HYGËIA

Essays Moral and Medical

Introduced by
Robert Mitchell
Duke University

THOEMMES CONTINUUM

Printed in England by Antony Rowe Ltd, Chippenham

HYGËIA

Essays Moral and Medical

Volume 1

Thomas Beddoes

With an Introduction by
Robert Mitchell

THOEMMES CONTINUUM

This edition published by Thoemmes Continuum, 2004

Thoemmes Continuum is an imprint of
The Continuum International Publishing Group Ltd

Thoemmes Continuum
11 Great George Street
Bristol BS1 5RR, England

http://www.thoemmes.com

Hygëia
3 Volumes : ISBN 1 84371 098 6

Reprinted from the 1802–1803 edition owned by the
Rare Book, Manuscript, and Special Collections Library,
Duke University

British Library Cataloguing-in-Publication Data
A CIP record of this title is available from the British Library

Publisher's Note

The Publisher has gone to great lengths to ensure the
quality of this reprint but points out that some
imperfections in the original book may be apparent.

This book is printed on acid-free paper, sewn, and
cased in a durable buckram.

INTRODUCTION

In 1802, Thomas Beddoes – physician, chemist, and democratic reformer – published *Hygëia: or Essays Moral and Medical, on the Causes Affecting the Personal State of our Middling and Affluent Classes*. Beddoes had already published a large number of books and pamphlets, on topics ranging from chemistry to educational theory and preventive medicine, but *Hygëia* was his most lengthy publication, running to almost 1500 pages. Published only six years before his death, *Hygëia* was also the most complete – though not necessarily systematic – account of Beddoes's understanding of the connections between medicine, politics, education, and consumerism. It was one of a group of 'radical' self-help medical texts published in the eighteenth and nineteenth centuries, designed to promote the Enlightenment goal of widespread diffusion of practical knowledge. At the same time, though, *Hygëia* was a particularly complicated example of this genre, for as historian of science Roy Porter has noted, it is a self-help book that is deeply suspicious of self-help books.[1] Beddoes believed that the popularity of the self-help genre was an index of the increasing commercialization of medicine itself, and in *Hygëia* he sought to reform both the public and the medical profession by outlining an almost paradoxical balance between medical authority and public education. For this reason, the text has been of especial interest to historians of medicine and education, as well as literary critics interested in the relationship

[1] For Porter's discussion of the paradoxical status of *Hygëia* as self-help book, see especially 'Plutus or Hygeia? Thomas Beddoes and the Crisis of the Medical Ethics in Britain at the Turn of the Nineteenth Century', in Robert Baker, Dorothy Porter, and Roy Porter, *The Codification of Medical Morality: Historical and Philosophical Studies of the Formalization of Western Medical Morality in the Eighteenth and Nineteenth Centuries*, in *Philosophy and Medicine*, vol. 45 (London: Kluwer Academic Publishers, 1993), pp. 73–91, as well as chapters 5 through 8 of *Doctor of Society: Thomas Beddoes and the Sick Trade in Late-Enlightenment England* (London: Routledge, 1992).

between Beddoes and authors such as S. T. Coleridge (with whom Beddoes had collaborated on political texts such as *The Watchman*). The importance of *Hygëia* for illuminating the intersection of politics and medical authority on the cusp of the early nineteenth-century 'professionalization' of British medicine has been emphasized by a number of historians and literary critics, and especially by Roy Porter in his cultural history of Beddoes and his context, *Doctor of Society: Thomas Beddoes and the Sick Trade in Late-Enlightenment England*.[2]

Unfortunately, after its initial publication, *Hygëia* remained out of print for almost two centuries, available only in a small number of research libraries. This reprint edition makes this important text available to a much wider audience of scholars and readers. The purpose of this introduction is to help readers situate *Hygëia* within the context of Beddoes' *oeuvre*, as well as his within his cultural context. It will also outline some of the medical assumptions that undergird Beddoes' claims in *Hygëia*, for while (as the book's subtitle suggests) it was written for a lay, rather than medical, audience, 'common sense' understandings of human physiology have changed drastically in the last two centuries, with the result that many of Beddoes's claims may appear unusual to a modern reader. Accordingly, the remainder of this introduction is divided into three sections: the first section briefly outlines Beddoes's biography; the second turns to a more focused discussion of the medical advice offered in this text; while the third considers the place of *Hygëia* in the context of early nineteenth-century British literature.

I. Chemist, Physician, Democrat: Beddoes as Doctor of Society

Beddoes's life has been the subject of three biographies. The first, commissioned by Beddoes's wife shortly after his death in 1808, was written by J. M. Stock, a physician and family friend.

[2] In addition to his book-length study, Porter has published numerous articles which deal with Beddoes in part or in whole; while some of these overlap with *Doctor of Society*, both 'Plutus or Hygeia?' and 'Consumption: Disease of the Consumer Society?' (in John Brewer and Roy Porter, *Consumption and the World of Goods* (London: Routledge, 1993)) are particularly useful. For a discussion of *Hygëia* in the context of educational debates, see Dorothy A. Stansfield, 'Thomas Beddoes and Education', *History of Education Society Bulletin*, vol. 23 (1979), pp. 7–14.

Stock's *Memoirs of the Life of Thomas Beddoes, M. D.* (1811) stressed Beddoes's career as a physician, and provided lengthy summaries of Beddoes's writings; however, this often came at the expense of an account of either his early interest in chemistry or his 'radical' political activities in the 1790s.[3] Dorothy A. Stansfield sought to address some of these deficiencies in her *Thomas Beddoes M. D., 1760-1808: Chemist, Physician, Democrat* (1984) by highlighting Beddoes's political and scientific work, while Roy Porter used Beddoes's life as a focal point for a more general investigation of 'the "sickness culture" of late-Enlightenment England' in his 1992 *Doctor of Society* (1992).[4] I refer readers interested in the details of Beddoes's life to these accounts, and provide here only a sketch of the major events in Beddoes's life leading up to the publication of *Hygëia*.

Thomas Beddoes was born in Shifnal, Shropshire in 1760, the son of a relatively prosperous tanner. In 1776 he began study at Oxford, where he focused on scientific courses (for example, botany, mineralogy, and chemistry).[5] He finished his B.A. in 1781, and then studied medicine at London's Great Windmill Street Anatomy School and the University of Edinburgh (both significantly more rigorous medical schools than Oxford in the eighteenth century), before returning to Oxford to take his M.D. in 1786. While in Scotland, Beddoes became a friend of Professor of Medicine Joseph Black (the chemist who had isolated carbon dioxide in 1757). Beddoes also employed his foreign language abilities to edit and publish translations of Italian and German experimentalists.[6] After completing his final degree at Oxford, Beddoes began his career as a 'Chemical Reader' at Oxford.

3 John Edmonds Stock, M. D., *Memoirs of the Life of Thomas Beddoes, M.D., with an analytical account of his writings* (London: Printed for John Murray, 1811). Thoemmes Press issued a reprint edition of this text in 2003.

4 Dorothy A. Stansfield, *Thomas Beddoes, M. D., 1760-1808: Chemist, Physician, Democrat* (Dordrecht, Holland: D. Reidel Publishing Company, 1984); Porter, *Doctor of Society*, p. 1. Porter compares his approach to biography with that of his predecessors in 'Taking Histories, Medical Lives: Thomas Beddoes and Biography', in Michael Shortland and Richard Yeo (eds.), *Telling Lives in Science: Essays on Scientific Biography* (Cambridge: Cambridge University Press, 1996), pp. 215–42.

5 Stansfield, *Thomas Beddoes M. D.*, pp. 12–20.

6 See Luigi Spallanzani, *Dissertations relative to the Natural History of Animals and Vegetables*, edited by Thomas Beddoes (London: J. Murray, 1789).

Beddoes also became increasingly politicized in the late 1780s and early 1790s, writing privately and publicly in favour of democratic reforms, a position that had unfortunate consequences for his career as a chemical researcher. It seems likely that had Beddoes been politically conservative (or had he at least remained undeclared), he would have been offered the Regius Chair in Chemistry at Oxford. However, Beddoes's explicitly democratic principles and pamphlet writing made this an impossibility. In 1792, he was listed as a 'disaffected & seditious' person by the Home Office, probably as the result of a letter from Charles Willoughby to Henry Dundas, Secretary of State for Home Affairs, in which Beddoes was described as 'a most violent *Democrate*'.[7] While Beddoes could not have known his Home Office status, he nevertheless was able to sense that his options at Oxford were becoming increasingly limited, and in 1792 he left Oxford – and the field of chemistry – to set up a medical practice in Clifton, near Bristol.[8]

While eighteenth-century physicians often had to endure a few lean years before establishing a profitable patient base, Beddoes seems to have found patients rather quickly. Richard Edgeworth was one of Beddoes's first acquaintances in Clifton, and it is probable that Edgeworth's connections helped Beddoes begin to establish a patient base.[9] Beddoes also began reviewing domestic and foreign medical books for *The Monthly Review*, which kept him well abreast of scientific and medical developments in Britain, the continent, and the United States.

Beddoes's interest in ameliorating individual health as a physician paralleled his continuing interest in promoting wider democratic social reform. Throughout much of the 1790s, the 'violent *Democrate*' authored a number of political pamphlets

7 Cited in Trevor H. Levere, 'Dr. Thomas Beddoes at Oxford: Radical Politics in 1788-1893 and the Fate of the Regius Chair in Chemistry', *Ambix*, vol. 28, no. 2 (July 1981), p. 65.

8 For a full account of Beddoes's relationship to Oxford and the Regius Chair in Chemistry, see Levere, 'Dr. Thomas Beddoes at Oxford', pp. 61–9. For a more comprehensive account of Beddoes's chemical interests, see Stansfield, *Thomas Beddoes M. D.*, pp. 31–59 and Jan Golinksi, *Science as Public Culture: Chemistry and Enlightenment in Britain, 1760-1820* (Cambridge: Cambridge University Press, 1991).

9 Stansfield, *Thomas Beddoes M. D.*, p. 103.

(many of which attacked Prime Minister Pitt), and collaborated with Coleridge on the latter's political newspaper, *The Watchman*.[10] In many of these pamphlets, Beddoes advanced an understanding of politics in which social inequity was tied to questions of health. In his satirical *Letter to the Right. Hon. William Pitt, on the means of Relieving the Present Scarcity and Preventing the Diseases that Arise from Meagre Food* (1796), Beddoes mockingly urged Pitt to combat diminished food supplies by doling out opium to the poor (perhaps even substituting it for salt), which would certainly reduce hunger cravings (and political resistance), but also would lead to a host of health problems.[11] However, like many of his contemporaries, Beddoes's output of political writing waned as the decade came to a close, partly as a consequence of an increase in governmental crackdown on 'seditious' authors, but also because Beddoes's hopes for moderate British reform diminished in the wake of Napoleon's rise.

In addition to the publication of explicitly political prose, Beddoes sought to encourage reform through more indirect means, such as the publication of educational tracts and the creation of medical institutions. Many of his educational texts were directed at specific segments of the British population. His most popular educational tract, *The History of Issac Jenkins, and of the Sickness of Sarah his wife, and Their Three Children*, for example, was designed to discourage the poor from excessive drinking, while his extensive comments on education in *Hygëia* were directed toward more affluent parents.[12] Beddoes's institu-

[10] Representative titles of Beddoes's political works include: *A Word in Defence of the Bill of Rights, against Gagging Bills* (Bristol: N. Biggs, 1795), *Where would be the harm of a speedy peace?* (Bristol: N. Biggs, 1795), *An Essay on the Public Merits of Mr. Pitt* (London: J. Johnson, 1796), *A Letter to the Right. Hon. William Pitt on the means of Relieving the Present Scarcity and Preventing Diseases that arise from Meagre Food* (London: J. Johnson, 1796) and *Alternatives Compared: or, What Shall the Rich do to be Safe?* (London: J. Johnson, 1797). For Beddoes's collaboration with S. T. Coleridge on *The Watchman*, see *The Collected Works of Samuel Taylor Coleridge* (London: Routledge and Kegan Paul; Princeton, New Jersey: Princeton University Press), volume 2: *The Watchman*, edited by Lewis Patton and B. Winer.

[11] Thomas Beddoes, *A Letter to the Right. Hon. William Pitt*, pp. 28–9. Beddoes also outlined a plan (complete with diagram) for the construction of machines that would extract nutriment from animal bones (pp. 18–21).

tional innovations were also oriented, though in several different ways, toward the poor. His most (in)famous institution, the Pneumatic Institute, was designed to be a both research centre and patient clinic, focused especially on the effects of gases (such as the recently discovered 'airs', oxygen and nitrous oxide) on diseases such as 'consumption' (a eighteenth and nineteenth-century disease category that included what is now called tuberculosis). The recent discovery of the role of oxygen in blood flow and respiration had encouraged Beddoes to see therapeutic potential in the delivery of gases into the human body, and Trevor H. Levere notes that '[a]s early as 1791, while visiting Davies Giddy in Cornwall, [Beddoes] had dreamed of a research institution to investigate the therapeutic effects of different gases.'[13] In Bristol he began to locate the money and individuals necessary for such a project, and by enlisting the aid of people such as James Watt, Thomas Wedgwood, and eventually Humphrey Davy, Beddoes was finally able to open his Pneumatic Institute in 1799. The Institute contained a laboratory, as well as a clinical facility in which possible therapies were to be tested on poor patients, and Beddoes hoped that the research produced there would enable '[c]ommand of the elements which compose the animal substance'.[14]

Unfortunately (and by Beddoes's own admission), the Pneumatic Institute was not particularly successful from a scientific point of view, and its focus on 'gases' became the butt of jokes (James Gillray's mocking cartoon of the Institute, which has one member of the Institute inhaling 'airs' from the posterior of another, has become almost an icon for Beddoes himself). Beddoes twice had to change the name and function of the Institute: first, in1802, to the Preventive Institution (which was a Dispensary and out-patient clinic, rather than a research site),

12 Thomas Beddoes, *The History of Issac Jenkins, and of the Sickness of Sarah his wife, and Their Three Children* (Madeley: J. Edmunds, 1792). Stansfield notes that *Issac Jenkins* was still in print in 1860; see Stansfield, *Thomas Beddoes M. D.*, p. 87. For Beddoes's discussion of boys' and girls' schools in *Hygëia*, see especially essays III and IV.

13 Trevor H. Levere, 'Dr. Thomas Beddoes (1760–1808): Science and Medicine in Politics and Society', *British Journal for the History of Science*, vol.17 (1984), pp. 191–2.

14 Cited in Stansfield, *Thomas Beddoes M. D.*, p. 145.

and again in 1804, to the 'Medical Institution, for the Sick and Drooping Poor'. Apparently the Institute was far more popular in its later incarnations, for in 1804 a reviewer of Beddoes's pamphlet *The Rules of the Medical Institution, for the Sick and Drooping Poor* noted that, whether or not the institution succeeded, '[i]t would appear that this Charity is a popular one, since 3533 individuals applied to it for relief in the course of last year'.[15] Yet despite the failure of Beddoes's original research-oriented plan, his institutional innovations should be seen as part of a more general fin-de-siècle interest in creating new institutional sites for generating scientific and medical knowledge and conveying it to 'the public'. So, for example, the young chemist Humphrey Davy left Beddoes's Pneumatic Institute to become part (and eventually head) of the fledging Royal Institution, which was very similar in structure to Beddoes's original Institute.[16]

II. Medicine, Sickness, and the Middling and Affluent Classes: Publishing *Hygëia*

By 1801, when Beddoes began writing the essays that were to make up *Hygëia*, he was far less hopeful than he had been a decade earlier that democratic change could be achieved through direct political action. The failure of the Pneumatic Institute, moreover, had cooled his hopes that the creation of new medical research institutions would provide the '[c]ommand of the elements which compose the animal substance' that he had earlier seen as the mission of medical science. At the same time, though, he had in his political writings begun to develop the notion that social inequities and biological illness were interrelated, which suggested that at least some diseases could be countered by reforming the interrelationships of individuals and groups. As

[15] Anon., *The Monthly Review*, July 1804, p. 328.

[16] For a discussion of the Royal Institution as a site for reformulating the relationship between science and the public, see Robert Mitchell, 'Here is thy fitting Temple': Science, Technology and Fiction in Shelley's *Queen Mab*', *Romanticism on the Net* (http://www-sul.stanford.edu/mirrors/romnet/21mitchell.html), Morris Berman, *Social Change and Scientific Organization: The Royal Institution, 1799-1844* (Ithaca, New York: Cornell University Press, 1978), and David Knight, *Humphrey Davy: Science and Power* (Oxford: Cambridge University Press, 1992).

Porter notes, Beddoes's understanding of disease in 1801 was significantly more complex than it had been when he first began his medical career: '[a]s a young researcher, he [had] regarded disease as an enemy which laboratory science could conquer, indeed would rapidly and gloriously vanquish'; a decade later, he was convinced that 'illness was not simply an adventitious foe ... [for] [i]ts incidence was [also] an index of certain social realities: it spoke about absolute and relative distributions of plenty and poverty, of power and oppression, of education and ignorance, succour and neglect'.[17] *Hygëia* was the first of Beddoes's texts to develop this position at length, as he used its eleven essays to probe the interrelationship between individual habit, social groups, and illness.

Beddoes published *Hygëia* as a series of eleven individually printed essays, which appeared monthly, and then were bound into three volumes following the last issue.[18] This relatively expensive distribution method underscored Beddoes decision to limit the readership of *Hygëia* to the 'middling and affluent classes'. The alignment of these two classes in the title was itself significant, for it indexed the extent to which the 'middling class' had successfully positioned itself as the corporeal equals of the aristocracy. Earlier exclusive self-help medical books, such as George Cheyne's *The English Malady* (1733), often had been directed to an exclusively aristocratic audience, operating under the assumption that aristocratic bodies and modes of living were so different from those of their social inferiors that the diseases afflicting the upper classes would have little relevance to the middling or lower classes. Yet Beddoes suggests, both through his subtitle and in the seventh chapter (on consumption), that the forces of social emulation had encouraged the ambitious middle

[17] Porter, *Doctor of Society*, p. 2.

[18] The precise details of *Hygëia*'s publishing history remain obscure. In *Thomas Beddoes M.D.*, Stansfield suggests that the text appeared in two separate editions (p. 276), but this does not appear to be the case, for the printing signatures on existing editions suggest that it was always intended to be bound in a three-volume edition. In *Memoirs*, Stock suggests that Beddoes thought of *Hygëia* as a 'periodical publication' (271), and Beddoes employed a number of different printers for the individual essays, but this simply meant that, like many books of the period, it was 'periodical' in terms of its printing, but it was nevertheless intended to be a three-volume publication from its inception.

class to attempt to catch the diseases of its social betters. And in this, Beddoes notes, the middle class had been successful, having now 'inherit[ed] a title to consumption' (2.vii.41).[19] By pointing to the apparent identity of disease profiles between affluent and middling class readers, Beddoes sought to justify his decision to address the book to both classes.

Yet Beddoes's exclusion of lower class readers from his intended readership did not signal any waning of his interest in social reform; rather, it indexed his growing belief that social change required a sort of leadership, or at least impetus, from above. He notes in the first essay that he has directed the book to those most likely to facilitate widespread social change:

> In my choice of topics, I shall principally hold the middle and more opulent classes in view. There only do we find leisure for study; there only the necessary degree of intelligence; and the means of carrying good counsel fully into effect. I believe too that whatever benefit these essays may be capable of conferring on my indigent countrymen, will be best conveyed through the medium of their superiors.[20]

The decision to focus on the health needs of middle and upper classes readers, from this perspective, was primarily strategic, for they could best facilitate the improvement of lower class health.

Most importantly, Beddoes directed *Hygëia* toward a specifically classed audience in order to emphasize one of the major points of the book, that the disease geography of fin-de-siècle Britain was inextricably linked to the dynamics of wealth and commerce. Porter has been most attentive to this aspect of Beddoes's text, noting that *Hygëia* is premised on a belief that Britain's commercial ascendency had resulted in new modes of disease.[21] *Hygëia*

[19] Nor was Beddoes alone in his suggestion that diseases had migrated from the aristocracy to the middle class: Thomas Trotter, in *A View of the Nervous Temperament* (London: Longman, 1807), made the same diagnosis. For a comparison of Beddoes and Trotter, see Porter, *Doctor of Society*, pp. 86–118.

[20] See this edition, vol. 1, essay i, pp. 8–9. In subsequent references to *Hygëia*, I shall employ the standard parenthetical citation method of volume-essay-page (e.g., 1.i.8–9).

[21] This is the central claim of Porter's *Doctor of Society*; for a concise summary of the major points of his argument, see chapter 6 ('Enlightenment and Pathology'), pp. 86–118.

partakes of the eighteenth-century critique of 'luxury' outlined by John Sekora and others, but Beddoes provides a particularly nuanced analysis of the mechanisms by which the pursuit of wealth was translated into pathology.[22] Beddoes points to the historically contingent nature of desire, arguing that '[i]n the social arrangements which have gradually formed themselves in Europe, WEALTH, the most general object of power, becomes the most general object of desire' (1.ii.52–3). Yet as Beddoes notes, what people really sought was to *display* wealth, a desire that they could accomplish only through the purchase of things. The result was a sort of positive feedback between consumer items and desires, for '[t]he multiplication of the roads to wealth unavoidably keeps pace with the multiplication of gratifications, and of contrivances for displaying accomplishments, which wealth can command' (1.ii.53). The result was a 'disunion of desire and of the means of satisfying desire' (1.iii.7), and this, Beddoes contends, was 'the deep root, of the tree of disease' (1.iii.83). The lust for things – and the increasingly dominant belief that the satisfaction of desire was inextricably linked to purchases – was a drive at best indifferent to health, and more frequently, led directly to illness.[23]

Yet of course *Hygëia* was itself just such a commodity – one of a whole genre of self-help books that promised to translate a purchase into health – and thus Beddoes took pains to emphasize those ways in which proper use of his book would undercut the very market that enabled its dissemination. In the first essay, he insisted that he would only provide his readers with a 'negative doctrine': i.e., his exclusive object was 'PREVENTION of mischief', rather than positive remedies or cures (1.i.52; 1.ii.13). He acknowledged that many of the readers would find this frustrating – hadn't they paid for a book of cures? However,

22 See John Sekora, *Luxury: The Concept in Western Thought, Eden to Smollett* (Baltimore: Johns Hopkins University Press, 1977), Christopher J. Berry, *The Idea of Luxury: A Conceptual and Historical Investigation* (Cambridge University Press, 1994), Maxine Berg and Elizabeth Eger (eds.), *Luxury in the Eighteenth Century: Debates, Desires and Delectable Goods* (Basingstoke: Palgrave Macmillan, 2002).

23 Beddoes does not contend that commodities necessarily led to sickness, but rather that the lust for things had increased without any attention to their consequences for health. Thus, he notes that '[i]n what regards commodiousness, considerable improvements may have taken place; – few in what regards health' (1.ii.75).

alluding to the original monthly format of the text, he hoped that his

> bookseller's powers of insinuation (to which the public has, at times, appeared not a little sensible) will ... prevail upon them [his readers] to bear with me yet a few months, before they give me up, in order that they may, at least, have complete evidence, that my notions are less correct than theirs. It is, I assure them, a question little short of life and death. (1.ii.13–14)

Thus, *Hygëia* is something of a paradox: a sort of prophylactic self-help book, one that sought to undercut the culture of self-help books. As such, it would not point its readers towards products with which they could medicate themselves, but instead would encourage modes of self-reflexive knowledge, by means of which individuals could reform the habits that were leading them toward illness.[24]

What sort of knowledge was this to be, exactly? After two introductory essays (addressed to parents and ministers, respectively), the remaining nine essays of *Hygëia* suggest three general rubrics of knowledge: (i) for parents, knowledge of the criteria by which they should assess and select educational institutions for their children; (ii) knowledge of the body's physiology, which would allow individuals to understand fully the corporeal consequences of poor personal management; and, finally (iii) strategies of 'self management' that would allow individuals to avoid habits that produced ill health.

Education of Children

Beddoes was particularly concerned about the health of children in educational institutions, and sought to develop for parents a set of criteria by means of which they could select 'healthy' boarding schools and schools for girls. Beddoes had written frequently on educational topics since the early 1790s, and had consistently adopted Lockean and Rousseauvian principles,

24 As Beddoes notes, '[w]ritings intended to warn against the destruction of health, can hardly be confounded with such as pretend to teach people how to restore it. To direct a stranger how to traverse slippery ground without injury, is one thing. To instruct him how to set his leg, should he break it, is quite another' (1.ii.35).

arguing, with Locke, that knowledge originated in sense impressions and associations between impressions, and, with Rousseau, that children could not be force-fed facts, but should instead come to knowledge through self-initiated contact with their environments. In *Hygëia*, Beddoes outlined the implications of his associational principles for health, arguing that the body 'stored' its early impressions, with the result that unhealthy impressions could continue to exert themselves throughout an individual's life. 'Any state of feeling or of action', Beddoes wrote, 'which has repeatedly taken place, may be re-produced after an interval of years, by circumstances that would otherwise have had no such operation... . [thus] whatever impressions are early made on the nerves, will recur on slight provocation, after a prodigious interval' (1.iv.11).[25] In essays three and four of *Hygëia*, Beddoes outlined in abundant detail the sorts of food, exercise, bodily temperature, clothing, and reading material to which children should be exposed.[26] In all of his comments on education in *Hygëia*, Beddoes sought to convince parents that their decisions about the educational environments of their children were at the same time decisions for or against their child's lifelong capacity for health.

Knowledge of Physiology

Beddoes also believed that an understanding of human physiology was vital for all individuals. He advocated the establishment of physiology lectures and demonstrations for children and adult men and women, for he believed that it was only

[25] Thus, Beddoes argued, 'a compleat investigation [of all the kinds of associations] would not only comprehend the whole history of human nature in all its states, but disclose the means of maintaining the system in its most desirable known condition, and perhaps of exalting it to a higher' (3.ix.71). Beddoes's belief that stored bodily impressions could produce pathology was relatively widespread at the fin-de-siècle; for a discussion of similar positions taken by some of Beddoes's contemporaries, including William Godwin and Thomas Trotter, see Peter Melville Logan, *Nerves and Narratives: A Cultural History of Hysteria in Nineteenth-century British Prose* (Berkeley: University of California Press, 1997).

[26] Like many of his predecessors and contemporaries, Beddoes was adamant about the dangers of masturbation (which Beddoes called the 'passion of the groves': see 1.iv.35–48). For a brief discussion of the relationship between Beddoes's critique of masturbation and those of his contemporaries, see Porter, *Doctor of Society*, pp. 104.

through such knowledge of their own bodies that individuals would truly come to believe that bad habits could result in illness.[27] While Beddoes did not believe that textual descriptions of the body's workings could supplant actual demonstrations, *Hygëia* nevertheless provided discussions of the effects of temperature on the skin, surface membranes, and inner organization of the individual (essay five); the symptoms that indicated the onset of scrophula, a disease that caused swelling and eventual destruction of the lymphatic glands (essay six); a description of the lungs and respiration, and the ways in which environmental irritants could engender consumption (essay seven); a detailed description of the physiological mechanisms of digestion (essay eight) and the nerves (essays nine and ten).

The image of the body that emerges in Beddoes's discussion is that of a corpus naturally inclined towards healthiness, but nevertheless fragile and capable of falling into disorder through relatively small causes. Beddoes provided a horrifying description of the ways in which seemingly minor events could cause permanent mischief, describing the effects of a drop of water on an exposed arm:

> from the rapid communication of the parts of the living system with each other, by which what comes in contact with any one will influence another, ever so remote, catarrhs are brought on by the most trivial seeming operation on a distant member. Thus leaving bare the arm for a few moments, if it have usually been covered with a glove, the slightest moisture finding its way to the feet, a quantity of rain not sufficient to wet a handkerchief through, shall produce an instantaneous sneezing, and a speedily following inflammation of the mucous membrane. This shall pass from spot to spot down the wind-pipe into the chest, and keep up an irritation to cough for many weeks. (2.v.67)

Through causes such as these, bodily constitutions can be permanently weakened. Yet Beddoes did provide his readers with some hope, insisting that through certain kinds of mental and physical

[27] See Beddoes, *Hygëia*, 1.i.36–48. For his description of the ways in which children should be educated about physiology, see 1.iv.51–3 and 2.vi.46.

exercises, damaged bodies 'will often be rewarded with a return of some portion of their lost vitality' (2.v.58).

Beddoes's understanding of the body owed a good deal to the work of John Brown (whose *Elements of Medicine* Beddoes had translated from Latin), though Beddoes cannot precisely be called a 'Brunonian'.[28] Brown had argued that disease was not innate to the body, but instead resulted from the relationship of the body to its environment, which could force the body into a state of over-excitement ('sthenia') or under-excitement ('asthenia'). In either case, health depended on moving toward the opposite pole, often through the use of stimulants (such as alcohol) or depressants (such as opium). Beddoes adopted Brown's basic schema of keeping the body between extremes. In a discussion of exercise, for example, Beddoes urged his readers to keep to the mean, insisting that both '[t]oo much or too little use renders the organs incapable of properly performing their office. And the mischief seldom stops at incapacity, but runs on to local or general disorder, and often terminates in the destruction of the entire system' (3.iii.40). At the same time, though, Beddoes did not endorse Brown's theory of medication, arguing instead that health often required that the body employ its own energies to locate the mean. So, for example, he argued against using external heat to warm oneself, instead urging that over-cooled parts should 'be made to warm themselves by their own action' (2.v.13).

Strategies of Self-Management

Finally, throughout all of the essays of *Hygëia*, Beddoes provided concrete suggestions designed to help individuals monitor, and improve, the state of their own health. Some of these were bodily exercises – for example, practices for weaning oneself from the need for warm rooms (2.v.53) – but many more were mental

[28] John Brown, *The Elements of Medicine of John Brown M.D.*, trans. and introduced by Thomas Beddoes, 2 vols (London: J. Johnson, 1795). For Beddoes's discussion of Brown in *Hygëia*, see 2.v.66. For a very thorough discussion of the relationships between Brown, Beddoes, and Coleridge, see Neil Vickers, 'Coleridge, Thomas Beddoes, and Brunonian Medicine', *European Romantic Review*, vol. 8, no. 1 (Winter 1997), pp. 47–94; see also Nora Crook and Derek Guiton, *Shelley's Venomed Melody* (Cambridge University Press, 1986), pp. 22–6.

exercises designed to bring to self-consciousness the corporeal bad 'habits' into which an individual had fallen. So, for example, Beddoes proposed that problematic dietary proclivities could be improved by recalling and judging the day's consumption: '[t]o pass our table-transactions in frequent review ... to dwell upon their consequences, particularly their disagreeable ones ... is our best moral preservative against the danger of becoming dyspeptic; and without this, our physical resources will hardly get fair play' (2.viii.60–1). Beddoes also outlined ways in which an individual's self-monitoring could become a node in the development of a nationwide informational network. He proposed that insomniacs and their relatives maintain personal logs of factors that might be affecting their sleep:

> Twenty journals drawn up by as many attentive and ingenious couples, would I believe, do more towards the art of procuring composed sleep, and consequently towards breaking the force of nervous complaints than an hundred volumes by physicians, who have no better opportunity of information than the ordinary course of practice affords. (3.ix.128)

Beddoes became increasingly interested in this aspect of medicine in publications after *Hygëia*, and as both Stansfield and Porter note, his methods were part of a growing interest among physicians and scientists in gathering 'statistical' information about the nation.[29] It is worth noting, though, that in *Hygëia* this distributed system of information gathering was a means towards *both* the relatively remote end of creating disease 'maps' of the nation, and the much more direct end of increasing individual awareness of corporeal habits.

III. *Hygëia* as Romantic Text

Beddoes's insistence on the importance of self-scrutiny for health provides a useful point of departure for assessing the relationship

[29] See Stansfield, *Thomas Beddoes M. D.*, pp. 180–82 and Porter, *Doctor of Society*, pp. 46–8. For discussions of the rise of statistical methods, see Ian Hacking, *The Taming of Chance* (Cambridge: Cambridge University Press, 1990) and Gerd Gigerenzer, Zeno Swijtink, Theodore Porter, Lorraine Daston, John Beatty, and Lorenz Kruger, *The Empire of Chance* (Cambridge: Cambridge University Press, 1989).

of *Hygëia* to 'Romantic' texts produced by contemporaries such as William Wordsworth and S. T. Coleridge. The relationship between Romantic medicine and literary production has certainly not been neglected, but treatments of this topic have tended to focus on the presence of medical 'themes' within creative literature, or on biographical relationships between Romantic era medical and literary authors.[30] However, the centrality of practices of productive 'self-management' to *Hygëia* suggest another approach to understanding the place of this text in Romantic era textual production. As Michel Foucault stressed in *Discipline and Punish*, psychological mechanisms of self-monitoring were central to eighteenth and nineteenth-century penal, military and educational institutions, and subsequent authors have highlighted the role of self-monitoring in many other forms of 'Romantic' era cultural practices and texts.[31] In the closing pages of this introduction, I shall focus on the relationship between self-monitoring and consumerism as a way of highlighting the points of connection between *Hygëia* and British Romanticism.

An increasing number of scholars have located the emergence of a 'consumer culture' in eighteenth-century Britain. This 'consumer revolution' had both an objective and a subjective dimension. On the objective side, eighteenth-century Britain was characterized by the emergence of modern 'shops' and understandings of 'shopping', a significant increase in the

[30] For particularly useful discussions of the relationships between Romantic literature and medicine, which include mention of Beddoes, see the following: Trevor Harvey Levere, *Poetry Realized in Nature: Samuel Taylor Coleridge and Early Nineteenth-century Science* (Cambridge University Press, 1981); Crook and Guiton, *Shelley's Venomed Melody*; Hermione de Almeida, *Romantic Medicine and John Keats* (Oxford University Press, 1991); Timothy Morton, *Shelley and the Revolution in Taste: The Body and the Natural World* (Cambridge University Press, 1994); Vickers, 'Coleridge, Thomas Beddoes, and Brunonian Medicine'.

[31] Michel Foucault, *Discipline and Punish: The Birth of the Prison*, trans. Alan Sheridan (New York: Vintage Books, 1979). For an application of Foucault's thesis to 'Romantic' educational institutions, see Alan Richardson, *Literature, Education, and Romanticism: Reading as Social Practice, 1780-1832* (Cambridge: Cambridge University Press, 1994). For an application to Romantic poetry, especially that of Wordsworth, see Thomas Pfau, *Wordsworth's Profession: Form, Class, and the Logic of Early Romantic Cultural Production* (Stanford: Stanford University Press, 1997).

number of individuals with disposable cash, and an ever-expanding range of relatively inexpensive consumer goods for sale (many, such as Wedgwood china, consciously created in fashion cycles).[32] Yet as Colin Campbell and others have noted, the eighteenth-century consumer revolution required more than just an increase in potential consumers and goods; it also required a fundamental 'subjective' psychological shift, through which the desire for goods was transformed from a vice to a potentially acceptable pursuit. Just as Max Weber located the origins of eighteenth-century capitalist production in modes of protestant self-scrutiny, Campbell locates the origins of eighteenth-century consumerism in 'an alternative Calvinist ethic' that sought proof of religious election in internal feelings.[33] This initially religious interpretation of feelings was partially secularized at the start of the eighteenth century through the development of philosophies of 'benevolence', which encouraged individuals to prove their moral worth to themselves by their capacity to take pleasure in the cultivation, and manipulation, of their own emotions. Over the course of the eighteenth century, Campbell contends, this strategy of emotional manipulation was tied to the category of 'taste', and through taste, to commodities: individuals could demonstrate aesthetic taste (and by implication, moral standing), by purchasing 'tasteful' products. As Jean-Christophe Agnew notes, Campbell outlines a theory of consumerism in which '[c]onsumer letdown and the longing that disappointment nourished were indeed consequences of manipulation, but of a manipulation ... in which consumers pulled their own strings.'[34]

[32] See, for example, Lorna Weatherhill, *Consumer Behaviour and Material Culture, 1660-1760* (London: Routledge, 1988); John Brewer and Roy Porter, *The Origins of the Consumer Society* (London: Routledge, 1990); John Brewer and Roy Porter (eds), *Consumption and the World of Goods* (London: Routledge, 1993). For a discussion of the relationship of new patterns of shopping to eighteenth-century literature, see Elizabeth Kowalski-Wallace, *Consuming Subjects: Women, Shopping, and Business in the Eighteenth Century* (New York: Columbia University Press, 1997).

[33] Colin Campbell, *The Romantic Ethic and the Spirit of Modern Consumerism* (Cambridge, Massachusetts: Basil Blackwell, 1987), p. 127.

[34] Jean-Christophe Agnew, 'Coming up for air: consumer culture in historical perspective', in Brewer and Porter, *Consumption and the World of Goods*, p. 25.

As noted above, Beddoes was remarkably aware of these psychological shifts that had attended the change in 'the social arrangements which have gradually formed themselves in Europe', and *Hygëia* provides an intriguing critique, and at the same time an appropriation, of these same mechanisms of self-management. In order to combat the regimen of self-manipulation that had enabled the consumerism of medicine, Beddoes proposed his own regimen of 'self-management' that he hoped would undercut consumerism itself. Protestant practices of self-scrutiny were the paradigm for Beddoes, though this mental discipline was directed toward factors that affected the body, rather than the soul; in addition to keeping track of food consumption, readers were to practice mental 'exercises' through which they imagined themselves in the place of the ill (2.vi.88). Beddoes made the link between economics and corporeal self-monitoring explicit, contending that 'the very same reason that makes it advantageous to be able, at the shortest warning, to call before us a summary of our *pecuniary* circumstances, applies with at least equal force to our *personal* circumstances' (1.i.29). Beddoes never questioned the need for self-monitoring, but he sought to transform a mode that led to commodity consumption into a mode that stressed the individual's *production* of knowledge about him – or herself.

Through its emphasis on the production of knowledge through self-monitoring, *Hygëia* bears an intriguing resemblance to more clearly 'literary' texts such as William Wordsworth and Samuel Taylor Coleridge's *Lyrical Ballads*. In *Wordsworth's Profession*, Thomas Pfau (drawing on Campbell) argues that what unifies much of early Romantic discourse was a shared, and often quite conscious, effort to use literary form to transform the 'consumer' of art into an 'imaginative producer'. Central to this emerging aesthetic, argues Pfau, was a relocation of 'production' into an imaginative register; that is, viewers and readers were to engage aesthetic products in a 'productive' manner such that the artwork 'stimulates "mind" to further, more productive displays of imaginative mobility'.[35] Pfau argues, for example, that Wordsworth's and Coleridge's *Lyrical Ballads* (1798–1805) was an attempt to 'devise for the emergent middle class a vigilant and adaptive

[35] Pfau, *Wordsworth's Profession*, p. 8.

sensibility, one whose cultured reflexivity *is* its sociocultural capital, one that has emancipated itself from the superstitions of a fading rural culture without, however, surrendering to the dissolute pursuits of the higher classes' (179).

By understanding Beddoes's text as 'Romantic' in this sense, one can locate the continuities between Romantic literary and corporeal practices. Where authors such as Wordsworth and Coleridge sought to use poetry to encourage in their readers the production of a self-reflexive, 'vigilant and adaptive sensibility', Beddoes endeavoured to produce precisely the same monitorial sensibility through attention to the practice of health and bodily regimen. In both cases, moreover, this 'cultured reflexivity' was understood as an alternative to consumerism, highlighting the way in which romantic art and practices of health were part of the same modern cultural discourse. Earlier 'lower class' self-help texts, such as John Wesley's *Primitive Physick*, had sought to create readers who were literal producers of their own medicines; Beddoes's *Hygëia*, by contrast, situated its middle-class readers as self-reflexive producers of knowledge about themselves.[36] We might usefully think of Beddoes's text as 'Romantic', not first and foremost because of the content of its medical knowledge of advice, but rather because of its form, which is itself a function of the attempt to understand, and critically restructure, the relationship between readers and consumerism at the start of the nineteenth century.

* * * * *

This edition of Thomas Beddoes's *Hygëia* has been an effort facilitated by the help of many people and groups, but I particularly would like to thank Philip de Bary of Thoemmes Press and Sara Seten Berghausen of Duke University Libraries for their invaluable efforts in making this project possible. I would also like to thank the Duke University Libraries for their generous purchase of an original edition of Beddoes's text, without which this facsimile edition could not have been produced. Many thanks

[36] John Wesley, *Primitive Physick: or an easy and natural method of curing most diseases. The fourteenth edition, enlarged (Twentieth Edition)* (London: 1781). Wesley's text includes recipes for home medicines that promise to cure everything from cancer (pp. 34–7) to the plague (p. 86).

as well to Jane Henderson of the Wellcome Library, Stephen J. Greenberg of the National Library of Medicine, and Dorothy A. Stansfield for their help on questions relating to the publication history of *Hygeia*. My thanks as well to Richard Sha, Tim Webb, Andrew Mitchell and Tim Fulford for their help in directing me to resources relevant to *Hygëia*.

Robert Mitchell, 2004
Duke University

HYGËIA:

OR

ESSAYS

MORAL AND MEDICAL,

ON

THE CAUSES

AFFECTING THE PERSONAL STATE

OF

OUR MIDDLING AND AFFLUENT

CLASSES.

By THOMAS BEDDOES, M. D.

VOLUME THE FIRST.

BRISTOL:

PRINTED BY J. MILLS, ST. AUGUSTINE'S BACK,

For R. PHILLIPS, No. 71, St. Paul's Church-Yard, London.

1802.

CONTENTS.

ERRATA.

Essay i. *p*. 51, *l*. 9, *for* prepared *read* compared: 58, *l*. 9, *for* suppositi*ons*, *r*. suppositi*ous*: 61, *l*. 1. *dele* all: 79, *l*. 11, *dele* out: 82, *l*. 23, 24, *dele* in an uncommon degree.

Essay ii. *p*. 5, *l*. 22, *insert* pretender to sanctity: 6, *l*. 23, *dele* some: 10, *l*. 22, *dele* after: 53, *l*. 9, *for* only *r*. not: 87, *l*. 2, *before* a young man *insert* you point out: 89, *l*. 17, *for* should, *r*. could.

Essay iii. *p*. 3, *l*. 9, *for* natural, *r*. national: *p*. 58, *l*. 17, *after* thus, *insert* we shall escape: *p*. 88, *for* food, *r*. cloathing.

Essay iv. *p*. 9, *l*. 9, *for* sufficient, *r*. insufficient; *p*. 12, *l*. 3, *for* example, *r*. examples; *p*. 16, *l*. 12, *insert* of scholars; *p*. 60, *l*. 20, *after* informed, *insert* in all the dangerous points; *p*. 79, *l*. 7, *after* our, *insert* schools and.

ESSAYS

ON

THE MEANS

OF

AVOIDING

HABITUAL SICKLINESS,

AND

PREMATURE MORTALITY.

ESSAY FIRST.

ΑΡΙΦΡΟΝΟΣ

ΣΙΚΥΟΝΙΟΥ

SΚΟΛΙΟΝ.

Υγιεια, πρεσβιστα μακαρων,
μετα σευ ναιοιμι
το λειπομενον βιοτας.
συ δε μοι προφρων συνοικος ειης.
ει γαρ τις η πλουτου χαρις, η τεκεων,
τας ισοδαιμονος τ' ανθρωποις
βασιληιδος αρχας, η ποθων,
ους κρυφιοις Αφροδιτης αρκυσι θηρευομεν·
η ει τις αλλα θεοθεν ανθρωποισι τερψις,
η πονων αμπνοα πεφανται,
μετα σειο, μακαιρ' Υγιεια,
τεθηλε παντα, και λαμπει Χαριτων εαρ·
σεθεν δε χωρις ου τις ευδαιμων.

INVOCATION

BY

ARIPHRON

OF

SICYON.

O Thou! among the blissful host
Deserving mortal incense most!
What yet of days remain to me
Hygeia! let me pass with thee,
And thou my willing inmate be!
If ought of solace coffered hoards—
Whatever progeny affords—
    Or sovereign sway
Exalting mortals to the Gods above—
    Or that sweet prey
Which struggles secret in the snares of love—
To toil-worn man, by Heaven's behest,
If balms unnamed refreshment bring—
All owe, benignant Power! to thee their zest;
Thou giv'st its radiance to the Graces' spring;
Nor but with thee may child of earth be blest.

I desire to be instrumental in diffusing a taste for the most useful species of knowledge, and in converting nations into HUMANE SOCIETIES. There is an art not suspected by the multitude to lurk among possibilities, and never yet cultivated by any people, although its honorable title was usurped by a system of intercourse, once established among the French. This is the ART OF LIVING; for whose reception men's minds can only be prepared by being familiarised with just ideas concerning animal nature; and whose precepts can issue only from the shrine of Hygeia.

Preface to Consid. on factitious airs. Parts IV. & V.

ADVERTISEMENT.

THE instruction of an audience, sufficiently docile from conscious ignorance, brings with it very little care except that of providing and arranging materials. But a body of popular information on health, where the author's secret purpose is not merely to fill a certain number of pages, constitutes a task of far greater anxiety and circumspection. Whoever undertakes such a task, may to a certainty calculate upon preoccupation. The just importance of every man in his own estimation, infallibly produces the desire of knowing what to think concerning the causes by which he is affected. But the judgment of the great majority — not excepting even those who have undergone the longest and most expensive education—being utterly unprepared for exerting itself with effect upon this subject, such notions are picked up as chance may offer: and what is thus lightly believed, is warmly recommended and blindly followed. This is a source of evil so copious, that whoever, by shewing the insufficiency of the grounds of common assent, could here introduce a prudent forbearance in thought and deed, would probably, in a single year, render his species a greater service than any one has been happy enough to render it in a whole life. So great a change, however, must be the fruit of a long course of years, and of much cooperation. No time, therefore, should be lost in preparing the way for the reception of just ideas; and considerations to this tendency will, I conceive, form no improper introduction to a work like the present.

But the pre-occupation of which I speak, will exist in different degrees. The inquisitive indeed out of the profession carry on their pursuits, pretty much as if there lay an interdict upon the ingenious against employing their talents, and the opulent their wealth, to promote the knowledge of themselves. Yet one man will have his reason better exercised, and his memory better stored, than another. Hence a writer in my situation finds himself obliged to fix upon an imaginary standard of capacity. This should doubtless be taken rather too low than too high; there being no comparison between the inconvenience to the intelligent from a few superfluous pages, and the evil arising to others, from deficiency of elucidation. If therefore, I shall be sometimes found adverting to errors, at the grossness of which common sense, assisted by a scanty portion of accurate information, revolts, I hope to stand excused; and no less if I descend to particular directions, which tolerable quickness of apprehension must discover in the very enunciation of the principle, which they are designed to carry into effect. It is enough that for want of the directions, serious mischiefs are in fact committed; and that the errors involve multitudes in misery and destruction.

If the reader, under any interesting emergency, have made the experiment of in vain searching the books, upon which he relied for instruction, let him recollect the acuteness of his disappointment; and then decide how far it is excuseable to be minute upon some points; and how far inexcuseable not to be minute upon others.

There is besides danger of misapprehension. When the judgment is neither previously informed, nor quickened by the urgency of distress, it easily ranks among minutiae, matters of decisive importance. In every part of life, there are *seeming* trifles, which if neglected, take the most severe revenge; and no seeming trifles are so vindictive as those relating to health.

In my choice of topics, I shall principally hold the middle and more opulent classes in view. There only do we find leisure for study; there only the necessary degree of intelligence; and the means of carrying good counsel fully into effect. I believe too that whatever benefit these essays may be capable of conferring on my indigent countrymen, will be best conveyed through the medium of their superiors. In some cases, the most circumstantial details would scarce afford security against misapprehension; and many prejudices must be expected to resist the dead letter, which example, sympathy and exhortation by the living voice, will with little difficulty overcome.

My plan by no means comprehends the whole of PREVENTIVE MEDICINE. For such a plan I am not enough acquainted with society in its present complicated state. I have not learned from experience the unfavourable effect of all the variety of human callings upon the human frame; nor can I point out therefore with sufficient certainty the means of counteracting each unfavourable effect. An undertaking, so great and useful, could only be accomplished by an union or concert of powers, not likely I fear soon to take place.

Writers, the better part of whose brain lies in the books out of which they copy, have in very few departments been more busy than in this. There is no department, however, where the compiler has been so constantly baffled. For unless the objects be examined and re-examined by the eyes of him who describes them, the most bulky volume will prove nothing better than a chaffy bundle, cruelly mocking the hope of those that expect it to yield a supply of wholesome doctrine.

Notwithstanding the limitation of my plan, I proceed in its execution with the most unfeigned diffidence. I have lately been employed upon a *general treatise of physiology*, for the use of all to whom their own nature is interesting. The treatise and its illustrative engravings, are in forwardness. But I do not, I confess, look with more anxiety towards the time when that work of comprehensive enquiry and arrangement is to appear before its judges, than I experience in sending abroad this smaller one of selection and exposition. Whatever may be the reason, I believe it is an unquestionable fact, that medical writings for the profession, have at all times been infinitely superior in their kind to those for the public. No wonder therefore that I should feel too proud to be gratified by a comparison, however to my advantage, with certain predecessors. But there does exist a model to which I am truly humble at the idea of being referred. It is *that*, which daily contemplation of domestic sufferings must have created in the breast of every experienced and discerning physician, whose feelings are not bound up in the profits of his craft.

ESSAY
ON
PERSONAL PRUDENCE,
AND ON
PREJUDICES
RESPECTING
HEALTH.

TO

HEADS OF FAMILIES,

Inhabitants of the

BRITISH ISLES.

IT is to you in particular, my countrymen and countrywomen, whose most severe afflictions it has long been my endeavour to trace to their source, that I address myself at the outset of my undertaking. In its progress I shall consider myself as addressing you more frequently than any other class. As parents and protectors of beings whose happiness is at stake, inasmuch as their constitution is unconfirmed, the majority among you have more than a single interest in the art which I attempt to teach. It is not upon yourselves alone that your want of proficiency in this art is visited. By unwitting violation of its prêcepts, you deal out misery where the thought of dispensing happiness (in such a case I will not hazard offence by speaking of *duty*) constitutes your prime delight. That your interesting cares are multiplied as your family receives accessions, you do not require to be told. But I know not if you are fully aware of the necessity of extending your views to a variety of situations and to a long course of events. For what will avail protection from early injury, should you fail to make a worthy choice, when you commit the objects of your solicitude (as I fear you frequently must) to the hands of substitutes?

Nor let us imagine the requisite parental foresight to end even here. All that has been so far done, though done ever so well, may tend but to the increase of moral and physical evil. If you have any difficulty in assenting to this proposition, keep your eye for a time upon a certain number among those young persons, who have remained safe while under superintendance, and whose condition appears the most prosperous at the moment they are ventured into open life. How many of them will you perceive dashed to pieces early in their course, or disabled by obstacles of unknown nature and position! And will you not put such examples to heart? and rank it among your principal concerns to equip your children, as they advance to that age when they must be left to their own guidance, with the most precise instructions respecting the hazards they will soon have to encounter?

In the accounts which have descended to us of the Grecian solemnities, you may have remarked that the most awful were introduced by observances, designed to awaken attention, and to purify the soul. As I must be desirous of communicating my own sense of the importance of my subject, I should be happy if by any means I could accomplish the double purpose of those preparatory rites of antiquity.

It is certain that in recommending to men the care of themselves, no voice has hitherto had half the influence attributed to that of the Hierophant upon the Eleusinian votaries. What possessor of a curious piece of machinery feels satisfied till sufficiently master of its construction to judge of its state; and who but carefully attends to the preservation of its substance and form?—Yet the usages of life, as they have come down to us, proceed upon the supposition that the human machine is made of materials so impassive, and so firmly put together, as to render superfluous that insight, without which vigilance is blind. No admonition, no example, no species of affection, no personal sufferings, have been effectual in rendering a contrary conviction general. Motives, which, upon a cool contemplation of the facts belonging to the natural history of man, one is tempted to pronounce singly irresistible, have failed when united. Otherwise innumerable hands would have been long since at work, and the world weeded of half its evils. But the heart and head have been heretofore filled with other concerns; and nothing beyond a casual thought, a fugitive feeling, a barren ejaculation, has been allotted to that which is doubtless the most important object in mortal life—HABITUAL WELL-BEING. Inattention is the ordinary state of mind. But there are few whose security some near alarm does not occasionally shake. Then opinion occupies the intervals of inattention; and, of the two, opinion is, I believe, much the more mischievous. For it is commonly both erroneous and strong—*erroneous* as being suddenly formed upon a difficult subject by the random application of maxims, in themselves very precarious; but *strong*, in virtue of the concomitant affections and of the defect of information on this particular subject, however different the case may be with regard to others.

Unhappily, the most effectual means of introducing a salutary mode of thinking, appear to be wanting in proportion as they are to be wished for. One principal topic of persuasion must be abandoned. The general excellence of the thing to be recommended is already felt with more force than it can be represented. It is a case in which, as the shortest sentence has a happier effect than the most eloquent oration, panegyric must be mute. For to the slightest hint of the advantages of health, all hearts rebound immediate assent, as well-disciplined troops start, with one accord, into movement at the voice of their commander. This, I suppose indeed, may be one of those truths of which the very obviousness is among the causes of their being neglected. By transferring to the means what belongs to the end, it comes to be imagined that the object itself is as easily secured as its value is conceived.

It is one thing, however, to acknowledge a principle upon a formal appeal; and another to be impressed with a sense of its importance, such as shall bring it before the imagination in its relation and bearings, with vivacity enough to determine our proceedings on the spot and without the delay of deliberation. This is genuine practical knowledge. It is this alone which has power to preserve and rescue. It may either be strikingly exemplified on a large scale amid those heroic scenes which History loves to paint; or, by an operation too gradual to be perceived by slight observers, it may procure affluence to its possessor, or the more enviable prosperity of a sound mind in a robust person. In all cases, its essential nature is the same. The manner of its acquisition, and the benefits it confers, remain unchanged. It consists in the clear discernment of good and evil, of cause and consequence; being an infallible guide, where faithfully followed; but leaving men exposed to the worst accidents of a road, infested by dangers, where it fails altogether, or sheds an unheeded light.

To bear in the mouth that *health is the first of blessings*, not only answers no purpose, but tends to create that sort of hypocrisy or self-deceit, which substitutes the repetition of a maxim for its observance. Habits such as will stand firm under difficulties and temptations, can be created only by taking up the means of securing this blessing as a study;—that is, by fixing the attention severally upon the modes in which it is forfeited, on the advantages that accompany its possession, and the consequences of its loss. If ever the Roman address of congratulation

Felix quem faciunt aliena pericula cautum,

be appropriate, it may with the greatest right be claimed by him who learns caution with regard to himself, from observing what brings infirmity upon others. In all conditions there occur too frequent calamities, fit to serve as lessons; and they should be carefully noticed by him who would acquire, and distinctly pointed out by him who would instil, an efficient sense of self-preservation. Towards our neighbour there is no humanity in making the good or ill that befals him the subject merely of loose, fleeting, unapplied prattle. But where the views are not yet formed, we may do infinite kindness by using it in illustration of those laws which mortals can seldom infringe with impunity. For in the order of nature, much more invariably than in matters of human regulation, does it obtain as a rule, that *ignorance of the law is no excuse for the transgressor.*

At the age proper for admission into the school of life, the feelings should first be addressed. The understanding will then take its turn; that kind of curiosity ever leading to most improvement, which is awakened by touches upon the heart. As long as there exists a newspaper to put the occurrences of common life into circulation, or indeed till Rumour herself perish, we can never be at a loss for copious matter of instruction. On minds, not preoccupied by vicious associations, the accidents to which humanity is liable, may always be so exhibited as to produce due impression. Such minds will not refuse themselves to a representation of the pain, the languor, the dependance, the incapacity for receiving or affording pleasure, which various kinds of indisposition produce. The history of few fathers and mothers is so uniformly prosperous as to be incapable of supplying materials for the present purpose. Where the education of one generation has been, in the usual sense of the phrase, *neglected,* tenfold exertions are frequently made in behalf of the succeeding; and it is well if more evil do not ensue to the one from superabundance, than the other has sustained from want, of tuition. But this anxiety seldom extends *systematically* beyond literary information. In more essential concerns, if another Fontenelle were to arise, he would still find that the experience of parents is without profit to children.* Yet if parents would take a little pains to analyse themselves, and to communicate the result at a proper season and in a proper manner, they need not fear forfeiting the smallest share of respect; and examples so near, would produce sensations, which no time could obliterate. Lasting trains may likewise be formed out of the history of persons, snatched in their prime from every object of customary desire, because, in a luckless hour, they slighted or overlooked some noxious power, which had planted itself in their way? Nor will those revolutions of the human frame, more tremendous than its demolition, in consequence of which, the activity of *many* is absorbed, for a series of years, in endeavours to soothe the bodily sufferings, or retrieve the mental alienation of *one*, be lost upon the youthful observer of his species and himself. For they

---

* Les sottises des peres et des meres sont perdues pour les enfans.

do not, any more than the phaenomena, which we behold with less emotion, spring from malignant supernatural agency; against which there is no defence; but may in most instances, be satisfactorily traced to the want of appropriate intellectual culture, and its genuine offspring, blind temerity.

In what I shall propose, I am not, to such a degree, apprehensive of objections on the score of impropriety, as of remonstrances on that of difficulty. The elements of every sort of doctrine have doubtless their difficulty, even where the means of information may be procured, like so many articles of commerce, upon order. And much more will the beginning of that study be difficult, where new means are to be provided, and a new scheme to be drawn up for their application. To *talk*, it is true, of bringing up pupils with a sacred regard to health, is nothing less than new. Not a school-bill but has the assurance to come forward with professions in this strain. Of professions, indeed, as an article without which his occupation will not thrive, no sort of quack is stupid enough to be sparing. But to bestow the talent of distinguishing the safest footing at every advance in life, and to inspire that true economy of existence, which takes ever so long a circuit rather than wantonly encounter a suspicious obstacle, and, whenever safety ought to become the secondary consideration, lays out beforehand for the best chance of escaping unimpaired, is an affair so different from pouring out vague promises, and is at the same time so unusual, that great cities and populous provinces, nay whole kingdoms may be searched, without the discovery of a single serious attempt to carry such a plan into practice. On the contrary, it may be hard to find examples where that which alone can spread sunshine over the years of a human being, is not of all acquisitions the least sought.—I speak not of houses, where every other wish is absorbed in that of decorating the young people, particularly the girls, with accomplishments, which may for a time attract the idle passenger, but soon come to be as little alluring as may-garlands that hang withering round their pole.—It is alas! common to meet with families, in which right feelings, respecting the culture of the youthful members, are accompanied by notions right indeed, but only to a certain degree right. Then education, considered as an apprenticeship to the art of securing the blessings, and escaping the miseries, by which we are beset, usually miscarries. And the miscarriage, traced to its inevitable consequence, produces somewhat of that painful eagerness of compassion, which is inspired by desperate efforts to snatch a beloved object from destruction, when they fail solely because the eye is prevented by some casual impediment from directing the hand. At the prospect of disappointments, so severe and so little merited, it is difficult to refrain from addressing remonstrances to those by whom they are about to be incurred, however small be the encouragement which experience holds out either to the moral or medical monitor.

"You are too discerning," it may without flattery be said, "to confound shew with "solidity, ephemeral with lasting attain-"ments. You are also too faithful to your "charge, to incur the guilt of neglecting "that which you approve. But your views, "though they are far from resting, like those "of so many others, upon the colours of the "surface, do not sufficiently penetrate into "the substance."

"If you would effectually teach that "parsimony of health, which in the illus-"trious De Witt was so closely united "with prodigality of life, you must no lon-

"ger suffer the impediments, accumulated "by the preposterous industry of ages, to "keep you aloof from human nature. Ap-"proach yet many steps nearer; and con-"vince yourselves that as the genuine hap-"piness of man has its source within, it is "his internal condition which demands your "earliest and most unremitting care.—Do "not imagine your error the less fatal be-"cause it regards only a single point. If "you neglect this point, it must be under "peril of casting over your declining years "a shade of sorrow, deep in proportion as "your affections are kind. By those affec-"tions, then, I conjure you to take the ful-"lest security which existing means can "give; and, above all things, in tracing "your plan, to beware how you employ a "scale, which can only be of service in fil-"ling up particular parts."

"Let it live in your recollection, that "*affluence to render all the arts its tributary,* "*and talents adapted not less to constant use* "*than occasional exhibition, combined with* "*the fullest self-approbation, can at best but* "*soothe the fatigues of him who is doomed to* "*bear through life the burden of a sickly* "*frame; and that the valetudinarian, though* "*he must indeed be less unhappy for being*

" *possessed of such advantages, can by no* " *means be accounted happy.* If you lose " sight of this one indispensable axiom, the " result is certain. You will give to the " superstructure the attention due to your " base. In which case, how vain your la- " bour! how much too like rearing an " edifice without tie or foundation, in the " face of a tempestuous sky!"

The reader, if he have large experience of the ways and means of life, will not require to be informed that an end clearly in view, however remote, is the best security against going astray; and that, without such an end, no one can say, whether he is, at any moment, nearer the attainment of his desires or further off. Hence I am in hopes that he will not disdain a familiar illustration, if it promise to be of service in keeping him upon the right track. What I could wish then, and what every one who has taken serious pains to follow up the most irreparable and most regretted of human sufferings to their origin, will agree with me in wishing, is:— that seasonable care should be taken to provide each individual with a set of ideas, exhibiting the precise relation in which his system, and the several organs of which it is compounded, stand to external agents, particularly to those with which he is likely to come most in contact; — that these sets of ideas be so placed in his head, that he may refer to them with as little difficulty as to the watch he wears in his pocket;—and that as by the one he adjusts his business to his time, so by the other he may be always able to accommodate his actions to his powers.

The distance at which we at present stand from such a consummation is no reason why future generations should, like the past, be abandoned to their fate. The relation of the animated machine to the powers by which it is put in motion, is unhappily not enough understood for the purposes of minute medical philosophy; but so far as it is understood, it constitutes, as I hope to shew in the sequel, a doctrine rich in lessons for common life.

There need be no difficulty, I am sure, for want of practice. Even where information is at the lowest ebb, and where the understanding is exercised with the least confidence, the process I recommend is quite familiar. Of this, on subjects of inferior importance, we have examples in the instructions, which young people are perpetually receiving on the improvement of fortune, the choice of connections, the maintenance of character, the arts of captivating vanity, and cajoling weakness. Nor is the labour lost. It is in this way men acquire what is vulgarly called *prudence*, but what is in fact the less essential part of that comprehensive virtue. There are few members of society who by such preparation have not obtained some facility in judging how they stand with the world: and the acquisition is in constant request. Few suffer a long interval to pass without bringing their affairs under distinct review; and the slightest tokens of a public or private change, are sufficient to induce them to enter, many times a day, into council with themselves how they may best meet probable contingencies. And in what way is this habit conferred? In teaching prudential combination, do we trust to premature generalisations? to mere maxims, and unimpressive precepts? Do we give only dim views, and transient glances, as if for the very purpose of perplexing?—Are not rather all possible particulars passed and repassed before the mind, till they have learned to rise at a word in their native shape and colours? Do we not devote whole years to arithmetic, to the rules observed in making bargains, the laws of commerce, the marks denoting the qualities of dead stock, and of live stock? Is not every case of ruinous misconduct pointed out as a warning? every opposite case as an example?—What is this but the *anatomy* and *physiology* of external prosperity? And as the pupil is more thoroughly initiated into the one, does he not proceed with more certain steps towards the other?

The very same reason that makes it advantageous to be able, at the shortest warning, to call before us a summary of our *pecuniary* circumstances, applies with at least equal force to our *personal* circumstances. To many it is necessary that the latter should be in a prosperous state, in order that they should subsist in *any* way; and is not the same thing necessary to us all, in order that we may subsist in a *desireable* way?

In forming the inclinations, so much pains are taken (and very justly as far as independence is concerned) to instil a lively regard to loss and gain, that one class of facts might have been expected to obtain more notice. For instances are by no means rare, where total prevention of success in life depends upon certain consequences left behind by disease, even where they do not involve the smallest degree of physical inability. This misfortune is too often accompanied or speedily followed by the destruction of internal peace; so that, by one and the same cruel operation, the frail edifice of our being loses support from within, just when it is most needed from the failure of that from without.

To have a distinct perception of the origin of these evils, it is only necessary to recollect how we are affected by the aspect of those who approach us. What impression, for example, does a sallow, unwholesome complexion, seams from the small-pox, scrophulous scars, the blight of beauty, and those marks which debauchery is apt to stamp upon the face, make upon the spectator? Is he not in general disposed to turn away in disgust from these appearances? or if politeness forbid him to give way to his feelings, do they not rise to a greater height for being dammed up?—And in what manner does the mind of those who perceive themselves to be objects of aversion react?—Instead of going for an answer to the theory, which explains how our habits are formed, I shall bring one of the most sagacious of self-observers to speak for himself. The late professor, J. George Busch, whose memory the city of Hamburgh is at this moment employed in honouring, tells of his having had the small-pox at nine years old, and though they were attended with no imminent danger, they left him badly marked. "I was "afterwards informed that previously to this "disorder I had a comely appearance. For "myself, I had never attended to the point. "But one thing I know well. After this "time, I perceived that those who visited "my parents, universally withheld from me "that kindness of attention, which with a "child is the first motive to render himself "agreeable, though they shewed it to my "well-looking brothers and sisters. All the "pranks that I, poor pock-pitted boy, to "whose feet no dancing-master had given "the right position, or drawn the head "from between the shoulders, played in my "vivacity, might indeed have been per-"formed with an ill grace.

"This too, I should observe, was the "period when the chief art of education "consisted in hard words and blows. On "me reproaches rained from all quarters. "When my parents, who alone treated me "with any degree of sense, carried me to "Haarburg, my grandmother and a brisk "grand-aunt so maltreated poor George, that "my father and mother were extremely "unwilling I should repeat the visit. In my "grandfather's favour I stood all the higher. "for he was stock-blind; so could not judge "of me by my exterior." From his subsequent history—particularly from the hypochondriacal complaint, into which he afterwards fell—it should seem that this man, active, wise and useful as he proved, never entirely recovered of the wound inflicted on his tender mind. What then can be hoped where fewer resources exist, and fewer favourable circumstances concur in riper years?

Unkindness is the natural provocative of malignity; nor will it often be in the power of Fortune to counteract, by any favours, the early union of painful feeling with a large mass of ideas. So important is it that nothing should intercept those sympathies of voice and countenance which are to children a source of such exquisite enjoyment, and a portion, by no means contemptible, of the discipline of humanity!

Unhappy feelings are said to spring from the consciousness of deformity. If the remark be just, these feelings should, in my apprehension, be ascribed to slight or insult rather than, as they commonly are, to envy. Wherever young persons, that have unwarily been suffered to deviate into crookedness of growth, are kindly treated, they do not, I am sure, yield in temper any more than in talents, to those of more erect figure. Malignity is not, I hope, supposed to have any particular affinity with dwarfishness. Yet superiority of stature must be felt as an enviable distinction. But the contrary defect, being precisely that which at an early age is least offensive to sight, offers the inconsiderate small temptation to adopt a manner, injurious to the disposition of children.

Doubtless when the study of our common nature shall have assumed its station among the pursuits, which no person of education can have neglected without discredit, corporeal blemishes and defects will seldom be suffered to take place; and where they do meet the eye, the spectator will have been tutored to too high a sense of humanity to degrade his countenance into a mirror, in which they are reproachfully reflected. Disgust is commonly the self-created torment of those who have no evil to plague them, *except* want of employment for their time. Few things, in fact, after having attracted rational curiosity, excite disgust. And curiosity, directed in the manner best entitling it to the name of *rational*, will gradually extirpate this factitious evil, and prevent every rising of unkindness towards innocence, oppressed by misfortune. Nor will it be slow in producing another effect, equally happy and closely allied to these. We shall derive from it sagacity to discern, and firmness to resist, pretensions founded on mere shape and air—pretensions which, when admitted, induce disregard of all worthier titles to esteem. Virgil has exhorted the Romans to spare the submissive, and to beat down the proud. But to encourage those who ought to be repressed, and to repress those who ought to be encouraged, still remains among the most common errors of mankind. That it in great measure originates in superficial acquaintance with the laws of their own existence, is an idea which may escape instantaneous condemnation, if these two truths be recollected;—that the more deeply the understanding penetrates into the productions of art and nature, the less does the fancy dwell upon the qualities that strike at first;—and that more accurate information often reverses the original effect of superficial qualities. I know not whether I may so far presume on my credit with the reader as to add that however the error arises, it requires correction scarcely less, as may be made to appear in a future number, for the sake of health than of morality.

When a person, tender of the feelings of others, once resolves upon an explanation, face to face, with a friend, he usually sets down the conversation at some distance, in hopes of finding a smooth approach to the disagreeable spot. I have just been practising a stratagem of this kind. For I am exceedingly apprehensive that what in strict order should have been introduced two or three pages earlier, will sound more forbidding than any thing besides in the whole course of these papers. To inspire my readers with resolution enough to venture on a path, which they may be apt to regard with dread, I have exposed beforehand to view, a small specimen of the advantages at which, after moderate exertion, they will be sure to arrive. But let them by no means be content with what I hold out. I invite them to fix their imagination, as strongly as possible, on the desirableness of existing in a state of constant preparation for the exigencies of thought and action. Of this state I neither dare attempt to enumerate the advantages nor the disadvantages of its opposite. For I should appear an insufferable dealer in common place.

Though scarce any one does, in reality, advert to this important matter, except in the gross, and by starts. Nor do I see how it can be properly comprehended or explained, unless that science, which *our* customs confine, to a particular profession, should be drawn forth from the corner into which it has been so undeservedly banished, and established with due honours in the midst of society. For by saying, " take care how you allow " yourself such an indulgence, as it will do " you mischief," I ought not to suppose I shall make much more impression than if I had uttered so many words in an unknown tongue. And in fact, the import of the words must be unknown till it is found in the dictionary of nature.

It will be perceived that the science to which I allude, is the science of the human structure and functions. A certain proficiency in this science, will give a firmer hold of that body of information, or those sets of ideas, in the right application of which *personal prudence* consists ; and without it we shall risque falling short of the perfection desirable in an endowment, which at any moment may be summoned into use.

One general rule applies to *physical* instruction. It should, in the first instance, be always addressed to the senses. By the help of books it can go on, but at its commencement, the very productions of nature herself ought to be presented. I know, indeed, with what abhorrence the processes and preparations which disclose the interior structure of animals, are considered by the unreflecting. And I have lately read of a country—not Great Britian—where two centuries are required to mature any highly useful proposal,—one century for discussing, and one for carrying it into effect. In Great Britian itself, though we see a new amusement adopted the instant it is announced, the more widely a plan promises to diffuse its utility, the longer interval must, in some cases, be allowed between its proposal and general approbation; and again between its approbation and execution,—especially where the profits are not to be reaped immediately *en especes sonnantes*—in money that you may ring on the table, or chink in your purse.

With these checks in full view, I shall venture to advance that nothing would be more easy than to establish in every one of our considerable towns, lectures for a *mixed* audience, on select subjects of ANATOMY. And there is, I should imagine, no considerable town where moderate encouragement would not induce some active member of the medical profession to engage in their delivery. To travelling professors of science it might by degrees answer, if to their philosophic apparatus they should add an assortment of preparations, models and drawings on a large scale, suited to popular anatomical lectures. That teacher of chemistry, who when treating of respiration, has been accustomed to exhibit any of the contents of the chest, if he did not excite a deep interest in his audience, must have been exceedingly unfortunate.

In asserting that much of anatomy will bear to be rendered popular, I hazard nothing on speculation. I have been happy enough to get the trial repeatedly made. It answered compleatly, and is likely to be persevered in. Having given an account of this interesting experiment in *a lecture introductory to a course*, delivered in Bristol, (*Johnson. 2d edition*), I shall now only state what a cloud of witnesses would be ready to confirm—that the elegance of many preparations afforded a pleasure not to be surpassed by any exhibition whatever, insomuch that taste itself found full gratification—that parents were struck with the obvious application of what they saw to the prevention of disasters, common in families—and that nothing took place which could give the most timorous female delicacy alarm. I might produce precedents enough from some foreign countries, where courses of *anthropology*, for so they are entitled, are common enough. But I foresee how easy it will be to object,—*Yes, such a thing will do very well abroad, but it does not suit the feelings of English people.*

This statement will, I trust, procure my scheme a few advocates among the ladies. But I have said enough to shew that I am not such a stranger to the prevailing modes of thinking, and to the difficulty of changing them, as to expect much immediate success. Innumerable mothers, though accustomed to consider themselves but as instruments for promoting the well-being of their offspring, and to act and suffer accordingly, will feel an unconquerable repugnance against paying any sort of attention to our internal confirmation, even though the means of instruction be placed ever so conveniently within their reach. That we are *fearfully and wonderfully made* they will piously believe, but every particular proof of this proposition they will consider as lying beyond their province. Most sincerely respecting their general tone of feeling, I renounce every hope of working to any purpose upon these matrons. And I shall but request that they will take into deliberate consideration a case, which I shall lay before them, as it arises out of the notes of a traveller, who has lately visited a community, separated by extensive deserts from the rest of Africa, and not included in the discoveries of Houghton or Park, though Hornemann may be expected to bring some farther accounts, relating to it.

The exemption of this people from wars, from intestine distractions and epidemic complaints, together with the abundance of every thing required by our grosser appetites, presents, in the original narrative, a most delightful picture. And this picture is heightened by one circumstance, upon which our traveller particularly insists—the universal force of parental affection.—It is not, he thinks, exceeded in any of our polished states. Yet a spot thus seemingly destined to be the seat of happiness does a single circumstance actually convert into a scene of general misery.

This circumstance is a degree of imbecility in parents, which seems to bear a near analogy to that species of insanity, where the patient is perfectly reasonable on all subjects but one. The cause of unhappiness begins to operate, and is extremely fatal, in early infancy; or, if not fatal, is then productive of injury, that seldom terminates but with life. Though a few, who, in their early years, have to thank their stars for an escape from what is considered as *proper* care, present the most pleasing association of self-enjoyment with the capability of enjoying external things.

"The peculiar cause of misery here," premises my author, "cannot be heard without a "smile, though I hope no one is capable of "relating its effect with a gay countenance." In short, an absurdity, inconceivable by *us* who manage so much better, pervades all the arrangements of common life. The meats though of excellent quality, are made offensive as well as pernicious, by bad cookery. The purest water is rendered brackish before it can be drank. Every infant has his limbs cramped by the national dress. And so uneasy are the very beds, that the sleepers are kept for ever tossing in distressful dreams. I pass over the strong remonstrances which our humane traveller hazarded.— "This very people," says he, "who take "so much pains about the instruction, mo-"rals and establishment of their children, "could not be persuaded to stretch out a "hand in their behalf against abuses so de-

"structive; though that would have been "sufficient for their abolition. They would "hear of no new practices. Their system "was founded, they said, upon immemorial "custom. The cooks, upholsterers, taylors "and other artisans of the country, had not, "as they were taught to believe, their equals "in the universe. In short, insight into "the things, to which I ventured to assert "that all ought to attend, was the province "of none, but certain persons, of whom "each had his peculiar department."

Thus far my narrative. An image of supineness so barbarous, will, I am sensible, shock my readers—my fair readers in particular. But it is by no means my desire to shock any one. I make the quotation solely, as I have said, to put a case.—Should there exist a nation equally infatuated in regard to the *internal* causes of pain and pleasure, by which of the two systems are the respective parents most likely to defeat their fondest wishes? Which set of children would a superior Being, who regards with an equal eye different human tribes, and their discordant manners, be disposed most to compassionate? Would it in his estimation avail the Africans to say that attention to such things as food, apparel and beds, was a deviation from traditionary practice, and in itself insuperably disgusting?

On the opposite side, could it be admitted as an apology that nothing imaginable was left undone, but that it was looked upon as a care, foreign to heads of families, to qualify themselves for judging of the interior condition of the young members? that such a qualification could not be obtained without trouble and disgust? that the prevention of that morbid sensibility, which brings with it continual suffering, and of that debility which renders the first disorder ruinous or fatal, to the constitution, belongs to *particular persons whose department it exclusively is*, and of whom people of other callings *are taught to believe, that they have not their equals in the universe.*

It can easily be conceived how the Being would argue with that delicacy, which shrinks from the contemplation of the most admirable works of creation, and combining with inconsideration, exposes those, who indulge it, to the misfortune of losing the most cherished objects of their affection, or even of seeing them pine under irremediable wretchedness. As to the pretended perfection of that art, to which the apology alludes, it would, in the supposed circumstances be a fine subject for raillery or indignation.—For what, it might be asked, is the result afforded by the clearest and most authentic documents, relative to this art? Do they, or do they not, bear, that above one-third of the infants brought into the world, perish before attaining their third year? that few reach the term deemed *natural*? and that of these few, a large proportion languish (perhaps half their lives) in sickness? and this, though in innumerable instances, the most *eminent* practitioners of the said art—men supposed to carry into every house they enter an atmosphere of security, as the anchorites of old did of sanctity—exert their utmost skill, with every possible advantage to second them?—Assuming a milder tone, our Being might illustrate the culture of the person by that of the temper. For who, with a view to ensure to any human creature an uninterrupted succession of placid feelings, would depend altogether on *external* causes? Does not uniform experience attest, that it is the *internal* difference almost alone, and not the difference of circumstances, which makes this man contented, and another the contrary? So it is in a great degree as to health. Unhappiness, it is true, remains unhappiness, whether it arise from within or from without. But in the latter case, it adheres loosely, and may be thrown off with the cloaths, left behind at the last inn, dissipated by the smile of a mistress or rolled into the lake of oblivion by a turn of the wheel of fortune. Is there any thing we can rely upon for giving satisfaction to the unhealthy and the morose? Do not both frequently find a sting in that which gives all others the greatest pleasure? And will not both, at times, be as ill at ease on a bed of roses, as Guatimozim on his red-hot gridiron? So that she who has secured to her progeny all the *external* means of well-being, may estimate her achievements at about a tenth of what every one must wish on the part of a parent; whereas to secure the *internal* means, is to go as far as nine parts in ten.

Leaving those who, without knowing its aspect, much less its applications, reject at first sight that which forms a most essential part of the preparation for a due discharge of the parental trust, I proceed to the suggestions which I have still in reserve for others. Well-conducted lectures on the construction and movements of the animal machine, will enable the hearer, as I have already intimated, to avail himself of other aids. It is in this view that lectures on the different branches of *physics* are capable of being rendered so exceedingly useful. Books and engravings alone may answer one good purpose. I shall hereafter point out the grossness of medical impostures, which frequently pass with persons who have enjoyed the most liberal education. They are so gross that they can often be detected by a quantity of precise information, which to the whole of physiology does not, perhaps, bear a much greater proportion than the first four rules of arithmetic to the whole science of quantity. Now more than this proportion may be obtained by reading. I regret my inability to name any English work, calculated exactly for conveying it. Treatises designed to guide the judgment of the physician, or the knife of the surgeon, it is obvious, will not answer. But until the author of these essays, or some one else, shall succeed in measuring out information according to the wants of the common reader, select portions of such works as we have, appear to be the only recourse. Most modern writers, to a description of organs, join some account of their functions: and I am not acquainted with any decisive reasons for preference. But not to leave the uninformed wholly without directions, I shall name *the compendium of anatomy, for the use of students.* 2 vols. 8vo. (Davies). Mr. John Bell's unfinished *anatomy of the human body* (Cadell) recommends itself, by a spirit, unusual in this class of publications, but effervescing occasionally with too great briskness. The learned reader may apply with most advantage to Dr. Soemmerring's classical volumes *de corporis humani fabricâ*, in which constant reference is made to the most accurate figures. A *compendium of anatomy*, in 3 vols. 4to. by the very accurate Mr. Andrew Fyfe, might have been mentioned as the best adapted collection, if attention to cheapness apparently had not caused the engravings, to be too little laboured. The *morbid anatomy* and the illustrative *series of engravings*, by Dr. Matthew Baillie, 4to. published by Johnson, will, both together, afford much instruction concerning changes of our organs, whose causes ought to be universally understood.—Though in many situations intelligent curiosity may not at present be able to obtain complete satisfaction, such helps as can be had ought not to be disdained. It is no barren course with but a single prize at the goal, which I am exhorting the public to traverse. At every distance will be found some recompence for exertion; and the recompences will continually increase in value, as further advances are made.

For all the labour that can be bestowed, sobriety of sentiment alone will be an abundant recompence. For opinion without information, is universally baneful: and where life is at stake, it must continually lead to fatal consequences. In questions of medicine, the most obstinate belief is usually coupled with the most profound ignorance. What we hear uttered with oracular confidence, in sick apartments and at convivial meetings, over wine and round the tea-table, when it is considered as idle speculation, can only provoke a smile. But if it be regarded as, one day, destined to decide the fate of infants and of invalids, it must excite pity and horror.

In tracing to their origin many generally received maxims, medical and diætetical, the unprofessional enquirer will meet with the most instructive lessons. He will find them derived from the most fantastic combinations. How else indeed could they have been framed? Let us imagine a savage alarmed at the sight of blood, gushing from his mouth and nostrils. Will he not grasp at the smallest chance of relief? and if a ruddy substance falls in his way, may not the idea of trying it flash across his mind? Does the haemorrhage by chance subside — behold a remedy in credit! But we do not seize analogies of colour so readily as analogies of form. When once rude anatomical ideas are acquired, roots or leaves are concluded, from resemblance in shape, to be medicinal for certain organs. This is the celebrated doctrine of *signatures*. Nothing, we may be sure, could exceed in caprice the philosophy of the golden age. Two events, of which one causes considerable sensation, fall out together. A conjurer, priest or witch in repute, connects them as parent and child. The people have faith. When the parent appears, the offspring is expected, and ten thousand to one, expected in vain. No matter. The people have still faith. The child ought to have appeared, but was perhaps seized with an undutiful freak, or fell lame by the way. The good people persist in believing as they had believed at first, and all along find themselves comfortable in their reliance on superior illumination.

Should any hesitation remain as to the justness of this deduction, it may possibly be removed by the following incident, which happened within these few weeks, and which I hope the reader will believe, because it is in character, and also because I can assure him that it comes to me from authority I cannot doubt.—A certain author of advertised medicines, whom rumour declares to have more solid reasons than any brother author whatever of the age, to congratulate himself on his publications, undertook a considerable journey. On the way, one of his companions turned the conversation to physic; and asked the compounder of panaceas, what was the cause of the late bilious flux at Liverpool. *Sir*, answered he, *I can attribute it to nothing but the prodigious quantity of flies this summer. Never were so many flies! All the meat was fly-blown!*

The doctrine of diseases gradually puts on a more scholastic form. The blood is supposed the life, because any considerable loss of this fluid is immediately fatal to animals, like man. The blood, therefore, becomes the subject of speculation; and is imagined liable to changes, which liquids more open to observation undergo. As wine turns sharp, and as oil turns rancid, so in different circumstances may the blood. The idea of its *ropiness* gave rise to that practice of drenching the sick with ptisans, which to the present hour subsists in France. On the continent grains of rye, enlarged by disease *(secale cornutum)*, have the name of *mother-*

corn. The name, whencesoever derived, (and perhaps the tumefaction suggested the analogy of pregnancy), seems to have occasioned its employment to facilitate labour. It is wonderful with what avidity prejudices in physic as well as in religion are imbibed, and with what obstinacy they are retained, by the untutored mind. And in whatever case premises have not been accurately prepared with conclusions, in that case must the mind be considered as untutored. On the subject of these essays, it would, in most instances, be treacherous complaisance to make much difference between the peasant in the field, and the philosopher in the closet. At this day perhaps, both one and the other sometimes set about to regulate their constitution, according to fancies, conceived under the shades of Dodona, and amid the groves of the Druids, blended with the crude ideas of Grecian speculatists, and of their Arabian disciples. For these remote fountains have all contributed to swell the stream of modern popular error. Opinions cast off, like thread-bare apparel, by the faculty, go each in their turn to the public. *In the politest and best informed circles,* according to newspaper phraseology, no day probably passes without its victim to some antiquated hypothesis concerning the *purity and impurity of the blood, the acrimony of the juices, the transmigration of humours, the salubrity of the air of this or that spot.*

Hence arises a distinction, most important to the unprofessional student of the laws of his own existence, between *positive* and *negative* doctrines. It is upon a knowledge of the latter, that all for whom I am writing, ought to lay the greatest stress. And so far as any one has informed himself respecting them, he may be said to have made progress in the art of self-preservation. The term *negative* will serve for a definition. Many suppositions reign for a time in the schools, become exploded, and afterwards make their fortune in the world. I need not say, that the proofs of falsehood or want of foundation in these suppositions, constitute the *negative* doctrines of medicine.—And it would be well if, every generation at least, a popular report of the reasons for their rejection were published by authority, could there in such a case be any authority, which mankind might safely trust. The use of the report would be to give fair warning against doing what is always the worst thing that can be done—acting upon notions, ascertained to be groundless.

So much is there to unlearn on the present

subject, that to reduce the mind to that *blank* state in which, according to Locke, it originally exists, would be no mean advantage to four out of five among those, who may take up these essays. The author is certainly accustomed to see invalids, for whom it would be happy if their whole mass of ideas—provided those were included that relate to the means of their recovery—could be abolished. No physician but must have found even disorders less difficult to manage than patients, and both disorders and patients together, less difficult than the friends of patients.

For so rapid are often what we call our reasonings, that we scarcely advert to the obscurest subject before we have *made up* our mind. Of a system it is still more true than of a fortune—come by it how we may, we must have one. In antient Greece a philosopher, who had never thought of ascertaining the constituent parts of any substance whatever, would single you out the element, of which the DEMIOURGOS fabricated the universe: and without troubling himself about the structure or properties of the stomach,* he

* Voluerunt Graeci statim esse philosophi naturae, antequam literas linguae naturae callebant. — Hoc modo de concoctione alimentorum disputarunt, antequam conformationem ventriculi noverint, quomodo hujus succus oritur.

Riegels Philosoph. Animal. I. 3,

would dispute with you all day long about digestion. Practice this convenient logic as much as you may on other occasions, it behoves you to take care how you suffer it to tamper with your person.

Those members of society upon whom the lot of others depends, being supposed in possession of the requisite information, how shall they most advantageously apply it? The question is of too much importance and extent to be slightly treated. And as the naked rule will be less instructive, I shall hereafter endeavour to teach by example the method of applying physiological knowledge to domestic use. At present I shall only observe, that writers, who are nothing wanting in zeal, when they inculcate the necessity of making all things else in the practice of education, subservient to health, have appeared to me to labour under the defect of too passive a plan. For what signifies it, how well you mould your mass, if every impinging body is capable of changing its form? Consider whether it be not your wish to turn a self-maintaining agent out of your hands. If so, manage your pupil as little, and let him manage himself as much as possible. The first care, it is true, must be of habit. But to habit I would not delay to join reason as an associate. Habit may be safe enough while under the wing of authority; but as soon as left to himself, bad example often quickly proves more than his match. With the knowledge of good and evil by his side, Habit will walk through a world of temptation unseduced.

But there is an idea which, as lying heavy on many a heart, it might perhaps have been incumbent upon me earlier to do my best towards removing. Is not the counsel I give in effect to hang a naked sword, by a single hair, over the head during the repast of life? or to waste that time which should be devoted to the free enjoyment of the pure scenery of nature, in exploring a gloomy labyrinth, full of spring-guns, where though you may escape by incessant caution, the fear of an accident is almost as bad as the reality.

Having in early life been altogether a stranger to medicine as a trade, I naturally acquired the habit of regarding it purely as a body of doctrine, productive of certain advantages to society, and as respectable or the contrary, in proportion to the amount of those advantages. For the same reason perhaps, my own first impressions remain fresh enough to enable me to sympathize completely with those who feel themselves heart-struck, when they come first into view of the host of human maladies. I may therefore venture with the greater confidence to pronounce that these terrors will immediately decrease in violence and soon vanish, if we do but maintain a steady countenance. Those who have felt them most strongly, will recollect how rapidly they chace one another across the field of the mind, till it is left quite clear. And on the solemn occasion when a medical student first faces his professor, to what class are we to refer him, if not to that of the ignorant? Are not his apprehensions those that attach to this condition? Do they not vanish with his ignorance? Can any one member of society flatter himself that, when once past the years of thoughtlessness, he shall escape the frequent sight of the forms, and the frequent knowledge of the effects, of disease? And will not those terrific phantoms, which once for all croud before the student and disappear, be flitting in view of others continually through life?

Let us also for a moment consider the general effect of knowledge. Is it not manifest, from the whole history of mankind, that the ascertainment of causes has scarcely been more beneficial in preventing real danger than in banishing false alarms? Is it to be believed that knowledge should produce this great benefit in all cases, except only in that of animal nature? The more closely we investigate, the less shall we be inclined to abide by a conclusion, so contrary to analogy. For where is the physician who has not perpetually occasion to observe, that the most harrassing and groundless apprehensions arise from want of acquaintance with the most obvious part of the structure and operations of the animated machine?—I now speak only of distress of mind. Another opportunity will occur for noticing the mischiefs, produced by blind efforts to obtain deliverance from imaginary maladies.

It is singular that the precepts of criticism should teach the same lesson with the history of knowledge. But so it is. There is not an author on the theory of fine writing, who does not tell you that to make impressions of dread you must avoid distinct images. But if you would shock your readers or spectators, till they are

——— distilled
Almost to jelly with their fears,

you should, on the contrary, only show a limb in the dusk and leave it to the fancy to make out the entire monster for itself: that is, you must give glimpses of your fictitious beings, just such as people catch of the ailments, that flesh is heir to, as they go along the beaten track of instruction.

Some philosopher remarks, that a life would hardly suffice to enumerate the vain phantasies of mankind. No title in the list would, perhaps, be more copious than that of suppositions and exaggerated diseases. One day you meet with an insuperable dread of dropsy, either because the body is at certain hours distended from indigestion, or because in consequence of casual sickness, a small quantity of bile is thrown up. How so trifling an accident can produce such dreadful consternation, every one may not be ingenious enough to divine. But the reasoning is fresh in my memory, for it was addressed to me by a person of no ordinary or unimproved understanding, since I began to commit these thoughts to paper. It ran thus. "Here I "have been vomiting bile. The bile is con"cerned in jaundice. After the jaundice I "may have the dropsy; then I shall die "the most odious and lingering of deaths." Again, the most insignificant inflammation shall occasion the shocking anticipation of a cancer. And this extreme of credulity shall meet the extreme of incredulity, in sensible and carefully educated people. I do not say *well-educated*, as in these very distresses I find a proof of something wrong in the mode of thinking, which in matters of important concern, argues a capital defect in education; and as I am in general convinced, that for entertainment and use, the study I recommend is preferable to the four greater and three smaller sciences—provided one must be excluded in favour of the other. I believe too, that this study would be the best preservative against that weariness of living, which is among the most hopeless of our miseries, and the surest forerunner of suicide.—It is no uncommon thing to hear invalids confessing that they have for a time withheld their suspicions, respecting their own state, because they could not bear to contemplate it steadily. And very often the practitioner of physic is forced to prescribe for the alarms of patients, when there is no call upon him from their danger. To treat their false fears lightly, would be attended with the certain effect of forfeiting their confidence. And then there would be a degree of danger lest they should find the terrible and sublime, which many a quack bill so happily blends with the soothing, more congenial to their feelings than unadorned truth.

But after all, is not the enquiry, whether people should or should not have their notions respecting diseases and remedies, fundamentally idle? Do we not find that they have and will have them? And then does it not remain, as the only question worth considering, *whether any pains should be taken to render these notions as accurate as possible in themselves and to give them for their boundary a full sense of ignorance?* To suppose any reader incapable of answering such a question for himself, would be too injurious.

That this sort of truth cannot be had, more than others, without the trouble of seeking it, may be proved by a profusion of examples. We have, in a considerable variety of cases, the good fortune to possess circumstantial memorials of persons, not merely well-informed, but distinguished at once for their diligence of research and the length to which they have carried their scepticism. And in these cases, the justness and general power of the understanding present a most humiliating contrast with its imbecility in that one thing, which concerns us above all others put together.—In what opinions an enquirer, whom Dr. Adam Smith (with the universal consent of all his learned and judicious friends), places "at the very head of the whole literary tribe, "existing in Europe," could acquiesce, we may discover from his own letters. Dr. Tissot "assures me that in his opinion, the moisture "of England and Scotland is most pernici-"ous, the dry pure air of Switzerland most "favourable, to a gouty constitution." So says Mr. Gibbon, (*Gibbon's memoirs and Letters, No.* 177.) But what says Mr. Gibbon's experience? In less than two months after he begins to respire this salutary air, his "*old acquaintance*" returns; and, in spite of Dr. Tissot, is brought on by the dry N. E. wind, (No. 171). After an interval, for which he perhaps was indebted to a greater degree of abstinence than he had practised in London, where, he says, "the late and long "dinners *would* soon *have been* prejudical to "his health," this *long-lost* acquaintance makes amends for his delay by a visit, in which he occupies him solely for three months (No. 182). Even then he leaves behind him the terror of his return (183), occasions "wearisome days without amusement, and "miserable nights without sleep," "discon-"certs all his plans" (No. 193—4), returns for a longer season and with more severity than ever (No. 203), quits the sufferer in a condition to fall a prey to an attack, seemingly not formidable, since "the medical gentle- "men expressed no fears for his life," and so not only brings a host of evils on his latter years, but prevents him from attaining nearly that age, which has been often reached by persons, originally not less feeble. "Let us now," says Mr. Gibbon, "drink and be "merry" (*Letter Dec.* 31, 1792). How far he was in earnest, a stranger finds himself at a loss to determine. Though he might not carry his observance of this precept so far as sottishness, I still suspect that but for the Madeira "improved by age," for the safe arrival of which he repeatedly expresses so much anxiety, he might never have had such a cluster of unfortunate reasons for renouncing his faith in dry air and Dr. Tissot.

The claim of Frederic II, to be accounted the most efficient character of his age, seems much less disputable than that of Mr. Gibbon, to rank at the very head of the literary tribe, existing in Europe. In sagacity the prince will be allowed at least to equal the historian. Frederic, who was so much happier in single sallies of thought than in a chain of reasoning, had, it is certain, the justest sense of the advantage of *true* self-knowledge, namely, of that which is acquired by comparing the structure of the inanimate body with all that the living man does, says, thinks, and feels. Even on subjects of medicine he occasionally took more than royal pains to form his opinions correctly. That he did not begin enquiry at the right end, but either neglected the only sources of genuine information, or failed to draw from them enough for any useful purpose, must be lamented both on his own account, and that of others. Else he who so widely influenced his contemporaries, might have given their pursuits a direction more favourable to humanity. There is indeed hardly any limit to the benefits, that might be rendered to the community of mankind, even by a rich private individual, who should lift his conceptions to our highest interests, and devote his means to their promotion. Alas! for what an infinity of sufferings would relief or prevention be at hand, if a fourth part of the sums squandered upon horses, and dogs, and weeds, were wisely appropriated to the service of the human species.

Mr. Garve, by whom his great king has been, in so many respects, successfully appreciated, is particularly struck with what he reckons an anomaly in this extraordinary character. Frederic, it seems, required that a disorder, of which he well knew the cause to be deeply rooted, and widely diffused through his system, should be immediately removed by a specific remedy.

"If a great personage of the common "stamp," observes Mr. Garve, "or even "if a person of reflection and of elevated "ideas, but utterly unacquainted with the "œconomy of the human body and with "medicine, had made such a requisition, it "might surprize, though it would not ap-"pear unacountable. But how Frederic—"he who among his other studies had not "been inattentive to the nature and origin "of diseases, and who has given out the "limits of art for much narrower than what "they really are—could seriously in his last "illness exact this from his physicians, is a "fact, in no way admitting of explanation. "Every invalid indeed makes such requisi-"tions in a certain degree, that is to say, he "has wishes of the kind. But he keeps "them to himself, well knowing their im-"possibility. And perhaps the king ex-"pressed them only to give a different turn "to his reproaches against physic and physi-"cians, on account of the little certainty of "the aid they afford."

"As little is it conceivable how a man,

" possessed of such self-command as to be
" capable of renouncing every comfort and
" every sensual enjoyment for the sake of
" public affairs, could be so much a slave to
" his relish for certain dishes, when he found
" the gratification of his appetite was fol-
" lowed by sufferings so severe. What! un-
" der the influence of motives, deriving their
" force from self-preservation, or from the de-
" sire of avoiding excruciating torture, could
" he *not* command himself? and yet *could*
" command himself, when the object was
" nothing more than to gain honour or set an
" example? Could considerations, seemingly
" so much feebler, counteract passion more
" effectually than the love of existence?
" The great man then, it should appear, re-
" strains, with more ease, a keen sensual ap-
" petite, when he has to attain certain ends—
" to execute an enterprize — to gain the
" opinion of the world — to fulfil official
" duties, than even when life itself is at
" stake!—So it is!—The finest fibres in our
" moral organization, are at the same time
" the strongest; and the coarsest, in some
" respects, the weakest!—It is so with great
" spirits in particular. In such, ideas prepon-
" derate over bodily sensations.—The inter-
" est of activity — that of useful designs

"prevails over the interest of self-preser-
"vation. — If none but this remains at
"a certain period; it concentrates itself
"within the interest of sensual gratification.
"*What,* thought the king, *is life, if there*
"*is nothing more to do, or to enjoy?*

"But equally extraordinary, during his last
"illness, is that association of vigour with
"imbecility—that freedom of mind which
"could occupy itself with other objects
"besides the disorder; and that capacity of
"prosecuting speculations in literature and
"politics, when bodily pain was so forci-
"bly soliciting the whole attention.—His
"concern for his guests during these last
"days, speaks as highly in favour of his
"heart, as the part he took in conversation
"in favour of his understanding.—And even
"the circumstance of his bringing on the most
"dreadful symptoms by eating, and then
"bearing them without complaint, exhibits
"a remarkable example of what philosophy
"indeed teaches, but what *such* cases only
"can impress—*that man is a strange tissue*
"*of contradictions.*"

Of moral difficulties, stated and relinqui-
shed by the greatest master of the philo-
sophy of life, that the most literary people
of Europe, and perhaps, Europe altogether,

has produced, I should not presume to attempt the solution, if a new approach did not sometimes give a clearer view of obscure questions.—Not to submit to regulation in the article of diet, is in character for a veteran despot. To suppress every inclination to complaint, is not less in the character of a man, who through a long life, has looked down upon human creatures, from an elevation too great to allow him ever to think of an appeal to their pity.—Here then, is really, nothing contradictory. We see effects produced by their proper causes. But these are irrelevant particulars. The principal phænomenon falls within the pale of principles, thrown out in the preceding pages; and it deserves to be put to heart by readers of every denomination.

In the soul of Frederic, let us reflect how large a space was occupied by the desires, which the estimator of his motives represents as constantly predominating over his propensities to epicureism. Did not the attainment of qualities, that go to the composition of the hero and the statesman, occupy him, almost entirely, during his best years? On other objects, what did he throw beyond a casual glance? For what purpose did he open the encyclopaedia of medical science, but to give his satire a finer edge, to add variety to his allusions, or to extract a ray of light for some obscure passages of his philosophy. The ideas, therefore, of power and glory occurred most frequently, and were interwoven with his whole body of thought and feeling. The ideas belonging to the causes of health; were fainter even than the remembrance of the pleasures of the palate. And, in the great king of Prussia, the force and frequency of these several trains determined (as in all mortals they determine) the order and preference of actions. Our moral organization is pretty much what it is made to be. The strength of its fibres may be said to depend principally on the hand, by which they are drawn out. This simple statement, I flatter myself, removes Mr. Garve's difficulties. Frederic, in his moments of suffering, wished for that which, in his speculative moments, he had considered as impossible. So loosely were these speculative convictions fixed in his mind, that the first tide of suffering could sweep them all away.

This illustrious example is, at the same time, one of the most instructive, with which I am acquainted. It determines with sufficient precision, the necessary quantity of instruction. And from the nature of the case, the quantity determines the mode.—In fact, if we would do full justice to the child, whose reliance is upon us, we should first teach him his duty, or interest, as a human animal:—interest and duty being, here at least, perfectly synonymous.—The interest or duty of that vocation, which makes him a human being of peculiar habits, ought to be but a secondary object with his preceptor. Of so much more consequence to happiness, are the circumstances in which we agree, than those in which we differ. Nor can any care, employed on the variable accessories, ever make amends for the neglect of the essential and common properties of the species. Unhappily, in general, the accessories absorb all anxiety, and the essential properties are disregarded.

It may sound extraordinary, if I say that valetudinarians and invalids, whose spirit is nothing refractory, every day experience sufferings, similar to those, which Frederic 11 brought upon himself by wilful indulgence. Enjoin a lively patient strict abstinence from such and such exertions. Obedience will be readily promised and sincerely intended. But, in a short time, some new impulse is given. The spirits flow at the old rate. The injunction is carried out of the thoughts.

The penalty is severely levied in suffering. The sufferer laments exceedingly—he really meant to forbear: but it happened so and so; and the caution was forgotten. What occasions this forgetfulness but the misfortune in early life, of not having had a particular class of ideas strongly impressed, and frequently repeated, so as to be completely blended with the rest?—It is not necessary in these cases, to have the excuse of vivacity. The dullest mortals are subject to the same default; and always for the same reason. If no trifling portion of human unhappiness be owing to the facility with which the healthy yield to every temptation, addressed either to the palate, or to any other organ of sensual gratification—if the opulent suffer from repletion almost as much as the poor from inanition,—we must still resort to the same want of particular and distinct impressions, as the cause, and to the communication of such impressions, for the prevention of these evils.

There exists scarce a person of weakly habit, or affected in a degree short of *bed-lying* illness, that does not receive injury from the inconsiderate usages of the world. Yet by these usages, which exact so much, and really obtain more than duty towards ourselves and those dependant upon us, a commerce is kept up, the most absurd that can be imagined. One party loses, and the other does not gain. Sometimes both lose—one while the penance imposed by civility lasts; the other both then and afterwards. "I know it will half kill me," said a lady the other day, on setting out upon a formal visit, "but it is a matter of *decency*, and I "must go." It is the glory of the modern code of politeness, to have set people at their ease. The boast is somewhat premature. Many burdensome observances have unquestionably been abolished. But one would imagine, that every body must wish the abolition to go on, till it sets those at ease, who have the fairest title to indulgence. It is not that the nuisance passes unperceived; nor is it merely the want of spirit to *abate* it, as in the quondam case of servants' vails. But the bad consequences are not fully understood; hence they are continually reproduced. I would not be so rash as absolutely to affirm, that complaisance destroys more lives than contagion; but even if a balance could not be made clearly out in favour of complaisance, I am sure that its separate account, if well looked into, would bring about an abrogation of conventions, indifferent alike to morals and enjoyment.

Wealth exempts from certain kinds of labour. But the wealthy stretch this privilege a great deal too far for their good. The vain forms, to which they subject themselves, and the consequent injuries, might suffice to convince them, that they cannot, with impunity, flinch from the task of learning how to take care of themselves. Of such exemption, I think it highly probable, that the nature of things will never admit. I am sure that, as physic and physicians stand at present, the most opulent derive from their fortune very moderate security. We see them every day undergoing what a great poet of antiquity reckons among the last of human calamities—that of consigning their children and their children's children, to a premature grave. If we reckon from the middle station upwards, it would, I believe, be more just to assert, that the unhealthiness of families is in the direct, than that it is in the inverse proportion of their wealth. Money, I perceive, can put all the members of the faculty and all its appendages into busy motion. But to what purpose? very frequently not to the purpose of deferring, one hour, the fatal crisis. If indeed, all this bustle console the sick, or the survivors, then must it be considered as well purchased, at whatever price. Otherwise, I often think the doctor's pomp and apothecary's mysterious hurry, fit only to figure, along with the efforts of the shopman and the vigilance of the nurse, in an *elegy in a city church-yard*, or *on the vault of all the Capulets.* So impracticable has it been found to reduce health to the state of a mere article of commerce, and so entirely does HYGEIA disdain to become the slave of PLUTUS.

It is obvious that what riches cannot do for their immediate possessor, they cannot do for those about him. Few families are happy enough, like Sancho Panza during his brief authority, to include in their establishment, a medical attendant. And if they did, there would still be calls enough upon the discretion of the presiding members. In health itself, there are moments when the destiny of an individual is decided by a single indiscretion. But in infancy, which may be considered as a valetudinary state, and in the course of a complaint of any hazard, it cannot but be that the conduct of the friends of the sick and those of children, should have the most powerful influence upon their fate.

The disposal of young people of an age to quit the paternal mansion, has always been an anxious and a difficult question. Notwithstanding our numerous treatises on education, we still want principles of sure application. When public and private education come into competition, health seldom misses the compliment of being mentioned. But it rarely receives farther attention, though it ought in so many cases to determine against all other motives. Sentence is passed by judges, who have little knowledge of the case, and less suspicion of their own incompetency. Some acquaintance with the progress of slow disorders, and consequently of remote origin, must be gained, before it can be conceived how frequently a large school is the antichamber of death. Masters of schools are not, in general, understood, nor do they I suppose, pretend, to have any accurate knowledge of the living frame, though that be the qualification most essential in a superintendant of youth. When a child is ill, their humanity, I know, will not permit a moment's delay in *calling in advice.* But how often does the internal enemy sap the constitution before he draws the eye of the preceptor, or alarms the pupil himself!

In these scenes, danger in a multitude of forms lurks round the puny and the young. A blow from a stouter companion falls unseen, and would be forgotten by him who receives it, if progressive blindness, or pains increasing with years, and rendering life uncomfortable, or terminating in a fatal disorder did, not keep the original *accident*, fresh in the memory. Probably, at a large school, a month as seldom passes without some such disaster, as a public rejoicing without a specimen of that destruction, which it is usually designed to celebrate.

I have heard the annual mortality in Bengal, from venomous animals and beasts of prey, estimated as high as some thousands. The numbers, ultimately destroyed by brute violence at schools, or spoiled for life, must, every year, be considerable. But it is not at schools only, where violence breaks out against the defenceless. In large families, it is by no means uncommon to meet with instances, where the faculty of feeling, or the vital organs, have received an irreparable shock, during some passionate or playful sally. The former seems to be more frequently the fate of girls, the latter of boys. An explanation of the manner in which the mischief happens, accompanied by expressions of just horror against the perpetrators, I apprehend to be necessary to an effectual system of prevention. Well-chosen facts, related minutely, but without exaggeration, would, doubtless, leave an impression as strong as can be desired; and such facts would usefully supply the place of prohibited or permitted fictions.—Ah! if writers of books for children *could* copy nature faithfully, they might give their imagination a long holiday.—The feelings, thus generated, would spread from family to family, till they became national, obviating many atrocities, and prompting infinite acts of kindness.

Hence we appear to derive one of the most solid arguments against corporal correction, or against its public infliction, at least. Propensity to imitation, and fondness for exercising power, will easily enable children, more advanced in growth, to discover a sanction in the example of parents and masters. At school, who does not remember to have seen a domineering boy adopt the mien and language of his seniors, when they proceed to execution; and, in this assumed character, exercise his cruelty with double satisfaction on some unresisting comrade, upon whom he forces the part of culprit?

Chills are sometimes irretrievably injurious: and the puny are for ever subject to chills at school. I know what may be urged on the subject of inuring to hardiness. Prejudice will still be in league with pedantry. But pedantry should be required to confer the Spartan constitution, before it introduce the Spartan discipline. Some boys and girls resist cold better than others; and these may with less injury, bear that tyranny in a stronger companion, which obliges them to warm his bed with their body: they may remain longer, without hurt, in water, or sustain long sittings in the cold of a school-room. But for whatever time a young person may resist such powers of mischief, if he is at last chilled, he is in danger. Either some violent attack will be the consequence, or the foundation will be laid for a fatal slow complaint. The less robust will become more and more enfeebled, till at last the powers of life fail; though not, perhaps, while they remain at school.

I have sometimes been struck with astonishment at the infatuation of the friends of children, who have nearly the same title to a mortal disorder, as to their patrimony. Let a child, whose ancestors, for one or two generations, have fallen victims to such a disorder, be himself delicate, still it is, not unfrequently, impossible to impress an effectual warning. Should the fatal symptoms not immediately threaten, (plainly because the season for their appearance is not yet arrived) the fate of the progenitors is disregarded. Unfavourable signs may even shew themselves to no purpose in the individual. For these omens bear an import, as the arrows in the quiver of Pindar resounded, only for the intelligent.* Merciless eagerness to secure the W—— —r or the E——n or some other *tic* hurries away the victim, tender in years, infirm in constitution, and perhaps, effeminate from ill-advised care. The aunt, the uncle, the guardian *hope* that the descendant will escape the fate of the family ; fools ! as if hoping against experience ever turned nature out of her course.

But I grow too warm. For it is really out of reason, to expect people accustomed, on all important occasions, to hire the use of foreign eyes, to see with their own, further forward than just to save the nose. Be it so. Let those, who have nothing of the commanding glance of a MOREAU or a LECOURBE, be excused from discerning the relative position of objects in a country, just

---

* ——— ωκεα βελη
Ενδον ενλι φαρελρας
Φωνανλα συνελοισι.

opening before them. But he must be an unobserving traveller indeed, who is at a loss with regard to districts, he has passed in a clear day, at his leisure. Yet it is certain that they, who are loudest in the behalf of public schools, are just such travellers. It is a point, you may easily satisfy yourself about. When they tell of the scenery of their childhood, that it was purely composed of limpid streams, verdant plains, and breezy mountains, you must hear them out without contradiction, to the end. Then question them quietly, and you will find that they have forgotten bottomless quicksands and pestilent bogs, though they had met with them, in every square mile of their terrestrial paradise.——Or should you prefer allusions to art, a little cross-examination will still supply you. If a certain author, for instance, be right in comparing a hypochondriacal body to the cask, in which Regulus was rolled to death ; you will find, that it has been in one of these same blessed manufactories, that their ailing encomiasts have had the longest and sharpest spikes driven through their sides.

To some young children, sudden violence is not more prejudicial than example is to others. Genius, in virtue of its sensibility, eagerly pursues the first pleasures of excess; and in virtue of the same sensibility, receives more injury than mediocrity, which, however, is carried along by the general movement, while it is the common sport to conspire to initiate dullness into the mysteries of debauchery.

For children, the frequent society of children alone, is nearly the worst possible society. It should be avoided equally for the sake of manners, morals, intellect and health. In their memorials, I find no circumstance so much in common to eminent persons, as familiar access, during childhood, to judicious and conversible adults. The moderns enjoy the advantage of the press; but they unaccountably relinquish the free intermixture of persons of different ages—an advantage, for every rational purpose of education, hardly inferior to what we derive from the press, and indispensably necessary to give this its full effect. There is, however, a feeling, kindling through society, similar to that which preceded the reformation of religion; and the coming generations will not be the victims of those popish errors, which have so long found an asylum in the recesses of our seminaries. Of late, it has happened that many persons, anxious that their children should attain robustness of constitution as they rise to maturity, and that they should, at the same time, be well-instructed, have consulted the author;—so many indeed, as to convince him, that an institution would, at this moment, find support, where glaring absurdities and ruinous abuses should be avoided, where the instructors should be companions to their pupils, and science, usefully active habits, good principles and personal prudence should obtain their due regard. Experience has taught him, that a much greater proficiency in antient literature, than usually takes place at schools, is compatible with all these objects. Nor can he ever advert to this subject without afresh deploring a friend, who clearly traced back the origin of his illness, to one of our greatest schools. From the fate of a parent, indeed, there was too much reason to fear for him. And on this very account, he should never have been trusted within those dangerous precincts: or at least, not till he had been confirmed in years and habits.

Though conspicuous among the parliamentary opposers of Mr. Pitt, (whom he considered, as of all politicians the most bloated with conceit, and as director of the longest course of clumsy experiments upon property and life, ever endured by an enlightened nation from one man), ministerial calumny wisely refrained from assaults upon his character. And highly as he was respected on the faith of rumour, he was still more esteemed wherever he was familiarly known. The more, concern for an increasing family caused him to revert to the consideration of his own experience, the less did he approve the system of education, followed in his own case, even when he viewed it without reference to the deep injury he felt that it had inflicted upon himself. Having unexpectedly emerged, for a time, into health, he actually began a course of self-instruction, and resolved, by the help of his ample fortune, to promote the useful arts. But an ill-advised residence in one of the few southerly countries of Europe, then open to the English, apparently brought back the danger it was meant to avert. In consequence, at an age little exceeding thirty years, the community lost a member, who united, in an uncommon degree, the best inclination and greatest means.

Through his illness, this excellent and amiable person followed the advice he deemed best for himself, as religiously as he had, at all times, regulated his conduct with regard to others, by the strictest principles of integrity. And whoever knows much of the retired sentiments of young people, particularly of the young women of the present day, must have been struck with their anxiety to act up to injunctions, which require the renunciation of the longest established habits and the constant practice of self-denial. And who but must be deeply affected, when such a disposition is of no use towards saving them from an untimely fate, though if well directed in time, it might have blessed them with a length of happy days ?

That *no one knows the value of health, till it is lost*, is among those paradoxes, which are oftenest repeated in a tone of lamentation. But where is the wonder ? If we come into the world without ideas, how should we expect to feel the value of any thing, till it is taught by express instruction, by observation, or by accidental experience ? And is it not a prerogative of the human species, and the glory of civilized ages, that they can avail themselves of the first to supply the want, or escape the injuries, of the last ? There is nothing, surely, in the nature of health, which should render a sense of its value either an innate principle, or particularly difficult to instil by proper means.

To be understood, it must, no doubt, be taught. Whoever will examine or observe, may satisfy himself, that we owe to instruction early begun and sedulously continued, those judgements concerning merit and demerit in actions, which prove, where they apply, such alert and faithful guides. To form a moral *sense*, for so the ready habit of certain artificial sentiments has been metaphorically styled, is not more practicable in itself, than to form a *sense* for health, or for happiness, which latter must be, in great measure, composed of the sense for health. It is, in fact, only necessary to proceed in the same manner as with regard to the moral sense. A broader foundation, indeed, if I am not mistaken, stands ready laid in our nature, to receive the sense for health. The labour would, therefore, be more easy and solid.

The individual, in fact, need only be taught to *sympathise* with himself, if that term may be applied to the feelings, excited by different possible states brought strongly into contrast. Consciousness of health, thus contemplated, will become just as much a source of pleasure as the consciousness of virtue. Of course, even while it subsists in its fullest vigour, its value will be fully felt. As to extraneous approbation and disapprobation, they can as steadily be attached to actions, affecting the person himself, as to those affecting others. Indeed, to say that they *can* be attached, is to speak too feebly. They *cannot but* attach themselves. The tendency of conduct, when clearly perceived, must involve censure or applause. That these have hitherto operated so much more as social than as selfish motives may, in part, be ascribed to the inevitable order of moral discovery. For it has been in this, nearly as in physical, where the desire of knowledge began at the remotest extremity of visible nature; and proceeding from the planetary masses to the inanimate bodies of our earth, fixes at present, with ardent and diffusive zeal, upon plants and inferior animals, but passes by the majestic species of man, as unworthy to be studied, but for the ignoble end of gain.

New suggestions justly excite scruples.—May not disagreeable and ridiculous effects be apprehended from putting mankind so minutely on their guard against noxious powers? What can we expect, but that plain old common sense should be turned out of doors, to make room for prudery in regimen; and that every generous energy should be crippled by coward caution?—But if no evil have followed from the pains, that have been so wisely taken towards putting us on our guard against deadly nightshade and against lead, why should any be produced by a complete reasoned catalogue of *poisons*, in the most comprehensive sense of that term? Lead does but occasion palsy; and whatever destroys activity and enjoyment, produces palsy too, of the worst species. For the advantage of sprightly vigour over pining sickliness, is greater than that of the enervated over the palsied, or of the palsied over the dead. Can we seriously fear, that if we suffer ourselves to be persuaded out of the use of poisons, both quick and slow, the feast of nature should not be various enough for a healthy appetite? Did any one, once master of their effects, ever pine for the berries of belladonna or the sweets of lead?

I see, indeed, that folks who *live by rule*, do not offer the most tempting example. Nor have I any intention of proposing it to the world. Just the contrary. There is nothing in my opinion, so much to be deprecated. Better, I sincerely think, not to live at all. But I deny that this is likely to be the lot of any one brought up with the knowledge of things hurtful, and in the practice of that knowledge. The difference between what I recommend, and you dread, is equal to the difference between a faquir and a stripling, enjoying in the presence of assembled Greece, the consciousness of powers, about to distinguish him as conqueror at the olympic games. For what are these *methodists* in meat and drink, whom we are all so justly averse to the idea of resembling? Are they such as prudential care set out in life with the science of good and evil? or have they, in ignorance, abused their animal powers till their sensibility acquired the unnatural property of deriving pain from the slightest impression, and could be no longer agreeably affected by any? Examine well, and see if they be not drooping, blighted creatures, frequently half-insane from a mixture of regret and remorse, with whom "the past is a scar, and "the present a sore," sensible only in the gross, from what quarter have issued the darts, by which they have been stricken, and so wild with fear that they hear the wing of the angel of death in the rustling of the gentlest gale of heaven.

These forms of misery have been laid to social refinement, as its genuine but monstrous offspring. Rousseau has proposed to annihilate the one, in order that the other might not deform the face of creation. It has, however, been rendered probable, that the unhappy singularity of Rousseau's temper and circumstances, drove him to this desperate expedient. The difficulty he found to satisfy common wants, the slavish occupations to which he submitted rather than owe obligations to bounty, and the bitter self-reproach, consequent upon some things he did in compliance with his intractable spirit of independence, all contributed to irritate him; till at last he perceived in the state of nature, the sole preventive for the diseases of society. So intolerable did he find these *relative* sufferings, (which, as a savage, he undoubtedly would have escaped), that the farther he receded from the institutions of polished society, the nearer did he seem to himself to approach happiness.

Knowledge of evil is the first step towards its remedy. We should, therefore have obligations enough to Rousseau, if the transcendant blaze of the beacons which he has set up in different parts of his Emilius, had not saved thousands from perdition, and if his insanity were no excuse for his extravagancies. But sagacity, greatly inferior to that of Rousseau, will discover that right does not consist in wild deviation from wrong. Nor does it require much acquaintance with what is related of the different modes, in which our species exists, to see that we should gain little, in respect to physical or ideal pleasure and pain, by the change proposed, even if we could disencumber ourselves of our habits.

To accelerate the progress of civilization, is probably our only practicable and only effectual plan. The sentiments of uncultivated tribes, as echoed by travellers, have, for the most part, a mournful accent. And of any such tribe, it would be difficult to prove, that they live in so much hope, and in so little fear, as to be worthy of the honour of our envy. If man is essentially a reflecting animal, the advice to push reflection as far as possible, might have been more wise than to renounce it altogether. It may, at least, be worth while to consider, whether the exertion of intellect cannot repair mischiefs, which it must be allowed to have incidentally produced, and whether that havoc of health and life, with which our civilization is justly upbraided, may not be the effect of the barbarism, which it still retains in its composition.

The effort to expel this alloy, as it is of private resort, may be a silent operation. In exerting it, there is no more occasion to wait for the consent, and no more necessity to incur the censure, of others, than in providing our apartments with more commodious furniture. Of our success, domestic felicity will be the sure reward. No firmer bond of family union can be imagined, than for one part of the members, in addition to existence, to have received from the enlightened love of another the faculty of enjoying existence.

Whatever one parent may suffer before a child is brought into the world, with all that both may do for some time afterwards, ought never of itself to be depended upon, for securing such a return as every parental heart must wish. These obligations can be known only, like the history of Gengis Khan, from extraneous testimony. The party obliged has no means of entering into them, but by reducing his dimensions according to a scale, not furnished by consciousness, and which it is seldom a peculiarly interesting operation to apply* to onesself. For kindness

---

* In this case, the imagination would be best assisted by portraits, or rather by family groups. I know not if it have entered into the head of artists, or their employers, to aim at promoting mutual endearment, by shewing a child his forgotten self, in some *interesting* situation with a parent.

to produce the desired effect, it must, at some time, be felt by the receiver as kindness. A surfeist of kindness causes a parent to be considered as a sort of importunate slave; and children are regularly ungrateful (*ungrateful?*) to those, by whom they are spoiled. The opposite sentiment as regularly arises, where parental care lays a train of gratifications along the path of life. With each gratification the idea of him who bestows it, occurs, and it has often the highest possible advantage,—that of being introduced by agreeable surprize. Such obligations, in regard to the warmth with which they are felt, must be the very reverse of those conferred on unconscious infancy. They must strengthen the bond of filial gratitude, at the very period when it otherwise will grow weak, or be dissevered. "I FEEL MYSELF INDEBTED FOR THE HEALTH I ENJOY, TO THE FORESIGHT OF A PARENT," is at once the most affecting eulogium, and the most permanent of motives for gratitude.

How encouraging this prospect! how full of despair, the opposite! What would be the feelings of that father, to whom, as he was contemplating his infant on its mother's breast, it should be announced by a voice, which he could not but believe prophetic:

That being, whose happy smiles thou art now watching, shall, one day, have cause to exclaim ; " Ah ! had not the affection of my parents " been as blind as it was strong, never should " I have been the wretch I am."

END

OF

FIRST ESSAY.

ESSAYS

ON

THE MEANS

OF

AVOIDING

HABITUAL SICKLINESS,

AND

PREMATURE MORTALITY.

ESSAY SECOND.

Printed by BIGGS and Co. Briſtol.

It is certainly right to take all proper care of our preservation, and were to be wished that the voluptuous and debauched would give more attention than they do to dangers, which threaten the shortening of their days. - - - - - - -

Life considered barely in itself, is a thing indifferent, neither good nor evil. If the kings of Colen, as legends pretend, slept 300 years in a cave, they must have been alive all that time, but were no more the better nor the worse for being so, than if they had been in a state of non-entity. - - - - - - -

We may look upon the shortening of life through our own negligence as a real evil; and if we make the prevention of this evil the object of our solicitude, it will keep us constantly attentive to our preservation.

A. Tucker on death.

DEDICATION.

TO

MINISTERS OF THE GOSPEL,

OF

EVERY DENOMINATION.

IT is said by Cicero, that nothing is too absurd for some philosopher to have advanced. In like manner, it may be asserted, that there can scarce be imagined an observance, detrimental to the constitution of man, but what has been enjoined by some teacher of a false religion, or added to a true one by its expounders, and practised by the bigots of both.

But YOU, my associates in the labour of mending mankind, do not so teach.—*You*, one and all, are ready to hold up your hands against such abominations. The human frame must, according to you, be a consecrated vessel, destined to traverse the ocean of life, till it shall be carried by such gales, as it may please its all-wise architect to dispense, into the secure haven of the grave. Unfaithfulness to the charge of conducting this vessel, you can hardly rate, as inferior in criminality, to the violation of any duty, enjoined by the decalogue. For how shall he, who renders himself incapable of fulfilling commandments, escape being reputed their breaker?

You, in virtue of your office, enjoy peculiar advantages for influencing the minds of common men. And almost every page both of civil and ecclesiastical history bears in pretty distinct characters, that your powers are superior even to your advantages. You hold, in short, in your grasp, that which I should infinitely prefer to the golden rod,* of which a mere glimpse could quell

---

* ——— Aperit ramum qui veste latebat.
——— Tumida ex irâ corda residunt.
——— Ille admirans venerabile donum
Fatalis virgae, longo post tempore visum,
Caeruleam advertit puppim, ripaeque propinquat.

the peevishness of Charon himself; namely, the key which causes the chords of the human heart to spring into placid or tumultuous vibrations, according as you supply it. But you do not, it seems, easily part with this precious instrument. You would not, for a moment, commit it in trust, for the use of certain *cheap* tracts, which did a scheme, proposed and exemplified by your humble servant, a very unexpected honour. You withheld it alike from the fair writer, and from her champions in a late (alas! not fair) transaction. How then may *one*, who has no pretensions to come into competition with these holy persons*, for your favour, hope to borrow it. Would to God! you could, at least, be prevailed upon to employ it yourselves, for exciting in the flocks you respectively tend, a sense of what is due to their mortal, as well as to their immortal, part.

---

* HOLY HANNAH! cries *Horace Walpole*. Unless the popes are belied, it is not the first instance of a pretender to sanctity, canonised by an infidel.

Of the topics, suitable to your public addresses, it is not for the profane to judge. But in private, you have endless opportunities of throwing out saving counsel, in regard to health. And such counsel you must allow, when effectual, to be the greatest *temporal* obligation, which mortal can confer on mortal. You perceive that WE members of the medical profession are, in this respect, precisely in the situation of the ladies. We dare not speak a word, till we are addressed. But YOU may, at all times, without indecorum, press the subject of timely care. To be able to form a probable judgment when delay is no longer safe, is the whole, about which you need be anxious. I would not have you go beyond your accustomed office of friendly admonition. Within this limit, you will be sure, that the good, resulting from your interference, will be as little alloyed, as the condition of human affairs may permit.

Some members of the establishment are, not unfrequently, impelled by benevolent zeal, to act also as bodily physicians to their poorer parishioners. To certain reverend authors we are even indebted for labouring to instruct us in our own profession. The active mind of JOHN WESLEY would not suffer him to leave his congregations without the best counsel he could provide against the season of sickness. But the *primitive physic* of this modern apostle is, I fear, fit only for a primitive age—that is, an age which wants none. In general, to caution the unwary against self-neglect, is the point at which you should make a conscience of stopping. Your own profession you must feel to be load enough for the shoulders of one man. The idea of encumbering yourselves with the additional study of an entire bulky art, may justly alarm you. I therefore only invite you to charge yourselves with a portion of general information, in itself agreeable, and capable of application to a most interesting series of observations. This, I am persuaded, in the present state of things, you will find as much more useful, as it is more easy, than to push your endeavours further.

It is, you may observe, by dint of cultivatng *your* talent of adaptation, that professional mediocrity has so constantly succeeded against genius and knowledge. By this (for it is the same principle at work under a different form) medical imposture is enabled to secure to itself retreats, as magnificent as that of the poet LUCAN;* in which, at its ease, it can contrive new arts of swindling, after the old have become stale. I have been credibly informed of an ingenious adventurer belonging to our profession, who was indebted for his first success, in one of the northern states of the American union, to a ragged coat.—And why not? What signifies a little difference in the objects of taste? Has a smart equipage never addressed itself with effect, in any of our great cities, to female favour?

The late pious editor of *Suessmilch's order of God in the changes of the human species*, declares that he has himself known many

* Contentus famâ, jaceat Lucanus in hortis
Marmoreis——

instances "where a cunning woman, or "mountain doctor extorted, from a sick "peasant, six, ten, or more dollars—a fee far "beyond what any city physician would "require. But the city physician is too "learned, too grand, and, in general, too "strange to inspire confidence. He may be "too much occupied to let himself down, "as far as could be wished, to the level of "the ignorant. But the parish priest easily "gains their good opinion. He knows their "way of thinking, and is more able to de- "scend, as it is fitting he should, to their "comprehension—"

If, therefore, as a body, you would take the part of the thoughtless against themselves; and of competent integrity against meddling imbecillity and obtrusive imposture; you would add more than you may suppose, to the obligations which society already owes you. The example of conduct, so beneficent and cautious, might check a scandalous abuse, which at present seems most rapidly increasing. Whether that vigilant prelate, my Lord of ——, means to encounter this abuse in one of those charges, in which he vies with the most vigorous General of the age, and shews what might be expected from some bishops, could they in these times take the field, cased in complete armour, I have not been informed. But I plainly forsee that, one day or other, if to the public detriment it continue unchecked, it will bring a stain upon your cloth. For look into almost whatever newspaper you please, you will find reverend names, dangling by dozens, to the tail of frauds at *half a guinea the bottle.* I would fain hope that the appearance of these names, in such a disgraceful position, does but add one item more to the advertisers' account of forgery. I, at least, for one, am unwilling to believe, that so many members of a respected profession can lend or let out their credit for a fox's paw, to rake the pockets of the poor; whose money, however ill they can spare it, is often the least valuable thing, out of which, from reliance on the judgment of the Clergy, in many cases they find themselves choused.

Wishing, among other blessings of a new year, that a seasonable hint may preserve you all from the misfortune of being either dupes or accomplices to these or any other blood-sucking villains,

I remain,

REVEREND SIRS,

Your most obedient servant,

THE AUTHOR.

Clifton, Jan. 1, 1802.

ESSAY

ON

PERSONAL IMPRUDENCE,

ACTIVE AND PASSIVE;

ON

THE INCENTIVES TO IT;

ON

ITS EFFECTS,

AND

USUAL ORIGIN.

IN my prospectus I have explicitly declared PREVENTION of mischief to be my exclusive object. From the specimen, produced by the preceding month, it will be seen that I am likely to be somewhat resolute in my adherence to the terms of this declaration. There exists, I am aware, a numerous body of readers, not belonging to the profession, but much interested about medical information, to whom this determination can be nothing less than agreeable. Of these readers, though it is impossible I should attempt to gratify their taste, yet I am unwilling to forfeit the patronage without an effort. But, under the dilemma in which I stand, I can perceive no resource, except in the reasonable expedient of an explanation, which luckily falls within the scope of the present number. This measure, reïnforced by my bookseller's powers of insinuation (to which the public has, at times, appeared not a little sensible) will, I hope, prevail upon them to bear with me yet a few months, before they give me up, in order that they may, at least, have complete evidence, that my notions are less correct than theirs. It is, I assure them, a question little short of life and death.

"*What is good against the head-ache,* "*Doctor?*" "Health, Madam." "*Well, if* "*you feel no interest about an old woman like* "*me—Marianne there, you perceive, has been* "*hacking all the afternoon. Do tell her of* "*some little thing, that is good against a* "*cough.*" "Health, Madam." *But are* "*you resolved not to give a more satisfactory* "*answer? In that case I shall take the liberty* "*of guessing why.*" "*Poh! Mrs. W.—*" cried a grave person in spectacles, from behind a full hand of cards—"*you should know that it is* "*the trick of these gentlemen* NEVER TO SPEAK "PLAIN, *as some great man says. And if*

"*they will not in a tête-a-tête, can you expect*
"*it before company?*—" I am not conscious "of having uttered any enigma. I am sorry "for the young lady. But I must still "answer—health, Sir—health, Madam." And so I should, were I questioned the whole nosology through, or as far at least as the enquirer is mistress of its nomenclature; still doubling my emphasis, as Simonides* did his demand for time, when called upon for an explanation of the nature of God. I do not know but I should add: "Madam, one has infinitely "less chance of becoming an adept in medi-"cal matters without regular study, than of "becoming rich without regular industry."

The lady, it will appear from the preceding little fragment, belongs to the corps, from which I am ambitious to enlist recruits. Many of these good people (for I will take no notice of those, who keep the physic left

* Roges me quid aut quale sit DEUS. Auctore utar Simonide, de quo quum quaesivisset hoc idem tyrannus Hiero deliberandi sibi unum diem postulavit. Cum idem ex eo postridie quaereret, biduum petivit. Cum saepius duplicaret numerum dierum, admiransque Hiero, requireret cur ita faceret: quia quanto, inquit, diutius considero, tanto mihi res videtur obscurior.

CICERO *Ernesti*, VI. 492.

by the last sick person in the family, for the use of the next) have, strong upon them, the passion for snapping up verbal and written recipes for every current *name* of disease. I have seen some of their pocket-books as full crammed as the cloth-hall at Leeds, during our dispute with the Northern Powers. Some treasure up these stores for family use. Others are public-minded, and know no greater pleasure than in drawing from their magazine for the benefit of every acquaintance within reach of a call or a note.

It is not that the former would grudge a neighbour a good turn. But where the affections are warmest, if a habit of shyness have been acquired, or an unpleasant intercourse with strangers have taken place, they are apt to contract themselves within the domestic circle. Then whatever is connected with home, has an *exclusive* sense of the agreeable, attached to it; and this is easily carried to excess. What others, for the sake of convenience, procure from abroad, these *exclusives* find the greater trouble, more than compensated by the greater comfort, of making themselves.—With what an air of satisfaction does one often hear it said: *We do such and such things within ourselves.* The phrase carries a happy union of oeconomy, independence, security and mutual affection. The dealer's profits are saved or believed to be so.—You feel above the necessity of trusting to out-of-doors supplies—What you have, you enjoy the more from being certain that it is genuine—All the time, too, you oblige, and are obliged by, those you love. What hedge-hog, rolled up round, tight, and warm in his skin, on a frosty morning, is more an object of envy than so close a family circle? So let those, who choose it, stitch, bake, churn, brew, and manufacture gooseberry wine at home.—They are much in the right. They have that, after which finer folks are eternally running, like so many wandering jews, but never find it. I advise them still to tend their sick without nurses by profession. But let them take care how they go so far as to take upon them the office of doctor. For by piling their comforts too high, they may all tumble down, one upon the head of another, like the houses which children build with cards.

The hoarders of single *infallible cures*, I am mistaken if I did not once designate, as the *pedlars* of medicine. There are others, whose larger style of dealing is enough to make me look up to them with reverence, and salute them with the title of its *hucksters*.

I might add, that these disdain to establish their concerns upon a paltry stock of soiled sibyl's leaves; that they lay in a thick close-printed publication, purporting to be a family compendium of the practice of physic; that they place it, as shelf-companion to the bible, and hold it next in veneration to the sacred volume; the one being the fund from which they draw and dispense what is necessary to happiness in the life to come, the other in the present life; but that it behoves them to examine well the authority for its genuineness before they depend too far upon their stock for profane use:—and I might go on to convey many agreeable and useful truths under these same emblems, if, at present, it did not suit my fancy more to introduce them to the public in a variety of transparent dresses, of different cut and colour.

I may, I believe, presume upon it as a pretty generally allowed fact, that in no line is there so large a proportion of bad hands as in that of book-making. There is certainly none in which so much cunning is employed in varnishing over flaws and imperfections; none in which the wholesale venders more frequently contrive to pass worthless articles upon the consumer.

Concerning the immediate makers of those books, which profess to distribute far and wide a capacity for discriminating diseases and applying remedies, I do not wish to propagate an unfavourable opinion. I place their intentions on a par with my own; and in order to shorten a question, on which so much seems to me to depend, I shall advance this single proposition: *such labour must, from the nature of things and persons, be either nugatory, or productive of disastrous consequences.*

Here let me beg the reader to consider the power and province of mere rules in practical affairs of the easiest kind. No one has, I suppose, yet come forward with pretensions to teach the coarsest handicraft by a book. But in the tumult of literary projects, amid which we live, scarce any absurdity being impossible, let us imagine some adventurer, sufficiently intoxicated to undertake to communicate the capacity for exercising one of our humblest, and most useful trades, without apprenticeship, by a tract on *domestic shoe-making*. Should any one, after studying this tract, conceit himself qualified to handle the awl and the paring-knife, I leave it to be imagined by the reader, how unmercifully the leather would be pricked and slashed, and what would be the condition of the poor toes, condemned to be lodged in the receptacle, prepared by these learned hands.—Does common sense spurn at the idea of efficacious instruction in such an art by such means? Are the qualities, then, of leather more complicated than those of the living body? Does the art of managing the former to most advantage require a long apprenticeship, and not that of managing the latter? Are the tools that lie within the compass of the shoemaker's bench, more easy to employ properly, than the articles of the materia medica? I see, indeed, one essential difference. The incompetent mechanic will soon be marked. No clumsy workmanship of his can pass. Whereas, in medicine, bunglers may go on, I know not how long, without disgrace. This chance of escaping detection is, no doubt, an encouragement for *private practitoners*,* such as nothing can countervail, if they be agitated by the same

* Of one of these dangerously absurd publications, the title announces, "*a dispensatory for the use of private practitioners.*" *Private* practitioners! what a winning thing is a gentle name! Why should not one be coined for the assassin, and some ingenious Italian manufacturer advertise the stiletto, as a weapon for the *private* gentleman soldier?

restless daemon, that possessed Lord Cheterfield's blood-letting peer. But I have no hope of effecting any thing, except with active, but misguided benevolence. Insanity must be differently dealt with; and wrong-headedness is scarce to be reclaimed by the plain dictates of prudence. Otherwise, a consideration, yet untouched, would be decisive! For the defect of the artisan, who leaves his work imperfect, can be afterwards supplied. But an amending hand may be vainly applied in case of omission during sickness, where it is often just as fatal to leave undone what is right, as to do what is wrong. What then shall we think of the defence, which conscious incapacity is so apt to set up by anticipation: *Very simple my advice is: but you may be sure if it does no good, it can do no harm?* Oh, yes, but if it does no good, it *can* do harm—all possible harm, provided in killing there be harm. It can arrest the rescuing hand, till the silent, but progressive finger of fate, move from TIME IS, to TIME IS NO MORE. There are plenty of occasions on which water-gruel, upon the harmless principle, will do a man's business, just as effectually as laurel-water. And what, I pray, does it signify to the killed, whether they come to their end by the sauce-pan or the still?—To the killer, the difference, we know, is all in all. Yet he who simply thrusts his ignorance between the sick, and the means of recovery, will really have done more mischief, inasmuch as he will have more largely accumulated pain upon death. And surely where law cannot interfere, the call is so much louder for public censure. It is by far too unequal a game to be allowed in society, where one party stakes empty professions of good-will against the other's existence.

Some dozen years ago, how many of us were seized with an invincible loathing against sugar,—because we thought, we perceived upon it spots of negroe's blood! For the same reason, ought we not to turn with equal aversion from drugs in hands, not taught their use? They may, in such hands, be regarded, without exaggeration, as swords, having two edges of unequal sharpness, but both most certainly stained with European blood—perhaps with that of our kindred.—Let us take care they do not get another crust from our own.

The strangest thing of all perhaps, is, that mixture of self-confidence and distrust of others, which many of these meddlers exhibit. Their presumption is always eager to discover itself, but I doubt if you can ever bring it to shame. When one of them presses his services in behalf of a sick friend, if you do not immediately reject the offer, but hesitatingly mention different practitioners, among whom it may possibly be expedient to consult one or the other; be assured you will not have proceeded beyond the second name, before a violent objection is started.

"But why?" you enquire, "has *he* any "marked deficiency of common sense?"

"*Oh! as to common sense! I do not say* "*that. But what is common sense! Is not* "*something more wanted, where life itself is* "*at stake?*"

"True. In that I perfectly agree. But "he has had, I am told, the usual oppor-"tunities for becoming thoroughly ac-"quainted with his profession. Did he "shew any particular negligence of dispo-"sition at the time of his studies? or has he "been a dissipated man since? Have you "heard any thing to that effect?"

"*I cannot say I have. But this I know.* "*For it came to me from one, who had it from* "*a particular friend to the family. In my* "*Lord Burroughdale's last illness, he cer-*"*tainly mistook a putrid for an inflammatory* "*sore-throat; and had one ounce more of*

"*blood been taken, his lordship, I am confi-*
"*dently told, must have been rotten in his*
"*grave, by this time.—Nay, the surgeon*
"*himself, it is whispered, hesitated about obey-*
"*ing the prescription for bleeding.*"

"Such mistakes, sure enough, are far from "pleasant."

"*Pleasant ! what security have you against*
"*one as bad now ? Such blunderers, really*
"*ought not—but I wonder how he ever got*
"*a patient after my Lord's narrow escape.*"

"Well then ! I thank you for your "caution. And I believe I must call in "Dr. ——."

"*What, Dr. ——! That notorious man*
"*of experiment, whom every body is crying*
"*out against. Did not he give iced water in*
"*a burning fever, though the whole world*
"*knows it is so hazardous to drink cold things,*
"*when hot ? Nay, by carrying the prescrip-*
"*tion himself, my very next door neighbour*
"*discovered at the apothecary's, that he had*
"*ordered for his daughter arsenic, think of*
"*that! under a different name. Such a violent*
"*thing for such a poor puny creature as that*
"*girl ! My God ! and are human creatures*
"*then, no better than rats, that they are to*
"*be enticed to swallow poison in disguise !*"

"Suppose then, I send to Dr. F. Nothing "like this can be alleged against him."

"*He! a mere Dr. Slop. I'll answer for "it, where something must be done without "delay, you will as soon trust to your "housekeeper's panada, as to Dr. F's medi-"cines.*"

In this manner would a determined private practitioner traverse the whole list of graduates, English and Scotch; discovering in each some quality, that renders him as dangerous as a beast of prey, to be turned loose upon a sick man. His own recipe, he would still persevere in pressing as infallible; never once suspecting that, without the interposition of a miracle in his favour, he must be liable, in a hundred-fold degree, to some one, at least, of the objections, which he urges as decisive against another. He may, indeed, easily contrive to unite, in the stile of his practice, the faults of different medical mannerists. For what! if disorders of opposite nature, but with some resembling features, *may* be confounded by a person of sense, aided by appropriate instruction, how can they *but* be frequently confounded by one, who probably wants the sense, and certainly the instruction? A fine chance, therefore, must a person in the situation of the lord, mentioned by the principal party in the preceding conversation, have from a recipe of his! — Again, should he have got hold of a *harmless* composition, will he not make as miserable work, (where something is to be done without delay) as the executioner, who cannot sever the head of an agonising criminal from his body, without repeated strokes of the axe? But, by way of amends, should chance have made him owner of one of those heroic medical weapons, which he will not tolerate in the hands of an experimenting brother, he may make as clean work with his patient, as the blade of the guillotine itself.

If it be the height of social imprudence to expose oneself to the operations of the hindmost in the train of justice, it is equally the height of personal imprudence to come under the hands of those agents of fate, who follow up the encouragement, they receive from the press, to officiate as *private practitioners.* How much would it be for the good of the suffering part of the community, if the odium, attached to the one, could be transferred to the other! To check troublesome kindness, must be always somewhat of a painful effort; and compunction may arise, on pushing back blind officiousness, when it lovingly aims a dagger at the throat. But is one to open the collar and stand steady, for such a reason as this? What is manslaughter, perpetrated under the cover of benevolence, but a pitfall, planted with roses? Trust me! few officious people have nerves so delicate, as to receive serious damage from a resolute repulse. The worst injury they can sustain, must, at all events, be infinitely less than that, which they are likely to inflict. Sometimes, probably, they have no claim to be treated with tenderness. They feel none.

Vanity—if I may venture to declare the plain truth—vanity is, ten times as often as humanity, the actuating motive. And, indeed, vanity, I take to be the grand principle of this contraband trade. Look into many strong characters, and you will find them tormented by an itch for being admired on account of trifles, the most remote from their real excellencies. The deep metaphysician shall strain to rival the Bond-street beau in easy, tripping, fashionable chat; and the sturdy sea-officer shall go about to display his taste by instructing ladies, how to distribute sprigs over the tambour frame. But if people thought it worth while to observe, they would find little minds still more infected by the same pitiful ambition. In our own times, did not a most reverend archbishop pride himself on being the best dresser of sallads in a great lettuce-loving nation?* The Roman history represents Nero as anxious to shine in the orchestra, and Commodus in the arena. If we see our young nobles on fire to emulate their grooms, why should not our gossips burn to be esteemed a match for the apothecary at least, if not for the physician? How delightful, by merely willing it, to outstrip in their own art, personages, who pique themselves on their profoundity! How charming, to be so clever and so useful on such easy terms! On balancing yourself against others, what a golden lump of secret merit, with which to sink your own scale, when all the world is imagining you must mount, light as an air-balloon, towards the beam!

Not that all look so far. Numbers circulate scraps of paper, marked with medical hieroglyphics, just as they circulate scandal,

---

* "Even M. Juignè had his talent. He was the best "hand at dressing a sallad! With such a qualification "one may be sure of reaching posterity."

George Forster's Schriften. VI. 408.

for want of other employment. But could the world be persuaded to give none of these busy-bodies the least quarter, they might find resources against idleness, which would render their own life equally happy, and have no tendency to abridge the term, which nature has marked out for that of their neighbours.

Of each species of recipe, it is worth while to remark the origin and authority. The most prized are, as may be expected, usually the most contemptible. These are the manuscript; which are really so old as to be mistaken by their collectors for new. They were originally transcribed from obsolete works, published during that barbarous state of pharmacy, from which this art has been of late much reclaimed. Of course, above ninety-nine times in a hundred, practitioners of medicine have, in common use, improved formulas, corresponding to the more efficacious. But in truth, a great majority of those, which I have examined (and I always encourage their exhibition) have proved the most roundabout contrivances imaginable for blending water, mucilage and sugar. —They are labelled—*never known to fail in coughs, colds and consumptions*: and, with these, our modern English *soeurs de charité*, soaring far above that chaste pattern of human kindness, Sir Charles Grandison's aunt NELL,* cloy such acquaintainces, as have no serious ailment, and, encouraged by their great success in these cases, proceed to treat others, till they make them more than a match for the best doctor's skill.

The remaining resource of uninitiated dealers in medicine, requires to be considered somewhat more at large. It consists of popular treatises on the practice of physic; against the utility of which, there immediately arises a strong presumption from this circumstance; that no such treatise has been attempted by any experienced physician, whose judgment had been approved by other productions. To write for the people usefully, requires the application of genius and judg-

* "He writes now and then to aunt Nell;".. she "folding up the letter carefully in its cover, puts it in her letter "or ribband-case, which shall I call it? For having but "few letters to put in it, the case is filled with bits and "ends of ribbands, patterns, and so forth, of all manner "of colours, faded and fresh; with *intermingledoms* of "goldbeater's skin, plaisters for a cut finger, for a chapt "lip, a kibe, perhaps for corns, which she occasionally "distributes very bountifully; and values herself—as we see, "at such times, by a double chin, made triple—upon "being not unuseful in her generation."

ment to experience, in a manner, which few have hitherto thought about in earnest. But waving this distinction for the present, I doubt whether any physician of avowedly large experience — (it is the trick of us medical authors, to *talk* of our large experience)—has ever undertaken to instruct the unprepared public, in the general knowledge and treatment of disorders. The almost total impossibility of ascertaining some complaints, unless the discriminating phaenomena are pointed out in nature ; the frequent necessity of accurately distinguishing the relative conditions of a given action, as of the pulse, or of a given organ, as of the lungs ; the darkness in which those must wander, who are destitute of the light of anatomy ; the changes, even to an opposite state, which the system undergoes during a course of symptoms, referred to the same denomination ; the transition, required by this as well as by other causes, from one mode of treatment to another, are among the difficulties which, upon the present average of apprehension in the most enlightened people of the globe, bid defiance to any mortal talents for exposition. And alas ! it is not readers, blessed with the greatest felicity of apprehension, any more than writers, the best qualified to assist it, who are forward to cope with such difficulties.

If we follow down the art of healing to the present hour, from the day, when pain or fear first incited to experiments on the sick, we shall neither find ourselves able to compute nor conjecture, how much our acquisitions have cost the human species. But possibly, wars, or attempts to destroy in the gross, have less contributed to dispeople the earth, than attempts to save by detail. Let it be considered, at what period endeavours towards the removal of disease must have been attended with the most unfortunate issue. Undoubtedly, at the most early period, that is, before experiments were sufficiently multiplied to introduce analogy and a spirit of caution—when the whole was hazard, both the judgment formed concerning the nature of the case, and concerning the effect of applications. Has not universal opinion established it as an axiom, that practitioners, the best taught, learn to cure their other patients at the expence of the first? Now then, where is the difference between the practitioners, who *formerly* went to work before any preparatory knowledge had been accumulated, and those who *now* go to work without that, which has been accumulated? If there be little difference, or none, every incentive to private practitioners, whether consisting of single recipes in writing, or a whole printed dispensatory, must surely tend to bring back the most murderous aera of medicine. Should our art, as M. Gilibert contends, have been prejudicial to society, is this for want of ignorance in its most ignorant, or for want of knowledge in its most knowing, professors? No person who meddles with the sick, uninformed of *any* thing that may be learned, can acquit himself of destroying, where another might have saved—much less then can the tamperer, uninformed of *every* thing. The imperfection of the healing art being an incontestible fact, to which not only every bill of mortality, but every assemblage of grave-stones bears witness, what is the obvious inference? Can the sick afford the smallest deduction from that, which may, at its most, fall short of their necessities? Who that wants to raise a load, and doubts the sufficiency of his greatest power, goes to work with a power much inferior?

I do not require to be told that certain adventurers have obtained a degree of public confidence by their endeavours to popularize the practice of medicine. But though it was in order, that they should supply the market with fresh editions, according to the demand for their performances, their *subsequent* experience has not enabled them to advance a single step towards conquering the impossibilities, inherent in the undertaking. Several compilations of this class seem to be entirely speculatious of their publishers. It is a scheme, perfectly in the spirit of our literary traffick. A bookseller, who feels his influence on the fortune of books, will not easily want the ingenuity to discover, that he may as well supply customers with an article of his own, as suffer a rival to pocket the publisher's profits. Now to him, who has only in view the making of money, a grammar, a gazetteer, a medical compendium will appear identical, provided they prove equally saleable articles. The projector of a new domestic medicine, meaning well by himself and the public, does what he can to exceed his predecessors in the quality of his goods. The predicament of *quantity* he can ensure. Without dreaming that his PANDORA's box is only the more pernicious for its bigness, and without the slightest suspicion of any mischief ensuing, (except to himself, from the possible failure of the enterprize) he resorts to the *house of literary call,** puts his job in hand, and has it got up, like a suit of mourning, by the day he appoints. The bookseller indeed might, in this case, be bookmaker too, were the saving of copy money worth his care ; but he must be a paltry architect, who can do no better than put together the materials, for which he has designed the plan.

Writings intended to warn against the destruction of health, can hardly be confounded with such as pretend to teach the people how to restore it. To direct a stranger how to traverse slippery ground without injury, is one thing. To instruct him how to set his leg, should he break it, is quite another. It is remarkable that, in departments so essentially different, the qualifications of authors should have been pretty nearly alike. It could not indeed well happen otherwise, as far as the projects of booksellers have in-

* Houses of call, may be found in most great cities. They are the head quarters of journeymen taylors, who there receive by lot the orders of the masters. None but competent workmen can come forward. A journeyman taylor, nearly cured of palsy, once told me, he did not think he could quite venture to answer at the house of call, because he was not quite equal to a fair day's work. I wish journeymen authors could and would be as considerately backward.

fluence. But neither could any thing be more easy than for persons, spontaneously engaging in an unexceptionable undertaking, to mistake themselves. It is a result of common experience that the more familiar you are with the dialect and manner of thinking, which prevail among the vulgar, the better can you adapt your knowledge to their capacity. Upon this just principle many have proceeded, feeling (what is also very just) that they themselves stood but in the next degree to the untaught. It seems as if it had not struck them that no one can deliver what he does not possess. And there is another maxim equally resulting from experience, and more applicable to the occasion: that dimness of sight is the worst of all qualifications in a guide for the blind. I could, it is true, quote a most sagacious adviser, who recommends it to those who have to walk in the dark to put themselves under the safeguard of a blind conductor. But then in our compilers on preventive medicine, and in directors, raw from the schools, we have neither the caution of blindness nor the security of clear vision.

What indeed can the greatest readiness at rehearsing certain general propositions, and at quoting the facts, of which such propositions are the abbreviated enunciation, avail in a case like the present? Persons, whether setting out or advanced on the road of life, but unacquainted with its dangers, can be compared to nothing more aptly than to children, who have an imperfect use of their limbs. Would you save them from accidents? You must examine the place where they are to essay their limbs, what shoes they wear, and, one by one, the manner in which they lift their feet. Nor is this all, by a great deal—(for to do any good, you must be able to discover what is amiss, to strike out improvements, and to explain and enforce your ideas so that they may be adopted)—but it is enough to shew, that without opportunities of minute information, all your labour will be lost. Yet we may see persons, setting up for instructors to a whole nation, without having had admittance to a dozen families, on such terms as can furnish the means of judging in any manner of their state, and of its causes. Yes, and what is more, some without troubling themselves about our capacity, would fain teach *us*, inhabitants of the British dominions, the alphabet of health out of horn-books, designed for foreigners, but, I fear, not very well calculated for any people.

On reconsidering the foregoing remonstrance against certain common incentives to the dangerous vice of personal imprudence, I do not find that I have any thing of importance to add. Whenever people are suddenly reclaimed from hurtful practices of any kind, it is (as far as I have been able to observe) by the shock of a sudden alarm for reputation, property or life. There is, however, an easy way, by which those who have been addicted to the collection and employment of directions for medicinal preparations may pass, by degrees, to harmless occupations. They may try to fix their affections upon papers, similar in size, shape and style, but which direct how to compound dishes instead of drugs. Of their treatises on the art of curing diseases, if they should be too tender-hearted to commit them at once to the flames, they may contrive perchance to barter them for *the art of cookery made plain and easy* by the notable Mrs. Glasse, or for the more promising production of that great proficient in the discipline of Apicius, Mr. Farley, sometime cook at the London tavern in Bishopsgate street. I may venture to congratulate them beforehand on the exchange. Many of them are vastly good-natured at bottom, and will, I am sure, experience greater satisfaction in serving up palatable preparations to their willing friends, than ever they found in cramming nauseous ones down their throat. Where the propensity proves incorrigible by reason, I flatter myself that those whom it would trifle with or destroy, considering what harm it may do, and what good it may prevent, will not merely meditate a civil escape from its persecutions, but have the courage to treat it, front to front, in a manner, corresponding to the enormity of the consequences, which it ever threatens to produce. Though the active party is generally a female, and females are entitled to more than civility, public opinion cannot be too severe on this article. I wonder, indeed, that decorum, of which the sex is so properly studious, does not interpose in favour of invalids. Were a lady, educated and endowed as ladies usually are, to produce, from her pocket, a case of surgeons' instruments, every body would feel confounded, and nobody, I suppose, would submit to the hands of the operator—be they ever so fair, and her probes and lancets ever so bright. I however defy any one to assign a good reason for supposing such a lady, better qualified to wield the equally dangerous tools of the physician.

There remains a kind of active personal imprudence, (or disposition to interfere, uncalled and unqualified, with the personal management of others) which, though partial in its operations, produces mischief enough to deserve to be pointed out. In the *shippe of fooles,* all without exception, who are actuated by such a disposition, belong to the berth, distinguished by this inscription: *they who take few circumstances into account, are sure to be rash in their decisions.* And single adverse occurrences, considered without reference to the whole sum of events, are perpetually provoking these hasty decisions. There will, for instance, be times, when patients in great jeopardy do not obtain the same beneficial chance from a surgical operation, or a new and powerful remedy, which others, in nearly equal danger, have derived both from the one and the other. Should such a failure take place in a distinguished family, what is the consequence? Its members, though from a proper feeling of incompetency, they had before made a point of refraining from giving, much more from obtruding, any opinion on questions, within the province of the physician and surgeon, shall now sally forth to all the winds of heaven, uttering imprecations against a process, which they have *once* † known to fail, though the strongest probabilities all united in its recommendation. Nay, if the treatment succeed beyond all reasonable expectation, but should an irreparable injury to the organization have previously taken place, the crime of domestic neglect is charged, without scruple, upon beneficent art. Thus, at a former period, when a long succession of ague-fits had shattered some of the vital organs, multitudes were deterred from the use of Peruvian bark by the senseless outcry of squires and knights and noble dames. In our days, half a city or a whole county shall be deprived, for years,

† "The vulgar are commonly very positive; thinking "themselves possessed of *absolute certainty* in almost every "thing they know. This happens from their weighing "their evidences *singly*—which will naturally produce that "effect. For we can judge of weights only by their oppo- "sition, because any one, thrown in alone, drives down "the scale forcibly. But the contemplative use themselves "to compare the judgments, as well of the sense as of the "understanding."

Light of Nature.

of the benefit of a valuable medical discovery by mandates, issuing from the mansion-house or the chateau. He who has but one or two dim appearances before him, and can but squint at these, shall set at nought judgments, deduced from hundreds of facts, carefully inspected with well-formed and well-exercised eyes.

No! I declare not a soul belonging to me, shall ever touch a drop of that violent thing, is a sentence which we physicians are doomed, every now and then, to hear pronounced against one or other of our most efficient instruments. It is also, every now and then, a capital sentence against those, for whom the use of the prohibited substance is proposed. Thus the very quality, common to all bodies worthy the title of *medicinal*, becomes a bar to the use of one in particular. And where others, more to be feared on the same ground, would pass unquestioned, this one, though alone perhaps to be depended upon, shall be inexorably rejected, unless it find admittance under an unsavoury disguise, thrown over it by the apothecary's art, as Falstaff, by favour of a heap of foul clothes, made good his way under the very nose of Ford himself.

I have often known letters of proscription transmitted from kingdom to kingdom, against some of the most valuable productions, which Nature supplies for the benefit of afflicted man. I have even heard the authors of these rigorous measures disclaim, *in the general*, all acquaintance with remedies and disorders. But we, inconsistent mortals! are most of us ready enough to infringe the very laws, under sanction of which we proceed, without scruple, to condemn every one, who holds up his hand at the bar of our judgment, for the self same misdemeanour.

This species of active imprudence, or ignorant interference, would lose its malignity, if the world at large could be persuaded to enter into the following very simple cousiderations. Between what are called poisons, and what are called medicines, there exists no fundamental difference; nor can they be separated by any line, drawn either by theory or practice. The one and the other are substances, which in small quantities produce strong effects; and this character distinguishes both, accurately enough for every useful purpose, from food. It includes, indeed, as I shall afterwards point out, certain things, which popularly rank among articles of diet. No substance, therefore, it is evident, can have the virtue, commonly required for the removal of disorders, unless the same be capable, when misapplied, of throwing the living system into disorder, and of extinguishing life. In addition, then, to what has been said above, those who become invincibly prejudiced against a drug, because it has occasionally exerted a powerful operation, should be referred to the writings, in which various men of sense have exposed the folly of arguing against a thing merely from its abuse. The occurrences from which such prejudices arise, are rather presumptions in favour of the remedy; and if there be room for censure, it should fall only upon those who use it improperly, either in respect to time or quantity. For persons, not minutely instructed, it would be most prudent to consider what common language denominates *medicines* with as much horror as *poisons*, and *poisons* with as much complacency as *medicines*. That the one happen to have been received into the pharmacopöeia of this and that college of physicians, is nothing to the purpose. Analogy or accident may, any day, bring to light an equally good reason for conferring on the others the same honour. In the last pharmacopöeia of the London College of Physicians, the writer of the preface, who does not seem to have perfectly clear ideas on the point, nevertheless observes, that "the an-"tients were miserably haunted with the ter-"ror of poisons; and it was one of their "principal occupations to devise antidotes. "How different has been the lot of poisons "in our days! The healing art, so far from "treating them as inveterate enemies, seems "now rather willing to enlist them as auxi-"liaries. In our catalogue we have inserted "some poisons, which we have ourselves "tried, and shall be ready to adopt others, "upon the faith of proper experiments."

Whether this representation admit of any exceptions, I pass over, as a question purely of idle curiosity. But that the exceptions are infinitely fewer than people, dextrous at splitting the hairs of argumentation, might imagine—and how unsafe it must be to act upon the assumption of their existing at all—will be manifest, if it be considered that there is no medicinal agent, of which the power can be taken as absolute, but that each is subject to endless variations, depending on the condition of the system, to which they are applied. Thus the food that does but support, the temperature that does but cherish, the wine that but seems to enliven, the healthy, may aggravate a precarious into a fatal distemper. The roast beef, which would render the labourer only more fit for his occupation, would, in some inflammatory diseases, operate to all intents and purposes, as perniciously as the rankest poison; and before the apprehension of mischief from these sources is dismissed, as a vain imagination, I hope that what I shall have to say, in some of the next succeeding essays, on heat and cold and diet, will be well weighed. We have here, however, without further search, two distinct considerations, which ought to influence the conduct of those, who have not been able to acquire medical information in as full measure as the existing sources can supply it; first, the general activity of valuable remedies; secondly, that change in the relation to the articles of subsistence, the most common and reputed the most innocent, which renders these, where improperly administered during certain states of human existence, as virulent as any of the articles of the materia medica, or as poisons, if the reader prefer that term.—In all points of view then, we must regard individuals, seriously ill, as tottering on the brink of a precipice, and the private practitioner, as one rushing to their rescue in the dark, and therefore full as likely to tilt them over as to draw them into safety.

Sufferers under the same grievance, be it moral or physical, find themselves relieved by communication. It is nothing uncommon with invalids to meet with dangerous delusions, where they are merely seeking this kind of comfort. From the presence of one, perhaps but grossly resembling symptom, the identity of two dissimilar diseases is confidently inferred; and whatever treatment seems to alleviate the one is, in consequence, concluded most likely to relieve the other. In spite of all remonstrances, the opinion takes deeper and deeper root, and at last is acted upon. Thus sympathy with the supposed fellow-sufferer, is more powerful than confidence in the physician, and it is not conceived that for any two patients to have a cough in common, no more proves them to have the same disease, than their having each a nose on their face, proves them to be the same person.

It is thus that ignorance of the most general and best ascertained facts, contained in the physical history of man, presents itself, at every turn in our path of inquiry, as the source of most mischievous error to the individual, and of embarrassment or failure to those who, at a period of danger, have him in charge. With this more especial cause, the habits in which we are, all of us trained, powerfully co-operate. Between both, it is impossible to say how much is endured, that might be compleatly avoided, if by the benefit of certain general principles, thoroughly inculcated, the will was made firm against temptations, to which, at present, it always yields. As these principles would influence the members of society but in their individual capacity, or, at least, could have no tendency to produce the smallest sensible alteration in their relative situation, there can be no circumstance in their operation to give the most timorous alarm.

From the custom of writers on the means of escaping ill health, the public has learned to expect scarce any thing beyond particular prescriptions; and to minds little inured to reasoning, scarce any thing else is acceptable. The commencement of the process for opening the sphere of intellectual vision, (chiefly, perhaps, from the unskilfulness of the operator,) has seldom been attended with pleasant sensations at any age. Children in understanding and in years, believe alike, that you are foolishly tormenting yourself, when you take steps towards a distant end: and with them, a first effort towards the same is apt to bring on a strained feeling, which generally causes them very soon to desist.—Nor do I doubt but I shall be blamed by many for idling away, at a distance, that time, which each will think I ought to have employed in spreading a plaister for his sore. This is unfortunate—for many sores do not admit of being healed, till the *habit* is mended, and if dried up, will break out afresh,—and our chief blessings are attainable only by casting our purposes to a sufficient distance from the passing moment. It is a law, to which none of our blessings are more rigorously subject than health.

Different navigators have discovered tribes, eager to barter their almost immediate necessaries for momentary gratification. It is not the disappointment, accompanying the fast approching hour of want, that so severely punishes this improvidence, as the baneful consequences, flowing from the unsatisfied calls of nature. In a state somewhat more enlightened, where exigencies are foreseen, but too dimly either to hold infant industry steady to the purpose of providing supplies, or to prevent inconsiderate waste, famine and pestilence return pretty nearly with the regularity of the seasons. Nothing more, one should think, could be necessary to rouse our speculatists from the dream, in which they connect happiness and uncivilized life. In all but the most favoured countries, something will be perpetually omitted to the destruction or detriment of the untutored inhabitants. In the happiest climates, something to the same tendency will be perpetually committed. Nor even in these, will disasters, the fruit of omission, be entirely unknown. It being impossible, in any situation, that human beings without the knowledge of cause and effect, should escape many fatal errors. But in the reveries, occasioned by our accounts of Otaheite or Tongataboo, it happens, I suspect, as it does when we amuse ourselves with the idea of returning to an early period of life. In conceiving how happy we should be, as rude natives of some *fortunate isle*, we forget to leave at home the foresight, which every inhabitant of a polished state inherits from the lessons, that his most remote ancestors have received in the severest school of adversity. If this be retained, there could be little dispute. To suppose, at least, that a condition superior to any which exists, may be compounded of the advantages, peculiar to each, can hardly be made matter of compliment, as if it were any notable discovery.

That at the expiration of some golden age, Astraea quitted the earth, must be numbered among the grossest fables of mythology. But from the records of medicine, which (though they have never been kept with a view to the great end, they are capable of serving) exhibit more exact, more interesting and improving views of the human race, of the effects of our appetites, passions, of our ill-laid and too short trains of thought, than the meagre chronicles or the puffy narratives, that are held in such unmerited esteem under the title of *history*—from these records, it appears that the improvident GENIUS of the savage state, carries on the same cruel sport with civilized society. Or rather, though he varies his proceedings, he abates nothing of his cruelty.

We read of barbarous tribes, whose deficiency of forecast is visited in public disasters, which consume at once a large proportion of their numbers. Polished communities, suffer perhaps in equal proportion, from the same cause. But instead of being swallowed up by tens of thousands (as is recorded of the Israelites) in one common gulph of misery, *we* are permitted to breathe the breath of anguish or of languor for a length of years, before death comes to our release. Calamity in the one instance, is *acute*,—in the other it is *chronic;* though the distinction scarce applies with accuracy, but to the higher ranks, where the distinction of ranks is established, The lower, like their tattooed and painted forefathers, suffer in the gross ; for (in addition to the consequences of their own misconduct) upon *them* bursts the torrent of evil from the crimes and the insanity of their superiors, collecting more exterminating rage, as it comes down from a loftier point.

The scheme of life is universally laid, (whether it be well or ill laid), in power. In all conditions, and at every age, voluntary action must be regulated by sensation. "We "never stir a fibre, but for the sake of some- "thing we desire to have produced thereby." Pleasant feelings we strive to perpetuate, and to escape from the painful. Our foresight, whether it be confined to an hour or stretch into eternity, is fixed upon enjoyment and upon suffering. To be masters of the sources of both is still the object of our cares.

In the social arrangements which have gradually formed themselves in Europe, WEALTH, the most general object of power, becomes the most general object of desire. Its active pursuit, and paramount administration, almost exclusively engage the stronger sex. In its preservation, a portion of the females bear an auxiliary part, as they exert themselves to effect given purposes, at the least possible expenditure. If property descend largely to some, it is a rare case. The majority can only hope to share its advantages fully, (to which they have by nature an indefeasible right) but by captivating its possessors. To this, female education principally tends; and to exhibit the advantages, which women have derived from nature and art, is one great end of the intercourse, established between the sexes. By attending to the drama of modern manners, we shall see that men engaged in the chace, or in the œconomy of riches, and women that attract, or that try to attract these men, are its heroes and its heroines. The rest serve but to fill the scene, and as they fall into the train of the leading characters, they necessarily tread in their footsteps.

The multiplication of the roads to wealth unavoidably keeps pace with the multiplication of gratifications, and of those contrivances for displaying accomplishments, which wealth can command; for it is by profiting from the sale of these gratifications and these contrivances, (that is, of the various luxuries of the table, of the toilet, of furniture, of equipage, of the fine arts) that more and more members of society grow rich.

Savage life is uniform and still: polished life makes a more varied and an ever-moving picture. Of the latter a miniature is presented by certain dances, in which all the parties join hands, and tread a ring; then dispersing, each in his several line, towards the different points of the compass, speedily rejoin and pace their round once more.

In the endless mazes of this dance, every movement carries us close by some obstacle, over which we may stumble, or against which we may strike. And we shall escape or pay the penalty of imprudence, according as we mark out our distances, and according as the eye guides the feet through the intervening spaces.

When an ELWES or a LAURENZIUS* de-

---

* Dr. Laurenzius lived some years ago in Leipzig. He was a jurist, noted for his opulence. "At home, he lived "like the poorest person, keeping neither man nor maid, "partly from thinking he could not maintain them, partly "from fear of being robbed. He lived in a building, attached to a large house of his own, in which he had a "suite of four rooms, through all which he had to pass

viates a little from common practice in the order, which prudence assigns to things here below, every body is struck by their folly, and

---

"on going out. He kept these rooms fast locked, that "thieves might be obliged to burst open four doors, "before they could reach his Mammon. He seldom sent "for meat enough for one meal, and on this, when he did, "he lived at least three days. He took neither beer nor "wine nor coffee. In short, his life at home was a con- "stant fast. Though when invited by his legacy-hunters, "he stuffed like a thresher, and toped like a canon."

"Under the most biting hunger (of which he actually "complained to me) he had not the heart to rob his coffers "of a single penny. He came to me, oftener than once, "as I was eating my breakfast, and begged for a bit of "roll. '*He felt a little qualm: otherwise he never, never eat.* "*A single mouthful was enough. More would be his death.* "*He would chearfully send for a whole roll, but, he vowed to* "*heaven, he had not a halfpenny at home—and it would be a* "*sin too, as all above a mouthful must be left to spoil.* "But when I forced upon him half a roll, he eat it with "the utmost glee.

"I have twenty times witnessed, when servants brought "him presents, how he would steal to the grated hatch, to spy, "if they were thieves; with what fawning devotion he would "draw his bolts, take the cake and wine into custody, and "begin: '*ah! my dear fellow, return a thousand thanks to* "*your master and mistress for the refreshment they vouchsafe* "*a poor wretch—ah! how glad should I be to give you some-* "*thing to drink—but, look you, may I never share the joys* "*of heaven, may I be cast into everlasting perdition, if I have* "*a farthing of money here within — but, be sure, tell them* all the righteous (I mean, all that go with the majority) exchange congratulations on the punishment, which these poor, preposterous sinners inflict upon themselves. But the troop, that hunts wealth, directly or through the medium of seductive acquirements (and it will make no difference, though some should strive after admired accomplishments, without further aim), sacrifice alike the end to the means.——Do not their grovelling views for ever rest upon certain outward helps? and hence do they not *underlook* that sole essential condition to happiness, the inward state? In what light can we view the designs of HIM, who endowed man with sensibility, understanding, and will to proportion his exertions to his active and passive faculties, without, at the same time, seeing that the ordinary devotees of wealth counteract those designs, as compleatly as the extraordinary. Is it not fair

---

" *in my name, I will remember them in my will — trust me,*
" *I will not forget them.*'

" Thus did the man forswear himself thousands of " times—and he was really an extremely religious man, " that believed, according to his catechism, in the devil " and hell, and in every thing else, wherewith the clergy " endeavour to scare people from premeditated sin."

Dr. C. F. Bahrdt's life. I. 339.

to take the estimate, by compounding what both parties do to embitter their existence, with what they do to curtail its span? I am sure I do not wish any one to rate the cold fits of terror, and the hunger-qualms of the tribe of LAURENZIUS too low. Were these withdrawn, the mountain of their misery would sensibly subside. But then, on the other hand, to lose that delight, which the very brutes find in existing! to be visited by the GOUT, the DROPSY, the PALSY—by the BLUE and OTHER COLOURED DEVILS lodged in the system, that compensates business by boisterous pleasures, and privation by gross indulgencies! Truly, when I think what is implied in the tremendous characters traced above, I am obliged to confess, that the followers of this system endure more intense torments than the miser. And durable they often are! And here too does not gold bring with it its plagues? Sensation, then, for sensation, and body for body (as long as each lasts), I can scarce find half the phalanx of a little finger to choose between them. But as to longevity—in the abstemious annals of griping avarice, I am struck by examples, not inferior to that of Cornaro himself—and in a lease for lives, I should have no hesitation to prefer the retired, self-denying hunks to his sociable, joyous rival. I say so of trades. I say so of professions. I say it of monied men, who follow neither trade nor profession, but contrive to split their piety fairly between Comus and Mammon.

But (as our actuating trains are most commonly composed) it is not only in what aldermen are said to love—though over-fondness for the *good things* of life is by no means confined to this worthy class of magistrates—that the appetite is found too strong for the mind, and for the body too. It is seen as much in the taste for more refined luxuries. I have often dwelt on a passage in Mr. Burke's *reflections on the revolution in France*. It is the only one perhaps, which the English nation never put to heart. And yet no passage was ever penned, which comes more home to our case. It is where, with too ingenious humanity, he defends the monks, (making them out to be but packs of harmless hounds, that devour their mess of horse-flesh and barley-meal, quietly in the kennel); and where, with equal humanity he glances at a much more numerous, an immured, and severely disciplined race—not indeed confined to British ground, but more abounding here, I believe, than elsewhere.—"The Monks," says he, "are

" lazy. Be it so. Suppose them no other-
" wise employed than by singing in the choir.
" They are as usefully employed as those
" who neither sing nor say. As usefully
" even as those who sing upon the stage.
" They are as usefully employed as if they
" worked from dawn to dark in the innu-
" merable servile, degrading, unseemly, un-
" manly, and often most unwholesome and
" pestiferous, occupations to which by the
" social œconomy so many wretches are in-
" evitably doomed. If it were not generally
" pernicious to disturb the natural course of
" things, and to impede, in any degree, the
" great wheel of circulation which is turned
" by the strangely directed labour of these
" unhappy people, I should be infinitely
" more inclined forcibly to rescue them
" from their miserable industry, than vio-
" lently to disturb the tranquil repose of
" monastic quietude. Humanity, and per-
" haps policy, might better justify me in the
" one than in the other. It is a subject on
" which I have often reflected, and never
" reflected without feeling from it. I am
" sure that no consideration, except the ne-
" cessity of submitting to the yoke of lux-
" ury, and the despotism of fancy, who in
" their own imperious way will distribute

" the surplus product of the soil, can justify " the toleration of such trades and employ- " ments in a well regulated state. But, for " this purpose of distribution, it seems to " me, that the idle expences of monks are " quite as well directed as the idle expences " of us lay-loiterers."

Had this great poet and philosopher, (who consequently was an industrious searcher too) ever made his way into the inmost recesses of the archives of humanity, he would have found records, that must have moved him as much for the fate of those who consume, as of those who manufacture, various articles of luxury. Aye and there are those who, disdaining the dull dissipation, that has not attractions enough to assemble its loitering throngs by the appointed hour, as the prospect of heartfelt enjoyment * always does, are nevertheless full as much entitled to Mr. Burke's pity. The things, upon which

---

* " One may see how lamely this method of employ- " ment answers its purpose, by the great dilatoriness there " is in going to engagements, by which means they have " been put off later and later, until the hours of amuse- " ment are run almost into midnight. Whereas where " there is a real expectation of delight, people are eager to " run to the place beforehand "

they are intent, are such as no wise man would neglect, but they have not the habit of referring their conduct to that one thing, from which no wise man will suffer his thoughts to stray, and which he will make the central point of all his movements. Hence even they, who escape uninjured by dissipation, suffer from self-neglect, and call it *chance*. But what Montesquieu observes of states, is more eminently true of individuals. Their condition depends still less on causes, placed beyond the sphere of human foresight and power. Whether it shall maintain itself in prosperity, or sink into adversity†, depends on the aptness and comprehension of the general plan, chalked out for life.

In estimation and in appearance, the natives of Guinea and of Great-Britain are certainly not more widely separated than the wretches, whose rescue Mr. Burke meditated, are from the whole or half men of pleasure and

† Ce n'est pas la Fortune qui domine le monde. . . . . . Il y a des causes generales, soit morales, soit physiques, qui agissent dans chaque monarchie, l'elevent, la maintiennent ou la precipitent.

de la grandeur et de la decadence des Romains.

the women, that in the temples of Fashion, celebrate the rites of that fawning, treacherous divinity. For these are the patrons of the dismal industry, exercised amid chilling damps and mephitic exhalations, in dens which an antient poet might have appropriately placed on the banks of his Acheron and Styx. But in the most essential traits of human nature, they often approach near enough to one another. In what the consumers are superior to the manufacturers, the former take care to display before every eye. But if we descend into the interior of each, we shall not find a corresponding difference. We need not think here, of the distress, sometimes experienced by the employed; it may stand against the want of relish so constantly felt, amid their superfluities, by the employers. Let us take, as a test, that annoying, uneasy sense, which, as it has held so large a space in our breast, ought not to have been so long without an expressive name in our language. Which of the two does this sense most frequently impel to provocatives, that render the succeeding listlessness more intolerable, and another forced orgasm more indispensable? till, at last, the sinking frame requires to be deposited—in the one case, on the iron steads of its hard, narrow, uncurtained, hospital-bed—in the other, upon an expanse of down, guarded round by chintz, profusely flowing in the mockery of idle state—both waiting their destined weeks or months of aggravated misery, before they are removed from the aspect of the sun into the common receptacle. The cells and conventicles of gentility, it is true, harbour beings, upon whose wearisome uniformity the tumult of pleasure never breaks in. But so do the cells and conventicles of labour, when it is fixed down beside machines, whose eternal rotation produces no greater variety of chearful sounds than the rattling of the turnkey's bunch of keys, or the creaking of the prison doors.

These classes, then, whatever be the distance between them, are inseparably linked together by the chain of destructive vanity: and though born under such different stars, they pretty equally share a lot in life, which no image can better represent than that of helpless crews, committed in frail barks to an uncertain sea, without chart to warn them in time of currents, shoals and rocks, and without skill in the manœuvres, necessary to steer clear of so many perils.

Polished society! how happily denominated! its habits being uniformly regulated according to the effect on superficial observers! and never, in any instance, tending to improve the texture or durability of the ground, on which the polish is laid! How perpetually do they run counter to that homely, but never-failing maxim: *hunger is the best sauce!* The hunger they leave to come as it can; and not only so, but it is the grand concern to provide such sauces as kill hunger by inches. The first circumstance is evident enough at first glance. And if we inspect this order of things a little minutely, we shall perceive whether the minor, but sacred, duties which it enjoins, fall in any degree short of a complete code of personal imprudence.

In the first place, what disposition do we find in those who give and receive the ton? What, in them, is the spring of those unpremeditated actions, whose aggregate fills up four-fifths of the whole space of human life? What can be the result of attention to the trifles, belonging to the capricious sway of fashion, but to produce an impotence of mind, and to preclude all proficiency in pursuits, requiring any degree of patient meditation?

It is the peculiarity of modish conversation (any body may see) to fly zig-zag, always making as quick turns as lightning, though it is not always so brilliant, notwithstanding, at times, it may blast as fatally. The ideas that pass through the head, will naturally strike the same angles, as those that come out at the tongue. And of course there can be no *esprit de suite*—no reach of thought. The place of those affections, that most largely distribute comfort over human life, will be usurped by a restless disposition for shopping and running about on morning calls. And a rage for *parties* will supplant the beneficent and noble passions, which, giving consistency to our proceedings, produce all that is good or great in the world. *As soon as possible!* is ever the watch-word of the spoiled, overgrown babies of genteel life, when they bespeak new toys on finding themselves, after half an hour's trial, tired of the *old* ones, for which, however, they were as impatient but the day before. *As fast as possible!* is their injunction on starting for a place, whence they will set out again with as much hurry as they came, to arrive at a third, where they feel no more satisfaction than in the station they originally quitted.

For they imagine dispatch a good thing in itself, and never have reflection enough to ask themselves; *what a foolish race must that be, where there is no goal?*

He who has an useful purpose, to pursue, finds minuteness in trifles, that have no reference to any such purpose, an intolerable burden. This punctiliousness, like the process performed by Aurora upon Tithonus, always reduces the understanding from its natural dimensions to those of an importunate, chirping insect. While the faculties are dwindling away under its influence, the spirit is necessarily debased, and the moral sentiments, too, become perverted, as in certain diseases, debility is attended by movements, contrary to the natural. There is nothing so unsteady, nothing so inquisitorial, as the laws that regulate the mode. They are more harrassing than the subtilties of casuistry, which once imposed so galling a yoke upon mankind, inasmuch as there is scarce a square inch about us, for which the said laws do not prescribe minute regulations, without a pretence of rendering to that part or to the whole the smallest service.

As to the standard of utility, they always keep that out of sight. A suckling, enraptured by elegancies, *just come out*, would laugh to scorn his own grandfather, who should ask *cui bono?* Your last year's coat may have fitted just as well, your carriage have run just as easy, your legs have felt no sort of inconvenience from the colour of the stockings or the pattern of their clocks. Yet where is the modish man, that would not feel twice as much compunction on appearing in these superseded articles, as from the violation of a clear maxim of morality? Crimes of moderate magnitude, do not excite so much repugnance as an oversight in any of these minutiæ. To judge accurately of expedience, is far from being always an easy task. On which account the habit of attending to a series of consequences cannot be too assiduously cultivated. But to be all anxiety, respecting one particular method of doing a thing, when twenty other methods would answer just as well (except in procuring you a momentary stare) is to abandon the exercise of reason altogether, and to take pains to become disqualified for forming a sound opinion in cases, where the unperverted judgment can discern no room for hesitation. Thus, from the very moment of its birth, caprice deserts the consideration of utility, and when it grows up, by way of shewing its courage, never fails to fly in the face of this motive. Hence legislators, as well as subjects, in the domain of modish caprice, ever have an insecure tenure of their most essential advantages, and in cases of emergency, turn out the most helpless of mortals. The ministers of *refined* luxury, as it is called, are scarce ever set to work but by caprice; and to speak very moderately, it is always matter of chance, whether any given article they fabricate, shall be injurious, indifferent or prejudicial. But from the incessant variations in the products of their labour, it is as impossible that the prejudicial chance should not frequently occur, as that a dye, perpetually thrown, should not frequently turn up a particular number.—The ferment of their operations being as brisk and lawless as the imaginary one of the elements in chaos, how is it possible that noxious combinations should not be formed?

But it is not merely from the total abstraction of the mind from all regard to the *retroactive* tendency of conduct, from which punctiliousness in trifles stands in constant danger. Toiling and thinking men, as I have stated at sufficient length in the former essay, are too deaf to the most urgent calls upon their prudence from within. But their occupations are prospective. They at least profess allegiance to utility ; and are not for ever striving to do for their interior, what Circe did for the exterior of all that came within the influence of her wand. They remain in communion with rational beings, by having respect to the benefit to result from their proceedings. But this is hardly the case with the followers of accidental, capricious, changeable usage, who may be conceived, as figures twisted by wrong attitudes, till the eyes are turned the contrary way to the movements of their limbs.

But it is not enough to shew, that devotion to fashion induces a disposition, that excludes any proper culture of the best gifts of nature, and all effectual care of the compages of bodily organs. Without direct injury to these, it seems impossible to live in the fret and fume of fashion.—Men, not soul-bound, by the halter of necessity, to the rack and manger, can hardly depend upon a healthy flow of mental complacency from any other source, than from a multifarious stock of ideas—which, in a situation, unfavourable to the acquisition of fresh ones, they can employ themselves in arranging, or which they can interweave with impressions, when they have the good fortune to discover a new world of objects. But it is upon an endless variety of ephemeral appendages to himself, to his retinue, or his mansion, that the heart of him, who means to figure in the beau monde, is eternally set. The passion, with which these things are pursued, is, I am sure, as insatiable as avarice, and, I think, more tormenting. What the miser acquires, however he may thirst for more, *lucro adponit*—he at least, stores up as solid and durable treasure. But, in the mart of the mode, every new pattern makes all the old purchases worthless; and (as I have sometimes noticed) what these have cost, creates a sense of vexation more violent, than a large stake, lost at the gaming table: the disappointment being almost equal to the loss. For should an expensive service have been laid in, against an occasion of high parade—and, upon the very heels of the acquisition, a new pattern appear—imagine what must be the mortification of the purchaser at seeing the whole glory of his taste and splendour fade away, at the moment he expected it to bloom in all the charms of the night-blowing CEREUS?—In no other respect, I confess, do certain periods of antiquity seem to me to possess a more solid advantage over modern times. Exempt from the miserable obligation of wasting their property and their thoughts upon alterations without the sense, or contrary to the sense, of improvement, they could enjoy a sort of repose, to which our labourers, and beggars almost, are strangers. Their faculties of both denominations, were free to make beautiful and vigorous shoots, whereas ours are stunted and warped by a force of perpetual operation; and in contemplating the *fair* and *good*, they were filled with a grateful feeling of human dignity, to which our beaus, I am afraid, do not derive any thing equal from the best stored taylor's card, or our belles from the milliner's shop, most in request at the West end of the town. *The trophies of Miltiades will not let me sleep*—burst from the lips of a youthful hero, aspiring to the admiration of Greece, and destined to overtake his great predecessor in the race of fame. Alas! what elicits equal sparks of emulation from the eyes, and induces agitations that equally disturb the rest, of thousands of striplings and of damsels, panting for celebrity in Britain?—Why, may be, a nosegay of artificial plumes, or a well stiffened collar.

All the world is melancholy, because all the world is in debt. The observation, certainly, will not account for the whole of the melancholy among us; and I will undertake, any day, to shew the lively author of the *Tableau de Paris*, upon the shoulders of English creditors, visages as long as he can find upon those of any French debtors. But it will account for a great many of those unwelcome visitations, against which no gaiety of apparel is able to protect many a snowy bosom. For how is it possible for those that have taken the infection of fashion to escape a disease of the heart, known to common mortals under the name of chagrin?— Before one set of desires is well gratified, new ones are kindled by the infinity of bright temptations, which ingenuity is daily displaying in the view of taste. In this dilemma, what can we do but take either disappointment or debt? Whichever we choose, we shall get a spice of mortification into the bargain. And mortification from debt will often be more sickening from envy of a more brilliant competitor.

So far moral and physical considerations go hand in hand, for I have at present nothing to do with the former, where the others are no way involved. But the body is not merely worn down by the perturbations of the mind. It is directly attacked. To the *circumstances* of different persons, the rage for shining in the gay world may be injurious or otherwise, (though it always absorbs sums that could be better employed) but to *themselves* it threatens injury, to whatever class they may belong. In some barbarous despotisms, we read, if I do not mistake, of inferiors feeling it, as an honour, to suffer and bleed under the hand of their superiors. I am confident that this is the case in the barbarous despotism of Fashion. Those who are abject enough implicitly to submit, and take a pride in submitting, to regulations framed by others, without any reference to reason (that is, to good and evil, to the causes of pleasure and pain, considered upon an extensive scale,) pass sentence of condemnation on themselves by the very act of submission. Those at the head of this department of luxury are but mortals; and if they are not inspired, how is it possible they should know any thing of what contributes to health—and they have always pretty well shewn, that they will not be taught on easy terms. Their requisitions are as imperious as those of ROBESPIERRE, and as tamely endured. A table of diseases and of deaths, directly and indirectly produced by these requisitions, with the manner of their production, (and I shall have occasion hereafter to furnish some articles towards such a table) might be inscribed: THE REVENGE OF THE SLAVES OF REFINEMENT UPON THEIR TASKMASTERS.

It is thought that Fashion, in some instances, is become more humane. But I am afraid this increase of clemency is only what an Emperor of Morocco would shew, in resolving to-day to strike off heads with a sharper scimitar than he wielded yesterday—a resolution, to be sure, which, in that potentate, would deserve the highest encomiums of his poet laureat, if it only gave security, that he would not return to the use of the old instrument to-morrow.

What may be considered as permanently gained, is to be seen well exemplified at the beginning of the *task* by Mr. Cowper—

Yet restless was the chair, the back erect
Distressed the weary loins, that felt no ease;
The slippery seat betrayed the sliding part
That pressed it, and the feet hung dangling down,
Anxious, in vain, to find the distant floor. - - -
But relaxation of the languid frame,
By soft recumbency of outstretched limbs,
Was bliss reserved for happier days. - - -

In what regards commodiousness, considerable improvements may have taken place;—few in what regards health, among which few the dereliction of hard, tight stays, has a right to be considered as the principal, *so long as it lasts*. But this has been more than compensated by modern barbarities in general dress, which have probably been as destructive to the tenderer sex in one year, as all the tight-lacing of our great grandmothers and grandmothers in a dozen: so that in these iron times, from sympathy or imitation, Fashion, like other rulers, may be said to have been far more madly prodigal of human life, than in the days of yore.

To some readers it must have sounded as a crude, wanton paradox, when I advanced that great part of the evils, ascribed to civilization, really spring from what still clings to our habits of antient barbarism—of its thick, ropy lees, deprived of all the spirit of wildness. I trust that, by this time, some are so far reconciled to the idea, as to be induced to make it the subject of reflection. No more consoling idea can be proposed to a nation, brooding over the reduction of its power, in the presence of a dangerous rival, by its own outrageous follies—a nation where pestilence, the fruit of war and famine, has so long ravaged the dwellings of the poor, and where the true, interior condition of the rich ill bespeaks that *prosperity*, with which knavish or ignorant men in power are perpetually dinning our ears—a *prosperity*, too nearly resembling the state of an animal, tormented by some spoiled child, till the awakened humanity of the mother causes it to be *put out of its misery*.

Let those who refuse to believe what I say recur to the evidence of their own senses. I entreat them to survey, with considerate eyes, a crowded opera-house, and to compare the gewgaws, so ostentatiously stuck upon many of the spectators, with what we know of the effect of glass beads upon various uncultivated tribes. Do the members of these tribes take pride in possessing shining trinkets—so do the frequenters of the opera as much. Do the tribes barter their most substantial valuables in exchange for trinkets; — the frequenters of the opera acquire and exhibit *their* trinkets at the price of comfort, and at the risk of their health, and often of life. Do the natives commit such mistakes in the appreciation of things, for want of having been trained up to set future suffering against present enjoyment (which is the true criterion of cultivation): but how does the (self-entitled) most refined part of the human species, differ in this respect from those, upon whom we look down as the grossest savages, though they doubtless pique themselves upon *their* taste, as much as we do upon *ours*? — Thus by looking a little curiously at the mere surface of things and persons, what instances may we find in confirmation of the sound old doctrine, that *extremes always meet!*

I have still much evidence in store to prove that the wretched objects, that moved Mr. Burke, have not such cause to envy certain superiors, as they and others may imagine. But this evidence could not be compressed into the present essay. It will find its place under particular topics, to be treated hereafter. Meanwhile, if there be any one, who trusts in himself, as capable of adorning a subject in the rich festoons of that eloquence, the pattern of which Mr. Burke took from the sweeping periods of Cicero, I propose to his ambition a minute, *probing* examination of the different classes, that act and re-act upon each other, within the compass and sphere of the British islands. Let him take them at their birth, and, having pursued them through the stages of their mortal existence, state distinctly, how much pleasure or pain each derives from the abundance or the want of spirits; from the sense of health, and from the gratification of appetite; from vigour of constitution or from infirmity; from the over-excited organs of feeling or the over-exerted organs of motion; from wilful or involuntary exposure to the elements; from the failure of desire or of the objects of desire; from pride of station or its opposite shame; from hope or fear. The lots, which I should most wish to see contrasted, are;

1. *that of boys, born to idle affluence.*

2. *of boys, brought up to professions: and to the land or sea service.*

3. *of boys brought up to trade.*

4. *of girls, genteelly educated, under which division are included all our females from the the daughters of the middling tradesman up to the princess royal.*

5. *of menial servants of both sexes, engaged permanently or for the job.*

6. *of the male yeomanry.*

7. *of the peasantry.*

8. *of soldiers and sailors.*

9. *of the men, women and children employed in manufactures of prime necessity and convenience.*

10. *of men, employed upon articles of refined luxury.*

11. *of the women and children, so employed.*

12. *of beggars.*

13. *of slaves, well treated.*

A comparative physical census of the population is possibly a leading step towards gradually and without disturbance removing that state-pestilence of polite luxury, which, although it parade in open day, and by the light of a thousand torches, contrives, unsuspectedly by the multitude, to mangle and destroy whatever it meets, and on both sides of its path. Could not science and rural œconomy absorb the labour of those, whom the poet terms the cankers of a quiet world and a long peace, and upon whom the political œconomist looks, as upon a race of two-legged cattle stalled in our manufactories, in order that in due season, they may be driven to glut the dogs of war*? I do not

* "In times of peace and tranquillity, the superfluity of "the produce of the land goes to the maintenance of manu- "facturers, and the improvers of the liberal arts.—It is "easy to convert many of these manufacturers into soldiers. "- - - - - - Those who labour in commodities, least "necessary to subsistence, must either enlist in the troops, "or turn themselves to agriculture, and thereby oblige "some labourers to enlist for want of business."——This liberal and enlightened author would have altered several things (*Hume. Part II. Essay* 1 *and* 2.), had he seen as clearly as Mr. Burke, the physical state of the numerous gangs of our manufactures. Politics is the art of producing individual good by general measures; and no man, who is not acquainted with the ultimate result, can either reason securely, or act without making havoc among his fellow-citizens.

think any one, who has reflected with care upon the possible improvements in science and rural œconomy, will hesitate to answer, that all this and much more labour could be turned to better account for all parties. Every possible improvement in these departments must, sooner or later, be realized. The individuals and the nation, destined to have the greatest influence upon the pursuits of mankind, present an union of means, tending to this end, without example in history. And some late spirited endeavours to diffuse physical knowledge at home, and to promote the culture of the earth, which, in spite of the counteracting policy of certain state-sots—implacably hostile to the Muses and Minerva — shedders of blood and wine in equal profusion — threw some chearing gleams upon the horrors of war, may now inspire brighter hopes for humanity. They suggest, at least, the reasonable expectation that the cultivation of truth, by experiment and observation, will finally accomplish what legislation would only have made itself ridiculous in attempting — if we suppose the opulent part of the community to have remained equally incurious of the world within them and without; and legislators equally unacquainted with the subject of national health, that is in reality, with the prime causes of individual well-being. An alteration in the first of these respects, at once, of itself, puts an end to the mischief. For why does the " despotism of fancy", before which Mr. Burke so profoundly humbles himself, distribute the surplus product of the soil in a way, " not to be tolerated in a well regulated state, except on the ground of necessity?" Why, but because, almost from their first hour, pains are taken with persons, born to wealth, in order that their imagination may become enamoured of articles, that owe their existence to " most pestiferous and " unwholesome occupations" ? In other terms, because those who determine the application of industry, are crammed full of desires, doubly immoral, that is, destructive to themselves as well as to others?

These reflections I began with offering to the clergy, because I felt that nothing could so much assist their labours in the cause of virtue, as a just and enlarged view of the consequences of personal imprudence.—*Where Prudence is present, no divinity,* says Juvenal, *will be missed.* It may, with equal truth, be asserted, that if disregard to the state of the bodily system were banished from society, a numerous train of vices would, by the same act, be sent into exile.— That I should have taken a range, so much more extensive than that of authors on the prevention of diseases, will not surprize any one, capable of entering into so very simple an idea; and to those, who feel any difficulty—I may remark that it is nothing uncommon for a domestic lecture against a particular fashion, to be a lecture on health. He, indeed, must entertain either a very high opinion of his own talents as a lecturer, or very inaccurate notions concerning the human mind, who can expect, by a few irritating words, to counteract a disposition, which it has probably been the labour of his life to form. For with how many parents is it a principal business, by direct precepts, or by conduct, in which the precept is implied, to fill the hearts of their children with anxiety, respecting the appearance they are to make in the eyes of others, instead of shewing them that it ought to be their first concern to ensure to themselves the longest possible succession of agreeable feelings; a purpose which modish manners. and the applause they procure, upon the whole go to defeat instead of promoting. There was a time, when the pulpit resounded, more frequently than the parlour may at present, with remonstrances against some piece of dress. But the zeal of the preacher seems to have been spent to as little purpose, as the fury of the wind in the fable. The louder each grew, the more closely did the beau and the traveller hug their garments. In fact, did not preacher, as well as wind, take just the wrong way to disrobe the subject of their experiments? And do not the *distinct* light of beneficent truth, and the warmth of affecting moral sentiments, stand the best chance of preventing that obstinacy, which rough treatment will but confirm?

To excite a dread of the consequences, produced by personal mismanagement, appears the only thing belonging to *this* world, which can be opposed with effect to inordinate vanity, to avarice, and to those other lusts, which, as the divines tell us, are always stealing away our affections from the *next*. Or rather, it is the only thing, belonging to *either* world, in which there is room to hope for any speedy and general concurrence of sentiment. In regard to the regions beyond the grave, people have always differed, and too often quarrelled and fought, the moment they began to develope their ideas. And in whatever system of doctrines any one shall acquiesce, he is sure to find himself in a very small minority. Religions and sects are not only divided by seas, rivers, and mountains, but a party-wall often bounds the territory, belonging to different creeds; and, this week, a set of lodgings are filled with the sounds, proper to the rites of one worship; and, next week, by those of another. But as to feeling, men are generally enough agreed, and not likely to differ, in regarding pain as what is to be avoided, and pleasure as what is to be sought. — All, that in quest of the latter, fall upon the former, coincide in principle with those more fortunate rivals, that contrive to shun pain, except where they are willing to pay it, as the price of more durable or more exquisite pleasure.

The contrary error, which when it is of a certain magnitude, produces the most poignant regret known to the human breast, and very often gives rise to dreadful reflections upon the memory of parents and preceptors, appears to spring from a very obvious source. This is the want of being, in time, impressed with proper self-regard, self-respect or reverence. As there is no eatable substance, so nauseous, but what some nation or tribe devours with keen relish, so there is no visible or tangible object, which has not been rendered an object of love or adoration. So fast and loose do the feelings of the human heart play, that there is no circumstance in life or death, (the rejoicings of martyrs and savages amid their tortures, bear me witness to the last fact!), with which pleasure or pain will not connect itself. The association depends upon education in the enlarged, proper sense of that term.

Should it be objected that tortured martrys and savages may personate a part, (an objection, upon which, I apprehend, none acquainted with the play of our ideas and sensations will insist), it must be allowed that there is nothing so vile, which has not been approached, with religious awe or rapture, in some part of the habitable globe. That is simple matter of fact. A Swedish author has published a learned work upon the various delicacies in request in various countries; which has had the honour of commentators of greater erudition than its author. But a catalogue of the multifarions objects of pious veneration would be far more useful. Logs of wood, beasts of prey, venomous reptiles, as crocodiles and snakes, have been among the number; and perhaps, in addition to the excrements of animals, toad's spawn and stinking fish might be enumerated. The nature of the rites is not a whit behindhand with the quality of the substances adored, As every one could furnish examples, it would be useless to tell at large of wretches suspended in the air by a hook, passed through the muscles of the back, to please one divinity, or mutilated in honour of a second, or murdered to appease a third. What would appear incredible, where the monster Superstition is forbidden the gratification of seeing wretched mortals writhe so dreadfully in her fangs, these instances place before the eye. Their use is—not to put to shame those who hold it but—to correct the opinion, that brutes have greater docility than man. But a swarm of examples, easily to be collected from travellers, depose in favour of the two-legged, unfeathered species—even though it be clear from home experience, that quadrupeds can be flogged into such an elegant accomplishment, as dancing. The eternal lamentation, therefore, that is poured out over *our* refractoriness, deserves to be treated as one of the lame excuses, to which incapacity always resorts. The fault does not lie in the stuff, but in the workman, who is paid to bring it into shape. Here is a creature, who will fall upon his knees, in the filthiest mud or upon spikes, before the foulest idol:

and he is said to have no religious instincts! —A young man, whose amorous propensities are such, that old age and deformity are not safe beside him.—And will you make me believe, that he would turn away, indifferent or disgusted, from beauty in its bloom?

Should any one enquire how he has to avail himself of the tractability, which I pretend to discover, for the prevention of personal imprudence, *Begin with inspiring early self-reverence*, would be my reply. Should it be asked again, how this is to be done, my answer is ready made. I have only to refer to processes in common use. In this country, as in others, we find a number of things, which mortals esteem sacred;—as spots of ground, buildings, habits, vessels, food, liquors. These though no way, in form or composition, excelling others in common use, most people would shudder at the idea of converting to any but to their own particular purpose. To do so, is *profanation*,—to steal such as are moveable, is not theft, but *sacrilege*: and behind a name they are perhaps more secure, than they would be behind bolts and bars.

On entering a church, a stranger to Europe may forget to take off his hat. He is reminded of the omission. He demands why he should shew, to one pile of stone and mortar, a sign of reverence, which is not exacted in behalf of another, though ten times as elegant. He is gravely told "that its claim lies in its destination." He is told so, perhaps, by one, that has never shewn or felt any regard for himself.—Ask the monitor, if a heap of dead matter can be consecrated by its destination, how it is with living, reasoning, eternal man?— Again—"the corporeal represents the spiritual. There are emblems, mementos, tokens that ought to be looked up to, on account of their signification."—But does not the light itself, unveiled and clear, deserve regard in preference to any reflection! —Now challenge the *property-man* to open the wardrobe of the church. Let it be examined in all sides; then let the *human person* be inspected; and say if, under every aspect, *this* do not meet the eye, as more august. Whether the destination or the tendency be considered, or the author, if *such* things can be accounted *holy*, the lord of creation is surely the *holy of holies*.

Should it, in so plain a case, be vain to distinguish or to deny, enlarge your ritual, Conscerate the child of your affection to himself. Employ, if you please, names, gesticulations, ceremonies, catechisms, invocations, precept, example; but, above all, employ facts. Such facts, as cannot fail to strike, whenever submitted to the senses, are alone sufficient, and alone safe. Let books, conveying these facts, and fit objects make part of your daily worship.—What consideration pretending to set the civil or ecclesiastical history of the Israelites, above the natural history of man, can for a moment, bear to be scrutinized by the prudent, or even by the pious mind. The Jewish history, you say, comes from inspired penmen. What! is not the other, the record of the power and the wisdom of the inspirer? That religious rites and codes of morality, have been without due influence, has been a perpetual complaint. Certainly. It should not be otherwise. They are imperfect and impotent machines. They want the main-springs and master-wheels. And this is the reason, why, as is truly said, they have never been able to lift any considerable portion of mankind out of the mire of vice.

How few rites of any church, but what are of arbitrary, human institution! Like other poetry and pantomime, like architecture and dress; these rites are mostly the progeny of *taste*. The taste of a barbarous, or of a semibarbarous, may ill suit a refined age.

It is at last widely felt as odious or ridiculous. Then ensues a long and bloody conflict. It is not the war of Pagan against Christian, of Mahometan against both, but of Christian sect against Christian sect. The sword is sheathed, but the hostile mind remains.—Nay, among frequenters of the same temple, a secret struggle is nothing uncommon. The functionaries will have no rite questioned. The congregation will not abandon their privilege as reasonable creatures. Whether the flock do, or do not, harbour any thing of hatred, fear abides with the pastor. But the sentiments of self-veneration, and the practice of self-management, as founded on the essential properties of the object to be reverenced and preserved, can never suffer from any versatility of taste. Fellowship cannot be weakened by animosity. Neither heartburnings nor jealousies can arise. Whatever, in consequence of better knowledge, is to be taken away, or to be added, will draw down no ruin. Of the rest, the old can never become obsolete, nor the new incapable of joining exactly to the old.

If we could not otherwise learn the principles of our own nature, and the duties deducible from those principles, it ought to be the first concern of every state, to maintain public teachers of this greatest branch of piety and morality. With zeal and ability, they would soon become the favourite instructors of the people. Look at the history of sects. There, more than in courts—*Dress makes the man.*—The tenets of sects are, in general, so absurd and so discordant, that the secret of their success must be elsewhere sought. This secret is easily found in the rude eloquence of their propagators, in gestures, looks, ejaculations and groans. Whatever the preacher brings, the audience receives. Mahometanism or Methodism takes root and flourishes indifferently, if the new cultivators of the mental soil do but work with more ardour than the old.

If prodigious, unnatural, destructive doctrines have easily made their way in the world, shall comprehensible, congenial, salutary truth fail to be received? Our children plunge headlong into vice and perish—our populace are debauched.—Why, then, do we not make haste to explain to them, in intelligible language, the laws, by which the Creator has attached certain punishments to certain crimes against self? Tricks, that decoy for the moment, we may relinquish to the juggler, the mountebank and the fanatic.

We need but be perspicuous and impressive. If no crouds, drunk with enthusiasm, reel to our standard, converts will incessantly come over with deliberate speed. We are the apostles of a faith, among whose disciples there never will be found one apostate. We aim at the gratification of every heart. We teach but the fulfilment of universal desire—how all may dwell in the gardens of pleasure with least hazard of being stung by scorpions, that lie hid in their bowers, or of being cut off by poisons, that mimic the fruits.

END

OF

ESSAY SECOND.

Printed by Biggs and Cottle, Bristol.

ESSAYS

ON

THE MEANS

OF

AVOIDING

HABITUAL SICKLINESS,

AND

PREMATURE MORTALITY.

ESSAY THIRD.

Printed by MILLS, St. Augustine's, Bristol.

" The first and capital article," observes a work of 50 years standing, " of town-effeminacy, is that of *Dress*: which, in all " its variety of modern excess and ridicule, is too low for serious " animadversion - - - -

" Every circumstance of modern use conspires to sooth a man " into the excess of effeminacy: warm carpets are spread under his " feet; warm hangings surround him; doors and windows, nicely " jointed, prevent the least rude encroachment of the external air."

" Vanity lends her aid to this unmanly delicacy. Splendid " furniture; a sumptuous side-board, a long train of attendants, " an elegant and costly entertainment (for which earth, air, and " seas are ransacked); the most expensive wines of the continent, " the childish vagaries of a whimsical desert—these are the supreme " pride of the master—the admiration or envy of his guests.

" Luxury is not idle in her province, but shares with " her sister Vanity in the labours of the day. High soups and " sauces, every mode of foreign cookery that can quicken taste, " and spur the lagging appetite, is assiduously employed. The " end of eating is not the allaying of natural hunger, but the grati- " fication of sordid and debasing appetite. Hence the most infla- " ming foods, not those whioh *nourish* but those which *irritate*, " are adopted." - - - -

" - - - - In the refined period, where manners and principles " are lost, the luxuries of life become necessaries among the " Great; and therefore will be as obstinately adhered to, and " quitted with the same reluctance, as food and clothing by the " poor."

" The poor *must* indeed increase their wages, in order to subsist, " yet this increase never takes place, till they are compelled by " the last *necessity* and *want*. - - -

ESSAY

ON

THE INDIVIDUALS,

COMPOSING

OUR AFFLUENT

AND

EASY CLASSES.

I. BRITISH CHARACTERISTICS.

Erroneous views of the points on which natural habits bear. Mischievous consequences of these errors.

OFTEN, as I go along, communing with myself, how I may raise into credit plain and apparently trite, but useful and disregarded truths, the fairest name in story, with its rich escort of ideas, rises before my imagination. And I grieve to think that the fairest cannot be saluted also as the happiest. Either the colours of private life, in no degree correspond to public splendour; or, alas for the generations that acquire inferior renown, or pass unrecorded away!

In spite of the miscarriages that, within thirty years, have tarnished the reputation of our country, and of the miseries, which have torn its bosom, BRITAIN still remains the second name to ATHENS; and perhaps is not second. From equal powers, profiting of the performance of a predecessor, much indeed is to be expected. But Britain has exceeded the measure of just expectation. She has successfully contended with rivals, as formidable as Sparta. The subduers and expellers of her domestic tyrants claim, equally with the Athenian HARMODIUS*, to have their memory cherished, and to lead an immortal life in the islands of the blest. Scions from her root of freedom are spreading wide, in ever-green vigour, over the new world; and grafts from those flourishing plantations have renovated the cankered and corrupt growth of the ancient world. What

* Φιλταθ' Αρμοδι', ου τι πȣ τεθνηκας'
Νησοις δ' εν μακαρων σε φασιν ειναι,
Ινα περ ποδωκης Αχιλευς
Τυδειδην τε φασιν Διομεδεα.

It must be owned, however, that Harmodius either deserved better company, or that he did not deserve to be in the islands of the blest.

does not hold of patriarchal SWITZERLAND, or of resolute HOLLAND, holds of Britain as perfectly as of Athens. Her genius has not been surpassed by her spirit. The greatest of poets are her's, and the greatest of philosophers, whether we regard the world of sense or spirit. On one part of the continent, a jargon as barbarous as any that ever reigned in the schools, affects to develope the operations, and to circumscribe the powers, of the human mind. Elsewhere, they pride themselves on garbled and disfigured transcripts from our HOBBES, our LOCKE, HARTLEY, and TUCKER—enquirers, who have almost exclusively the merit of discovery in the obscure regions of intellect, because they almost exclusively pursued the track of genuine observation. There always have been, and there always will be minds, in which no modern production of reason or imagination can excite more overpowering sensations than were felt by the artist, when he cried out—*I, too, am a painter!* But NEWTON, and SHAKESPEARE still farther, seem to have left the most aspiring hopes of humanity behind.

Our moral sentiments have been equally honourable to us, as a people. No-where has distress more readily found relief, than in the openness of British generosity; no-where accused innocence a more certain safe-guard, than in British laws, administered by British juries.

Among the most striking facts in history, we may consider the rank, which some communities obtain above their native means. Neither in this does BRITAIN yield to ATHENS. No community has acquired so great a command of the articles, which foreign soils and foreign industry yield to the requisitions of sense and fancy. No-where has domestic ingenuity so largely contributed to the same demand.

The most interesting question, however, still remains. Is all this, solid benefit or empty boast? What revenue of health and comfort has our vast dominion over art and nature yielded? Have we borne our faculties more wisely than the Athenians? One may account the pride of power as the same thing, in effect, to the body politic, with intense, enervating pleasure to the body natural. Has Britain wasted her vigour in the debauches of glory? Has she bruised her breast by straining to grasp unwieldy masses? Have lives innumerable been sacrificed in vain trials to extend her controul over a huge continent? Has she stained with blood every shore, accessible to her navies, using her own miserable children, as the destroying and destroyed engines of ambition in her statesmen, and of rapacity in her merchants? Is the multitude exhausted by toil and want, that their superiors may pine under the effects of intemperate or effeminate luxury? Do many drudge, like Helots, that a few may revel and repose, like Sybarites?—Before the rule of law is established, the victor savage obtains what he strictly wants and truly enjoys. But may not the original state of war of all against all, subside into a latent civil struggle, wherein both parties come off worsted; because each unconsciously deprives the other of one of the requisites to a happy existence; and because natural good is converted into artificial evil, by the disunion of desire and of the means of satisfying desire?

Wherever this unfortunate separation shall have taken place, it will be sensible in its effects. But these effects are to be looked for in private life. There only, will they be at first discernible, and there will they be always most striking. In the chain, that suspends the order of created beings to the throne of the Creator, the councils of the cabinet are but a few links from the condition of the members of a state; and the genius of a people has always decisive influence upon the genius of its diseases.

About the middle of the century, lately closed, a lively picture of the reigning manners* is well-remembered to have produced a diffusive and powerful impression. But the impression was transitory. And like all unsuccessful projects for altering public opinion or practice, it served but to confirm possession. The author connected certain accidental disgraces of our arms, with the spirit, which had gradually insinuated itself into our superior orders. And he calculated upon those disgraces, as the first members of a series, about to end in our subjugation by a foreign enemy. He forgot that misconduct in matters, relating to the person, was most likely to be first visited upon the person. Like many other explorers of the human species, he took the wrong instrument into his hand. It is the telescope, which brings within our sphere of vision the stars, that form the nebulæ of the heavens, and ascertains their relative position. But the micro-

---

* *Estimate of the times, by the author of Essays on the Characteristics.*

scope makes us best acquainted with the minute parts of single, near objects. And, respecting ourselves, it is the information, to be procured by the microscope, of which we have the greatest need and the smallest store. The calamity, that befel this bold delineator of his age, contributed to deprive his representation of the credit, which it really deserved. Nor was it considered, that the intellectual eye is often unusually illuminated before it is dazzled. What commonly happens to our kindest monitors took place in the present case. They clearly perceive mischief in preparation. But for want of sufficient insight into the nature of things, they know not exactly what the issue will be. Yet they will be rash enough to fix upon something specific. The prophecy so far fails, and brings thenceforward, all salutary caution into disrepute. For such mistakes are turned, without any great degree of dexterity, to the advantage of evil habits. They occasion the justest grounds of apprehension to be confounded with false alarms.—In the least numerous class among an enterprizing and powerful people, many baneful propensities strike an observer. Seduced by the force of his just forebodings, he seizes upon the first public miscarriages, however slight and fugitive, forces them into an unnatural conjunction with the propensities that excited his fears, and ventures, upon the strength of an accidental coincidence, to predict the speedy national annihilation. But the millennium of misery does not commence, at the time and in the manner prescribed. And the example shall be produced as a ready-made refutation of those wary enqurers, who presume no farther, than to trace the connection between sin and suffering in the same individuals.

PARTICULAR CLASSES.

THE duration and use of every production of nature and art, depend upon the manner in which it is managed, and preserved.—There are, probably, several, which would keep longest in a receiver, void of air. Others must be, at least, laid up in glass cases, the delicacy of their construction requiring that treatment, which the elder HAMLET seems, more fondly than judiciously, to have observed towards his queen—

> When he permitted not the winds of heaven
> Visit her face too roughly ———

Whether the question respects things fabricated by hands human or divine, the rule still obtains. In order that they should last long, and properly perform the services, to which they are destined, the nature of their materials, and the peculiarities in their construction, must be religiously regarded, either when they are employed for any purpose, or put aside for future occasions. Of this attention the necessity is as great, in respect to human beings, as to machines of any other description.

Some human beings are worn down by labour, without mercy on the part of those, at whose disposal they stand, and without being suffered to exert any will of their own, as to the kind of occupation or the season of repose. Even where slavery is not allowed, the measures, that enhance the price of necessaries, establish a condition, scarcely less constrained, and in some respects worse. Others have more latitude. In no country is there so large a proportion of inhabitants, with such liberty of choice, as in Great Britain.

During the first period of life, while they continue dispersed in families, it remains impossible to treat of these privileged beings collectively. The diversity of plan, pursued by parents, is one obstacle; and the difficulty of obtaining exact information, presents another obstacle, much more serious. I shall, therefore, prefer speaking of the management, most proper, during this early stage of existence, under distinct heads;—leaving it to those, whom it may concern, to compare their own proceedings with my counsels.

At a certain age, we find children of both sexes assembled in bodies, under the care of persons, whose qualifications do not very widely differ, and in whose proceedings there obtains a considerable degree of similarity. In extreme cases, common discretion will be able to judge without assistance. If otherwise, what is true of prevalent practices will apply, with still more cogency, to the greater number of exceptions.

OF SCHOOLS FOR GIRLS.

The keepers of boarding-schools are among the persons to whom the physician often deems it adviseable to render particular homage. Medical men have few better patrons. Perchance, even I should hardly dare to breathe sounds that may prove displeasing to the ear of those, whom my interest might incline me to court, if I did not feel courage from having, so far, made my way, in spite of impediments more formidable than any, which the lady abbesses of our temporary nunneries can oppose to me. And is it not better that some one, at his private peril, should bring to light, or cause to be brought to light, universally interesting facts, than that the truth should be unknown, or that no discussion should arise?

My situation offers me an endless diversity of opportunities of enquiry. For among the invalids that resort to this place, females belonging to families in some degree of affluence, form the largest proportion. Of these, an immense majority have been educated, in part, at boarding-schools. They are accompanied by relations, and visited by friends, who have been trained in the same manner. It is easy, therefore, by turning the conversation to these seminaries, to become acquainted with the lot of an infinite number of our young women; and the comparison of many accounts will obviate the danger of being misled by the inaccuracies of individuals. But the essential points tax the memory and the judgment so little;—and they afford such very small temptation to exaggerate;—that full reliance may, I suppose, be placed upon the answers. As a basis for the observations I have to make, I shall produce, in sufficient quantity, specimens of the information which I have collected. I have held in view particulars, relative to food, fuel, exercise and clothing. It was easy to put the parties upon relating the school occurrences. When I had thus ascertained the uniformity which actually exists, I declared my object, and procured in writing, or took down myself from the mouth of the person I was questioning, a relation of the circumstances. It was understood that I had determined sacredly to abstain from local or personal allusions ; and I found every one, to whom I addressed myself, ready to give me all the satisfaction in her power. To contribute information, that may eventually be serviceable to equals or to juniors, is doubtless more than allowable, when no expressed or implied obligation is violated. Indeed it is a subject, on which there ought to prevail the greatest publicity. If legislation, when it has condescended to extend its regard to helplessness, had not generally been capricious and desultory, because it followed no fixed principle, there might have been as good cause for requiring houses for the reception of children, to be licensed and visited, as houses for the reception of lunatics. For both, it is either nugatory or necessary. But law being as full of fictions as poetry, it will be easy to feign a reason for interference in the one case, and for forbearance in the other. No good reason, however, can be deduced from the comparative number or situation of the parties. To protect the young against the ignorance of relations, is perhaps still more requisite, than to protect the insane or flighty against their rapacity. And how lunatics are more likely, of the two, to be sufferers from incompetence, is past the power of an unenlightened imagination to conceive.

To complain of any thing in the following statements, unless misrepresentations should have happened to creep in, would, I imagine, be deemed no presumption in favour of the complainant. Where they exhibit a genuine picture of the interior of those institutions, as I have no doubt they do in numberless instances, it would be better to make the necessary corrections, than to complain. But there is a foible, even more common among the instructors of little people, than among us authors, who pretend to instruct the large public. Both take much greater pleasure in giving than in following good advice. Both would rather find faults in others, than amend their own. Hence, to tender exceptions to what the one practises, or the other produces, is little less than to expose oneself to be treated as an invading enemy.

It will be enough to give one set of queries entire. The others I shall abridge.

What time did you rise?

In summer at six, in winter at seven.

What time did you breakfast?

In summer we walked an hour before breakfast. In winter we had seldom time. We never had our breakfast in less than two hours after rising.

Of what did your breakfast consist?

Of bread and butter with tea.

How many hours did you sit before dinner?

Three hours.

Were you and the other girls cold in the winter, particularly in the feet?

Yes: extremely.

Could not you go and warm yourselves?

We might go to the fire—but we never did for more than a few minutes; and hardly ever till we were very cold indeed. Besides, there were too many who wished to go, for any to get much good this way.

Were the chilblains much among you?

Very much: both in the fingers and feet.

After morning school hours, what did you do?

We walked, about an hour, before dinner; (as we did also again before supper, in summer). In foul weather, we played about within doors.

What was the interval between breakfast and dinner?

About five hours.

Had you nothing to eat between?

No; unless we bought it, or were supplied by our friends.

Where, and how did you walk?

Always backwards and forwards along the same ground—and two and two: but never arm in arm.

What time did you go to bed?

About nine.

Were stiff stays worn?

Either stiff stays or back boards. Stiff stays were recommended for the larger girls; and generally worn.

No. II.—Rise at six in the summer—a little after in winter—work and read till half past eight—never go out before breakfast.

As soon as breakfast is finished, go into school, or else, in winter, walk for an hour and half—school lasts till one—when some walk on a terrace, which the greater part are not fond of.—Dinner at two—afterwards school till five.—Then, if the next day is lesson day, learn for an hour, or hour and half.—Dance two evenings in the week—go to-bed at half past eight—chilblains very common—feel very cold in cold weather—may go for a few minutes to the fire.

Stiff stays recommended, and worn almost without exception.

No. III.—Rise at six in summer, at half past in winter. Breakfast at half past seven—on tea or bad milk—school in half an hour—school over at twelve—dinner between one and two—school again till near six—tea in the evening—for supper, bread and cheese, alternately with potatoes and salt; beer to drink—go to bed at half past eight.

A small play-place behind; but all were indifferent to it—scarce walked once a week.—

All uncomfortably cold in cold weather—chilblains very general.

No. IV.—Rise in the summer at five, and in the winter at six.

Tea for breakfast, made by the governess, at eight o'clock.

Dinner at three. Slice of bread and butter at one.

Tea at half past six, made by the governess.

Sit in school five hours—learn tasks in the

mornings and evenings, which takes from three to four hours. Go to bed at eight.

In cold weather all are almost starved with cold feet, or rather benumbed, and many have chilblains.

Many girls have stiff stays, which the governess approves. Back boards not much worn; but ribbon braces very uncomfortably tight.

No. V.—Must be down by seven in summer, and by half past seven in winter—breakfasted about eight, dined at one, supped at seven—went to bed at eight.—Sate about seven hours in school; learned and practiced besides—walked little—wore stiff stays.—Chilled in winter time, and much troubled with chilblains.

No. VI.—Up by seven—breakfast at half past eight—might buy cakes at twelve—dine at two—drink tea, made by the girls, often strong. Supper afterwards—go to bed at nine—stiff stays are common and back-boards worn.—Chilblains common—greatly pinched by cold.

No. VII.—School in a very damp situation—chilblains common in the house—many girls never had them till they came to school—cold feet—rose about six o'clock in summer, and in winter as soon as it was light—

breakfast about an hour and half after—sate in school about seven hours a day, besides learning and practising.

No. VIII.—Another from the same reporters. — Hours and other circumstances much the same as in the last; except that chilblains were not *so* frequent or *so* bad.

No. IX.—Rose in summer at six—in winter at seven—breakfasted in an hour and half, on tea and bread and butter—dined five hours after.—School three hours in the morning; paraded two and two, for an hour, twice a day, in fine weather—in foul, used dumb bells, or played about—chilled in winter—and chilblains common—evening tea made by the girls, and often made strong. Stiff stays much worn.

No. X.—Monday before breakfast, which was always on the table at eight, the younger scholars rose at seven, and attended the governess in her bed-room, to read the psalms and repeat their prayers—the elder not obliged to rise before the breakfast hour. The writing master attended at nine, and remained with us one hour; then we had prayers and a moral lecture by the governess; —we then dressed for the morning, during which we were neither allowed to speak, work, read, or move, except when engaged in the dances. We seldom walked on Monday.—At one we dined on a plain joint of meat and pudding afterwards.—We had then one hour to ourselves, in the course of which a fruit or pie woman attended, with whom we were allowed to spend a certain sum.—We danced again from three to five. We had then an hour more to ourselves, at the end of which we drank tea, or supped, for we had only one meal after dinner—this consisted of tea and bread and butter, that we eat of as much as we chose. After it was over, (which was generally at seven) we employed ourselves in learning for the next, which was our first "lesson day," of which we had three in the week; Tuesday, Wednesday, and Thursday; on those days every thing was conducted the same as on the preceding, except that we read aloud, worked, and repeated our tasks of geography, grammar, spelling, &c. instead of dancing. At twelve school ended, and, if ordered, we prepared for walking, which we generally did, when the weather permitted, and which was always in the country, though we very rarely varied our walks. In these we were attended by two teachers, who were strict in not suffering us to walk *more* than three abreast: even this number was an indulgence, for we usually were limited to two.——Our punishment for mischievious tricks, buying eatables clandestinely, &c. was, being obliged to take milk and water, (sometimes gruel) instead of tea, and being placed during that meal on a form apart; generally accompanied by a severe lecture. For more serious faults, such as lying, insolence, &c. we learnt different and appropriate pieces of prose or poetry; and for carelessness we were disgraced, bearing the article in which we had offended, it being pinned to our shoulders when we walked. Those who learnt drawing never walked on Tuesdays or Thursdays, our drawing days.

Friday almost precisely as Monday. Instead of writing on Saturday, grammar-exercises and repetition.

In some chambers were four, in others, two beds, each of which, (if not very small) contained two young ladies. We went to bed early, in both summer and winter.

On Sunday went twice to Church, and before tea, learned a task from an explanation of the catechism; after repeating this, we sate till bed-time, round a table, to hear a sermon. This day we had not any bread and butter at tea, but were permitted to purchase as many biscuits as we chose. Our tea and sugar we always provided ourselves;

and selected our own parties, which generally consisted of five or six. We all felt the warmest affection and the highest respect for our governesses. The slightest illness was treated with the most particular attention, so that when we had a particular wish for a holiday and a black dose, the smallest hint was sufficient to procure both. There was school held in two separate rooms, in one of which those always attended who learnt french, therefore our french lessons never interfered with our walking or playing, as drawing and music did.—There was a person on purpose to dress the chilblains with opodeldoc—they were very common.

No. XI.—Comprises, as I find, by comparing different accounts, the situation of a large body of females; almost all the circumstances being common to more than one establishment.—Rose at six in summer, seven in winter—all who learned music, practised in winter for two hours before breakfast—in summer all, except the puny and sick, walked for an hour—six hours interval between breakfast and dinner, but each, in one school at least, had a large piece of bread at twelve. Out of school hours, in fine weather, an hour's walk, two and two, before dinner, and another before supper—in foul weather, used the dumb bells, or played about in the dancing room. The rest of the time out of school hours, practised the lessons appointed by the several teachers. Feet generally cold in winter; there were constant good fires in the school room, but the streams of air from the school room doors, were chilling. All wore stiff stays—" it was the fashion,—" so says one reporter, who was excellently situated for knowing the fact in numerous schools,—" the custom universally; the " friends, in general, expected or approved it."—Chilblains common, particularly on the fingers.

None of the preceding statements extend farther back than six or seven years. Some come up to the present moment. And along with a number of others, which I do not find to differ from them in any material respect, they include schools from one end of the kingdom to the other, especially those in or near great towns.

It has always been my endeavour to trace back, as far as possible, the constitutional history of invalids. I have been, in no instance, more able to do this in a satisfactory manner, than in young women about the age of puberty. In the case of these patients, every thing is favourable to accuracy of information. The life they lead is exceedingly uniform. What has passed, where they have not been under the immediate inspection of their parents, they can themselves supply. The period at which their health has begun to decline, has been frequently well marked. It has been often within a year, or a year and half, after their removal to school. Every one without exception has complained of being uncomfortably cold or quite benumbed, during school hours in winter.

CHILBLAINS.

I know not by how many mothers, I have been told that their daughters had no chilblains before they went to school, or that they became much worse there. According to indubitable information, when girls of the same family have gone to different schools, this has happened to them all alike. The following is one of the memorandums, which I have sometimes thought it right to request, that the deliberation of writing might correct any errors in the first recollection.—"None of my children had chilblains at "home. The eldest first went to school at "thirteen: the second at eleven, the third

"at twelve, where she is at present. Each
"got, the first winter, chilblains."

**** ******

It seems, therefore, safe to assume, that this complaint of the extremities, is much less prevalent in private families, except among particularly spirited adventurers in the snow; and that, at schools, it seizes on new-comers, with as much certainty as the ague upon strangers on their arrival in the fens, or as the yellow fever upon Europeans, when they set foot in the West Indies. It is an artificial malady, in general produced by the application of sudden heat to extreme parts, benumbed with cold. The previous coldness is an essential circumstance. In different habits, different degrees of previous cold and subsequent heat, will produce chilblains. But where they so frequently occur, the first indispensible cause must operate severely. The general feelings corroborate the inference from the effect on the hands and feet. The fact has been well remembered, whenever I have made the enquiry. In winter, the nine, ten, or twelve tedious hours of school and of preparation for school, particularly the former, were, in a greater or less degree, an uncomfortable, chilly, shivering season; leaving upon the memory, a strong, painful impression, and a worse upon the system.

It is difficult to imagine, that either this gross mismanagement, or its consequences, should not have totally escaped governesses and parents. But however that may be, they may be sure that there is not a single species of disease, indisposition, or incapability, prevalent among women in this country, which it is not fully sufficient to induce. Not a constitution but it will help to undermine. I have heard it related by many mothers, that after a residence of some months at school, a periodical function, peculiar to the female frame, has ceased altogether, or run into excess. Either irregularity will follow the action of continued cold. And I will venture to propose the prevalence of chilblains, as a simple and indubitable criterion of improper treatment. In children the most puny, and the most disposed to chilblains, they may be prevented by a very little care. They are, it is true, suffered to take place, to an equal degree, in private families—sometimes because attention is totally absorbed by vain acquirements—all acquirements are vain, if health be not secured;—and sometimes because the very desirable quality of hardiness is sought by a preposterous method. But a large proportion of young women, who have suffered long from chilblains, whether at home or abroad, will always be found to fall into fatal disorders about the æra of maturity, or to become miserable valetudinarians for the subsequent part of life.

MEALS.

Of the two, I am disposed to think a full supply of temperate warmth, more essential to the present generation than a full supply of food. But I should be sorry to drop a hint, from which any one might take occasion to think or say, that *both* are not most indispensable necessaries. It is a general opinion that children at school are less stinted in their diet than formerly. Every thing I have been able to learn, induces me to concur in this opinion. And so happy a change is probably an indication that strong public and private remonstrances from medical men, if they be continued half a century, will have a real influence on the condition of their fellow-creatures. It would, however, appear, that further improvement is still desirable. At a variety of schools, much too long an interval is suffered to pass between the hour of rising and breakfast, and again between breakfast and dinner. All will suffer from these fasts; and the puny will be materially injured. The girls are driven to bribe the maids, if possible, and at an enormous rate, to furnish them with eatables; sometimes with eatables as substantial as bread and cheese. But when the stock of money is exhausted, this contraband trade must cease, and the want of food in due season will be dreadfully felt. I have heard young women, that had lately left school, describe the faintness, internal sinking, and consequent loathing of food, which too long abstinence has induced. At some schools, (I hope not at many,) the fragments of the week are worked up into pies for Saturday's dinner. In these, the baker's unpunctuality is apt to make the members of opulent families feel, for one day in the week, what an incredible number of poor families have been universally feeling, almost every day, for these two years. But in the few schools, where children are not subject to the misery of protracted fasting, ease and cheerfulness, it is to be feared, are banished from their meals. In some, a stiff pudding is the detested prelude to a second course, carelessly dressed, and uncomfortably served up.

Elsewhere, feeding is rendered almost as much an act of penance as learning, as if it had been concerted on impartial deliberation, that the mental appetite should have no good reason to envy the corporeal.—An author, who has been strenuously endeavouring to soften the distresses of our times, devotes a chapter to the *pleasure of eating, and to the means that may be employed for increasing it.* And on the commonest grounds of humanity, it seems as if we may require of governesses, that they should provide their pupils either with more inviting dishes, or with a keener appetite.

In addition to these animadversions, I find myself obliged to state certain objections to one article in universal use. I mean

TEA.

In books on regimen, we find fruits, meats, gardenstuff, liquids, sauces and seasonings, regularly mustered from apples down to yams. Much is sagely said of *roast* and *boiled*; and the difference between the effects of *beef* and *pork* is laid down, point by point, as though the authors had been present, when these important articles of sustenance were compounded out of their elements. All this is very well, *if* it have the least foundation in accurate research — a condition, respecting which, I admonish the reader to demand full satisfaction, before he adopts a single article of the long diætetical creed, set before him.—

The full catalogue, has, however, a fine orderly, martial air. But the whole is mere shew. The service it has rendered, is next to nothing.—To many methodical, nicely reasoned, political treatises, it has been with perfect justness objected, that they are, at best, useless, because they treat of government *in the abstract.* So it is with treatises concerning things wholesome and unwholesome. To frame regulations is, in either case, a vain labour, without defining the circumstances, in which they are to be observed.

Foregoing, therefore, any apparent advantages I might derive from a specious arrangement, I shall introduce the articles of diet, entitled to a separate consideration, when I am speaking of the classes, to which they are most hurtful, or by which they are most abused.

If we look round the globe, we shall discover that the inhabitants of most countries have been ingenious enough to discover substances, capable of heightening the feelings, or of producing a temporary flow of spirits. The fabulous origin of the tea-plant is founded upon the supposition of its possessing this quality. For DARMA, having fallen asleep in violation of his vow, cut off his offending eyelids, and cast them away: when from the spot, where they fell, there sprouted up a tea-plant, the use of whose leaves quickly dispelled every cloud from the mind of the repentant saint, and recruited his system, exhausted as it had been by long abstinence, by watching, and other exercises of devotion. The Chinese seem in the gross to be well aware of the destructive effects of excess in tea-drinking. But their knowledge, probably, serves them to as little purpose as that of the Europeans in respect to the abuse of spirituous liquors. In Holland, of which the example is so much in point, as they drink tea and coffee so freely, there is a wonderful unanimity among the medical writers. I would undertake to quote the authority of twenty Dutch physicians, but I shall content myself with that of Dr. JAMES PETENSEN MICHELL. In that section of his late treatise on nervous disorders, where he speaks of tea and coffee, this physician affirms, that "*all* "who have written on the endemic disorders "of his country agree in this conclusion. "On which account," says he, "it would be "superfluous to enter more at large, into the "pernicious operation of this abuse. Only "this I will remark, that the abuse is almost "peculiar to the female sex, which must of "course principally suffer. - - Hence various "nervous complaints, which were *totally* un-"known before these liquors were introduced. "Hence a species of difficulty in swallow-"ing, which is peculiar to our country; and "hence it is, that in every place, where this "abuse particularly reigns, disorders of the "nerves are more abundant than else-"where."

Excellent observers of their own feelings report of themselves, that in certain states, they experience, even from moderately strong tea,* a sense of intoxication, with enfeebled faculties, uncomfortableness and languour

---

* Ipse memini, ante plures annos quum sensilitate nervosa singulari et debilitate ventriculi correptus essem, Theam viridem pretiosiorem me bibisse, quæ, licet modice nec saturate valde sorbillata, paullo post tamen ebrietatis sensum, memoriæ imbecilitatem et languorem insuetum cum displicentia rerum obversantium per tempus in me creabat. *Murray App. Medicaminum.* iv. 255.

afterwards. Dr. Whytt, formerly a celebrated professor of medicine at Edinburgh, says; "I once imagined tea to be in great measure "unjustly accused; and that it did not hurt "the stomach more than an equal quantity "of warm water; but experience has since "taught me the contrary. Strong tea, drunk "in any considerable quantity, in a morning, "especially, if I eat little bread with it, "generally makes me fainter before dinner "than if I had taken no breakfast at all; at "the same time it quickens my pulse, and "often affects me with a kind of giddiness. "These bad effects of tea are most remark-"able when my stomach is out of order."

In children unaccustomed to its use, tea of no uncommon strength, or fineness, makes an impression, of which it is impossible to suppose, that it can be often repeated without a degree of injury. I have observed it to occasion a wild vivacity, equal to what is excited in grown persons, by a considerable portion of strong wine; excessive flushing; a full quick pulse, with watchfulness, followed by deep sleep and heaviness next day. Having formerly, like Dr. Whytt, imagined that many things had been wrongfully laid to the account of tea, I was led by my observations on children, and by some other considerations, to have recourse to certain experiments, of which I shall quote an example. They were many times repeated; and the result was always the same. The experiments were made on frogs and toads; and it is proper to inform the reader, that several poisonous substances have a much more sudden and powerful effect upon *cold* animals than they have upon the human species. A weak infusion of opium introduced into the interior of a frog's heart, instantly destroys its movements, and renders it incapable of acting in any respect, as a living organ; whereas it would move many minutes after being treated in the same manner with water. The heart of a warm quadruped will beat long after having undergone the same treatment. The same organ, in a frog, will continue to beat for a considerable time, after being internally wetted with an infusion of Cayenne pepper, when it is so acrid as to give almost intolerable pain, if dropped into the eye, and to occasion a very disagreeable heat, when tasted.

It was first ascertained by a number of trials with a variety of preparations from vegetables, that laurel-water, infusions of opium, of digitalis, and green tea bear equal rank, with regard to their destructive effect upon the hearts of frogs and toads—all rendering them instantaneously incapable of pulsation. Infusions of hop came near these. Infusions of quassia, of peruvian bark, ipecacuanha, cantharides, and cayenne were less, or scarcely at all, poisonous; and stood nearly in the order in which they are here enumerated. We then introduced equal quantities of the following preparations into a hole, made through the skin and muscles of the belly in twelve different frogs:—into No. I. a saturated infusion of the tanning principle—into No. II. of cantharides—into No. III. an infusion of opium, diluted till it was just strong enough to destroy spontaneous motion in the heart—No. IV. decoction of quassia — Nos. V. & VI. infusion of hop—No. VII. decoction of digitalis—No. VIII. laurel-water—No. IX. infusion of ipecacuanha—Nos. X. and XI. of tea—XII. decoction of tea. The animals were placed in separate vessels, and observed. The other effects the present is not the place for describing. But Nos. X. XI. & XII. died within twelve hours after the tea had been poured under the skin. The next that died, were Nos. V. and VI. or those treated with hops. The appearances perfectly corresponded to the order of mortality. For in the organs of those that died first, the marks of inflammation and disorder were most striking. So that in all our experiments, tea proved as quickly poisonous as laurel-water, opium, or digitalis; and in some more so.

These experiments indisputably prove green tea to possess a most formidable quality. From a few others with the black species, there is reason to suppose that it is less suddenly destructive; which, if it should hold in trials to be hereafter made, would agree with human experience. But even black tea, with the usual additions, must affect the irritable habit of children too violently to form a proper part of their daily diet. It might still be said indeed, that a substance may be exceedingly deleterious to frogs, without being in any degree so to the living human frame. And the consideration always deserves attention, when experiments on brutes have any view to the benefit of man. But could any one, in prudence, overlook such strong facts, even if there were only room for the slightest suspicion, that the result was at all applicable to himself and his own species?—But there is room for much more than slight suspicion. The greater susceptibility of cold animals to certain noxious powers only exhibits, in a sudden and striking manner, phænomena that take place more faintly and slowly in animals of a higher order. Strong spirituous liquors immediately destroy the mobility in the heart of frogs, and kill them very soon, when poured under the skin of the belly, or into the stomach. And after the multitude of miserable instances which common life exhibits, will any one be so sceptical as altogether to deny the injurious effects of hard drinking, because they are in our species progressive?

The condition of the system in young people appears plainly to be such, as to cause them to be more strongly affected by tea, than persons of more advanced years. It is so with opium, with fermented liquors, and with that whole tribe of substances which, whether they find admission into the cellar, or into the apothecary's shop, or are thought too dangerous to be received into either, frequently exhilarate first, and depress afterwards. Though in some circumstances, without any previous exaltation, they bring on lowness, tremors, head-ache, and a state of sensation in which it seems to him, who has made free with them, as if he were out of his element. They should all, without distinction, be banished from the regimen of the young and healthy. How far they may suit the sickly, is a question foreign to these essays, though it is certain that, by the habitual use of tea, and of the products of the vinous fermentation, we unwisely deprive ourselves of remedies, as precious as fox-glove and opium.

I am far, however, from imputing the puny habit, and the habitual sickliness of our opulent classes solely to tea. I am also far from wishing this liquid to be universally relinquished. I only regard it as one among many concurring powers, that operate perniciously upon the constitution of children. Were they to follow a course, materially different from the usual one, in some other respects, they might perhaps drink tea with impunity, unless they should be allowed, as is the case at some boarding-schools, to make it excessively strong. To such immoderate strength, it is probable that elderly people often owe the obstinate continuance of various disorders.

EXERCISE.

In their earlier years, whether spent at school or at home, it unfortunately happens that the generality of our females are never affected by any cause, capable of counteracting tea-drinking. All perhaps without exception, that can be distinctly perceived to affect the health in any manner, concur in their mode of action with this. Among the rest, their very small bodily exertion may be fixed upon as the principal; and it is also sufficient to rob them of all sense of pleasure in themselves for the remainder of their days.

One part of the animal system we know to be destined to the purpose of feeling; another to the purpose of moving. But the due exercise of these functions is restricted within certain limits. Too much or too little use renders the organs incapable of properly performing their office. And the mischief seldom stops at incapacity, but runs on to local or general disorder, and often terminates in the destruction of the entire system. But there seems a remarkable diversity in the consequences, that result from suffering the respective parts of the system to remain in a dormant state. If a person labour moderately, and observe a temperate regimen, without ever experiencing a greater quantity of sensation than inevitably attends the animal appetites, and their gratification, he may attain the utmost term of human life without suffering. He may be as contemptible a member of society as can exist; but he will, in all probability, escape the never-ceasing and multifarious miseries of the valetudinary state. He will find himself free alike from the plagues, that wait upon ladies' indolence, and gentlemens' debauchery.

But the reverse, by no means, holds. Human creatures, made to think assiduously and to feel acutely, but debarred from the use of their muscles, will have but a questionable advantage over those, that are forced to slave in unwholesome confinement, without any adequate allowance of food or sleep. The one and the other may continue to breathe for years. But what consolation is that? when the charm of existence is gone—when the spell, cast by the Creator over his works, is broken, before its time, by the temerity of unthinking, or impious mortals. The collection of human creatures, prevented by ill-advised, unfeeling, barbarous art, from attaining a condition much above the brutes, which Britain could furnish, is beyond calculation great. Were our boarding-schools, and our cotton-mills, and other manufactories where children are employed nearly as in the cotton-mills, discharged into Salisbury-plain (where our Saxon ancestors may often have assembled to hold their wittenagemot), could the inhabitants of the respective mansions be readily distinguished?—*Yes, doubtless, at the distance of gunshot. To suppose otherwise were an affront. How could they fail to be distinguishable?*—I understand.—By their garb and their air.—But were they to exchange dresses, would the superiority of the young ladies in solid, personal advantages, at first glance, strike a stranger through the disguise?—Who durst swear that it would?—I wish it were in our power to summon into the field a third troop, that had been accustomed to enjoy their fill of sun, and air, and food, and warmth, and exercise, without being cramped, or distorted internally or externally, by unnatural garments, and without being harassed by odious tasks. The contrast, I feel well assured, would be instantly decisive. Deference to custom and aversion to thought—formidable as they are in their league against reason—would not be able to stand their ground against this show. However inferior in unimportant attributes, the recruits from the houses of labour might seem to those from the houses of instruction, both would stand at nearly an equal distance, on the scale of health and happiness, below their equals of the third division.

The greater part of the movements of girls

at boarding schools, no one, who has any regard to propriety of language, will characterize by the name *exercise* : and if we reckon upon them as such, we shall grossly impose upon ourselves. The hours, allotted for parading two or three abreast, must be deducted from the account of time, set apart for health. They scarcely contribute to it at *any* season. In the cold of winter, more inconvenience is, perhaps, occasioned, and more harm done, by this portion of discipline, than during any other part of the twenty-four hours. The tolerably robust must not measure the sensibility of feeble girls by their own. Yet, in the tolerably robust, to pace in solemn procession along the sides of a square, or through a public walk, would be a very hopeless expedient for exciting that chearfulness and glow, which can be the only rational purpose in sending young people abroad.—Play-grounds are usually too confined to allow of interesting amusements; and the majority of girls, with whom I have conversed, preferred staying within doors during the free hours. Not one has described herself as feeling any eagerness to repair to the play-ground. What shall we think of the gardens, cut out, at some country schools, after the model of a Lacedæmonian letter? For the best ends of education, I place great reliance on the nomenclature of botany, if connected with practical instruction in vegetable physiology — a science always fascinating, and now, from year to year, enriched by an accession of the most curious discoveries. This, with the present gaiety and lasting benefit of moderate, joint labour, would materially help towards forming a series of pursuits, advantageous in every point of view, to young people of both sexes. At all ages, but particularly in early life, those species of instruction are greatly to be preferred, which exercise the body along with the mind. But to be set to cultivate little plots of ground, without the inducement of reference to any kind of use, or to information in natural history, will infallibly be felt as what it really is—a paltry substitute for an occupation, in the highest degree, interesting and salutary, when properly carried on. It will, accordingly, soon come to be neglected as such substitutes always are.

Follow with your eye a mixed party, moving over an uninclosed space. What a difference in the manner of advancing of the individuals, of which it consists! The children, *if their organization have not already suffered under mal-treatment*, will always prefer the most circuitous, untrodden and difficult way; while the seniors will follow the easiest and most beaten. Truly, were the frequenters of our routes and drums made to lead a boarding-school life, I should see little subject of complaint.—To learn, every day, so many verses or so many questions and answers by rote—to practise so many tunes—to sit upon benches (provided the rooms were well warmed) so long in the morning, and so long in the evening, with so many *dead marches* between—would be a manner of filling up their time, nearly as useful, and quite as wholesome as what they now follow.—I do not see how their heads would be rendered more giddy, or their faculties drawn more awry. As to their megrims, their indigestions, their inappetence, their unrefreshing sleep, their morning mawkishness, and their all-day-long ennui; an ingenious speculatist might be at a loss to find a substantial reason for believing, that these must be aggravated. Nor, after the first awkwardness of change was over, would they themselves, or others, in any respect, suffer from their being immured in a new place of confinement, and performing their daily round of penance in a manner, somewhat different.—But to condemn, to the dry husks of words, children, all whose senses are hungry for the impressions, which fresh objects afford; and to debar them from giving a loose to those inclinations, by indulging which, they knit their sinews, swell their muscles, and harden the whole frame against the vicissitudes of the air, is a kind of severity neither indifferent to them, or to those about them—for the present or for the time to come. It is a system, during its continuance and in its consequences, as detestable as that of the *first* continental despot, into whose hands the ill-starred Baron TRENCK had the misfortune to fall; and it produces sufferings more to be deprecated than that of the *second*.

Like every other organ, the palate and the stomach, when left inactive and unirritated, lose their original faculties. Repeated fasting is fatal both to appetite and to digestion. Sedentary occupations gradually destroy the desire, and impair the power, to exert the muscles, more particularly while they are forming.

Of a climate, uncertain like ours, it is among the most serious evils, that it supplies a standing inducement to a close, retired, chamber life. It is evident (other circumstances being alike) that the inhabitants of milder regions must be more sociable and joyous, as well as more healthy. Where the seasons return with never-failing rigour, the inhabitants arm themselves with equal regularity: and art gives what nature denies. Of all nations, the English has probably shewn itself least ingenious in the œconomy of artificial heat, and most imprudent in the article of covering. Hence, and perhaps on some other accounts, we treble the disadvantages of an atmosphere, boisterous, wet, and inclining to cold. Bating the difference in elegance between caves, and their fixed or moveable hiding places, may not our women of genteel rank be considered, as having converted themselves into TROGLODYTES? Their influence has rendered their imbecillity an example, and thus it comes to be deemed reputable to shun, as much as possible, the light of day. By their indolence they in great measure lose (and we are partners in the loss) the beauty of the mild season, and by their tenderness, they are entire strangers to the sublimity of the severe.—For all, I believe, who are capable of enjoying the varied aspects of nature, and can bear to contemplate, at leisure, the stern features of winter, will agree that the poetry of THOMSON is not romance——

VAPOURS and CLOUDS and STORMS! be these my theme;
These! that exalt the soul to solemn thought,
And heavenly musing. Welcome, kindred glooms!
Congenial horrors hail! With frequent foot
Pleased have I wandered through your rough domain,
And sung of Nature with unceasing joy - - -
NATURE! great parent, whose unceasing hand
Rolls round the seasons of the changeful year,
How mighty, how majestic are thy works!
With what a pleasing dread they swell the soul!
That sees astonished, and astonished sings——

An aversion to move, and a horror for any temperature below spring heat will, I imagine, be found to reign in boarding-schools. When not acquired, these qualities, that properly characterize old age, are always confirmed there. To my questions on the subject of skipping, playing, and using the dumb bells —*we would always ten times rather sit by the fire—we never exerted ourselves when we could avoid it—it is amazing how indolent we all were*—have been the perpetual answers. To those who are sensible what such answers imply, they tell the story of the school as plainly as if it were laid out in a whole volume.

CLOATHING.

For more than one generation past, the injury to important, and even to vital parts, from stiff stays, and from tight braces of whatever denomination, has been perpetually echoed aloud in the public ear. Nothing therefore surprised me more than to find that those who have the charge of girls, so generally persevere in their use—thus long after women had *apparently* laid them aside. I say *apparently*, because the old plan of severe constriction, much oftner than is suspected, lurks below the free Grecian flow of the external habit. The different parts, I suppose, of the boarding-school system, by their united operation, reduce the happiest conformation nearly to that ricketty state, to which Pope was born: and stays, in most cases, may become necessary to keep up the sinking frame to its daily discipline: just as soldiers, fainting under the application of the lash, are brought to their feeling by having stimulants forced upon them. The insane desire of a shape on the part of parents, no doubt, frequently concurs.—There is a process, through which the whole human species is doomed to pass, and upon which a scattered individual here and there may have entered. This process is *the consolidation of the agreeable with the useful, and the subjection of taste to truth.* Now to a person of information, whose judgment the above-mentioned process has set above his feelings, the sugar-loaf scull of the Indian must exhibit nearly as pleasing a spectacle as the sugar-loaf bust of a British beauty. Considering *what lies beneath* in both cases, I hold it probable, that the measures, by which the head is made taper, are the less detrimental. But after what has been already written, it would be vain to bring together the strongest general terms. And by readers, not conversant in anatomy, a minute disquisition could not be understood without helps, which do not fall within the scope of these essays. If there yet remain a chance of bringing the public* to reason, it must be by calling in the eye to the aid of the understanding: which I hope to be able to do in the *Treatise on Physiology*, already mentioned. Mr. JOHN KING,

* That is, such part of the public as can command this essential necessary of life. It is remarkable, that what has been observed as to food, may be justly observed as to cloathing too. One part of the public cannot procure enough. The other injure themselves by its abuse.

surgeon, who unites the talents of a skilful artist and of an accurate anatomist, and who has considered the origin and effect of fashions with much minuteness, has undertaken to supply a set of drawings, exhibiting various dresses, as they act upon various parts. In respect to shoes, as the English reader may know, this useful design has been ably begun by the late Dr. CAMPER, and with respect to stays by Dr. SOEMMERING, whose work has not been introduced into our language.

I have not yet done with *boarding-schools for young ladies.* But what I have further to observe has a wider application.—I object to lessons by heart, as a practice by no means indifferent to health, and highly injurious to the understanding, where it is carried to any extent. But I content myself at present with stating a naked opinion, because the arguments would lead to a disquisition too prolix for this place, and too abstract for a popular work. In his tract on education, Mr. LOCKE, without tracing the mischief to its source, has said enough to convince every unbiassed mind of the unfitness of learning by heart for any rational purpose. Whoever will make the experiment, will find the reason, the imagination, and the *memory* more powerful, the less this method is called into use.

Female schools of high fashion, where the conductors—angels defend us!—threaten to compleat their pupils in accomplishments, and to give them *finished manners,* I take to be the most destructive. The croud of masters, some of whom are meant as *ministers of grace,* overpower their tender pupils. The rapid succession of lessons cannot fail most materially to damage the *mental organs,* not yet certainly arrived at their state of greatest fitness for constant service, and at all times liable to injury, without the grateful interchange of bodily exertion. Injury of this kind is positive disease, slow but consuming. It is one of those sorts of blight, that most cruelly despoil the spring of life. An explanation of the manner, in which *cramming* the head with dose after dose of heterogeneous ideas, before the first have had time to settle, operates, would seem perplexed to the general reader. But the fact itself is what concerns him to know. This can hardly be doubted by any one, who has ever experienced that confusion and dizziness, which the hurried exhibition of a multitude of objects, for *once only* and for a short time, leaves behind—And here we have the attention stretched and distracted several days in the week, and half the day together.

MUSIC.

Music, for a tedious succession of months, is a severe study, before it approaches towards the nature of a delightful recreation. In both predicaments it is objectionable. Nothing perhaps from first to last, is more prejudicial in education, as it is at present cultivated. Even when it charms, it co-operates with weights, already suspended with too little consideration upon the nervous system, and all pulling in the same direction. The good or harm of music is, doubtless, a question upon the case; and indeed discussions concerning health must be either unavailing, or they must be solutions of such questions. The SAVOYARD rustic, who carols as she trudges, is, I can well conceive, all the better for her elegant accomplishment. But the English Miss, with whom already almost every occupation is sedentary, and every pleasure passive, must, I fear, be the worse—the worse for the acquirement of the art, and for the delight it yields, when acquired. Those who are not acquainted with particular facts—that is, who pronounce without any sort of evidence—may smile. But I will not suppress my suspicion, that the largest pack of hounds we have, turned out *mad* upon the country, might possibly have committed less ravage, than that rage for excelling in music, which, of late years, we have seen invading families, and imposing the necessity of such strictness of application upon the girls—But this abuse is by no means proper to schools. It is not even most inveterate there. Unless the wheel of fortune throw up more salutary examples, it will increase with the increasing preponderance of private over public education. And this, for one sex, must become continually greater, till we shall be acquainted with schools as we are with nunneries—only by report.

Sickness any where is to be deprecated; and especially for children. At a boarding-school, it is double misery. Not from any fault of governesses. It is indeed out of possibility, that governesses, however large of soul, should feel for a host of pupils as much anxiety as parents for a few children; or that the more indifferent party should watch over *many*, as effectually as the more deeply interested over *a few*. For this reason, and for this reason only, insidious disorders will more frequently make progress at school than at home.—As to the rest, both parties labour under a common disadvantage. Both are destitute of the sense—a necessary, but an acquired sense—by which danger is perceived on its first approaches. Without suspicion of the foe, they can have no inducement to look abroad for an ally. But when the foe shews himself in full march, and they can no longer delay calling in assistance, it is —— —— ——. —— —— —— —— —— —— ——; —— —— —— —— —— ——. —— —— ——, —— —— —— —— (—— —— —— —) —— —— —— ——. —— —— —— —— —— —— ——, then the drenching scheme appears upon its grandest scale. It is into the stomach of these poor creatures, unprotected and helpless on their backs in bed, and hardly daring to draw a wry face at any thing offered them, that the wash-tub of the apothecary is *then* most unmercifully emptied.

I deliver these reflections to the press, under a full conviction of the impolicy of overcharging my statement. I am aware that in such cases, a detected inaccuracy in a single item discredits the whole account. Has not popery itself, in the course of the French revolution, under all our eyes, been persecuted into popularity? I am sensible that there may exist seminaries of an opposite spirit;—where more mischiefs are prevented than I can point out, and wiser regulations prevail than I can suggest:—where the culture of the interior is carried on by the hour, and that of the exterior by the minute:—where no jostling crowd of builders run up a flimsy piece of plaister-work, which in a few seasons, is to sink, with all its streamers faded, into the mire—the scorn or pity of all that pass by. There may, in short, be seminaries, where accomplishments, information, morals, dress, diet, comfort, air and exercise are so happily combined, as to produce rude health without rude manners, or frivolity of mind:—And where a full provision of that happiness, to which the *girl* is competent, ensures the same to the *woman.* For these two objects can never be disjoined.

Such seminaries my censure cannot touch, and my applause is at the service of their superintendants, as soon as they choose to make themselves known.—Of others, I had determined to speak more mildly than I should think. But facts speak for themselves; nor ought I to stifle their voice. The partial existence of some of the facts above adduced, I had learned in the intercourse of life, but did not suspect they were so common. Of their combination I was as little aware, as I at this moment am of the sum of discoveries, which the opening century is to produce.

The most judicious advocates for boarding schools, reduce the pupils to a number, not double that of some families. The more, undoubtedly, the worse in all essential respects. I am persuaded, from observation and experience, that few persons can do justice to five or six children. Where the proportion is greater, it will be necessary to have recourse to a mode of preventing harm, similar in principle to that which is practised in the case of maniacs. Restraint is much the same thing, whether imposed by fear or by the straight waistcoat.

ADVANTAGES FALSELY ASCRIBED TO SCHOOLS.

School-education is said to derive an advantage from the propensity of children to imitate those, who are a little older than themselves. But it probably derives more disadvantages than advantages from this source. Bad examples being apt to operate more powerfully than good; and being also more detrimental in their kind than the good are beneficial.—Emulation is a stimulus of exceedingly disputable quality. There is too much cause to fear, that it frequently generates a disposition, tormenting to him, who harbours it, and mischievous to others.—The rudiments of the knowledge of life are said to be best acquired at a school. But home-education, by no means, implies seclusion from equals. What is there to hinder the heads of private families from bringing their children together as often as can be for their good? Nor will it be difficult to put them in a situation like that, in which the advocates for free-will, suppose our whole species to be placed by providence: namely, at liberty to act, but their every action noted. And thus the double inconvenience of too much constraint at one time, and of too little observation, when the removal of that constraint is followed by its natural consequence, a disposition to practice all sorts of licentious pranks.——Home-taught girls would have more opportunities of mixing with their seniors. Their judgment, therefore, must be better exercised as far as depends on diversified objects of comparison; and there can be nothing at school, which can do half so much towards cultivating the moral and prudential sense, as the conversation of judicious parents.—I place more confidence in these reasonings, because I have seen them confirmed by actual trial. Children under my eye, (without example, emulation, or confinement,) have made thrice the classical attainments, which others, apparently of equal ability, have made at the most celebrated schools. By proper associations, it was practicable to prevent the very elements of knowledge from proving disgustful. Further advances were attended with increasing delight. So rapidly did the voluntary power wax in strength, that boys of eight or nine years old would compose latin verses, attempt an English story, dialogue, or play, or solve a question in fractions by the head. They went on without painful effort, and without decrease of the unaffected sprightliness, so becoming in children. The constitution improved at the same time. The improvement of the constitution, indeed, was the province delegated to me; but a fair occasion presenting itself for a noble experiment in education, I was tempted to draw out a plan for conducting it, and, as far as my more urgent avocations would permit, to inspect its due execution—Now a result, obtained at so early an age, must, it is evident apply to one sex as exactly as to the other.

I have never, I own, felt the propriety of huddling children together, in order that those, that suit may feel each other out in the throng. It is a way, in which I should imagine bad connections much more likely to be formed than good ones.—When the chemist desires to affect an union between two principles, does he set about, in the dark, to shovel all manner of materials into the same vessel, for the chance of a combination to his purpose? Or, does he mix, according to art, such substances as contain the elements, that have a mutual affinity?—To secure the formation of advantageous intimacies, I should rather trust to the superintending, but unseen influence of the parent, than to the untutored feelings of the child. No advocate for a public education, will surely offer the heads of private families such an insult, as to say, that they have not management enough to place young people within the sphere of each other's attraction. And shall we only concern ourselves about what is united by the process recommended? Shall we not also look to what is separated? If living together establishes friendship between aliens, living apart more often occasions alienation among friends.

Domestic duties, the grand concern of

every female, will always be best acquired, when the pupil is unable to say when she began to learn. They cannot be taught at large schools. At those upon any scale, they are, perhaps, nearly as seldom taught as any solid knowledge of foreign languages:—the common regulations for the latter purpose having, I am told, no more effect towards forming good linguists than the statutes of the University of Oxford towards providing knowledge, suited to the times. How far young ladies may be best instructed at schools in the disposition of their ostrich feathers before, and of their trains behind, I am not able to conjecture. I should, however, imagine that there will be, every where, stimulus sufficient to save them from the danger of inattention to dress; and, every where, examples enough to direct taste.

If boarding schools incontrovertibly possessed the alleged advantages, and others still superior, will any one who knows the difference between good and bad health, say that the destruction of the constitution ought to be risqued for their sake? But there are some, who count instead of weighing arguments. I therefore, thought it allowable to stop, while I examined whether certain arguments add any real weight to that side of the question, on which they are usually thrown in.

A mother, perhaps, here and there, (looking with more anxiety towards the means, by which a husband may be won, than those by which he is retained) will entertain fears lest girls, brought up with a sovereign regard to health, should not be delicate enough for the present demand. In consequence, it appears, of the superiority of the Grecian mind, the taste of that extraordinary people, in the article of beauty, has been adopted in all the countries, which the Grecian literature and Grecian art has reached. But if we have adhered to the cast of the *features*, have we not departed from our model in respect to the *person?* Were not the Grecian beauties of a structure, somewhat more firm? The remains of antient art seem to say this with regard to the Greeks, and to the Romans also;—at least, before the degenerate period when the Romans had dwindled into helplessness under the continued influence of luxury, and "the fierce giants of the North broke in "and mended the puny race!"—But whatever be in this opinion, there is certainly hazard in leaning too much towards enfeebling measures. Who would be rendered *needlessly* dependent upon the humanity, which she might indeed experience, and which she might also be very ready to exercise towards another? And is it impossible to conceive husbands, whose feelings shall not be of the most kindly character, when, with a fate, more tantalizing than that of Ixion, they find they have embraced sickliness instead of delicacy?

DUE CARE OF GIRLS' HEALTH.

When it is impracticable for parents to bring up their own daughters, by seeing that the following particulars are carefully attended to, they may prevent injury to the constitution. The same directions will apply, in a great measure, to a home-education.

The limited number of some few schools should become general. The number of pupils in the same house, should never, perhaps, exceed a dozen. But that properly qualified governesses may be tempted to undertake the charge, affluent parents should consider that a good education is the best part of a child's portion; nor should they be afraid of lessening the patrimony, where it is at all considerable, in behalf of so important an object.

The young people should rise early in summer; and in winter as soon as it is light. But, in both seasons, some agreeable occupation should be the inducement to call them out of bed rather than compliance with a strict rule. As soon as they rise, they should be offered some bread, or bread and butter, or a small draught of milk. That child is unwell from a permanent cause, or from transient indigestion, who refuses such offers soon after rising. An hour is too long for a child, not robust, to remain in the morning without food.

The interval of an hour should be devoted to some play or exercise. A breakfast of milk should succeed. Prepared milk or broth or beef tea should be the substitute of plain milk, where that disagrees. The liquids used at breakfast should not be taken above blood-heat in any case; and where there is no considerable or perspiring heat, they should, in warm weather, be taken of the temperature of the atmosphere. The previous exertion will prevent any one from sitting down chilly to her meal.

Genuine milk should be secured at any expence. It would be an essential circumstance in favour of that school, where cows enough for the family could be kept, or where it could be satisfactorily shewn that an unadulterated supply was provided.

A *moderate* walk of twenty minutes should succeed. Much exertion in feeble subjects might disturb digestion. But even much exertion with agreeable feelings, will have a less bad effect in this way, than reluctant occupation in a posture, by which pressure is made on the stomach. We see boys and peasants, using violent exercise immediately after meals without inconvenience.

Two hours of sedentary occupation might fol ow—but not more than two hours—in a temperature, as little as may be, below sixty degrees of the themometer, commonly used in this country. Of what they read, young people should give the substance in their own words, sometimes in writing and sometimes in speaking. It is better to give nothing to be learned (except, perhaps, the multiplication table) in order to be repeated *verbatim*. Getting passages by heart should be trusted to a sense of their beauty. Without this sense, the passages will soon be forgotten; and children are fretted and injured by frequent *tasks* of this nature. The suffering of the body through the medium of the mind, is a thing every day exemplified in the most sudden manner. But the *gradual* operation of this cause is not less manifest to close observers.* And it assists in accounting for a phænomenon, which the female sex, I believe, more frequently exhibits at an early age. The phænomenon I mean is the inversion of the character or of the temperament. The members of the medical profession, who act as *confessors* to the unfortunate among mankind, perpetually find those very women most labouring under depression of

---

* It is endless to relate cases in which the organization is instantly or progressively affected by the feelings.—The following instance I noted down from the lips of the patient. ——A healthy lady consulted a medical person about her son's eyes, and was bluntly told that he would lose his sight. That very instant, being extremely shocked, she felt something burst within. On driving home, she spit blood, and from the same moment lost her health. The organs of respiration and digestion all partook in the mischief.

To-day, (*Jan.* 14, 1802) I have seen a person, who, five years ago, neglected a sprained ancle. And now on feeling the least chagrin, the ancle becomes painful and swells; which it never otherwise does, except when twisted by a false step.—Thus then, if a girl brings a weak vital part to school, and is teazed and saddened by the life she leads, may not the unpleasant state of mind weaken and disorganize the part?—There is no doubt but it may; and that without local pain, as is actually the case in some of our most fatal complaints.

spirits, whose infancy and childhood were distinguished by an extraordinary degree of sprightliness. This change, I believe, very seldom begins at home; not unfrequently at school, and sometimes not till after marriage. It arises from repeated vexations—rarely from any single misfortune. A human being, endued with an extraordinary degree of sensibility, will, of course, be more than commonly elated amid pleasant scenes. And where the feelings are so expanded to impressions of either species, how can they but sometimes receive a blight from the unkindly discipline and diet of many boarding-schools?

The most chearful and healthy sort of instruction is that, which I have already recommended, as to be acquired directly through the senses. The different departments of *natural philosophy*, in the most extensive sense of that term, will afford an endless course of agreeable and wholesome instruction. But to be good for body and mind, the lessons must be short. One or two experiments will generally be enough. The general inutility of lectures of all sorts, and the injury they occasion to children in particular, depends upon their excessive length. The multiplication of elementary treatises will prevent any one, who is in earnest, from ever being at a loss for matter. Every place will afford persons, willing to instruct a teacher of girls in certain simple manipulations. If the practice of some mechanical arts could be introduced into female schools, the advantage would be great indeed to the constitution. It would, it should seem, be practicable to introduce some of these arts; such, for instance, as the use of the lathe. Both arms should, as much as possible, be equally exercised, not merely for the sake of harmony and power in the movements, but to obviate distortion.

In rather less than three hours after breakfast, food, in small quantity, should be distributed. Three hours is the very utmost, that a child, particularly if at all weakly, (and the great majority of the offspring of parents in genteel life answer to this description) should go without sustenance. Every quarter of an hour, after hunger begins to press, adds its share to the mass of mischief, which a variety of causes has most commonly conspired to accumulate.—An hour for work and study may intervene before dinner. To dinner, and to some entertaining reading or communication on the part of the instructress, during a simple desert, an hour and half may with propriety be devoted. In winter a brisk walk or cheerful play should succeed: and then school-exercises for an hour and half. But during all the sitting time, the kind of employment should be twice or thrice changed. Some grateful preparation of milk, as custards, or blanc manger, or orgeat, or a piece of fruit-pie, should be served in place of tea—I particularize fruit-pie, because I often see over-anxious educators deprive children of innocent and agreeable articles *for their stomach's sake.* Individual peculiarities will soon shew themselves. Otherwise, few plain eatables disagree with children, especially active children. To prevent needless privation is of some importance. It is of more to prevent attention from being wasted on frivolous precautions. I have heard many mothers cry—" my dear, don't eat *that*; it will " make you *sick!*"—It would not signify much though it should. And I wish I could hear a few say with equal emotion—" don't do " that, it will make you *sickly*:" or rather manage with constant reference to this idea. —The refreshment in place of tea should, in fine weather, be followed by a botanical excursion; at other seasons, by an active, in-doors game, in which as in every thing else, the governess or an assistant, should join. Then supper, and bed. But at every meal, a properly qualified person must be able to bring forward something, which will answer the purpose of a pleasant and healthful education. A piece of bread and cheese may be made the introduction to every art and science.

At all times, exposure to damp must be guarded against. In cold weather, the pupils should be felt and examined, that proper precautions against nightly chills may be taken, when required. In very cold weather, a fire should be made in the bed-room.

Under twelve years of age, then, it should be an invariable rule, that the hours of application should never exceed those of amusement and exercise. SWIFT observes that in the arithmetic of the customs, two and two often make one instead of four. The same holds in many cases of instruction. The children that have made, within my knowledge, the quickest progress, felt the deepest interest in knowledge, and retained their acquisitions most firmly, were never detained at their books above an hour at a time; and seldom above half the time. So perpetually true is it, that the other most valuable objects are best secured by the very measures, which regard to health enjoins.—*Proficiency* in music and drawing should be given up to a maturer age; and indeed till the system has become hardy, and the pupil is confirmed in active habits. Even when a love of bodily exertion, a facility in seizing ideas, and a power to resist the inclemency of the weather have been acquired, some plan should be devised, by which they may be retained. One day in the week, for example, every pupil should abandon her bench, her book and needle for a long excursion. A school of twelve would divide well into two parties, which might sally forth in turn. For this, and for very many other reasons, these seminaries ought to be conducted by a married couple, and not by women only.

The worst quality, which a female can carry out of childhood into puberty, is that languor, which, as I have said, it is hardly possible a boarding-school, according to the usual scheme,* should not create or fix. It

* While this number was printing, the following statement was received. The plan has many and great omis-

is the foundation of the worst, and of the most disagreeable, in the long catalogue of modern female diseases. When the constitution of a daughter has undergone such a depravation, the parents cannot make too much haste to correct it. The process should be the most gentle possible; but it should be carried on daily.—Women grown up, and become their own mistresses, need not despair

---

sions, but in some circumstances bespeaks an excellent spirit.——"In summer and in winter, the young ladies are expected to be in the school-room by seven o'clock; for which purpose they must rise about half past six. They breakfast at eight, and go into school again at nine. As soon as the younger girls have finished their lessons, which is usually about eleven, they are sent to play in another room, or in the garden if the weather be fine. School is over at twelve. They dine at half past one, and go into school at three, where they remain till five. Between five and six they drink tea, which then as well as at breakfast, is made for them. They sup at half past eight, and go to bed at nine. In winter they walk in the middle of the day, in summer in the evening: and in spring and autumn from four o'clock till tea-time. On all occasions they are encouraged to take as much exercise as possible; and as they do no work in school-hours, *they are not under the necessity of sitting to learn their tasks during the time, appropriated to play*——Chilblains are less frequent, and observed to be almost peculiar to scrophulous habits. A separate bed for each pupil."

of effecting a very salutary change in their own persons. The author could name several, who by dint of resolution, or in consequence of a change in the circumstances of the family, have got rid at once of their helplessness, and of their ill-health. Instead of falling into one of those fatal or lingering complaints, of which they stood in continual dread, and indeed in continual danger, they have recovered equally the tone of body and mind.

PERIOD OF YOUTH AFTER SCHOOL.

Were the young woman subjected to powers of a nature, opposite to those which acted upon the girl, the force of habit would probably, in most cases, prove insuperable. But the powers which act during the succeeding period are, if possible, still more unpropitious to health. Some sedentary occupations, begun by the direction of superiors, are more sedulously pursued from the natural desire of being approved or admired. It is now that the victims, which Pædìa had prepared, are sacrificed, in countless numbers, at the altar of Fashion. The monuments of antiquity generally exhibit some struggle on the part of the animal, before he bleeds. But in the victims of Fashion, there is either greater thoughtlessness, or greater indifference to their fate.

We wonder at the infatuation of devotees, who at the festival of a certain Hindoo goddess, throw themselves under the wheels of her heavy car, and are crushed to pieces. We have no right however, to call the ardour of these poor wretches *infatuation*. For a reason, not at all better, a much greater number among ourselves come to the same end in a more cruel and lingering manner. And in the majority of instances, they are of the sex, most exquisitely alive to pain.

FASHIONABLE DRESS.

Many women, in the morning, muffle themselves up to the chin and ears, and go about half naked all evening, braving disease and death The cold of our climate is sufficient annually to cut down thousands of females, who, having been tenderly brought up, will not guard themselves by sufficient covering against it. But the ladies are not content with the havoc, committed in this manner. Among death-bed confessions, I have heard of a practice of *damping* the cobweb garments, which otherwise would hang about the limbs too loosely. By this means, the killing rigour of an inclement atmosphere, is most materially assisted. The desire of obtaining a celebrity, equal to that of some monuments of antient art, which have been lately transferred to Paris, appears to me quite as good a reason for risquing life, as many of those, which have conferred mortality on the adventurers. But the beautiful followers of the *attitude* or *statue* fashion unfortunately do not stiffen into figures, like the Venus de Medicis, or like what they would themselves be in good health. The state, to which they are often reduced, is one from which every one turns away with horror, except the stealer of dead bodies and the dissector.

The last years of the century have seen the land, more deeply covered with mourning, than any of the preceding. War and pestilence have obviously had a large share in promoting the distress of families: but neither war nor pestilence a larger share, perhaps, than the prevailing modes of female dress.

The mothers of those young women, who

are still under controul, should observe whether they seem chilly, and examine whether they are cold to the touch. The internal flame can only be raised to a proper degree by exercise or labour. But it must be cautiously raised; otherwise the feeble spark of life itself will be extinguished. Meanwhile, warm covering should be applied. Hot rooms do not answer the purpose. They rather increase the evil. I have read, that Dr. Herschel, when he observes the stars on a cold night, has his feet covered by fourteen pair of stockings. The example of this astronomer should be followed, as far as is necessary. It is probable that the ladies of Holland, where errors of regimen are as gross as in Britain, owe their greater exemption from some fatal maladies, to their greater care in preventing chills.

READING.

Our literature is no inconsiderable source of our effeminacy. Books, which act perniciously upon the constitution, are prohibited at schools. But it is not by prohibition that bad practices are to be prevented. The governess, that has contrived to give a decided aste for that knowledge, which is most suitable to both mind and body, has done more towards preserving her youthful pupil from the various bad consequences of improper publications than could be done by a law of DRACO.

As to the sort of reading, most injurious to young females, I cordially assent to the opinion of almost all men of reflection. NOVELS, undoubtedly, are the sort most injurious. Novels render the sensibility still more diseased. And they increase indolence, the imaginary world indisposing those, who inhabit it in thought, to go abroad into the real. For useful lessons, doubtless, no vehicle is fitter than fictitious biography. It may exhibit select scenes only. But those scenes should never swell beyond the compass of life. Fiction begins with being extravagant. But its extravagancies become ridiculous. Its tone then grows gradually lower and juster; and it satisfies the longer, the less it attempts to surprize. Our novelists, however, are still guilty of that, with which SOCRATES was reproached. *They tread ether and look down upon the sun.*—It is probably easier to perform some feats of activity than to maintain a graceful carriage.

—Passages of Le Sage, Richardson, Marivaux, Sterne, of J. P. F. Richter and their equals, will remain. But as soon as science is a little more improved, and morals a little better understood, many admired novels will be banished in a body to the same shelf with *Jack the Giant-killer* and *Tom Hickathrift*.

The common love-stories are justly regarded as abominable. They relax soul and body at once.—But there is a class of novels—the mushroom growth of the present unhappy age—undeservedly considered with more favour. Certain projectors, totally unacquainted with the data for calculating individual happiness, set about to draw up schemes for the happiness of nations. These schemes were received with clamorous applause by second-rate reformers, as shallow and presumptuous as their principals. The principals by degrees incurred universal odium or contempt; and were deserted by their satellites. From the dead and putrid carcases, however, of their conceptions, there has issued a swarm of novels, as noisome as the mass by which they were engendered. The authors of these novels rank among the basest flatterers of corrupt opinion. They are clumsy enough to be obliged to overcharge what they decry; and manage their ridicule so, as to recommend existing abuses, not less baneful than any crudely concerted plans for innovation.—It does not belong to the present plan to pursue these ideas further. I shall only add that parents, to whom the health of their families is dear, should be less severe against pining love-stories than against tales, in which fashionable frivolity is directly or indirectly represented as amiable or respectable. Such tales, besides their other mischiefs, are to all intents and purposes codes of slow suicide.—Of modern narratives, I consider Sandford and Merton as the one, conceived in the most salutary spirit. Authors, whom we see darting thier literary shears at every shoot of generous ardour in youthful minds, are almost certain to impair the vigour of the plant, and may clip it to death.

Independently of novels, too exclusive addiction to the belles lettres must hurt the weaker sex, as at present constituted.—A couplet, eternally quoted from Ovid, recommends the *ingenuous arts*, as softeners of ferocious manners. The prescription is excellent; and for that very reason, unfit for the opposite case. That must be a wretched practice, which applies *emollients* to parts already too flabby.

HIGH AND MIDDLE LIFE AMONG WOMEN.

It has been supposed, that in a country, where food and fuel would be unnecessary, and which should produce sustenance at all seasons, the inhabitants would do little more than lie under the trees, and open their mouths to receive the dropping fruits. Now, surely, it makes little difference whether it be the gifts of nature or those of art, that can be had in this passive manner. Hence in the metropolis of every great empire, there always comes to be a number of females, virtually in the condition of the torpid natives of our imaginary fine and fertile climate. But in a great commercial state, the number may be much larger than in any other. For, in respect to the means of luxurious enjoyment, the extent of the empire must be equal to the sphere of commerce; and nearly the whole of the commercial country may, in this respect, correspond to the metropolis of an huge, but not commercial empire. This possibility we see realized at home. British consumers are supplied not only by British producers; but the labourers of almost every other country administer to the sensuality or to the indolence of our richer classes. This cause, assisted by others already in part explained, produces an effect the more to be deplored, as it appears progressively increasing. If we abstract from certain tropical islands, and from particular tracts in Europe, where the air and the soil are charged with pestilence, our own will, I believe, be found the most ailing country in the world; and of the two sexes, and among the different orders of society, women above the necessity of labour suffer the most. They also appear to partake the least of the most lasting pleasures of human nature; though these lie completely within their reach.

In London, if information from different quarters be not fallacious, many women of the higher ranks are up till two or three, and often (particularly in spring) till four or five o'clock in the morning. They take no exercise or air, which they can avoid; but constantly pass from hot rooms to close carriages. By this course, even inattentive observers can perceive how materially they are affected.—In the country, though the manner of living be different, and indeed too various to admit of a general description, it must be considered but as a return from worse to bad. It is still without activity enough.—The same fault descends through many gradations of wealth. In the lowest of those that keep up a shew of gentility, women sit too much, confusing their heads and fretting their sensibility by a rapid perusal of novels. They practice, besides, every thing that has been stated, as tending to weaken the moving part of the frame, and to render the feeling part, unfit as well for the ordinary as the extraordinary conditions of existence. The *slighter* consequences are, constant uncomfortable feelings, complaints which delicacy declines confessing, disgust in matrimony, the mortification of miscarriages, incapability of discharging the first office of the mother, or a state of atrophy, if it be attempted. The *graver* are to be found in the bills of mortality; and will be particularly described hereafter.

The evil has spread wide, and is, most probably, every day extending. It spreads; and those about to be consumed are unmindful of its progress, as is found to be the case, when the plague has raged for a time in a great city. He that proposes to take a circuit round its boundaries, will appear to wander wide of the mark. In a treatise on the preservation of the public health, the most essential parts will be disregarded as preliminary or digressive. And persons of sense, inadequately informed, will be impatient, if they do not see you lopping the branches, when you are delving to get at the deep root, of the tree of disease. They will overlook the inseparable connection between moral and medical topics. Nor will they be easily persuaded, that moral treatises have been of little or no utility, because the authors have torn asunder in their speculations, what nature has joined in the actual world. We have had, and we shall have, writers on morality, capable, like the Grecian sophists, of ringing a thousand pretty changes upon words. But unless they precisely understand, and can make others feel, how immoral practices affect the sensations and existence of him who commits them, and those of others, the audience will be nothing the better for all the music, when once it is over. A course of lectures upon the bodies of only ten young people, murdered by the preparation of articles of luxury, and of ten others, murdered by the use of those articles, might more improve the present age, in its individual and relative capacities, than all the elementary treatises of morality, that have ever been composed. But every useful observation in those tracts would find its place in such lectures; and here, in its proper place, it might have a chance of operating upon conduct. At least, since it is clear, that plans of education and of life, not calculating upon the physical nature of man, as their foundation, contribute nothing to health or happiness, it may be worth while to try, whether strict attention to this would not give a more desirable result. Perhaps, if there could be found, in our several cities and counties, as many families willing to accede to this proposal, as the angel required persons to save Sodom and Gomorrah, we might live to see the female sex, emancipated from constraint and maltreatment in youth, and from debility and sickliness in riper years. This will be clear gain to human happiness. The encouragement of manufactures, that is, the creation of a miserable and sickly population is a paltry excuse for lying to macerate in the stew-pan of luxury, till we become miserable and sickly ourselves.

END OF ESSAY THIRD.

Printed by MILLS, St. Augustine's, Bristol.

ESSAYS

ON

THE MEANS

OF

AVOIDING

HABITUAL SICKLINESS,

AND

PREMATURE MORTALITY.

ESSAY FOURTH.

Printed by B. M'Millan, Bow-Street, Covent-Garden.

Animus adeo a temperamento et organorum corporis dispositione pendet, ut si ratio aliqua possit inveniri quae homines sapientiores et ingeniosiores reddat, quam hactenus fuerunt, credam illam in Medicina quaeri debere——DESCARTES.

So much does the mind depend on the temperament, and on the state of the bodily organs, that if it were possible to render men wiser and more ingenious, the method must, in my opinion, be sought in medicine.

ESSAY

ON

THE INDIVIDUALS,

COMPOSING

OUR AFFLUENT

AND

EASY CLASSES.

PART SECOND.

TREATMENT OF BOYS.

BOYS, I presume, are more frequently educated from home than girls: and schoolmasters diversify their methods more than governesses. Some circumstances, however, of the greatest importance, and immediately relating to the subject of these essays, are very general, if not absolutely universal. These require to be considered at large, especially as their operation is often slow and silent. Peculiarities may be pointed out afterwards.

SEDENTARY CONFINEMENT.

We are determined by their descent, rather than by their qualities, to shew people neglect or respect. In the case of *persons* this is ill-judged. But of *customs* we may often pronounce upon the value by knowing their founders.—The hue and cry against modern philosophers has been loud enough to bring to every ear notice of the design, with which they are charged, of subjecting society to a new sort of discipline. Ages ago, a set of religious fanatics in the East, equally mad with their philosophical brethren of the West, and infinitely more mischievous, succeeded to a certain degree, in this monstrous undertaking. It was their maxim, that waging war against themselves was the most effectual method of rendering service to their Maker. They did not willingly see other necks free from the yoke, with which their own was galled. Over a considerable portion of the globe, their authority became so great, that none durst openly resist it. With the *men* and *women* of the world, indeed, they condescended to enter into a compromise. These, on condition of verbally renouncing its pomps and vanities, were allowed to partake of them, till they were glutted. But it fared so much the worse with the poor *children.* Those of the male sex were, in process of time, completely delivered into their hands; nor to the present hour have they been emancipated from the cruel processes, to which they were subjected by the descendants of these fanatics.

In the porches, gardens and gymnasia of those very ancients, whose works furnish *liberal* education with a principal part of its occupations, study and exercise relieved each other. The teachers often delivered their lessons, while enjoying the free air amidst their disciples. Ridiculous excess, perhaps, in attention to the body and to its members, suggested to Plato those mystic passages, which have contributed to produce so much folly and misery in after-ages. It is, at least, certain, that the founders of cathedral and abbey schools modelled them according to the rules of the monastic orders. From their own habits, they were led to number seclusion, constraint and inactivity among the chief elements of piety and virtue. It was impossible that these men should choose chearful places for the scenes of youthful instruction. It was equally impossible, that they should not array knowledge in a forbidding garb. Schools, planned by friars, could be no other than cells upon a large scale. The inhabitant of the monastery was enjoined to pass hours in lonely meditation. The school-boy was accordingly condemned to his bench for an equal space of time; solitude and the distance from wall to wall making the only difference. Long forms of prayer, and vain repetitions, directly contrary to the express commands of the founder of christianity, made a part of the essential business of every order. How could the pupils hope to be exempted from the odious mummery, by which the masters were harassed?

To the watchmen of religion, the night could not be allowed as a season of repose. Therefore it was decreed, that the slumbers of childhood, and almost of infancy, so indispensable at these periods within certain bounds, and so sweet, should be broken. ROLLIN speaks of himself, as trudging to school under a load of books, as early as four in the morning. We find, indeed, that this austere rule has been mitigated. But in our own country, hundreds of boys are required, even in winter, to appear at six. Let not any one suppose that they are so early collected to be early set at liberty. No; it is for a reason more in character;—to lengthen the hours devoted to *wholesome discipline* and *sound learning*.

He who reads history for pastime, if the historian be such an one as he ought to be, will often be roused from his idle mood by a representation of what mistaken piety has led men to impose and to suffer. But let no such reader dream that the representation is foreign to his own age and nation. The castigations, the penances, the restraints, which superstition has devised in her direst fits of insanity, may have reached himself and his family. His children, at the moment he is reading, may be pining under their rigour.

There have been sometimes delineated characters of inexorable rancour, where the expiring breath has been spent in imprecations, and the closing agony has been an effort to destroy. If the last monks, with whom our island was infested, may be supposed to have harboured such deep malice, they have been amply gratified. The spirit, and almost the substance of their worst institutions have survived; and over the descendants of their extirpators, they have celebrated a continued posthumous triumph. —We shall hereafter see, that there is good reason for supposing that their regulations are at present even more baneful than when they were first introduced. It entered into their plan to *mortify*. They succeed no doubt, more than ever, in the literal, if not in the metaphorical acceptation of the term.

It may be observed here, as in the preceding essay, that too long application, in a state of confinement, defeats its own purpose. More would be learned, if the learner were to have his book before him, just as long as his faculties are fresh, and no longer. But the principal loss is sustained by associating painful feelings with the means of instruction. Nor is it only a loss that is sustained. We have here a grand source of positive mischief. It is strict confinement, which renders both the frequent and the long holidays of many schools, so injurious both to mind and body. Children withdraw themselves and their thoughts as far as possible from their irksome labours, the moment they quit the scene where these labours are carried on. They consume their activity at first in sports, that have no beneficial effect beyond mere exertion; and frequently afterwards in pursuits of a positively baneful tendency. Learning, industry, and virtue being perpetually brought into a disadvantageous contrast with ignorance, idleness and misconduct, a very large proportion of those, to whom their circumstances allow a choice, prefer one or other of the latter;

many, all the three. In others, the desire of obtaining the distant advantage of wealth is far from sufficient to counterbalance the odiousness of instruction and the charms of dissipation. In all these instances, it is evident that the constitution will be in danger, and frequently be damaged or destroyed. Between too long sittings and intervals of insufficient exertion, we have found that girls acquire an indolent turn, frequently incompatible with tolerable health, and almost universally, with a joyous state of being. Boys of tender age and weak fibres, are injured in the same way. But those farther advanced and of more vigorous temperament are forced into the evils, which follow from an accumulation of animal power, where there is no sort of discretion to watch over its expenditure. They have, in fact, sensations similar to those of seamen, stung by long privation. To the prisoners discharged from school, perpetual opportunities pretty well make up, in the whole account of mischief, for the more head-long precipitation, with which prisoners from on board ship rush into excesses.

The rule of health for *sedentary* instruction is very simple. From two and a half to four hours in a day, is the longest time that children, from six and a half to nine years of age, should be required or rather induced to sit. But through this period, as well as previously to it, health and proficiency equally require that the principal pursuits should be of an active nature. There is an easy test, by which we may judge, whether they are at once good for the constitution, and proper in all other points of view. This is the case when they are followed with alacrity; when the curiosity of the pupil expatiates on all sides, and when he occupies himself, during his free hours, in combining the ideas, which he has gained during his hours of study. If, on the contrary, he is vexed and depressed by his lessons, he is most assuredly laying in the seeds of misery and disease. They may or they may not germinate and produce their proper fruit, according to future contingencies. But the habit of feeling, induced in this manner, during childhood, is often most severely felt during the later seasons of life. Causes which without such a preparation would have produced no bad effect, now operate with irresistible violence. The inevitable storms of life shake the system to its foundation, and threaten its demolition. Adversities, which should but furnish the occasion for a dignifying trial of courage, at once bear down all opposition. To execute one of the measures, prescribed by common prudence, and even to perform one of the offices, required in the usual intercourse of society, is felt as an Herculean labour. In some instances, the very light of day is shunned as loathsome; and at the sight of a strange countenance, a panic terror throws the whole host of animal spirits into confusion.—The reason is perfectly familiar to every person, conversant in the study of animal nature. Any state of feeling or of action, which has repeatedly taken place, may be re-produced after an interval of years, by circumstances that would otherwise have had no such operation. If you teach a child the sounds of different languages, and drop them till he grows up, his pronunciation will be much more perfect, than if he had not practised it early. Exactly in the same manner, whatever impressions are early made on the nerves, will recur on slight provocation, after a prodigious interval. In the delirium of a fever, a Prussian peasant surprised his attendants by repeating verses of Homer, which on enquiry he was discovered to have learned when a child, with his lord, 40 years before: in the mean time he had totally forgotten all greek. There are endless examples, similar to this upon record. Every man's history would, perhaps, supply examples, if accurately known.

That incapability, and those horrors, which seem so incomprehensible to persons who have not felt them, and for which there is no help in wealth, learning, or wisdom, shew themselves in an endless diversity of degrees. It would be unjust to ascribe them to any one origin. But there is no source, from which they more frequently spring, than the perverse discipline of seminaries, derived in a right line, from the establishments of the monks, and bearing in every feature a token of their descent. They would not perhaps occur once in an hundred instances, where young persons were not foully dealt with, and where the faculties should have full, proportionate play during the first ten years of life. This implies an almost uninterrupted and an immediate intercourse with the powers and objects of Nature, and such occasional communications from seniors, as shall render that intercourse safe, wholesome and instructive. Under favour of these advantages children would learn to brave little dangers, to despise petty mishaps, and to profit, for their next plans, from the failure of the last. The arteries would acquire a vigour of pulsation, the muscles an elasticity, and the nerves a firmness of tone, which scarce any single calamity, or series of crosses could enfeeble or unhinge; and which are the most precious ingredients towards courage, resolution, efficiency and health.

When little children are immured six, seven, eight or nine hours a day in a comfortless place of confinement, and when the business of this place continually damps the heart, and produces incessant aversion and anxiety, it does not require any preternatural intepreters of the book of fate, like the weird sisters of Macbeth, to understand, what many among these children shall be hereafter.

Mamma! I cannot bear it—Oh! how I do hate it! is language very common in the mouth of boys, just put into their latin grammar, when they dare express what they feel, and can hope for sympathy. But these cries of distress bring no relief. Those, who utter them, are still doomed, for a child's eternity, to the despair which this language bespeaks. The *immediate* cruelty of such treatment is obvious. The lasting mischief it occasions is vouched by every law, by which our machine is regulated, and by the actual condition of a multitude among the quondam pupils of grammar-schools. It is wonderful that parents as well as school-masters will not see, that to begin with rules, and to keep learners long to rules alone, is also the slowest, the most preposterous and ineffectual of all possible methods of teaching languages.

That species of sickliness, which consists in almost continual uneasy, or overpowering sensations, is at present one of the most common ailments among the opulent. Its formation is generally remote, and requires to be particularly pointed out, because it is unperceived by those, under whose eye it takes place. The principle, on which the disposition towards it may be prevented, is the following. *Take care that your children be not frequently subject to depression—and with despondency let them be entirely unacquainted. If they have small difficulties to struggle with at first; and if the contention be short; their resolution will be strengthened by the proximity of the prize, and they will be able to comprehend the causes of failure and success.* It is thus that firmness to act well their part in life will be acquired; an advantage, which many intelligent, but perhaps not sufficiently scrutinizing observers have ascribed to great schools; but which, if ever they possessed it, they seem in our times (as will be particularly stated below) to have lost —in consequence of the introduction of a system of manners, awkwardly aping the indulgence and dissipation of the great world. The present mildness of parental conduct, which under a better plan of instruction, would be an inestimable benefit to the rising generation, is utterly inconsistent with the original spirit of our ancient seminaries; and must operate so, as materially to aggravate the evil, and diminish the good, deriveable from those establishments.

NATURE OF THE CONFINEMENT AT SOME SCHOOLS.

In illustration of this important point, I shall relate the case of a boy of fourteen, for whom I have been consulted in the course of the present winter. I saw him on his coming from a considerable school, affiliated to another still larger. Among other things, his heels were dreadfully chopped with chilblains, a complaint which he had never experienced before, and which had risen to this height before the late severe weather. He and his companions suffered in the most cruel manner from cold.—" Their feet," he said, "were dead, so that they could scarce "walk, at first going out of school." The school-room is about an hundred and twenty feet long. It has no ceiling. The roof, walls and windows are in very indifferent repair. The rain beat in early in winter, so as to wet and chill the boys excessively. There is no fire-place, nor any sort of contrivance for artificial warmth. "They kept almost all "their cloaths upon their backs." Whoever had a great-coat and spencer wore them both. The whole number of scholars was nearer two than one hundred; and of these, 35 were ten years of age or under. The master (upon whom it did not depend to put the school into proper condition) would ask the puny or smaller boys, if they were cold; but sending them to a fire in another room for a time was of little service; and in the room, where they sat out of school-hours, some boys would not go near the fire, *because they should miss it so much afterwards*. They swallowed their liquid food almost scalding hot. The breakfast was insufficient to satisfy the appetite; and they would have been sorely pinched with hunger, when they had not money of their own, *if the shop-keepers had not been accustomed to give them credit*.

They went into school at six in the morning, and staid till nine, when they breakfasted: they returned into school at ten, and staid till one; and were again in school three hours in the afternoon. They had two afternoon holidays (on which days they were *but* six hours in school), and one whole holiday.

At one of the greatest schools in the kingdom, there is a wise and humane regulation for keeping the boys in the school-room no longer than while they say their lesson. But no inconsiderable part of the preceding horrible history holds of more than one public school: particularly the long confinement, and the want of fires. There the distressing chills oblige boys to have recourse to various expedients to relieve their painful feelings, such as fixing their feet, pressing back against back, and waving for a long time backwards and forwards in this position. Such practices must be considered as bespeaking excessive distress. Children of both sexes will endure cold to a very injurious degree, without effort or complaint, when they are cut off from play. And it particularly deserves notice, that the children most liable to injury from cold are precisely those, who sit most quiet, when they are intensely affected by it.

It requires very little information or discernment to perceive, that more disease and destruction must arise from this rigorous treatment now than formerly. At the æra of the institution of monastic schools, all the habits of mankind were more rugged. There was nothing like that contrast of tenderness and severity, which at present exists between school and home. Boys accustomed to apartments with cold floors, loose doors and windows, and walls which did not exclude the most cutting breezes of the sky, made scarce a perceptible transition, when they assembled in the school-room.

They were also infinitely better protected against the inclemency of the element, by which they were surrounded, being clad in coarse woollen. There was neither calico nor linen next the skin. The gowns, worn by the college-boys at Eton, are said to remain as a specimen of the apparel of those times. Thick and shaggy as they are, I am told that they are expressly ordered by the statutes of the finest texture, into which the materials could then be wrought. In an Essay on Consumption (2nd ed. *Longman* p. 89) I have shewn, from the most unquestionable authorities, of how much importance it is to persons exposed to the wet and cold of our climate, to be dressed in woollen altogether. Since the publication of that essay, I have had opportunities of collecting other information to the same purpose. In Glamorganshire, for example, and indeed in other Welch counties, where the dress of the common people consists of a flannel shirt and a flannel jacket (with a coat of homespun when they are not at work), I am assured by attentive observers, that the disorders which cripple *our* hempen, linen and cotton-clad peasantry, and which materially assist in inducing that premature old age, of which they are so frequently seen to bear all the marks, are scarcely known. Our elegant cottons have therefore probably reduced to the most miserable state, and cut off, still greater multitudes in the wear than in the fabrication. Among their victims may, no doubt, be reckoned many boys, confined in schoolrooms, not heated at all—*more majorum*—or not sufficiently heated, according to a common modern practice, of which the author perfectly remembers the severe effects.

INJURIES FROM VIOLENCE.

In the newspaper lists of sick or disabled persons, carried to infirmaries, we find a certain number, regularly classed under the head of *accidents*. Were a register annually published of our schools, and especially of our large schools, there are few where this title would not appear. Boys out of school are wild animals, subject, without their keeper, to fits both of wanton and sober cruelty. There is no injury, to which the arms, with which nature has furnished our species, are equal, that is not actually committed. To enumerate the several species would be impossible and fruitless. Obliging the weak to exertions beyond their strength, before their bones can be solidly formed—using them, for instance, as horses—is particularly shocking. At those great seminaries, where a regular system of vassalage is established, the practice of frequently keeping the little slaves out of bed till a very late hour, has appeared to me to deserve enumeration among the most odious oppressions of a tyranny, founded not on want of feeling, but of knowledge.

SCHOOL-MASTERS AS TO ONE GRAND POINT.

Every human being may be considered as the centre of action, or as the butt, of a certain number of surrounding powers. And it entirely depends on the spot in which he stands, whether he shall be struck by many of the fatal shafts. That he may distinguish those, which it most behoves him to avoid, his feelings must be appealed to in some degree. But this appeal is principally useful, in order that his understanding may receive, from the accumulated experience of mankind, impressions sufficiently powerful to determine his habits. The art of making those impressions is worth all the other qualifications of a preceptor put together. But at our schools there is not, in general, the smallest provision for bringing this experience to the ear of youth, at moments when it is most open to admit the practical result. Nor indeed does there seem, among parents or instructors, to exist the smallest suspicion that any such provision may be necessary.

I own myself, however, at a loss to conceive how it is possible that the heart of a father or mother should fail to melt within them, if they pursue the most obvious train of reflection upon this subject, and put together a few of the most incontrovertible truths.—“ Here is a human being, whom I “ have reared with infinite care and anxiety. “ He is capable of a long course of enjoy-

"ment, but is still, for reasons peculiar to "his age, extremely liable to sudden destruc- "tion, and particularly so to lasting injury. "The far greater number of the causes of "such calamities have been explored, and "stand registered in full. He is perfectly "capable of comprehending their operation; "and was he furnished with the knowledge "of it, he *must* regulate his conduct accord- "ingly. But by the most flagrant of all "prescriptive absurdities, it will be the "same thing to him, as though neither his "capacity nor his susceptibility to injury, "nor that accumulation of experience existed. "He will be in the condition of a ward, de- "prived of the most precious part of his "inheritance by the incompetency of his "guardians. For of all things human and "divine, it is acknowledged that the persons, "to whom he is about to be entrusted, under- "stand the least, what is principally to be "deduced from the fate of past generations, "for the benefit of the present. If so, what "a cruel mockery to commit him to their "care! As well keep all nautical instruc- "tion from one, whose future security is "entirely to depend on his skill in naviga- "tion; and by way of preparing him to "encounter the perils of the ocean, put him "apprentice to a fidler!"

WISDOM FROM SCHOOL-EXPERIENCE.

Something of the tactics of life is acquired at a public school. Who will deny it? In what situation, except in a dungeon, is not something of this kind to be acquired? By mutual collision, children will communicate to one another more or less of the valuable qualities of fortitude and enterprize. Even in the worst scenes of depravity and wretchedness, a few of the actors may benefit, in the same manner, at the expence of the rest.

Activity is alike essential to success, to utility, and health. But where is the necessity that activity ahould be exerted in destroying the produce of labour, and in inflicting pain? At school and at home, the *spirit of mischief* seems to be contemplated with improvident satisfaction. Whoever fosters this spirit in a young person, may rest assured that, whatever he does with regard to others, he will, some time or other, exert it against himself with effect.

If we consider what it is that ought to be gained towards the welfare of the rest of life, during the years between childhood and maturity, we shall find that schools, according to present usages, are lotteries with an infinite proportion of blanks; that the prizes themselves are of low value; and that a beneficial chance must often be purchased at the price of one's all. Of the present and future consequences of what they do and suffer, it is never any thing beyond a gross knowledge of the former, which school-boys attain. They find out which can give and which can take the hardest blow. They settle an intellectual precedence, which is often afterwards reversed. For the sum and substance of what is cultivated at classical schools stands, both in its applicability and in its influence on the faculties, so much below what is neglected, that if a boy be casually thrown into almost any other method of study, or meet with an accidental incitement, he may, in a single year, surpass a school-fellow, who has been, for seven years, his superior in mere Greek and Latin. —The forethought, which boys learn to exercise at school, seem to border too often upon the nature of cunning. And if so, it must contribute to their future disadvantage rather than their advantage. They delight to shew their wit, in eluding the vigilance of the master, and their courage in breaking the rules of the institution. They lay plots to obtain prohibited articles; and scruple not to obtain them by theft. Their secret thoughts are intent upon sensual gratifications, which become continually more dangerous, as those that indulge in them advance in years. And they know no bounds but those prescribed by physical necessity. There is, indeed, hardly a possibility that they should lay themselves under any restraint. Their proper business does not sufficiently engage their feelings; nor are they, once, perhaps, in ten thousand times, *duly* apprized of the tendency of their excesses. Hence, were another ASMODEUS to disclose to us the history of our tombs and sick-chambers, he would be often obliged to begin with the arrangements and omissions at places of education. But also, if he could plant a window in the breast of living multitudes, or render the feelings audible, schoolmasters would need no other inducement to qualify themselves for making *that* an essential part of instruction, which is most essential to happiness. It is to be considered, that they themselves have not been more fortunate in their education than their pupils. They do not wilfully reject the good. Their eyes have never been opened to it. FORTUNE is represented as blindfold herself. CUSTOM might be represented, as blindfolding all the world.

PECULIARITIES OF SOME GREAT SCHOOLS.

In some of these places, it appears ridiculous to expect that mind or body shall be invigorated by competition. The scholars frequently feel no interest in their studies. The poor substitute of compulsion is either not resorted to, or is unavailing. The stimulus of indigence may act upon some. A master may be weak enough to conduct himself differently towards boys of different description. To the miserable industry of the active, writing verse in a dead language is proposed as the supreme excellence. Of others, the faculties would rust, if they had no motion but what they derive from literature. But they are not unoccupied. They employ themselves in schemes for getting rid of their time and their health. They drink, row and ride. Little boys, if I am not grossly misinformed, guzzle at their meals so strong a fermented liquor as porter. The bigger frequently super-add wine; and not in very sparing quantities. The large sums they procure from home, and the more extensive credit they find at inns and shops, enable them to follow their seniors with no very unequal steps. There is nothing to check their career, except the failure of these resources. Two circumstances, which operate elsewhere as a considerable restraint upon vicious propensities, are wanting here; namely, difference of age, sex and occupation; and more or less acquaintance with consequences.—Where such an order of things prevails, it is idle to stop to settle the account between cause and effect. Every reader can do it for himself. It is obvious, that life must gush out in full streams, as from Seneca in the bath. And to judge by the state of the person, which should always be taken as the grand criterion between mortals, the richest heirs, in consequence of debauchery begun at school, often turn out the poorest devils in the country.

The debauchery that begins at school, increases at college. And those who before escaped, now enter into it. I speak from observation in the University at Oxford, at different periods, during eighteen years; and in so simple a fact, I have no apprehension of being contradicted. Drunkenness was said to be on the decline; and, I believe, most truly. Each college had not, probably, above one drunken party. But it was an unreproved practice, among a vast proportion even of the youngest members, to drink some glasses of wine daily. The poorer students contented themselves with ale. All competent observers are agreed, as to the pernicious effects of fermented liquors. And this operation is generally more pernicious, as he who drinks them is younger. By persons below five-and-twenty, they should never be taken but as medicines; and it would be better not to take them but as the cordials of old age, that is, always medicinally. It is impossible that three or four glasses of port wine—a quantity which usage miscalculates as small—should be regularly swallowed by the under-graduates of our universities without curtailing life, or making it miserable; and indeed, without doing both in a tremendous number of instances. The fact is evidenced in the most ample manner by the spleen, the gout, the palsy, the lunacy, the premature decrepitude, so prevalent among the gentry, the clergy, and others, who have been unfortunately deluded by example and by ignorance of the laws of their own nature into this habit, at an early period. There are, unquestionably, (whatever the zealots of temperance may say,) counteracting causes. Some of these can be assigned. And in a subsequent section, I shall consider how far there is any thing in the reigning manners, by which we may hope to be frequently preserved from the effects of this consuming poison.

DEBT.

Paper-money is said—

> To lend corruption lighter wings to fly.

The universal facility of credit in this country is an admirable accommodation to those students, who are in haste to ruin their constitution. If it acts indirectly by bringing the physical causes of disease into motion, it frequently also comes to operate directly as a moral cause. I have known several students, who being nothing inferior to their more opulent comrades in animal spirits and pride, embraced the opportunity which our universities so liberally afford of incurring debts, which the revenue of a curacy did not enable them to discharge. They afterwards suffered greatly in mind and body from chagrin on this account. The articles they procured on trust, brought on the horrible sensations, known only to persons, afflicted with hypochondrasis, and with a disposition to melancholy madness. In this situation, the dread and the persecutions of unsatisfied creditors, contributed to bereave them of hope or of their reason.

With regard to food, the schoolmasters, in general, of the present day, deal with their boys in the most liberal manner. There is reason to believe, that during the late scarcity, their humanity frequently induced them to forego all profit. There appear, however, to be exceptions; the school, of which some part of the history is related above, forms a remarkable one. There, and at some other schools, it has been a standing custom to serve a three-farthing* roll for breakfast. The letter of the regulation has been observed in cruel opposition to its spirit. The allowance, even during the late dear season, being determined by price, and not by quantity. The customs, as well as the rules of certain old institutions, which the master has no power to amend, inevitably drive boys into debt from stress of hunger. It is no-

---

* "Our allowance was neither a penny or a halfpenny roll, but a three-farthing one, during the plenty and scarcity; also the dinner was not sufficient, or supper."

Note in the author's hands.

thing uncommon, for example, to hear the parties aggrieved complain, that a kitchen perquisite, too paltry to be named, acts as an inducement with the cooks, to overdress the meat till it is unfit to be eaten. Nor do I repeat this upon hear-say, but after the examination and cross-examination of young men, by no means disposed to exaggerate the hardships they have to sustain. I am well aware that such a propensity does often exist, and have, consequently, always endeavoured to guard against being misled by it. Again, there is sometimes only provision for half the mess. The weaker division must, therefore, be content with the part of the inferior beasts, who had hunted down a deer in society with the king of the forest.

The calls of nature, however, will be satisfied. The pocket-money is not always equal to the provision of the necessary means. When that is gone, what resource is left but credit ?—The habit, thus begun in necessity, will often be continued for the sake of indulgence, and may terminate in wretchedness and ruin.

BOYS DISGUSTED INTO NEGLECT OF HEALTH.

These and other imprudences, whose effects future contingencies determine, in some instances, to the mind, and in others, to the body, can only be prevented by timely precautions. *But upon what sort of precautions can we place reliance?* Those, who feel a practical persuasion that health is the best provision they can make for their sons, have no call upon their understanding more urgent than the study of this problem. Its solution cannot be said to require great acuteness, but it requires perfect impartiality. And, in deliberations concerning the points, that most nearly touch human nature, impartiality is a quality more rare than acuteness.

From the practice of the world, we might collect a considerable list of precautions; but it would be almost wholly negative. The annals of private families, of our schools and universities, unite to testify against the sufficiency of the measures, commonly adopted. Advice is wormwood. Reproach in its first operation is cantharides. It not only makes the mind sore, but leaves it callous, if frequently administered. Exhortation and intreaty will rouse the nerves for a few times. But like other volatiles, their effect soon goes off; and, by repetition, they altogether lose their power. None of these articles of the moral materia medica appear fit to form part of a constant wholesome regimen, though in emergencies, if judiciously employed, they are capable, no doubt, of producing the happiest effects. I forbear to particularize other means, upon which mankind have long placed implicit reliance, and which only fail, perhaps, because they are seldom adapted to the case of the young and the unwary.

Infinite good may doubtless be done by oral addresses. But it must be considered that this is an affair, in which nothing is indifferent. What does no good, must do harm. Whoever treats the subject improperly, spoils it for another. To give disgust for the knowledge, that belongs to the proper care of health, is to labour to its destruction. This may be often seen. Advice, reproach, punishments and lectures induce such a soreness of sensibility, that whenever a topic, connected with self-preservation, starts up in the way of the party advised, reproached, punished and lectured, he will fly from it with as strong signs of loathing, as if it were an adder or a toad. Do you endeavour to open his eyes to the horrors of that fate, to which he is hastening? He will, perhaps, with inattentive complaisance, suffer you to proceed. Or, if you be one, with whom he does not deem it necessary to stand upon ceremony, he will cut you short with a familiar oath: *Oh! aye, right—just so—you have it.—But I have heard a d—n'd deal too much of that sort of thing already.*

In the preceding numbers I have said enough to point out the principle, on which I think that what is vainly attempted by repulsive measures, may be attained with the utmost certainty. It is now time to illustrate the principle by example. For which purpose, I fix upon the most delicate article in the whole scheme for creating a disposition favourable to the care of the constitution.

GRAND SOURCE OF UNHEALTHINESS IN THE MALE SEX.

The classes, whose present lot in life I am considering, are injured *by continued indolent or passive enjoyments*, *by abuse of stimulating liquors*, and *by mismanagement as to heat and cold.* From none of these sources, however, does there spring a larger quantity of bodily and mental affliction, than from the PASSION OF THE GROVES, as it has been denominated, in a particular tribe of animals, by one of our poets. In Europe and Asia there is observed, towards young people, by their friends, a treatment almost as different as the religions which predominate in those two quarters of the globe. In many parts of Asia, free and permitted indulgence reduces boys to a condition, which it is difficult to contemplate, without wishing that so horrible a system of manners may be abolished at almost any hazard. The difference of European manners, and particularly of those, which prevail in Great Britain, need not be particularly stated, nor their origin, which is distinctly laid down in history, related. There can, undoubtedly, be no comparison between the two methods of proceeding, but practitioners of medicine, and innumerable heads of families, have reason to know that the better is not good enough to guard against evil to the extent that could be wished.

Were it possible to keep young people in entire ignorance, nothing further would be necessary. A passion must, at least, be formed, before it can be injurious. In regard to the passion, to which the present section refers, the first object is to retard its formation. The second to hinder it from attaining uncontroulable violence, and to give it a proper direction. But, who can hope sufficiently to prolong the state of ignorance for either of these ends? Is it not in vain that parents maintain a religious silence? Will not curiosity be raised by occurrences, inevitable in any situation? and will it not be gratified by communications, almost equally inevitable? In a multitude of instances, will not desire be awakened before its natural season, and raised to a height, which shall never permit the body to attain its full vigour, and shall keep the mind in a state of constant perturbation, after the bodily powers are worn out?

This process must blast every hope of the enjoyment of health, and, on all accounts, it deserves to be a little more narrowly inspected. Almost every child is led to consider how he himself, or how an infant brother or sister came into the world. He cannot satisfy himself, and therefore applies, where he has always been accustomed to resort for intelligence in his difficulties—Is he put off with a ludicrous answer, accompanied by a corresponding expression of countenance?—He snuffs deceit, and feels humiliated: or afterwards detects the imposition. In either case, this, as we shall immediately see, may become the first step towards the various mischiefs, that lurk in the train of early sexual intemperance. For half a century past, parents have been properly advised by writers on education to exempt themselves, betimes, from the obligation of answering all the interrogatories of children. To recur to this practice in the present case, would be attended with much less danger than giving a false reply; but as most families are circumstanced, it will still be hazardous. The striking answer, recorded by Rousseau, could apply only with full effect to a child, that had itself suffered from the passage of a stone.* And, even so, it appears to deviate from that indispensable rule—*truth or silence.* †

Since then, in all but very uncommon situations, it is impossible to suppress information, it should be gravely, decently and early imparted. Anticipate servants and

---

* *Mon fils,* respondit la mere sans hesiter, *les femmes les pissent avec des douleurs, qui leur coutènt quelquefois la vie.*

† Je n'ai pas besoin d'ajouter que vos reponses doivent être vrais—Un seul mensonge averé du maitre á l'eleve ruineroit â jamais tout le fruit de l'education.

loose companions; that the great secret may not be presented in seductive colours. It is very seldom possible to raise a fence round a family, which shall prevent impurity from ever coming into contact with its younger members. But it is very practicable to follow the surer method of rendering them superior in knowledge to those, most likely to be their corruptors. But this superiority will answer no end, if it be founded on declensions and conjugations. It must include domestic œconomy and natural history,—particularly that of man. It is astonishing, what influence a very few ideas on this subject give a grown person over the mind of a boy. The wisdom, that extends beyond a child's sphere, he supposes to be boundless. Artists have confessed, that there has been a time, when they have received more delight from the painting on a sign, than from the works of the greatest masters afterwards. Philosophers, who can recall their early impressions, will be sensible, that there is no human intellect, to which they have rendered more profound homage than to that of the nurse and the footman. The communications of domestics are likewise the more captivating, because they are made up of particular facts, and because they are unaccompanied by any of the forms of rhetoric, noticed in the last section. Not to add that participation in secret, * incommunicable doctrines has, in all ages, been a bond of union, stronger than the ties of blood.

The following seems to me an irresistible argument in favour of full disclosure. The mischief, which privately lays waste society, arises from partial information. It is not any thing exhibited to sense and made plain to the understanding, but a mysterious cast of the features, nods of intelligence, and half-sentences, that usually act as the first ferment, and set the imagination to work, till it becomes corrupted. Amorous poetry and prose, to which this sort of kitchen tuition may be considered as preparatory, operate upon the same principle.—I doubt whether Hogarth himself could have managed the subject with decorum. Otherwise, a design, representing the origin of the generality of our novels, might have an excellent moral tendency. In the Latin accidence, there is shewn, for the encouragement of industry, a tree, loaded with the fruit of learning. The pupils are picking it up and stowing it in

* τα απορρητα.

their satchel. In like manner, I would have a tree hung with novels. Its stem should be just seen, dividing at the ground, into two main roots; and these roots—but I will not complete the sketch, for fear of drawing down upon a faithful emblem, the odium due to the thing signified. I should be sorry to excite a feeling, under favour of which many productions of this class should escape the reprobation of parents, intent upon rearing a family, unpolluted in mind and person.

The impressions, which are thus formed and thus strengthened in the unprepared mind, lead early, as is well understood of the stronger sex, to destructive practices* of

* After having seen persons, deprived of health, of the faculty of pleasurable sensation, and nearly of their understanding, by a vice, which has been the subject of many admonitory medical tracts, I can never doubt of its frequency, nor of its destructive effects. The one and the other rest upon authority as good as any set of facts whatever belonging to human nature. Since TISSOT, SALZMANN, OEST, WINTERFELD, CAMPE, VILLAUME, BAUER and others have collected a body of evidence, which may indeed well disgust, but which must convince, where there is no predetermination to reject all testimony on the subject. Though there are, it should seem, various things, destined to be disbelieved, because they are offensive to the imagination.

Many sensible people have conceived that the evil lies in the alarm, engendered by quack advertisements. These more than one kind. The ravages of a par-

---

advertisements may add to the despair of the poor wretches, who read them with trembling faith. But it is not possible that a diseased condition, marked, when it has arisen to a certain degree, by symptoms as clear as those of a fever, could have been so produced. There is proof enough that this is not their genuine origin. In a variety of examples, children have been reduced to the most deplorable state, without having any idea of the manner in which they were injuring themselves. I shall abridge an instance from a respectable author, and could add others from my own observation.——" At length, but ten years too late, " Tissot's tract fell into my hands—My eyes were now " opened all at once. - - *I was already excessively enfeebled* " *and reduced.* Every body said, *he is far gone in a decline.* " Yet I had never had a single suspicion of the real cause " of my waste. I now learnt it with horror. *Oh*, thought " I, *what culpable parents, teachers and friends! not to give* " *warning against this vice! What destructive ignorance* " *prevails in the world as to its consequences*—I resolved to " abandon it. This I found difficult, but not impossible.

" My present situation is as follows: My faculties are " greatly impaired. My apprehension is become blunt: " continued thinking is impracticable: my memory is " weak, or rather entirely gone - - - my body is helpless, " mere skin and bones. The sight of me excites horror— " Oh, that all ignorant and thoughtless sinners against " themselves were by to look on - - - But not only am I " disabled; I constantly feel the most violent pains, par- " ticularly where I have offended most - - - Add to this a " restlessness and a melancholy, that exceeds every thing— " Often should I have yielded to the temptation of putting " an end to my wretched existence, if the motives of reason ticular disorder, and of its specific remedy, amount but to a trifle in comparison with that misery, for which no advice is sought, except when it arises to visible disability—though in much inferior degrees, it extinguishes all present pleasure, and all hope as to the remainder of life. Into whichever habit, any one falls through ignorance or is misled, its baneful effects will, I apprehend, be in proportion to his youth, to his sensibility, and the rate, at which his excesses are carried on. In either case, there are certain signs, which do not indeed occur universally, nor early enough to prevent much damage, but which ought to raise suspicion in those, who have the care of persons, *almost ever so young*.

The eye loses its lustre and no longer sparkles with the fire of youth. Vision, in process of time, becomes dim. The eye-balls sink and the sockets grow hollow, as in atrophy or consumption. The features acquire a flabby appearance. A look of age supervenes. The complexion sometimes changes from

---

" and the doctrines of the most beneficent religion—which " is my only friend and treasure—had not restrained me. " I am indeed however convinced that in my case there " is little moral transgression, as I never had the least " intimation of the consequences of this practice."

clear to a muddy paleness; and frequently, an eruption of a livid hue breaks out. Briskness of motion and vivacity are lost. A morning heaviness, (almost a sleepiness) after dead sleep in the latter part of the night, comes on; and there are great complaints of being relaxed or overpowered in warm weather, and of being proportionably chilled in cold. Imagination and memory decay. No continued effort of mind or body can be borne. The ears incessantly ring; deafness ensues; and epileptic fits or atrophy. Suicide is a common subject of meditation; and is sometimes actually perpetrated. The sensibility grows excessive; but the feelings are all of the painful kind; and moroseness and habitual dejection follow in due gradation. The nerves are perpetually a-jar. There is within, an incessant discord of flats and sharps, of which, without having seen and heard persons tormented by it, no adequate conception can be formed. When animals, having cold blood, are made to take opium, they will lie stretched out, almost dissolved and perfectly motionless for a long time, but if you touch them with a feather, or strike the table, they are instantly thrown into convulsions. This experiment affords ocular illustration of the interior state of those baby debauchees, in favour of whose constitution their parents do not effectually provide for the observance of that rule, to which Tacitus* wisely ascribes the *unexhausted puberty* of the ancient Germans of both sexes.

Girls are, by no means, strangers to a similar species of ill-health.—The manner, in which the passion I am speaking of, and the objects of it, are presented to their imagination, produce effects, much the same as those last described in the case of boys. In boarding-schools, it is scarce possible but that there should be a few, to whom the mysteries of this passion should have been revealed in the way, most likely to kindle a dangerous flame; and even from *one*, it cannot fail to be communicated to a great number of school-fellows, and to be kept up among succeeding races, like the eternal fire of Vesta.

---

* Sera juvenum venus: eoque inexhausta pubertas: nec virgines festinantur; eadem juventa; similis proceritas. Pares validæque miscentur; ac robora parentum liberi referunt.

I say *baby* debauchees. I will not needlessly sully my pages with documents to shew the propriety of this term. But I could produce them in numbers.

The power of certain ideas to irritate certain organs, after the association has once been formed, is a fact that requires no illustration. Upon this principle, there requires nothing, but mere circulating-library literature to account for a great deal of the sickliness we find existing in society. The sensations, to which all these melting tales immediately give rise, and the voluptuous reveries, which they leave behind, may, without injustice, be regarded, as a part of the concealed fountains, from which the NILE of female unhealthiness derives its origin.*

---

* I disapprove as much as any one, of the clumsy practice of distracting attention by long notes. But I purposely throw certain matters out of the text, in order that those, who choose, may pass beside them.—A popular writer on the prevention of diseases would deal unfairly, were he not to apprize mothers of the possibility of the appetite of puberty being prematurely excited, and occasioning misconduct, much more active than any wanderings of fancy. As I had rather go abroad for illustrations of such a subject, I shall produce the following cases and observations from a writer, some of whose works are well known in this country. I mean the late Dr. Zimmerman. "A particular vice, (says he) is more dangerous in the "other sex than in ours, though it is less known, as being "carried on more in solitary chambers, and in the darkness "of night. There is no complaint, by which beauty is so

These, and other considerations, may per-

"soon blasted. Nothing so soon robs youth of its fresh-
"ness, and all enjoyments of their relish. Hence in young
"women sickliness so often without any disease. Hence
"that debility, which before and after marriage, is the
"cause of so many nervous ailments.—

"Girls, during the period of childhood and of perfect
"innocence, before the possibility of any propensity to
"voluptuousness is suspected, fall into a pernicious clan-
"destine practice."

Case 1. "A father consulted me in behalf of his daughter,
"only five years old, on account of a nervous weakness
"of the kind, that quickly passes into epilepsy. The child,
"he told me, took a particular pleasure *when alone*, in
"leaning over a stool or upon the stairs, and throwing
"herself into violent agitations, till the sweat dropped
"from her forehead. He was in hopes, *as she manifested*
"*an understanding far exceeding her years*, that he should
"be enabled to put a stop to the habit. He added, that he
"had discovered to his astonishment that the child had
"acquired it at school. The governess was an old, simple
"person. From mere ignorance, she had mistaken a vice,
"very frequent in her school, for a harmless practice, and
"even supposed it a symptom of worms."

Case 2. "A girl of thirteen has complained, for a
"twelvemonth, of a violent head-ache, which was origi-
"nally attended by dizziness, now intermits but few hours
"in the day, and at its greatest height brings on a burning
"of the eyes. She is, at the same time, heavy, often ex-
"tremely dejected, and apt to fall into tears without any
"external provocation. Her faculties are upon the decline;
"her look observantly timid. When a child, she was

suade the governors of some families, that

" remarkably lively; but now all the joyousness of youth " has vanished. Her former physician looked upon these, " merely as symptoms of an approaching change of con- " stitution.

" At length, the elder sister, who had read Tissot, " avowed that, *at six years of age*, the invalid had fallen " into the vice, of which that author treats. *She never went " to bed without practising it, till she fell asleep from exhaus- " tion. Though she herself had never received any intimation, " that this was improper, yet it had struck her as so. She had " therefore remonstrated in every possible way, but in vain. " At length, she had adopted the custom of holding both the " other's hands, till she fell asleep. This had gradually " weaned the younger from her bad habit.*

" Four years ago, however, the elder having married, " and the younger being left to herself, an irresistable " urgency impelled her to her old transgression. The idea " occupied her whole soul, and left no vacancy for any " other thought.

" She appeared, continues Dr. Zimmermann, exces- " sively pale, and had a livid circle round the eyes, whereas " before she had been particularly florid. Other conse- " quences gradually followed. She grew so weak as not " to be able to support herself on her limbs. Then came on " the head-ache with its accompanying symptoms; and so " excessive was her sensibility, that the smallest alarm made " her shake violently.—

" It will be said that worms occasion a feeling in girls, " that leads to those means of alleviation, which I consider " erroneously as an infamous gratification. But the objec- " tion is groundless. In these, as in other similar cases, the most violent, and consequently the most dangerous, principle of action, which nature has implanted in man and woman, ought not to be left to the direction of blind hazard. This being admitted, it follows as a necessary consequence, that the check must be placed close beside the moving power; or, in other words, that the understanding must be enabled to restrain and guide the impulses of

---

"there were no worms. And even should worms prove "the first incentive to this habit, what difference does that "make as to its consequences? A child that is led by "worms to discover a method of exciting a pleasurable "sensation, will renew it without that stimulus; and so "it falls gradually into a destructive vice.

"But it may be said that from various causes, there "arises, at a very early age, an itching in particular parts, "and should the method, that children take to procure "relief, be branded in such harsh terms?

"By no means. But much sooner in life than is com"monly supposed, voluptuous ideas will arise in girls "from any accidental irritation, however simple and "destitute of images they may appear at this period. In "a temperament of great sensibility, these ideas get pos"session of the imagination. And as soon as the plea"sure is sought without any bodily irritation, I should "call it by the offensive name. The frequent rubbing of "particular parts, whether to allay itching occasioned by "worms or by any other cause, is not the vice itself, but "leads directly to it."

appetite. For in the relative situation, in which mankind find themselves, and in which they will, probably, for ever continue, scarce any individual can attend to another with so much vigilance, as to secure him from the various and frequent dangers, to which the universal passion of youth and manhood must expose him. No one, certainly, can regulate the imagination of another, except by the communication of ideas and feelings.

Anatomy and physiology must be resorted to for the knowledge, necessary to prevent that corruption of mind and manners, of which the date is frequently so early, and the consequences so lasting. And it was with a view to this, more than to any other single application, that I stated an acquaintance with the sciences, as indispensable to parents, who are in earnest to secure to their children the internal means of happiness in full measure. There may be a variety of ways of applying the science of the human body to this purpose, not only without disparagement to a parent's dignity, but so as to form a close and indestructibe bond of family union. I shall delineate the outlines of that method, which appears to me easy and safe. If fathers and mothers, who may have the good fortune to be convinced, should not feel themselves instantly qualified for the execution of the plan, they may find, almost every where, books and persons to assist them. And half a year's application will not be thrown away, if, with respect to themselves, it should terminate in enlargement of their stock of ideas, and with regard to their children, in preventing

> The deep damnation of their taking off.

The first thing I have to recommend, is but the extension of a practice, happily now become very common. The number of elementary books on natural history, already composed for the purpose of education, familiarize children with the form and the manners of animals. To this, I would add, the natural history of vegetables also. From the exterior, it is easy to allure curiosity farther. The admission of air into the bones of birds, by which they are so admirably qualified for flying with ease, would be one of the examples of anatomical instruction, most easy to the preceptor, and most agreeable to the pupil. The different conformation of different birds in this particular, would add greatly to the interest, which the subject could not but inspire.

Digestion will form another copious source of entertainment, if the astonishing power of the stomach in some animals be properly unfolded.—By degrees, the circulation of the blood might take its turn; and this would lead to the difference that subsists in animals before and after birth. Hence there would be but a step to the reproduction of the various tribes of organic beings. I should advise taking the first instance from among the amphibia, as being remote from the human species, and therefore incapable of raising improper emotions. Spallanzani's dissertations on natural history, will supply the requisite information; in spring, the frog may easily be found, bringing forth its spawn. A female should be killed, as is instantaneously done by immersion in very hot water, and then opened to shew the origin of the eggs. —It should be a rule to explain this whole subject by word of mouth. A collection of facts and engravings, relative to this part of natural history, for the use of those, who have the care of young people, is one of the works, at present, most required by the state of society.—It would be the best substitute for popular lectures in anatomy.

The examination of a hen, when full of eggs, presents no difficulty. At least, a dissector, sufficiently skilful, is to be met with every where. And after such an examination, the thoughts might be cast off, to great advantage, in a variety of different directions. The superior fertility of oviparous animals, might be exemplified from the pigeon, that hatches two eggs, and the hen, that hatches a dozen and upwards, to the herring that lays a thousand at once, the carp, that lays several hundred thousands, and the cod, that exceeds a million. These facts would introduce an account of the tribes, that people the *multitudinous* ocean. And thus would innocent associations be formed, and, what is one most desirable object, the mind be preoccupied, to the exclusion of those trains, that secretly prey, like a canker, upon the constitution, or bring it to ruin before complete puberty.

It ought to be contrived, that a child, so far instructed, may be present at the labour-pains of a domestic quadruped. Afterwards, he might be led to consider the difference between the oviparous and viviparous classes. The latter might be represented as hatching within the body of the eggs, which the others exclude. The severe sufferings, which attend pregnancy, and the dangers of child-birth, if exhibited in just colours, would be productive of much direct and collateral advantage to both sexes. In our's, it would give rise to a disposition, favourable to the happiness of females; and how often does the contrary take place in consequence of the improper manner, in which the knowledge of the sexual relation is imparted! And to women, it would give a deep and effectual warning against those modes and habits, which affect them so injuriously in the character of mothers.

In all this course, as much of the *history* of diseases should be introduced, as consists in well-ascertained and intelligible facts. To the account of digestion, for instance, there should be subjoined an account of the injuries, which the organs of digestion so frequently sustain from free indulgence in fermented liquors.—When the functions of the brain and nerves are explained, (and for the purpose in view, they are sufficiently understood,) the explanation should be accompanied by a correspondent detail, This would be full of lessons, the most useful to the present and future generations. It would comprehend the convulsive startings and shocks to which the causes pointed out in the preceding Essay, render women for ever subject. It would shew how the most trifling chagrin can, in a moment, unman him, whose youth has been haunted by school-terrors, and his maturer years by the furies of ambition or of avarice.— It would go on to the general imbecility, and final loss of sense and motion, induced by the early habit and eager repetition of those intense pleasures, that shake the nervous system from the root to its slenderest twig.

The end of the proposed instruction being altogether different from that of lectures for students of medicine, the reader ought not to be surprised at the difference of method. Nor should the intermixture of subjects be regarded as disorder. The intention is, by variety of examples, to habituate the mind to the reference of the structure and use of parts to good and evil. When this has been done in a variety of other cases, it will be not only safe, but the greatest means of security, to enter minutely into the peculiarities of the sexes. This will be best begun by the physiology of vegetables. The leading ideas may be communicated, and nearly all the terms explained, in relating the experiments upon the parts of fruitification in plants. And I have heard this doctrine delivered to a mixed audience, without offence to delicacy or scandal.

The calculous concretions, that are formed in the gall-bladder and in the intestines, particularly of horses, may afford the instructor an opportunity of entering upon the formation and passage of gravel and stone in the human species. He might then speak of the greater exemption of the female sex from the sufferings, which men experience in consequence of the difference of dimension in the canals, which these hard and rough bodies have to traverse. Here would be an opportunity of shewing figures of the parts, or (what, if practicable, would be better) producing preparations of them from animals or from man.

The circumstance by which an acquaintance with the organs in man would be best introduced, is the immense length of the convolutions of the seminal arteries in the testis. An accident, in which this tender part should receive a blow, would furnish a fair occasion for taking up the subject. The delicacy of the organization would be rendered particularly striking by a drawing or exhibitions of the convolutions of the arteries, when unravelled. The injuries, to which these bodies are liable, would then be easily conceived. The fire and strength of the bull and the stallion would illustrate their importance to the system. The Italian *castrati* would shew that the same law extends to man. Indignation would kindle at the idea that, in civilized and christian Europe, there should be suffered to exist public signs and inscriptions, offering the performance of this male mutilation at a *reasonable* price, and that there should be parents unnatural enough to subject their boys to it. A lively picture of the feebleness, incapacity and frequent depravity of the beings, thus mutilated, would produce the most salutary impression. For it might be transferred, in full force, to practices still worse than mutilation; since in addition to their enervating effect, they induce a train of diseases, some of them the most painful and hopeless that affect human nature. Here it is obvious the various symptoms of the malady, falsely said to have been introduced into Europe from America, would find their proper place. Enlargement of the prostrate gland, strictures in the urethra and schirrus of the testicle would follow. And to elucidate these fearful ailments, the parts themselves, figured or prepared, might be brought into view without any danger of exciting one lascivious feeling. On the contrary, as without overcharging a single particular, it would be apparent that, nineteen times in twenty, men bring upon themselves all the mischiefs which have their seat in these delicately constructed organs, the excitement of such feelings would be prevented by so guarded a kind of representation.

Among the means of preserving young people from imprudence, it has been advised to carry them to the institutions for patients, labouring under the effects of indulgence of the sexual passion. The advice is said to have been followed with success. "I once," says a writer of good credit, "carried several of my pupils to the *Charité* at Berlin, "after having spoken to them on the origin "and nature of the diseases, to which that "hospital is appropriated. And I am satisfied that the impression of horror, excited "by the objects, can never be erased from "their minds."—I have made a similar experiment with the same result. It was in the case of a young man, whom I more than suspected to be entering, in full ignorance, on the career, which has secretly ruined the constitution and peace of mind of so many others. Without preparation or explanation, I took an opportunity of shewing him a debauchee, whose organs were half consumed by corroding ulcers. The sight, the smell, and perhaps a flash of light, suddenly reflected upon his own mind, rendered him immediately faint. On his way from the habitation of the mangled and offensive invalid, he proposed a variety of questions, suggested by the spectacle. I gave him to understand that the situation he had been witnessing, was, by no means, one of the most deplorable, to which the same appetite reduces the imprudent. He took up the remark with the utmost interest. It was easy to confirm it by a variety of facts. These had the effect of inducing a confession; and the vice, into which this young person had been plunged by evil communications, was without difficulty, and for ever, abandoned.

I should consider that child as perfectly shielded against external corruption, and against the more dangerous pictures, formed by the imagination, who should derive his first knowledge of the circumstances, constituting the distinction of sex from anatomy.*

---

* Since the above was composed by the printer, I have met with the following fact, in a writer on education. "When a youth of nineteen, I happened to be present at the "dissection of a female subject, when part of the doctrine "of generation was explained. The whole struck me "much. The person who delivered the lecture spoke with "such dignity, that I have often blessed him for it after-

It is not possible to conceive emotions more opposite than those excited by the aspect of a dead body, and by licentious conversation; or even by the jests, from which, unfortunately, many persons of reflection, and, in other respects, of decent conduct, do not restrain themselves in the presence of boys. No power would afterwards be able to deprive the ideas, impressed by such a sight, of their serious complection.

It is in vain that we dissemble to ourselves the eagerness, with which children of either sex, seek to satisfy themselves concerning the conformation of the other. No degree of reserve in the heads of families, no contrivance, no care to put books of one description out of sight, and to garble others, has perhaps, with any one set of children, succeeded in preventing or stifling this kind of curiosity. No part of the history of human thought would perhaps be more singular than the stratagems, devised by young people in different situations to make themselves masters or mistresses of the secret. And

---

" wards. My ideas were brought into good order by this " lecture; and I am certain it suppressed many loose " thoughts - - - - - - - Such lectures would assuredly " always be lessons of virtue."

every discovery, due to their own enquiries, can be but so much oil, poured upon an imagination in flames. Of course, every such discovery will endanger their innocence or their health. Of so great importance are the feelings, with which information on this subject is *originally* accompanied.*

The disclosure of the whole structure and functions of the sexual organs having once been made, there will be little difficulty. An enquiry into the periods of life, at which nature has made men and women in different regions, capable without injury of continuing their kind, will lead from natural history into morality and law. Here will be an opportunity for many discussions, hardly suscep-

* It cannot be too strongly inculcated, that the manner of introduction of this species of knowledge, makes all the difference in its operation. It may be perpetually observed that young men, who certainly are as well informed in all the dangerous points, as it is possible by means of words, without the application of the senses, persevere in baneful practices of more than one description. But the truth was not *originally* delivered in the same tone, which is thought necessary with regard to every other species of information connected with their welfare. They have worked their way to the notions they have, under conduct of a heated fancy; and hence these notions operate as handmaids to that corruption out of which they arose.

tible of levity. And the more all these topics are presented in their true and natural colours, the less will the pupil relish the common-place jokes of ignorant and inconsiderate people.

Of these jokes the whole zest seems to depend upon the impression, that has been left by the success of ancient fanaticism in sanctifying the self-denying system of celibacy.* Hence, in process of time, it came to be an offence (even in the bosom of families and on occasions, requiring the difference to be taken into account) to think or speak of men and women, as differently constituted. In consequence, at this day, by the magic power of custom, hardly a parent or preceptor will allow himself to enter into explanations, (however calculated to prevent the most serious disasters), upon which the most senseless of mortals affixed a mark of turpitude, many centuries ago. Children are, therefore, left to make their own way precisely there, where

* One abuse led to another as great. Just abhorrence of the libidinous manners of declining Rome (which certainly had not their origin in any diffusion of the physical knowledge of human nature) led the institutors of monachism o conclude, as is always the case with the ignorant and the passionate—

——Farthest from thee is best.

a guide is most needed; and, in disregard to every thing that has been observed respecting human conduct, it is required that discretion and ignorance should lodge together in the same bosom. Hence it is, that so many young people have been busy, by deed or thought, as occasion has served, in revenging upon themselves the fault or rather the oversight of their relations and tutors. The rest of the world *appear* to keep aloof. The wit only ventures on the prohibited ground, and the rather because it is prohibited—

Nitimur in vetitum semper petimusque negata.

There are analogies which might render comprehensible the future reward of youthful continence; and by a proper use of these, the preservative course would be completed. Sickness from an accidental surfeit, will help a child to conceive the folly of those, who, by a life of surfeits, destroy their appetite and bring on themselves perpetual mawkishness or even sickness itself. It would, therefore, appear in order, that the abuse of the sexual function, as of the other functions, should be followed by some peculiar penalty or loss of enjoyment. In pursuance of the same analogy, it could be stated—so as to counteract unseasonable and immoral desire—that hunger and thirst, are urgent and painful calls of nature. But as the design of providence is to protract human existence to a certain period, the satisfying of these calls has been connected with pleasure. Were eating and drinking indifferent or disagreeable, many would allow life to sink for want of support. As in fact, many really suffer ruin to come upon them from neglecting a thing which they are sure would avert it, but which is not attended with any immediate gratification, or which happens to be disagreeable at the moment.—It might be shewn, that the same providence, intending the perpetuity of the species, has connected its reproduction, both with transient and durable enjoyments; all of which are totally lost or lessened to him, who gives into the irregularities, against which this body of information is intended to guard inexperience. Pity for infants, that turn out short-lived or sickly, because their parents were not happy enough to receive timely cautions, might here be forcibly excited by an exact statement of facts. A dextrous monitor might afterwards introduce motives, derived from the various circumstances, constituting domestic happiness.

That these lessons might have their full effect, *some particulars* should be delivered in the most confidential manner, that is, to each child by himself. The idea of the trouble, which this would occasion, may strike persons, not disposed on any other account to object. But on reconsideration, I flatter myself, that such persons will not persevere in the objection. The end, supposing it can be obtained, will not fail to fill the heart with courage to encounter the means. And in a scheme for a school, which is never to lose sight of present and future health, I shall almost immediately point out the practicability of the thing proposed. Indeed, as to trouble, the frivolous decorations, attached even by the moderately affluent to their person, equipage and house, are attended with a large share. And it is trouble too, which costs a parent more than the purchase of the fairest chance of a continued happy state of being, in favour of a child. In fact, if we look merely to money, cheap education will always, upon the average, turn out the most extravagant.

A graver objection will be started. Apprehensions will arise for the conventional arrangements of society. Young people, so instructed, will be supposed incapable of restraining themselves from blurting out things, not to be mentioned in mixed company. Experience, however, by no means warrants any such apprehension; and we happen to have diversified experience, directly applicable to the point. In the first place, there are certain necessities of more than daily occurrence, with which children comply in silence, only because they are told, that to speak of them openly would be offensive. It is astonishing how easily they are trained to this observance.—But what is the case with regard to the subject directly under consideration? Where is the school-boy, the apprentice, the attorney's clerk, that is not proficient enough in the mysteries of Venus to be able to throw into confusion every mixed company into which he goes, if he were to communicate a hundredth part of what he has learned? Which of their acquisitions furnishes more towards table-talk, when they are among themselves? What they have acquired is not, indeed, the genuine order of nature in the most wonderful of her processes; but a mass of crude opinions, fallacious maxims, vulgar terms and obscene jests, from which last the conversation always takes a turn towards levity. Whence it happens that disorders, very severe in themselves, and, in all their circumstances, extremely fatal to young men, produce laughter instead of amendment; and the buffoonery, that originally attached itself to the subject, has the unfortunate faculty of rendering the roughest correction of adversity useless— We may, therefore, lay it down, that by multitudes of boys, these ideas are treasured up with greater fondness than any others. They also circulate with more activity. Yet we find, that the dealers in them take good care not to produce them where they will not pass.

How then, are we to reason upon these facts? Shall we conclude, that a method, the most favourable to modesty and consideration, shall produce indecorum of behaviour?—And this, when daily experience proves that the method, most certain to produce immodesty of mind and flippancy of tongue, can be effectually counteracted by respect for our social regulations? Do we, I pray, perceive, that persons, conversant in anatomy, are particularly apt to violate the rules established for the meeting of the sexes? Is this the case with medical men in general?—What is of as much consequence—do we find that their acquaintance with the structure and functions of the human frame renders them more dissolute in their own conduct? As to the date of their information, their general lot does not appear superior to that of others. Their trains of thought and habits of acting are all previously formed. But every fact, which has come to my knowledge, induces me to believe, that when their professional studies have brought them to the point, at which they, in common with the rest of the world, ought long before to have arrived, they frequently acquire, and never lose, controul over the appetite, most fatal to youth.

It will be remarked that I do not here enquire concerning the best plan conceivable, but the best plan practicable. Could the thoughts be diverted, till the age of puberty, from all attention to sex, I do not hesitate a moment to acknowledge that this would be by far the most desirable alternative. But the austere regulations of the monastery failed; and in failing to secure that purity, which was their object, they induced, or rather obliged, the transgressor to join other vices and crimes to his impurity. It has several times fallen in my way to communicate with invalids from families, where education is carried on almost with the regularity of a commercial concern. In some cases, it had struck the parents, as most advantageous in point of health and morals, to exert every effort to preserve the happy state of ignorance. And they flattered themselves alas! with having succeeded, when the surgeon could have informed them, how compleatly they were deceived in this belief. For a passage in Homer, a verse in the bible, a casual expression, the sight of a person in the state of pregnancy had introduced, in a moment, that turn of thought, which so many endeavours were employed to obviate. A longing desire to be fully informed soon afterwards came on—Then it was that the hidden purpose of so many precautions struck the mind. Henceforward, in proportion as one party was solicitous to keep the veil drawn, the other exercised all their cunning to creep behind it. Whether the young people ventured, or did not venture, to seek intelligence among the domestics, they never rested till they were, in some measure, satisfied. One consequence inevitably followed. Towards the authors of the concealment, a sense of alienation and distrust settled in the bosom, and destroyed the benefit of all future lessons of virtue and prudence.

However some may suppose themselves able to controul external things and inferior persons, there is a limit to their authority, which they will not refuse to acknowledge. The most skilful medical magician cannot flatter himself that, by his art, he can at all times adjust, precisely to his wish, every movement within the body; and if he cannot, ideas and sensations dependant on different movements, will arise in despite of all watchfulness and tuition. Were every existing improper idea, sensation and action banished from among young people, processes within the system would be constantly introducing them again. It seems a common opinion, that the sort of fever of the feelings, which I am proposing the means of preventing, always arises from infection. But this is a fatal mistake. The fact being perfectly ascertained, it ought to be universally made known to parents, that there are an immense number of instances, in which dangerous habits arise, independently of footman, nurse, or novel. And, if this be so, what further can be wanting to convince us how little IGNORANCE can be depended upon as a safeguard for INNOCENCE. * It may con-

* Let it be considered too that, as matters stand at present in regard to early instruction, the children of middling and opulent families are born to loathsome studies, of which

tribute to the removal of needless scruples, if we remember that a given kind of information cannot affect the untainted mind in the same manner as an opposite kind affects the tainted. And it is impossible to conceive any two things more opposite than the information here recommended, and that which young people hunt out to inflame passion, or as matter for voluptuous waking dreams, when they have not opportunity or power for more active indulgencies.—It is the lot of hundreds of parents to have to consider, in the anguish of their heart, the case of boys, who are wearing away their constitution in a course of unbridled pleasure. This case is surely less difficult. You can tell them nothing of a bad or even of a suspicious tendency. They have found out the worst. They have learned the road to licentious pleasures. They

---

the repulsive effect favours every accidental attraction towards loathsome diseases and loathsome habits.—I repeat that I can furnish *numberless* instances of the excitement of a propensity for hurtful gratifications, by causes acting within the body or upon the body, perfectly independent of seduction, and such as nothing upon earth could prevent, unless it be early, proper and particular addresses to the understanding and the feelings.

know it by experience. They do not know, minutely and in a manner to work conviction, that, by the evidence of the experience of thousands, the same appears to be the road to lasting pain, and to the loss of the faculty of feeling pleasure. On no principle, therefore, can the most ample communication be regarded as hazardous. But will it be beneficial?—This question can only be resolved by the testimony of those, who have learned the whole truth, as they stood upon the brink of danger, or had waded some way in it. Now, among the valuable materials, which various modern writers have collected towards a history of the sexual appetite in refined societies, there are actually to be found the most encouraging confessions; and they bear the stamp of the most perfect authenticity. Indeed, as they are rather humiliating, it would be difficult to imagine a motive for falsehood. In some instances, it appears that books, containing a plain anatomical and physiologicial exposition, have preserved or restored those, about to be the victims of their untutored appetites; in others, it has been an explicit account from the mouth of a respected instructor.

SKETCH of the Principles, according to which a HEALTHY SCHOOL can alone be established.

Many things under the same head in the former essay apply to boys. These I shall not repeat.—But it is more reasonable to require of school-masters than of governesses, that they should be conversant in the doctrines of animal nature, since men more easily find access to instruction. This instruction should be drawn from nature, not altogether from books. Reading is not sufficient to qualify a person for taking care of others' health. After perusal of the best works, he may have little of a practical eye for detecting mischief to the constitution in its rise; and he will be in danger of doing or suffering what may, immediately or after some time, destroy his pupils. To require so much can be nothing unreasonable, since it is only taking school-masters at their word.—They profess the care of health. But is it not one and the same thing to take care of health and to prevent diseases? How then can he, who has cultivated no acquaintance with the nature, origin and early phænomena of diseases, presume to offer himself for a charge, to which, even by his own avowal, if he has any meaning, an acquaintance with these things is necessary? It is in vain to profess, and even to feel, the most perfect readiness to apply to the medical practitioner, when necessary. To judge when this *is* necessary, often demands proficiency in the knowledge of the appearances, indicating ill-health, nearly equal to that of the practitioner, himself. And to adjust diet, exercise, and study to the exigencies of different constitutions, requires this same kind of knowledge, applied in a way, too little customary even with medical practitioners. For, surely, it is not enough that a boy continue to exist through the usual period of education. The seeds of future diseases must be destroyed; and since it is extremely easy, it would be but doing him justice to start him into life with a stock of vigour, which shall keep him in good plight thrice as long as is usually the case at present with the individuals, belonging to the superior classes.

The studies, therefore, of the preceptor ought in part to be the same with those of the physician. Inattention to this principle has been attended, but in an infinitely greater degree, with the consequences, ascribed by Homer, to the anger of Achilles. It has dispatched to Pluto, before their time, and to the inconsolable grief of their parents, innumerable souls of those, that *might* have turned out heroes or philosophers, and that certainly *would* have been happy human beings and useful members of society. Sometimes blows directed in the ignorance of anger, towards a vital part have done the business. It is more often silently effected by ill-adapted discipline, by parsimony in respect to food* and fuel, and by preposterous schemes for conferring hardiness, or permitted for want of

---

* Among the communications, relative to schools for boys, with which I have been favoured, there is one from which I think it worth while to abridge the principal circumstances. The author was at three different private schools, to all of which he informs me that his account of the third is, in the main, applicable. He was previously too at a child's school, where he suffered much from hunger.

" In ——shire, not 30 miles from London, still flourishes " the academy of ———. It is near a fine common, " well adapted to youthful sports. A broad, clear, yet " shallow stream affords the benefit of bathing without " danger. In such a spot, and with a master not inferior " in benevolence to the Man of Ross, what parent would " not have anticipated the improvement of his child? what " child could have disappointed his parent?—In describing " this seminary, I cannot help reflecting how many hun- " dreds like it, are crowded by the rising generation.

" Our number was about 40. There were two classes. " The parlour-boarders had the privilege of eating toast penetration to perceive the powers of mis-

---

"with their breakfast, sitting by the parlour fire out of "school-time, and drinking a glass of raisin-wine on Sun- "days. The other boys were, in all respects considered "as inferior; they were limited to the school-room or "kitchen; at dinner they were indulged with an extra- "ordinary allowance of *fat* to their meat; at other times "they lived on brown bread and milk. We all rose early "enough, in summer and winter, to be in the school-room "by six. After two hours' application we had our break- "fast, which we had scarcely finished, when we were "again summoned into the school-room. Here we con- "tinued till twelve, and often, if deficient at our lessons, "till dinner. At this meal, a large portion of Yorkshire "pudding was administered, before we were entitled to "animal food. At two we were dispatched to digest our "pudding over our greek and latin. At five we left school. "But there remained a task, which required at least an "hour to be completed with credit. A half-holiday on "Saturday gave us just breath enough to renew our labours "on the Sunday" [at the two other schools there was another half-holiday on Thursday]. "The Sabbath was no day of "rest to us. The moment we arose from our bed, we "began to pray, and excepting the time of meals, we per- "severed in praying, either in church or out, till we went "to bed again—

"How far so much confinement contributed to health, "I leave to be determined. But it will be clear that health "was little an object of attention, when I add that it has "frequently been my lot, for three hours at a time, to sit "with my feet chilled by damp stockings, my fingers "aching, and my nose blue with frost; nor could a par-chief during the earlier stages of their operation.

" tial glimpse of the fire, or rather its reflection from the " warm visage of the master alleviate the distress of my " situation.

" Instead of wasting our few hours of relaxation in " childish exercises, many of us fed on the chaste precepts " of C——n M——, or the beautiful sentiments of F—ny " H—ll. There were times indeed when we partook of " substantial food, and drank of intoxicating liquor. Both " were easily procured. A day-scholar felt honoured by " purloining a duck from some neighbouring farm, or " smuggling a bottle of brandy from the public house. " With these acquisitions, a select party would retire to a " room, *hired* for the purpose, and there celebrate the " orgies of Bacchus with heartfelt devotion.

" When I first became a member of the society of parlour-" boarders, I was unworthy of the rank. But, like a petty " thief after residence with more wholesale villains, I " became as accomplished as the rest. I drank, I gamed, " I ran in debt. I engaged in gallantry; and to complete " the measure of necessary qualifications, I wore my leather " breeches so tight as nearly to stop the circulation of my " blood.

" It may be supposed that the indulgencies above-men-" tioned, were interrupted by the officiousness of the " master. This did not often happen. The usher, on " adding a guinea to his *new-year's gift*, was ever our firm " friend—There were indeed *casual incidents*, which for a " time deranged our plans. But here again the assistance " of the usher, or his friend' the surgeon, *set all things to* " *rights.*——But it would be endless to recount the

Among the ablest students of medicine there will always be a proportion, to whom the exercise of their laborious profession is not suited, or who fail in the competition with their inferiors, because they are above playing certain tricks, that take with the directors of public opinion. The warm discussion, which the various topics of education have undergone since the leading and, with all its faults, the still unrivalled work of Rousseau, has induced a sufficiently general wish for improved schools. Now with the inestimable advantage, which persons, conversant in the science of health and diseases, must have above all others, there can

---

" qualifications which such a system of education yields.
" Here, as in similar institutions, the dullest genius, before
" he is sixteen, may acquire those habits and that know-
" ledge, which will qualify him either for — —, — —
" or N — —te."

Let parents, whose children are to take the chance of schools, attend to this account.—It comes from one, who has had not only various experience, but extensive means of enquiry, and upon whose accuracy I perfectly rely.—Let them hope, if they can, that their boys are to be kept from the knowledge spoken of in the preceding section, or that they will acquire it in a manner less dangerous to their morals and constitution than a true and natural exposition, divested of all the provocative tricks of rhetoric.

be no doubt of sufficient support, if such persons, not neglecting inferior endowments, were to engage in the task of education. It is common to hear incurable valetudinarians exclaim, *Ah! if I had known the consequences in time, I never should have brought all this upon myself.* There is scarce any one, who if he accurately reflects upon the occurrences of his life, but will feel himself entitled to hold some such language. He will not have occasion to reflect long, till he discovers that his sweetest consolation will be in conferring upon his successors the advantages, which the thoughtlessness of his predecessors has denied him. And the resolution to do so will diminish his regret on finding himself cut off from that ease or those enjoyments, towards which he has so often cast his looks with a sigh of despair.

That schools should cease to be fatal to all the valuable qualities of mind and body, parents must, on such considerations, become liberal, or more properly speaking, they should be œconomical. For money spent upon useful and engaging acquisitions in youth, if we reckon only upon saving in money, will be found to bring the most ample interest. The poet, says Horace, does not easily become a miser. In like manner, it may be asserted, that he, who when a boy, has been initiated in those theoretical and practical sciences, which the labour of ages has accumulated, will seldom prove a spendthrift, a gambler, or a voluptuary. On the danger of parsimony during the early part of education, the most general fact in the modern history of our schools and universities, and of most other places inhabited by the equals in age to our university students, must decide the opinion of every prudent father of a family. The fact I allude to is this:—young men of fortune, and those also of no expectations at all, commonly dissipate more money in a very few of their last school years and after quitting school, than would suffice for the education of two or three boys upon the most expensive plan, which any reasonable regard to instruction or health could suggest. There is, perhaps, scarce one individual, whose education upon the whole of the present plan, is not more costly than it would be upon the whole of the plan proposed. The mind not being agreeably engaged by the ideas it has acquired, and the vacancy of feeling being intolerable, nothing remains to produce the desired excitement, but some species of debauchery. And the temptation is always greatest, and yielding to the temptation always most baneful, to those, who have received from nature the happiest endowments. In such circumstances, no regard to the purse or to the constitution would probably be an effectual restraint. But for want of the least degree of precise information concerning what is due to the constitution or even to the purse, the youthful debauchee is destitute of the advantage even of this poor chance.

Such an inducement may be supposed capable of determining every considerate father of a family to co-operate with all his means towards preserving the patrimony and person of his offspring. That these means may be properly applied, the preceptor, in addition to the qualifications already mentioned, should be one of those rare persons, to whom, after the prime of manhood, brisk exercise is neither impossible nor irksome. He should have force of mind to break that habit of sedentary study, which is by far the greatest evil that the art of printing has produced among mankind. His own knowledge of the properties of human nature would furnish him with motives, powerful enough to lead him to forego and to banish from his table the use of every sort of fermented liquor, from small-beer inclusive. After a month's perseverance, no one, as the author can attest both from experience and observation, would feel this as a sacrifice, unless his constitution had been before irretrievably hurt by a vicious regimen. For the same reason, stimulants like tea and coffee, which grown people might use, without disadvantage, in a certain quantity and of a given strength, would be abandoned. By help of the master's example, but not in opposition to it, what he would be able to communicate from his stores of physiological and pathological knowledge, would set his pupils above danger of being seduced by the usages of the world. On the contrary, as bad examples are always salutary to those who consider them as bad, the feelings with which such boys would look on the unfortunate slaves to the dreadful habit of swallowing poisonous drinks, would strengthen them in their better resolution. And effects, which the master might place before the eyes of his scholars, by mixing ale or spirits for a time among the food of domestic animals, would confirm the impression of his doctrines. Such experiments, while they furnished an interesting employment, would secure the health against one of the greatest dangers to which it is exposed in the intercourse of modern society. Mathematical, philosophical, chemical, botanical, and *technological* instruction, judiciously intermixed, but at the same time carried on in conformity with a system, clearly made out in the mind of the preceptor, would keep the physical and moral faculties of the children in perpetual and proportional progress.* The thoughts would never stagnate; the heart never prey upon itself, and intemperance of every kind would be abhorred according to the nature of its consequences.

The plays, usual among children should make a very small proportion of the general

---

* There exists a school in Soho Square, London, at which, from the literary and philosophical attainments of the master, his acquaintance with the arts of common life, and his active and well-regulated habits, grown boys, and for such, the school, I believe, is designed, will doubtless find that grand security against destructive vices, which rational knowledge affords. I hear too of a few schools for little boys, where they are let loose as soon as their work is done, and not fixed to their benches while the tedious finger of the clock sweeps a certain portion of its circle. But these schools are unfortunately merely *classical*. Now this one circumstance cuts off all just expectation, that the boys shall grow up robust men. Nothing can ensure this greatest of earthly blessings to our opulent youth, but such scientific pursuits, as shall lead early to wholesome exercise of body and mind.

scheme of exercise, or be altogether disused. It is not that I would wish them prohibited; but active occupations, connected with future life, should be adroitly substituted in their stead. It is the misfortune of school-sports that they must be dropped as soon as boys quit school.—The exception of cricket is hardly worth naming; and the morning benefit of this game seems more than counter-balanced by the carousal, which commonly follows it in the evening.—But all the employments, to which the practice of physical science leads, could be continued in the more advanced stages of life. Thus there would be no intermission or languor in the pulse of thought or motion. Nothing like that distress, which the outer frame feels on the unexpected stopping of a carriage as it is bowling along a smooth road, would be known to the intellect. The leisure, which men of letters and of business devote to dangerous or doubtful recreations, might be employed, with much increase of individual and social happiness, in the continuation of these pursuits. The expence attending them would be less—it certainly could not be greater—than that of the surfeiting entertainments,—perhaps not greater than that of the insipid parties—to which, for want of having been trained to something more agreeable and wholesome, we are obliged to have recourse.

As an indolent turn and the enervating indulgences, necessary to render indolence supportable, principally occasion the early debility and disorders that prevail among the opulent, it will scarce be enough to connect the leading pursuits of youth with bodily exertion. Where the circumstances of the parent will admit of it, a habit of activity should be secured at whatever cost. The preceptor should be empowered to carry his pupils on distant excursions; and as much on foot as can be done without over-fatigue. Botanical and mineralogical information, the inspection of manufactures, the measuring of heights, would be objects delightful in themselves; and they would give to home-pursuits, a relish, which the wretched prisoner, obliged to pore, day after day, over his grammar, dictionary and lexicon, can scarce ever feel. On the summit of high mountains, we are sensible of a degree of freedom and elevation of thought, felt in no other situation. Here we have a commanding view of the labours of men and of the productions of the earth. We seem to have it in our power to reach at a few strides whatever spot of the diversified scenery most invites the fancy. These sensations cannot fail to inspire a taste for the charms of external nature; and the acquisition of such a taste is among the chief means of maintaining the constitutional vigour to the longest allotted term, allotted by his Maker, and of raising it to the highest pitch—and

Qui fait aimer les champs, fait aimer la vertu.

In the confidence, which would necessarily arise amid participation in active pursuits and in little incidental hardships, the delicate subject of the preceding section could be treated in common with other interesting topics. Many appearances in nature would occur to introduce it. Many human beings would also be accidentally met with, or might easily be found upon search, whose condition would give irresistible force to admonitions, or rather to prudential inferences from a detail of facts. This is, doubtless, the single point of most importance in male education. The want of information properly conveyed and properly illustrated, may possibly deprive a majority among the sons of rich families, before they are five-and-twenty, of that elastic force of mind and body, which ought to bear them light and chearful over the difficulties of life, to more than double that age.

The regulation of food will have nothing difficult, where the free, but not excessive, employment of the limbs secures appetite and digestion. The diet should be nourishing, but in no article stimulating. Bread and milk, divers preparations of milk, and fruits with broth or soup occasionally, will supply all the meals except dinner. Dinner should consist of a moderate portion of butcher's meat with plenty of vegetables. It is only in the case of puny children, that an additional quantity of animal food with beer will be necessary. This is a medical consideration; and is best determined and proportioned by a medical man, having constant opportunities of observing the boys, that require a medicated regimen. Such things are hardly ever, in fact, adjusted by a practitioner, called in for the purpose. His presence is only desired when diseases are actually formed. Some evil might be obviated by submitting schools once a month, to the examination of a physician, who has paid particular attention to the first symptoms of slow disorders, and to the tokens that indicate vices in the constitution. But still a constant observer of young people would have an infinite advantage over an occasional inspector ever so skilful.

It would be adviseable to commence establishment for healthy, for scientific and moral, in addition to classical education, with children, fresh and untainted from their own home. A single preceptor should not engage with above six boys; and it would be better not to bring above six together as permanent inhabitants of one house. But the public has been so cajoled with the idea of the members of a large body, disciplining one another to a greater expertness for the warfare of life, that with the addition of two or three well-qualified assistants, twenty might be received. Such a number of superintendants would afford the inestimable advantage of breaking the twenty into parties. The whole should never be left alone together for an instant.

It may he reasonably hoped that the rich will procure to the public the benefit, that would result from the example of such an institution. They can hardly hesitate, if they will consider with a little attention what it is impossible not to see, but what is perceived to no purpose. I mean the utter insignificance of one or two thousand pounds' difference in a fortune of eight or ten thousands.

—Should we even suppose, that the most advantageous culture of mind and body will necessitate such a diminution, of two boys, equal by nature, would not ten thousand pounds go farther in the hands of him, who should be brought up with a practical insight into men and things, than twenty thousand in the hands of him, who should be subjected to the blind-folding and debilitating routine? Many schoolmasters, I doubt not, wish in secret for a more active education: and six such seminaries might effect a gaol-delivery of the innocent, misused, little prisoners throughout the kingdom.

It will easily be credited on the faith of common observation, and I can assert from experience, that when the understanding is kept in constant but various exercise, and never overstrained or glutted by excess of the same study, progress will be much more rapid in this and every other line. The pleasure from difficulties overcome will give ardour to cope with new ones. We observe the sickened and spiritless pupils of our celebrated classical schools, frequently unable to summon up resolution enough to master the first rules of arithmetic. I have known some scores of college youths, regularly lectured in Euclid, but not one in twenty came to understand the first six books of that author. Whereas, boys of nine or ten, if kept in health and spirits, will pass the asses' bridge without boggling, and advance continually with augmented alacrity. Nor will they find the very chips and saw-dust of learning too dry.

MODE OF LIVING AFTER MATURITY.

The diversity of trades, professions, and callings creates a much greater difference in the habits of the male sex than takes place among females. But it is principally in the pursuits, in which men spend a *part* of their day that they differ considerably from one another. In the manner of passing the evenings, those who find themselves in easy circumstances or in affluence, are nearly uniform. Many inhabitants, however, of towns, and in the country not a few of the clergy and of the smaller gentry, pass their time without much greater exertion than women. A walk from the counting-house to the exchange, and perhaps, from the counting-house to the parlour, will satisfy some merchants for a week together. Others ride ten or a dozen miles between town and country, and flatter themselves that this is sufficient to keep them stout, and to counteract powers perpetually at work to curtail their existence. In no country are the men so fond of lounges on horseback as in England. In none are the limbs exerted so little. These customs depend upon the character, which the British constitution has gradually been acquiring, partly from hereditary disposition, and partly from the same causes, which produced the disposition in the father at first, continuing to operate at advantage upon the son, who has brought it with him into the world.—Of the causes of such a disposition, some are, it is true, of long standing, and beyond the reach of human memorials. But by those of later origin, the nature of men has been so much altered, and they have been rendered so much more susceptible, that the experience of our forefathers will delude us into dangerous errors, if we be inattentive to the disparity between them and ourselves. Nor is the result to be found by merely adding together what each cause of degradation might be supposed to contribute, if it were to act singly. Several are rendered more powerful by the concurrence of others. The sum, therefore, would be more nearly obtained by multiplying the single products by one another, if in such things we could approach towards the accuracy of a numerical statement.—But the fact will be clearly illustrated by bringing the more efficacious causes under view at once, and by adding a few observations upon the influence of one cause, not yet considered, as it operates in combination with the others. Were a perfect table of this kind constructed and proper explanations annexed, each individual might find his station on the scale of health, as readily as he could the class, order and genius of a plant in the most convenient botanical arrangement existing. He would, at the same time, be able to perceive what had placed him where he stands. He might make out a method too for raising himself higher, if the elevation were possible;—at least, he might see how to prevent the yet entire members of his family from sinking so low. —I need hardly say that I over-look, for the present, that part of the foundation for ill-health, which is laid before the school-going period.—

1. *Cold at schools,* in our days more severely felt.

2. *Vexatious discipline, and restraint in the use of the limbs,* more pernicious as the habit of body is weaker, and because their odious confinement precipitates the little prisoners, when let loose, into enervating gratifications,

to which access is more easy—while public opinion on certain sorts of imprudence has abated of its severity.

3. *Sexual passion* prematurely excited and inordinately indulged.

4. *Early intemperance* in wine; also in inferiour stimulating liquors.

5. *Insufficiency of food* at some schools.

6. *Example* of indolent habits, already established.

7. *Lazy literature,* or the host of modern writings, calculated, if for nothing worse, only to relieve listlessness for the moment, prompting to no one exertion, and therefore eventually tending to increase the languor, that overwhelms so many of the present generation.

8. *Ravenous desire of gold,* incited by eagerness to participate in the general luxury of the times.

9. *Disappointment and success in regard to this desire*—both necessarily more frequent than ever from the increase of competitors; and alike productive of habitual anxiety, or of absolute hypochondriasis or insanity, because several of the preceding causes will have reduced the nerves below the tone of simpler ages.

10. In females, and probably in some feminine males, *application to music and other sedentary accomplishments*—concurring, with the tumultuous perusal of light books, to injure the perceptive and associating faculties, and by consequence to spread havoc through the nervous system.

11. *Inhabiting warm, close rooms.*

12. *Half-nakedness* in women, or too thin clothing in men.

By these causes, or by as many of them as may chance to act together or in succession, the term of existence is so frequently abridged, that there are few large genteel families, where an example of such operation does not occur. But it is not on their power or the duration so much as upon the condition of existence, that I would have my readers fix their attention. *A short life and a merry one* is a wish, of which I shall not contest the rationality. But the expression is much misinterpreted, by uniformed boys, when their blood boils within them from impatience for enjoyment. Or, as the fiery temperament rarely occurs now-a-days, I should rather say, that it is misinterpreted by young rakes, who feeling already shattered, are glad of a seeming *philósophical* excuse for resorting to provocatives of merriment, without which they would no longer know what it is to spend a joyous day. But nothing is more erroneous than the principle they profess. Life is not to be raised into a bright blaze that shall go instantly out. It never is more languid towards the last than with those, who labour to force it into extraordinary vividness. Nor does the faint, wavering flame ever more tediously, than with them, linger in the socket. In cases of gout, spleen, apoplexy, dropsy, and in all other diseases of intemperance, there is often an interval between a predominant comfortable state and the grave, as fearful as that which Shakespeare plants between the formation of a plot and its execution, and always infinitely longer. Before the loss of feeling and motion from the palsy, what years of wearisome, stagnant being, of stretchings without relief! what uneasy sense of the flesh melting away from the bone! Very similar is the unnoticed transition to regular and accepted disease in other instances!

Among the *modern* devotees of Bacchus, few are allowed to escape this sad purgatory of suspended animation; and indeed but few of those moderate, but regular wine-bibbers, who look upon the worship of Bacchus with abhorrence. The reason of the more pernicious effect of the use of fermented liquors at present, I seem to myself clearly to discover in our luxurious indolence. In *nervous* females, during their best health, I have known a glass of wine and water produce giddiness, depression and all the *narcotic* effects of opium or foxglove, without the smallest sign of previous exhilaration. The same thing will happen to most modern gentlemen, if wine be taken, at an unusual season, on an empty stomach; but it does not happen to active, muscular persons, happily ignorant of *what nerves are.* In regard to the ultimate poisonous effect of wine, these persons seem to stand in the same order, and at perhaps nearly equal distance. In some districts of Ireland, the change from a rugged to a soft mode of life, which in England required perhaps above a century, has taken place within a few years. The fathers drank and rode from fair to fair, and hunted, and suffered the winds of heaven to blow upon them within doors. And they commonly reached the extreme term of human existence, unimpaired. The sons drink rather less; but from the change in the state of culture, they differ almost as widely from their progenitors in exercise and expence as our silk-mercers from our butchers. They are in consequence often worn out and diseased at thirty. Robinhood and Little John, the alert and airy inhabitants of the green-wood shade, could take liberties with the bottle, which are very unsafe for us, under our close roofs and between our stuccoed walls. The adroit adventurer who negociated the submission of prince Heraclius to Russia, and who was better acquainted than any other mortal with the rude tribes of Mount Caucasus and its precincts, relates facts, that strongly favour such a supposition concerning the foresters of Sherwood. Speaking of Iberia, he says—" The mild climate secures

" the native from rigorous cold and oppres-
" sive heat. In his cool shady woods, beside
" clear waters, amid exhalations from odori-
" ferous herbs, with the most exquisite wild
" fruits and wholesome wines, produced with-
" out toil or art, he is in general unacquainted
" with distempers, and attains to extreme
" old age. I knew an Iberian, named
" DADULO, who was 110 years old. He enjoyed
" perfect health, and every day drank two
" pattmans or sixteen pints of wine, to which
" he had been accustomed from his youth
" up. For 75 years without interruption, he
" had performed a pilgrimage on foot to the
" holy sepulchre of the female martyr Nino
" in Georgia, and although the distance is
" 280 wersts, he never complained of fatigue."
Sir John Chardin relates similar feats with consequences apparently not more disastrous. And these narratives would find confirmation among our own hardy mountaineers. Although it would appear from the destruction of the American Indians by spirits, that the habits most powerful in counteracting excess, will not allow the generality of men to go beyond a certain length.

I have thus delineated the causes, why the sluggishness, dejection and moroseness of old age is so common among us at 25 or 30, and why vigour and alacrity are so rare, at 60 or 70. With the advantages, which a moderate fortune can command, this order of things may be immediately reversed in a multitude of families; and a gradual return to a desirable state of existence may be secured in almost all. But if the inherited tendency to early decrepitude, to fatal or severe disorders, and mental alienation is to be checked, the necessary conditions must be complied with. A child is ridiculed for thinking that he can *eat his cake and have it.* Are not grown people equally ridiculous, if they complain of ailments at the time they are doing every thing in their power to bring them on? —There can be no hope of avoiding them, unless we be filled by a *sentiment* of the value of health, at least as lively as that which animates us to the pursuit of any other temporal blessing. Whoever can inspire the public with this sentiment will do more good than the discoverer of twenty specific remedies.—Specifics will only put away the disease. In all cases they *may*, in many they *must*, leave the patient sickly and wretched. An acquaintance with the influence of external powers and of internal feelings upon the living frame, with a correspondent conduct, is the sole specific for health. This knowledge excludes all reliance on physic and physicians for reproducing that condition of existence, which has been destroyed by long perseverance in enervating usages. To just *general* views, and the ruling sentiment required above, a popular account of diseases and of diseased constitutions in *particular*, will be an useful complement. By itself, it can be but of partial and trifling avail.— To trust to any thing like this is the same sort of policy as laying a garment to rot, with the view of patching it hereafter.— What Iago says of Othello, once infected by jealousy, may with some change of terms, be applied by the Dæmon of luxurious dissipation to the individuals, into whom he has instilled his insidious poison—

—— Not poppy, not mandragora,
Nor all the drowsy syrups of the world
Shall ever medicine thee to that sweet sleep
Which thou ow'dst yesterday.

END OF ESSAY FOURTH.

Printed by B. M'Millan, Bow-Street, Covent-Garden.

HYGËIA

Essays Moral and Medical

Introduced by
Robert Mitchell
Duke University

THOEMMES CONTINUUM

Printed in England by Antony Rowe Ltd, Chippenham

HYGËIA

Essays Moral and Medical

Volume 2

Thomas Beddoes

THOEMMES CONTINUUM

This edition published by Thoemmes Continuum, 2004

Thoemmes Continuum is an imprint of
The Continuum International Publishing Group Ltd

Thoemmes Continuum
11 Great George Street
Bristol BS1 5RR, England

http://www.thoemmes.com

Hygëia
3 Volumes : ISBN 1 84371 098 6

Reprinted from the 1802–1803 edition owned by the
Rare Book, Manuscript, and Special Collections Library,
Duke University

British Library Cataloguing-in-Publication Data
A CIP record of this title is available from the British Library

Publisher's Note

The Publisher has gone to great lengths to ensure the quality of this reprint but points out that some imperfections in the original book may be apparent.

This book is printed on acid-free paper, sewn, and cased in a durable buckram.

HYGËIA:

OR

ESSAYS

MORAL AND MEDICAL,

ON

THE CAUSES

AFFECTING THE PERSONAL STATE

OF

OUR MIDDLING AND AFFLUENT

CLASSES:

By THOMAS BEDDOES, M. D.

VOLUME THE SECOND.

BRISTOL:

PRINTED BY J. MILLS, ST. AUGUSTINE'S BACK,

For R. PHILLIPS, No. 71, St. Paul's Church-Yard, London.

1802

CONTENTS.

CONTENTS.

ERRATA.

Essay v. *p.* 13, *l.* 19, *for* ten *read* five: *p.* 35, *l.* *after* duschi *a colon*: *p.* 37, *l.* 3, *after* -stance *insert* who can: *p.* 41, *l.* 3, *for* girl *read* girl's: *p.* 46, *l.* 20, *after* -sion *insert* to have: *p.* 52, *l.* 20, *for* former *read* wanted: *p.* 60, *l. last*, *after* overcome *insert* by heat: *p.* 82, *l.* 19, *for* ation *read* ance: *p.* 92, *l.* 18, *for* 5152 *read* 51,52.

Essay vi. *p.* 14, *l.* 5, *for* more *read* once: *p.* 28, *l.* 25, *for* this *read* his: *p.* 42, *l. last* & 43, *l.* 1, *dele from* and *to* perceptible.

Essay vii. *p.* 11, *l.* 10, *dele* are: *p.* 22, *l.* 13, *for* 5,070 *read* 570: *p.* 23, *l.* 8, *for* 365 *redd* 290: *p.* 50, *l.* 18, *for* influence *read* inference: *p.* 54, *l.* 5, *for* knots *read* knot: *p.* 68, *l.* 4, *after* right *insert* side: *p.* 88, *l.* 26, *for* ascertained *read* entertained.

Essay viii. *p.* 19, *l.* 10, *for* thing *read* things: *p.* 32, *l.* 14, *after* not *read* of: *p.* 99, *after* pumpernichel *insert* (a coarse black bread): *p.* 110, *l.* 19, *for* scorn *read* seem: *p.* 115, *l.* 27, *for* bon *read* bons: *p.* 117, *l.* 23, *after* as *read* to: *p.* 130, *l.* 19, *after* different *a comma*. *p.* 137, *l.* 20, *after* male *read* sex.

ESSAYS

ON

THE MEANS

OF

AVOIDING

HABITUAL SICKLINESS,

AND

PREMATURE MORTALITY.

ESSAY FIFTH.

Printed by MILLS, St. Auguſtine's, Briſtol.

——Now have we many chimneys, and yet our tenderlings complain of rheums, catarrhs and poses. THEN we had none but reredosses, and our heads did never ache. For as the smoake in those days was supposed to be a sufficient hardening for the timber of the house, so it was reputed a far better medicine to keep the good man and his family from the quacke or pose, wherewith, as then, very few were oft acquainted.

Harrison's Description of England.

——Without meaning to boast, I may say of myself that my armour was to me as easy as the softest down: and such is my custom that when I now go the rounds of my district, I never take a bed with me, unless I happen to be attended by strange cavaliers. In which case I do it to avoid the appearance of poverty or penuriousness. But by my faith, when I have one, I always throw myself on it in my clothes. Such it is to be a true soldier.—Another peculiarity I have is, I cannot sleep through the night, but always awaken and get up in order to contemplate the heaven and stars. And thus I amuse myself, walking backwards and forwards, as I used to do when on guard, for a good space of time, without hat or cap:—and glory be to God! I never yet caught cold, nor was a jot the worse for it. And this the reader must pardon me for mentioning, it not being from vanity, but that I wish him to know what kind of men WE, the true-bred soldiers and real conquerors of Mexico were.

Bernard Diaz del Castillo.

ESSAY

ON

TEMPERATURE

AND

HARDINESS;

WITH

REMARKS ON DIET.

THERE are scarce any topics, which popular writers on regimen labour so little as those of the present essay. Yet there are none better entitled to their care. A hundred pages of plain and appropriate instruction how to secure the advantages of temperature, and how to escape from its mischiefs, would be of more service than a hundred volumes of such discourse as we commonly meet with on the comparative merits of fish, flesh and fowl. There are, in reality, very few persons who are not the worse for their own want of information on this head, or for that of others. Erroneous practices at one time give rise to the most tedious, and at another to the most severe illnesses.

What subject of contemplation is more melancholy than the uncertainty of life during its first stage? What oftener intrudes upon the purest and most placid domestic enjoyments, than the alarming recollection of this uncertainty? What plunges families into deeper distress than the occurrences, from which it is deduced by political arithmeticians? It seems, however, clear that the knowledge and application of a few simple principles, would prevent the greatest part of this alarm and distress. Of two equal lots of infants, I do not entertain the smallest doubt but the mortality would be less by half, in that where these principles should be steadily followed. I will venture to say the same thing of sickness, at this and at every subsequent period; and not of sickness only, but of that comfortless state, which wants a name, but will be well understood by the tenderer part of society, on recollecting their feelings in sharp weather, their greater weariness on waking than on going to sleep, their oppression after meals, and their sufferings on many other occasions, where the rest of animated nature finds solace.

During those epidemic pestilences of the dark ages, which were so destructive as to obtain notice in general history, the people and the medical faculty agreed to bewail the malignancy of the atmosphere. The atmosphere was accused in cases, where he who ran might almost have seen, that the calamity was disseminated by the grossest kind of contagion alone. But in acquitting the medium, by which we are surrounded, of producing one class of evils, its real injurious influence, if not absolutely overlooked by professional men, has never been sufficiently explained to the mass of those, on whom it is exerted. It is my intention, at present, to do what my own observations, assisted by those of others, will enable me towards supplying this great deficiency. — I shall separately consider different ages, constitutions and circumstances: and thus I hope to furnish most of my readers with data for determining the mode of conduct, best suited to themselves, and to those dependant on their discretion. I only expect that they, on their part, will bring with them common prudence to the consideration of common matters, and not suffer themselves to imagine, that any tricks of medical legerdemain can suddenly repair the ruin, brought upon the human machine by long negligence or gross error.

INFANCY.

If there were safety in the multitude of books, parents would have little to apprehend for their progeny. The literature of Europe abounds with productions on the diseases and the management of children. We have many such in our own language. A reputable living author has collected from his predecessors, with a degree of judgment. He has added some things from genuine observation. It is to be regretted, that he should have unintentionally given his sanction to the arts of common imposition by the needless employment of language, connected with exploded hypotheses.* It is

---

* For instance—"*Every species of this eruption is produced by the same cause as the thrush, but can scarcely be termed a complaint, being a kindly exertion of nature to throw off some* ACRIMONY".—Again "*otherwise the rash striking in, the* ACRIMONY *will fall on the first passages.*" Every body, who has taken the least pains to examine the grounds of medical opinion, knows that there is not the smallest evidence of any kind in favour of the existence of *acrimony*, or of its migration to and fro in the body. And such shallow nonsense from a regular physician to common readers gains credit for quacks, whose advertisements usually run in the

still more to be regretted, that by entering into the treatment of diseases, he should have thrown a temptation in the way of mothers and nurses to intermeddle with that part of medical practice, which is precisely, on many occasions, the most difficult, on account of the obscurity of symptoms in children. It would have been better to omit such dangerous matter, and in place of a few perplexed, indiscriminating sentences to have devoted the space that would thus have been gained, to a copious investigation of the effects of heat and cold. For there is nothing so essential in the whole compass of infantile regimen.

I hope I shall not be thought too trivial if I recommend, that care be taken to provide a fit habitation for the expected little stranger. The numerous opportunities I have had of becoming acquainted with the causes of suffering among the poor, first impressed me with a sense of the importance of this precaution. I have seen severe catarrhs, the

---

same strain. In this consists the mischievousness of this particular nonsense. Though indeed it is hardly possible for nonsense and falsehood to be innocent, when it is to be reduced to practice upon infants by the ignorant.

most violent inflammations of the eye, and other complaints, produced by a breach in the window, or a by fissure in the wall, opposite the spot where the infant was usually held. In some instances, indeed, the parents, inattentive as the poor usually are, have made the discovery themselves. The same thing has occurred in opulent families. The shrinking of the sash-frame, of a board in the floor, or of some other wood-work has admitted an unperceived, but strong current of cold air, by which infants have been seriously injured.

But it is not only necessary to guard against such mishaps. The fire-place should be so contrived, that even in cold weather a steady temperature of about sixty degrees may be kept up, for the first four or five weeks after birth. The air of the nursery should never be suffered to be below fifty degrees. And I would advise every parent to ascertain this by a thermometer, constantly kept in the room, at the risque of being thought ridiculously minute. No hot-house is suffered to be without an instrument of this kind. New-born babes come from as warm a climate as any exotic plant. They are as tender. And who will say they are not intitled to as scrupulous care? The directions, with which Count

Rumford has favoured the world, will assist in fitting up an apartment so as to be properly warm. And I suppose, few fathers, once convinced of the propriety of a steady, mild temperature, will hesitate to dedicate their most suitable apartment to the health of their offspring. Small nurseries cause many complaints, and many bad constitutions. Leaving the door open for a short time, fills the room with frosty air in winter: and a fire, somewhat too brisk, produces a heat, that will aggravate some disorders, and greatly enervate a habit, constantly immersed in it.

The membrane lining the nostrils and air-pipes is, at first, so susceptible, that the slightest impression will often produce a very sensible effect. Brisk motion, for example, against the air of a room sufficiently warm, will produce both coughing and sneezing, but more certainly the latter. And it had better be avoided. The frequent renewal of this irritation tends to keep up the excessive susceptibility of the membrane. A habit is established, which in some few instances, lays the foundation of asthma at a very tender age. I have seen this disease extremely violent in young children, and have traced it with a considerable degree of probability to causes, that kept up an original diseased dispo-sition of the surfaces, swept by the air in respiration. Where no striking evil is occasioned at the moment, future complaints of the same parts may be rendered more frequent and severe: and, perhaps, a propensity to catarrh, which is undoubtedly one of the greatest plagues a person can be troubled with in this climate, may thus be generated. There are a number of nurses, I know, who will laugh you in the face, if they hear you express any anxiety about an infant's taking cold. But these ladies do not appear to have been favoured with any peculiar illumination, as to the latent causes or remote consequences of indisposition. And I have known contemners of petty cautions, whose success in rearing children could not be said to add particular weight to their sentiments.

The use of the cool bath, at first, and afterwards of the cold bath, is the safest and most effectual method of taking off the too great sensibility of the membrane in question, and of habituating the whole body to variations of temperature. For a new-born infant I should prefer instant immersion in water at eighty degrees, to washing. There is no occasion to wet above the neck. The infant should be taken up into warm flannel which should be rubbed hastily, as it lies spread over the infant's body, with the hand for a short time. Another warm flannel should then be substituted for the first. I believe there will be no occasion for adding salt to the bath. But it may be done, if the child do not appear sufficiently comfortable from fresh water. In a week the temperature may be reduced to seventy-five, and progressively to sixty degrees, at which point we may stop for a twelvemonth.

"To see a little infant of a few days old," says a writer on infancy, "washed up to the "loins and breast in cold water, exposed for "several minutes, perhaps in the midst of "winter; - - - - itself in one continued scream, "and the fond mother, covering her ears "under the bed clothes, that she may not be "distressed by its cries, has ever struck me "as an unnecessary piece of severity."—But it is not only this degree of severity that calls for reprobation. A weakly child may be easily washed, by ever so nimble a hand, into irrecoverable debility, if the process is regulated by the clock, and not by the state of the subject. It is this, and similar inattentions, which occasion convulsions to be so formidable an article in our death-registers. Irritation will often be followed by convulsions, or otherwise, according as regard has been had to temperature.—In a hot, dry state I have, at different times, for many years past, ordered the trunk to be washed with warm water, to the apparent comfort of these young creatures. But when the heat arises from inflammatory disease, it will be for the medical attendant to direct.

Some infants are born with a defect in the membrane of the nostrils and air-pipes, or of the *mucous* membrane, as I shall hereafter call it. They sneeze frequently. They are heard to rattle in breathing, or to wheeze. Mucus appears in too great abundance at their nostrils. They are liable to appear more chilly than is usual, even at this age. At the time they breathe hard and have a superfluity of mucus, their hands and feet will be covered with a cold, clammy perspiration. These infants demand the most particular attention as to the temperature, in which they are kept. This attention, and perseverance in the following simple measures, will generally bring the mucous membrane to a healthy condition.

When the extremities are cold and clammy, they should be rubbed before a moderate fire with the hand, or gently with a soft flesh-brush, till they grow dryer and warm. No pains should be spared. The rubbing at first may require to be repeated many times a day. Merely holding the hands and feet to the fire will not answer. To any one, who has paid the least regard to the nature of animal motions, I need not explain the difference between the effect of external heat and of that excitement of the parts, by which they are made to warm themselves by their own action. The salt cool bath, and in its turn, the salt cold bath should be used instead of the fresh. *They should neither of them be used, in any case, when the child is cold, but rather when it feels above par as to warmth.*

In the clammy, perspiring state I have described, a little chicken broth, or gravy diluted with water, or diluted milk in which spice has been soaked, should be given at the temperature of about an hundred and ten degrees; that is, sensibly above blood-heat. And this in whatever way it is nursed, if the child will take it.

A small vessel for a warm bath should be part of the furniture of every nursery. Warm bathing will be particularly useful in cases of weakness of the mucous membrane. At times, during the chilly state, the infant may be placed in water at the temperature of ninety-six degrees, to which salt has been added.

This will be particularly proper if there should be pain of the bowels at the same time. And, in general, weakness of digestion will accompany the other weakness. The time of continuance in the warm water should be a quarter of an hour or longer, if there be signs of ease and comfort for a longer time.—The warm-bath will be a resource during such sufferings as arise from the thrush, and from teething as well as from indigestion. And it may, in case of emergency, be repeated four or five times in twenty-four hours.—A sloping *bath-bed,* which shall keep the body in a posture between sitting and lying, will be found convenient. A proper frame and soft covering will easily be contrived without a particular description. A small basket with its bottom and side lined, in which the water may just come up to the chin, answers equally well.

When the skin feels dry and hot, children, whether weakly or robust, should be carried into a room without fire, till the skin returns to its natural state. This, when there is no feverish disease, will soon happen. Neither coughing nor sneezing, as I have frequently experienced, will be provoked by this treatment, even where there is the greatest disposition to it at other times.

Nurses, in general, have an exceedingly vicious habit of holding infants at particular times near the fire, without regarding whether they are previously hot or cold. In their feverish complaints, there is the same want of care to accommodate the temperature to the state of the skin. Though perhaps it is not altogether want of care, but sometimes the effect of those mistaken medical opinions, which, in former ages, induced physicians to stove their patients, ill of the small-pox and pleurisy, in hot, close rooms.

It is, by no means, so impracticable as many would fain make us believe, to eradicate these and other prejudices from the minds of female attendants upon infants. But from the experience of the world, it is clear that it can very seldom be done by threats or by commands. Indeed there is much less disparity in the power of apprehension of different ranks, than the superior members of society are willing to flatter themselves. If men and maids have fair play, they will often reason as logically as masters and mistresses. And having once been made to feel that we can only arrive at an acquaintance with the nature of things by experiment and observation, they will with perfect readiness defer to the opinion of those, whom they believe to have had superior opportunities for observing. Orders and facts are so very different in themselves, that they can never produce the same operation upon any human mind. But obedience will always be more chearful and steady after a reasonable explanation. And this, in the case of infants, can always be given without derogation from authority or any other inconvenience. The interest, which these helpless and tender beings inspire, produces a sense of equality, which is every day expressed in the conversation of the nursery, and from which no danger arises to family subordination. The same interest will ensure conformity and vigilance, when the means of averting evil are clearly pointed out.

I have known an incident as simple as the following, put an entire stop to a system of mismanagement, from which much immediate suffering had arisen, and which threatened serious hurt to the constitution. In the middle of the night, during a smart accession of feverishness, one of the parents finds the nurse sitting up, the child upon her lap, flushed and with a burning skin, and in this condition exposed to the full blaze of a brisk fire:

Parent. *How has she been?*

Nurse. *Oh! very restless. She has tossed and tumbled about ever since bed-time, without one comfortable wink of sleep.*

P. *Bless me, how hot!*

N. *She has been like a fire-coal these two hours.*

P. *I wish there were some cooling medicine to take off this extreme heat.*

N. *Yes, indeed, I wish there were. The poor thing is excessively feverish. How she burns!*

P. *But since she burns in this manner, must not being so near the fire in this hot room be extremely bad for her? If she wants* cooling *medicine, would not* cooling *air do her good as well? And is not this hot air making her burn more?*

N. *But might not she catch cold? For it always sets her to sneeze, when there is the least breath of air?*

P. *Did you not hear Jane Saunders, this very week, say how much ease it gave her to be washed with cold water in the rage of her fever; and how eagerly she cried out,* MORE, MORE WATER? *And there are countries where persons in the small-pox are always washed with water; and they are now, here and every where else, kept in a cool air. So I shall take her into the next room, and do you put out the fire. I do not believe it is possible for her to sneeze, or to take cold there with this heat upon her; and if the hands grow cold sooner than the face and body, we will put on her gloves.*

The child grew comfortably cool in a quarter of an hour, and slept with perfect composure during the remainder of the night. Nor was there afterwards the smallest reluctance on the part of the nurse to take her out of the cradle or out of bed, even in the midst of a sound sleep, horrid as this piece of cruelty would have seemed before, when her skin felt particularly hot and dry. And she entered cordially into the plan of rubbing the hands and feet, when cold,—as also, of giving food, either below or above blood-heat, according to the state of the skin at the time.

Children, who have either the abovementioned defect of the mucous membrane or general constitutional debility, are particularly subject to indigestion—that common, but by no means necessary plague of our earliest years. I shall, therefore, step aside for a few moments from the subject of temperature, in order to point out a method of obviating one principal cause of that distress, the heart-piercing expression of which is made by Virgil,* with so much truth of feeling, to strike the ears of Æneas, when he arrives on the verge of the precincts of Pluto. The cause I mean is the perversity of the women usually about children. For it will be found still more easy to convince them out of their prejudices in the case of diet, than in that of temperature. For here the facts can be brought much more directly home to their senses. I have heard a variety of mothers complain that sugar, bread, biscuit and cakes disagreed in the most evident manner, and yet, that it was impossible, by any injunctions, to prevent the one from being made a part of the food, and the other from being given to stop the hiccup, or to produce a sensation that should suspend crying for the moment. After a tedious contest, these mothers have declared themselves obliged to abandon the point altogether. Now it is well known, that perpetually recurring complaints of the stomach and bowels arise from mere *sourness*, and the parties, by whose mistaken kindness or by whose delicacy of ear they are occasioned, are perfectly informed so far. It

---

* Vagitus infantum in limine primo;

remains only to carry their knowledge a step farther. Respecting the juice of the sugar-cane, it is a very striking particular, that in the West-Indies the purest sort will scarce keep a quarter of an hour in the receiver, without turning sour. This can only be told.---The acescent nature of bread, of sugar, and of the various compositions into which flour and sugar enter, may be shewn. For this purpose, it is only necessary that a solution of sugar in water should be made into vinegar. In like manner, bread and sweet cake, steeped in water, should be placed in a heat nearly equal to that of the human body, and the servant be put to taste the infusion when it has become acid. By an address suited to the object in view, there will surely be small difficulty in giving these simple experiments all the effect that can be desired.

I shall very contentedly allow the childless wit to laugh at me for the whimsical idea of tutoring nurse-maids in chemistry. I have a balm at hand for any wound the shafts of of ridicule may inflict. Considerate parents will avail themselves of so practicable an expedient; and many little sufferers will escape the consequences of an improper regimen. And these are, probably, far more serious even in respect to the future than the present. For it clearly results from a contemplation of the manner, in which human feelings and ideas gain their connection, that frequent discomposure of the stomach in the morning of life may be instrumental in overcasting its meridian and its close with a cloud of misery, such as neither skill nor fortune can disperse.

Among some of the common people, who deem themselves adepts in the art of managing children, it is a prejudice that those, fed by hand, and probably that others, will never thrive without more or less of solid food; and if the mistake be not corrected, it will induce them to persevere in administering acescent substances to the infinite torment of many nurselings. But the example of young animals, when distinctly pointed out, has proved sufficient to remove this apprehension. And the understanding being once satisfied, there is little hazard, in persons of common intelligence and humanity, lest the will should continue intractable. Need I add, that such only as are possessed of these qualities in a reasonable degree, should be entrusted with the care of beings, so extremely liable to receive irreparable injury from injudicious treatment?

Many writers on education have laid it

to the heart of mothers, not to commit to hirelings the task of *nurse.* Such writers have not, in general, been members of the medical profession. By these, the recommendation will be found a great deal too indiscriminate. And professional men have the best opportunity of knowing the consequences in different instances, and the fairest pretensions to decide, when an exception ought to take place. This interesting office should, undoubtedly, be always undertaken by the parent herself, when it is not likely to prejudice either party. But I am afraid that the reasons for *occasionally* declining it have become continually more and more cogent from a pretty uniform increase of debility among females. And I suspect them for some generations back, to have been somewhat more cogent than has usually been conceived.

The milk, it has been alleged, *disagrees with the child's stomach.*—No doubt it frequently does. But could not the reverse have often been as truely alledged ? Does the stomach never disagree with the milk ? That is to say, would not the disparaged milk afford perfectly wholesome aliment, where the digestive organs should perform their functions healthily ? In such case, the child, undoubtedly, should be weaned *for its own sake*, if not for that of its mother. And a plan of diet should be pursued under the direction of a medical practitioner, not less conversant in chemistry than in the vulgar routine of our art. The former is a qualification, which nothing can replace. The want of it sometimes, perhaps, levels the practitioners, most busy about ladies in the straw and the inhabitants of the nursery, with the most prejudiced of those women, whom we see busying themselves, with equal confidence in their own talents, in the same department. I shall only add, that preparations from animal substances and cow's milk, with the occasional addition of alkali, saturated with carbonic acid, and by no means always of magnesia or chalk, will form the basis of a regimen, proper under such circumstances. However inexpedient it may be to stimulate the young stomach in its robust state with the extract of flesh, this may be almost the only admissible aliment in a given condition of the digestive organs. Woe be to the attendants upon childhood, when they will not move in correspondence with the aberrations of nature!

But, in truth, these and other cases, where all the assistance required could be easily rendered, are far from being submitted to the judgment of any medical practitioner, whether competent or the reverse. The subjects of them perish from an imagined incorrigibility of their constitution, or else after struggling out of a first into a second *slough of disease,* they are lost in a third, without the solace of a single smooth or flowery interval in the whole journey of life.

These calamities would occur more rarely but for the two following causes. Parents, in the first place, remain in a state of torpid inattention to temperature, and in some measure to diet, as if the operation of temperature and diet were inconsiderable, because it is not always striking to the inobservant !—Nor is it remembered what the chances are against the persons ordinarily employed, managing in the best manner possible, in these essential respects.—It will not, I trust, be said that the very prolongation of existence does not at first demand the best possible management; whatever difficulty there be in extending the views to the sequel. I do not, as I have said, think lightly of the duty incumbent upon one parent to give suck, where nothing forbids. But it seems to me a duty, infinitely more incumbent upon both, to make themselves acquainted with the incessant and mighty influence of the vicissitudes of the surrounding medium upon the human frame, that they may guard their tender progeny against the ills, which some of these vicissitudes are so apt to bring upon it, when unsheltered.

On giving this whole subject due consideration, few things will, perhaps, appear less comprehensible than the practice of *nursing out of the house*, or confiding solely in the uninstructed and the undiscerning. I can scarce, indeed, conceive that this practice has ever been followed by any parent, who had once sat seriously down to consider, what are the requisites most essential to happiness in every stage of mortal life without distinction. If we abstract from certain uncommon instances of adoption, what shall we repute the ordinary condition of foster parents to be? What is the condition of those, who are sent abroad to be fostered? Is not the latter evidently such, that if Quintilian properly advise to *commit a beginner to the most skilful to initiate into learning*, the harder task of initiating into life ought not to be committed to the most unskilful?

As to foster-parents, it would be wasting words to describe their deficiency in means, information, expedients, vigilance and penetration. With scarce the ability to perform the easiest offices, relative to human beings, they are charged with the most difficult. Whatever be the number of ill-starred infants, that may have passed through their hands, it is impossible, without more than a miracle, that their qualifications should have received any material increase. It is the misfortune of mankind, that experience itself can instruct none but those who are prepared for its precepts by art, or by some very unusual bounty of nature. Upon the latter chance, nobody in his senses will reckon, and the former sort of preparation is always wanting to the persons in question. What is the common history of poor families? Why, that a large proportion of the members perish or become diseased, in the rearing.—To what are we to ascribe this? Surely, not to penury alone. Ignorance claims its share.—Nor must those faithful attendants upon ignorance be excluded—heedlessness on the present occasion, and rashness on the next. Does the history of those opulent families, whose heads divest themselves of the superintendance of the members during their most precarious years, differ very widely from that of the poor? I doubt it extremely. Were an inquisition taken, their bill of mortality and their bill of health would, I am persuaded, agree in a similar result. And whenever a business, capable of being conducted, in a great degree, according to ascertained principles, shall cease to be carried on at random, fathers and mothers, who have it in their power to provide for the proper accommodation of children, will less frequently have to lament either their loss, or the irreparable personal misfortunes of their infancy.

GENERAL PRINCIPLES FOR A MORE ADVANCED AGE.

A nursery upon the principle of regulated warmth might, in its fitting up, afford much scope for usefully minute ingenuity. In such a nursery, children, born in winter, would suffer nothing even during the most severe weather. Thus, one standing and frequently fatal evil would be banished from society. But in cold seasons the removal of children into the open world should not be precipitated. It is better to be content with habituating them to those variations of temperature, which different rooms in the same house will supply. They should, at first, not be carried into the cooler room but when rather warm, and then for a short time. In conveying them through stair-cases and wherever they are likely to be met by streams of air, the whole head should be covered with warm calico, in which it is easy to leave air enough to serve, till their arrival in the apartment, to which they are destined. They should be rendered hardy by exposure to a cold calm atmosphere for a short time at first; and at no time, till they are chilled. They should never encounter rude blasts till they can move briskly enough to produce warmth from within; and then only at intervals, successively prolonged. When they return home cold, the precautions to be mentioned below under the head *catarrh* should be employed. I speak all along as if the objects, whose security I have in view, were by nature something of invalids. But the parents, who feel a just confidence in the robustness of their offspring, will do no injury by proceeding with a measured pace. I intend that all shall in time become hardy. But we have access to no river Styx, in which one immersion shall render us invulnerable by the elements.

One fatal error seems to be entertained by parents and by many founders and conductors of seminaries of learning. The correction of this error will afford a rule, applicable to the healthy of all ages and to those who have not deviated from health. It was observed, that many persons, accustomed to be buffetted by storms, so much exceed the inactive fire-side *tenderling* in spirit, vigour and health, as to have a fair claim to be deemed a superior variety of the human species. The effect of certain cold mediums in giving hardness to some inanimate bodies, and in bracing others, or bringing their parts closer together, was joined to the preceding observation. It was also evident to feeling, that the stoutest men have their muscles most braced or most capable of resisting compression.

These several particulars being taken together into account, no wonder it should have been concluded, that to endow a young person with the most desirable qualities of body, there was nothing to be done but to expose him sufficiently to cold. This mistaken system was, indeed, infinitely more prevalent some years ago than it is at present. And many constitutions must have been suddenly destroyed in consequence. Others, originally unable to stand rough treatment, must have received an instantaneous shock, which could never be recovered, to whatever length life might be afterwards protracted. This seems to have been the case with Rousseau, whose compliance with the custom, once fashionable, of drinking freely of cold water in the morning to benefit the health, manifestly injured his nervous system to a degree, which I am tempted to suppose had its share in producing his subsequent unhappy state of mind.*

The more any one can resist internal and external cold, the more hardy may he justly be reputed. Nor is there any surer criterion of hardiness. But whatever perfection the constitution may, in this respect, have attained, *long-continued and repeated chills will, in the*

---

* After giving an account of a course, during which he drank two bottles of cold water fasting, he relates the particulars of a seizure, which must be deemed a partial paralytic stroke, since it instantly and for ever impaired his hearing. —" Dans ce meme temps, il m' arriva un accident aussi singulier par lui-même que par ses suites, qui ne finiront qu' avec moi. Un matin que je n'etois pas plus mal qu' a l'ordinaire ----je sentis dans tout mon corps une revolution subite et presque inconcevable --- Mes arteres se mirent a battre d'une si grande force, que non seulement je sentois leur battement, mais que je l'entendois meme. Un grand bruit d'oreilles se joignit a cela; et ce bruit etoit triple ou plutot quadruple --- Ce bruit interne etoit si grand qu' il m'ota la finesse d'ouie que j'avois auparavant, et me rendoit non tout-a-fait sourd, mais dur d'oreille, comme je le suis depuis ce temps-la."

first instance enfeeble, and in the second, bring on a susceptibility to the operation of the powers, that superinduce violent diseases. The true principle, therefore, is gradually to inure the habit to cold; then it may continually be enabled to bear a longer and severer application of it; but as soon as a chill comes on, be that whenever it may, to stop the process, and to take proper measures for returning to the natural condition without injury, Of these measures I shall say something below.

A middle course must therefore be pursued. It will be sufficiently understood from the facts, detailed in the preceding essays, that between the cold and the sedentary confinement at boarding-schools, girls must almost universally be chilled into debility. But were the school-rooms kept of a comfortable temperature, though much misery would be prevented, the end would not be attained. This class of young persons would not be rendered efficient, happy, and healthy. In their education every circumstance tends to unfit them for such a climate as ours. Such are the facts,, and such is the inevitable inference. I am aware that many parents will fully admit both, but that instead of attempting the emancipation of their daughters from so miserable a fate, they will content themselves with saying—*It is but too true. But all this signifies nothing. The thing cannot be helped. Girls must and will be sent to these places. What else can we do with them? We are obliged to employ all our care in scraping a little something together to put them out in the world.*

The celebrated traveller, NIEHBUHR, relates what he terms *a specimen of the firmness of the Arabs in misfortune (Reisebeschreiburg,* 4to. I. 354.) "One day," he says, during his residence at *Beit el Fakih,* "a "house an the south side of the town took "fire; and the wind blowing fresh from the "S. W. the greater part of *Beit el Fakih* "became a prey to the flames. Every thing "had been so parched by the extreme heat, "that the houses, which were covered with "grass, and surrounded by dead fences, "ran into a blaze the moment the fire "reached them. The Arabs were perfectly "cool upon the occasion. Not a cry, not "a scream to be heard in the streets. But "when we condoled with them on their "mishap—*It is the will of God,* was the "reply. We lodged in a house of masonry, "on that side where the fire never came. "We mounted the roof, and observed almost "all the roofs of the houses of the same con-

"struction as ours, covered with people, "composedly observing the conflagration. "A Fakih, or poor scholar, who used to pay "us frequent visits came this day, after hav- "ing secured his scanty furniture, and with "the greatest sang-froid imaginable, pointed "out his own house, when it began to be on "fire.—What an Arab loses by such a cala- "mity," continues Mr. Niehbuhr, "is not, "it is true, of equal value with what an "European loses.—However, though it be "but a hut, this must always be regarded "as a considerable loss to a poor man." All this is remarkable enough. But ten thousand such anecdotes would never prove the Orientals to exceed us Europeans in fortitude or in resignation, call it which you please—unless indeed any one should contend that a cabin of Yemen is more worth than a virgin of Great Britain. We can look on, while our daughters decline as coolly as the Arab can upon his perishing habitation. Our—*it can't be helped*—has not so pious a sound as the Arab exclamation; but is not the sense the same? And does not the conduct of the respective parties shew that there are as true fatalists under the banner of the *cross* as that of the *crescent*.—Nor indeed do I well know what can be said on the other side of the question: except that the wealth which we have been all the time labouring to accumulate for our childrens' sake, may still, if they die, afford us the consolation of erecting over their remains a monument, which every church-goer in the parish shall admire. Or if the machine holds together beyond all calculation, by help of this same wealth we can conceal its rickety state beneath trappings, which a whole metropolis shall eye with envy. And a troop of menials from Parnassus will always think themselves honoured alike in being put to bear cypress at the funeral, or myrtle at the wedding ceremony.

Besides the regular severity of temperature, which slowly undermines the constitution in so many schools, boys are subject to intense and suddenly pernicious chills. In the preceding essay, I have related an instance, where certain effects of cold were brought on during the milder months of the passing winter. This patient informed me that immediately after the rain broke in, as there related, several boys were carried home ill. The accident brought that suddenly to bear, which in the usual course proceeds gradually.

But a prime incidental cause of disease and

debility, I take to be too long continued swimming or bathing. In many, indeed in most school-boys, we may be sure that there may be thus produced any sort of slow complaint, according to the existing pre-disposition, or any sort of inflammation, according as imprudences shall be immediately afterwards committed. No human frame can be tempered so as to resist the effect of water, applied for a certain period, and at a certain temperature. The bosom of the earth hides multitudes, who have been suddenly cut off in their pride of strength by its operation, or whose hairs have been slowly brought down to the grave, before they were grey. Greater multitudes have been reduced by the very same agent to crawl upon its surface, or lie continually groaning under an inability to crawl. This fate befals the mountaineer of the Highlands and of the Alps. It equally befals races of men, in comparison with whom the Highlander and the Swiss would seem to deserve to pass for feeble and uninured to hardships *. What then is not to be apprehended

* See what Ehlich (under the name of Dr. J. Reineggs), says of the Lesgae and Duschi tribes inhabiting the very summit of Caucasus.

" From natural appearances one should imagine that there " were no inhabitants. Astonishment is produced at that from the tenderly reared offspring of parents not particularly robust, if they be allowed to dabble in water at discretion? I do not see how we shall avoid the opposite inconveniencies of renouncing wholesome practices altogether, or of carrying them to a hurtful

---

" temerity of rude art, by which the edges of immeasurable " gulphs are joined by bridges; dangerous but passable " roads conducted along precipices, and streams forced into " a course not their own. The exterior of the houses produces horror. They seem only reclined against the rock. " --- Both tribes sow their fields in peace. But seldom does " a harvest pass without hostilities. --- The stronger obtains " the advantage, and reaps where he has not sowed.—Therefore each guards his field at the hazard of life. ---- The " care of the Duschi for their flocks is extraordinary, and " their vigilance indefatigable.—A young shepherd was feeding his sheep, when the approach of evening, and a violent and continued fall of rain, obliged him to drive homewards. But he met with a stream, which he was obliged " to pass, so swoln and impetuous, that his sheep could " not swim it. However, the dread of becoming the nocturnal booty of the Lesgae, or of wolves, urged him to " seek a place of security. He did not therefore refuse the " toil of carrying two hundred and thirty sheep across this " stream. Relieved from apprehension, and content with " the issue of his efforts, he did not seem to trouble himself " about the incurable cramp, which the cold of the water " had brought upon his feet. I saw him in a state in which " his whole lower extremities were immoveable; and now, " at the end of twenty years, he serves as a pattern for young " shepherds."

excess, without a superintendant who can judge of the effect in every individual instance, regulate exposure according to the resisting power of each constitution, and put an immediate stop to any threatening mischief, brought on by youthful heedlessness or impetuosity.

It will be obvious, if there be any truth in the preceding observations, that no person, already chilled, is fit to encounter a more chilling medium. What can be safely done in a heated state, may call for a little farther explanation. It is clear, from a hundred instances that, when persons are heated, the sudden use of cold water, whether external or internal, to a certain extent, will destroy life and health. In the case of dry heat, however, it appears that * water may be drank and applied as cold, as nature furnishes it, safely and with the advantage of refreshment. But if the application be external, it should be but for a short time. And the quantity drank should be sparing: and the draught should be discontinued the instant there is the least sense of chilliness in the stomach, or rather before any such sense supervenes. This caution is the more neces-

* See Dr. Currie on Fever. Liverpool, 1798.

sary to those who have lived in habits of indulgence. Such persons chill faster than others, and they suffer more from frigorific or cold-producing agents, too freely applied. Our ladies should, for these reasons, be particularly on their guard. Experience teaches them how little they can venture to take liberties with themselves at *certain periods.* But in early life many become victims to this species of indiscretion. I have at present one example under my daily inspection. And a lady of observation has favoured me with the following account of an occurrence at school, from which she has been a sufferer these sixteen years. "A. B. having excessively heated herself at play with a girl of nearly the same age, they heard the dinner bell ring. Being afraid to make their appearance in that situation, [*mark the delicate discipline of these seminaries*], they both dipped their faces into a tub of cold water. They both felt extremely warm at the moment, had not ceased from play, and were perspiring. A. B. was then for the first time in a state that made more than usual care requisite. The almost *immediate* consequence was a violent inflammation in her eyes, and a fever which confined her to her bed for ten days. This was followed by a cutaneous eruption on her face, which has remained ever since, and has never in the slightest degree appeared in any part of the body. The periodical indisposition did not take place again for three months."

"The other young lady sustained no apparent injury at the time. She was soon after separated from A. B. who did not hear of her again for two years, when she was sent to the Hotwells for a decline. She however grew better, and had one child, but eventually died of consumption. Whether there is any reason for supposing the imprudent act above-mentioned to have had any share in bringing it on, A. B. has no means of determining." Unless indeed some indisposition speedily took place, connecting the first attack of pulmonary complaint with the act of imprudence, I should not suppose that these stood in any relation to each other. There needs nothing out of the common course of things to throw a young woman into consumption; and without the help of forced interpretations, there are warnings enough against the free use of cold water, when people are heated. In a small medical tract for the poor, (published ten years ago, and of which six thousand copies are dispersed among the public), I quoted from Dr. Franklin an authentic instance, in which four young men, who having worked at harvest in the heat of the day, plunged into a spring of cold water. Two died on the spot, a third the next morning, and the fourth with great difficulty recovered. The effect of the Cydnus upon Alexander the Great has been long celebrated as an example of the danger of sudden cold in an overheated state. An ingenious and experienced writer has endeavoured to prove, that Alexander was exhausted, and perhaps cool from perspiration. But it is too much to flatter oneself that the closest investigation of the case of Alexander can assist in removing any of the difficulties that attach to the particular point of perspiration. Can it be thought likely that any, even of the immediate attendants on the person of the Macedonian monarch, could have given an account, in the least satisfactory to the modern physiologist, of the state in which he plunged into the mountain stream? How much less then can the relation of historians be assumed as the basis of conclusive reasoning? Such speculations, however, must be allowed one important use. In the event of future unpleasant consequences, they will induce an attention to catch and retain the previous circumstances, before they vanish from the memory.

As an essential part of a system of security, I recommend it to mothers and governesses not to make such a crime of a girl overheating herself. She will not then take any hazardous step to conceal her guilt.—During perspiration, when there is as strong a sensation of warmth as is compatible with the perspiring state, and where the system has not been exhausted by fatigue, we may, I think, venture to pronounce it dangerous to apply water of the temperature which springs afford in our latitudes, either to the whole surface of the body, or to a small part. With regard to the whole body, the facts preserved by Dr. Franklin furnish as strong evidence as can be desired.

As to a small surface, the following case, into which I have had the fullest opportunity of examining, appears nearly decisive.—Feb. 4th, 1802, M. A. D. aged 26, applied to me on account of an eruption on her face of long standing. On questioning her as to its origin, she ascribed it to immersion of that part in cold water, when she was hot. Her answers to particular queries, which were extremely distinct, yielded the following information. She was at school and perfectly well. She had never been troubled with any species of eruption, or complaint of her skin. About midsummer she was thrown into a perspiration by play, and felt very hot, but not at all fatigued. She immediately dipped her face into a bucket of water, just fresh drawn from the spring. She is certain she had no general or partial chill after; on the contrary, her face in a very few minutes felt excessively flushed. Her skin from that time forwards grew more and more coarse, and in the space of a few months an eruption of pimples took place; and they have continued ever since, that is, for about ten years. The patient has also ever since been subject to the most violent flushings after dinner. The skin of the face is evidently altered in its whole texture.

But although in the *glow of perspiration* it seems dangerous to take general or partial liberties with cold spring water, Dr. Franklin is of opinion that "during the great "heats of summer, there is *no danger* in "bathing in rivers, which have been tho-"roughly warmed by the sun, however hot "we may be." And all subsequent experience seems to vouch for the opinion of this great observer. But the continuance in the river should be short. Only when *the chill of perspiration* has come on, then undoubtedly every medium, capable of depriving the body of warmth, and much more if it conduct heat rapidly, menaces danger. The moist garments should be laid aside, and the body rubbed before a fire. This seems the only safe plan.

The same distinctions ought to be recommended to notice, as to drinks. In the warm state, a moderate quantity of liquid, not less than thirty or forty degrees above the freezing point, may be admitted. But cold water, and still more ices, should be strictly avoided. The chill requires liquids, (as wine and water) above the temperature of the human body, and indeed as warm as can be conveniently swallowed. In case of a chilly seizure from the unwary use and application of cold water, very hot liquids, taken till the contrary sensation arises, would probably prevent all injurious consequences. And at those orgies of courtly gluttony, where we hear of the serving up of ices in such profusion, and where, every now and then, it is the fate of an ambassador to eat himself to death, almost on the spot, I am of opinion that there should always be at hand an urn, full of warm cordial beverage. By a draught of this, at the decisive moment, such serious diplomatic accidents would, I will venture to say, be prevented even more frequently than persons are recovered from drowning by the measures of the *humane society.* Nor am I acquainted with any contrast more miserable than that which is formed by the ingenuity of the professors of the culinary art in providing poisons, and their stupid neglect of antidotes. And this where the existence of grave personages, charged with the interest of nations, is at stake! Nor would it be amiss, if sovereigns, in selecting their representatives, were a little mindful of the power of the stomach, as well as of the head!

It may somewhat assist the judgment, if I observe that the effect of any cooling application should be considered as acting through successive moments, or by successive quantities. If a person in a heated state drink half a pint of cool liquid, that may not sensibly reduce him below the natural healthy state. But if he pour down a double quantity at once, the last half pint may be regarded as operating upon the system, reduced by the first, and sinking it into a dangerous chilliness. So immersion for a moment in a sunny river may strengthen and refresh, as many pedestrian travellers have experienced; whereas delay in the water would be attended with great hazard, on account of the continued operation of a heat-abstracting medium upon a system, sufficiently reduced in its temperature and in its actions, by the first plunge.

Every precaution, deducible from our present imperfect and inaccurate knowledge, should be observed with double fidelity by females, and by the weakly in general. Perspiration is a process in itself cooling. In the dry heat preceding perspiration, the thermometer in the mouth or at the arm-pit, will stand higher than after the skin has become moist; but in vigorous habits, particularly under the continuance of exertion, the infinity of small but brisk living movements, supply the heat nearly as fast as it is taken off by the oozing moisture; and a chill will not so soon be induced, as where the corresponding movements are fainter.

There remains another most material point, not sufficiently examined by the physiologist. It is one which, to avoid perplexing the common reader, I should at present pass entirely over, did I not apprehend that he might imagine the subject to be more simple than it really is, and reasoning from imperfect data, come to suffer for the very justness of his logic. In the instance of cold water applied to the hot perspiring face, as I have above related it, the injury to the skin certainly did not arise from any chill. The intense glow, which so immediately succeeded, sufficiently shews this. In fact, the consequence, whether direct or remote, seems to have been an *inflammatory* affection of the skin. A transitory inflammation is always felt upon coming into warm or calm air, after the face has been exposed to a cold wind, and it commonly supervenes in a still slighter degree over the whole body, after the use of the cold bath. It is probable that the continued greater action of the rest of the exterior, and of all the interior, parts of the body, contributed to the inflammation that was here produced. I at least doubt extremely if any inflammation would have taken place, had the whole body been dipped in the fresh drawn water. And I feel almost assured, that so obstinate a complaint of the face would never have ensued, supposing it, after total immersion or affusion, been in any shape inflammatory. It is important to put the reader in mind, that such accidents manifest to the eye, in what manner cold draughts or ices in certain circumstances will produce a fatal inflammation of the stomach. And in the same manner it will be understood how pleurisies and catarrhs (colds) are often caught by breathing cold air in the situation in which M. A. D. found herself when she dipped her face; the air producing upon the mucous membrane that lines the air-pipes, the same chilling effect as the water did on her face, but more slowly.

The *rate* at which the cooling process takes place upon these occasions, probably occasions a material difference in the result. It is most likely that the greater the rapidity the more the danger, whether it be inflammation, or some other bad effect that follows. I do not here touch upon the disasters which are experienced, when persons under the influence of mercury expose themselves incautiously. I suppose that the system, at the time the imprudence has been committed, was in a more heated and excited condition than during health. And in many circumstances where disease is produced in a healthy, but heated state, I take it for granted that it depends upon the kind and degree of excitation, and would not have occurred, though the system had been even under par as to temperature and action, when it was subjected to the refrigerating power.

There are conditions of disease, which confer prodigious power of resisting cold. I do not speak of insanity, which seems scarce to belong to the consideration of the action of cold mediums in heated conditions of the system, unless indeed the opposite testimonies we have from observers of unquestionable faith and accuracy may be reconciled by supposing, where maniacs bore intense cold with impunity, that they were in a hot, febrile state; and when they appeared as sensible to a low temperature as other mortals, that the skin was cool. Whence it would result that the power of resisting cold, which has been so frequently thought to constitute a peculiarity in the deranged, does not proceed from the derangement, but from fever, accidentally conjoined with it. For it is febrile action, in some of its periods at least, that hardens the frame so remarkably against refrigerating applications. On a journey into the Highlands in 1787, the author happened to pass two nights at a small inn by the river Tummel. A low fever raged throughout the country; and among several persons ill in the house, a maid servant was distinguished by the violence of her symptoms. It was the sixth day of the disease. The author, after examining her situation in the morning, ascended Schehallien, and on his return was surprized at hearing that this girl was nearly well, all but weakness. In her delirium she had uttered a desire for cold water, which being withheld, she had crawled during the absence of attendants to the brink of the river, from which she immediately perceived a herd of cattle with the drovers at some distance, on their way to cross the bridge. The sight induced her to make for the water, in hopes of concealing her nakedness. She waded up to her middle, and leaned against a fragment of rock. Nor was it till one of the drovers turned his horse towards the inn, that she was discovered in this position; and it was believed that she had occupied it not less than five minutes. Her delirium was gone; and the symptoms of fever had quitted her. Dr. Baynard, among other equally singular and apparently authentic examples, relates of a young man ill of the small pox that he escaped into a pond, and continued in it and in his wet shirt, while an apothecary was fetched from three or four miles' distance. Two physicians to the army, Dr. Sinnot and Dr. Sutton, have attested that during the rapid retreat of the British troops from the banks of the Waal to Embden, in the dreadful winter of 1793 and 1794, the fever-patients, though carried in open waggons, universally recovered. And by the accurate researches of Dr. Currie, it has been established that in the dry heat of fever, the affusion of cold water has a most salutary effect; and that, on repetition, it seldom fails to subdue the disease, if employed during the first stage.

I introduce these facts to give weight to the following rule, respecting the cold bath There are very few circumstances, in which cold bathing will not be safest and most advantageous, if the bather, immediately before immersion, exerts himself enough to produce a gentle sense of warmth. There are no circumstances where an unprofessional person should omit this precaution, without the express injunction of a practitioner of medicine. Weakly people every day destroy or injure themselves by going into cold water, when they are not sufficiently warm. The robust themselves probably bring down the tone of the system by the same inadvertency. For this reason, I can see no sense in the common notion, that it is best to bathe with an empty stomach. I would not recommend the cold bath to a person full gorged. But it is desirable that the stomach should be in a state of gentle activity, as well as the external surface of the body.

It has appeared to me, in one or two instances of grown people, that a diarrhœa was induced by the cold bath, in consequence of the stomach being empty and inert at the time of its use. I am certain that in feeble infants the same effect has followed, when the water has been even up to seventy degrees. When this happens, the temperature should be raised, and a small quantity of warm and spiced food should be given shortly before the infant is dipped. The sense of greater alertness and greater power to resist both cold and heat, are the surest signs that the water has been at a proper temperature, and that the whole process agrees. In infancy there is often, in a very short time, an evident increase of ruddiness, and a heightened complexion indicates, with considerable certainty, increased vigour of the whole system, as could be easily shewn by entering into physiological discussion.

The use of cold air is more ready, safe, and agreeable, than that of cold water. It is to the mere abstraction of heat, and not to any recondite quality that we owe, nine times in ten, the refreshment we experience on going abroad into the open air, when its temperature does not verge towards either extreme. It is from the same source that the few, who spend much time in the open air, without being chilled, or overworked and underfed, derive that freshness of feeling by which they are distinguished from sedentary mortals.

Were it not for field sports, and for some morning use of the limbs during a town residence, our gentlemen would be nearly as puny, and as liable to diseases of debility, as our ladies. The manner in which most advantage may be drawn from cool air, will be understood from the first part of the following section. I leave it at present to writers for the profession, to consider whether cold air would not answer in those advanced stages of fever, where affusion with water may be unsafe or ineffectual.

HARDINESS.

Hardiness, the most enviable of all the attributes of animal nature, can neither be acquired nor recovered but on certain terms. There are indulgences, and unhappily of the most common kind, with which it is utterly inconsistent. The gratifications of enfeebling luxury may possibly be preferred by some, from a mistaken idea that they can find new means of rousing the blunted senses after the former stimuli have lost their power, and that they can always shelter themselves against those natural agents, that seldom fail to bring disease upon the tender. If, indeed, they make this choice on deliberate comparison, all argument is at an end. It is however but fair to apprize them at what cost they purchase their gratifications. Let them only not look to the enjoyment of incompatibilities. A man may as well expect to break both his legs and be able to run a race, as to weaken the tone of his frame, and be able to encounter the rigour of the elements with impunity, or to be subject to the ordinary exciting causes of sensation without pain. After this explanation, let those who will not or cannot renounce practices demonstrably ruinous, lay aside the remainder of this essay. Others may avail themselves of the succeeding pages, to preserve what of their fund of vigour they have left, and to enhance it, if unimpaired.

One of the causes that contributes most materially towards the reduction of persons, living at their ease, below the standard of hardiness, is the dependance they place upon external warmth for preserving a comfortable state of sensation. Thousands of experiments made for the express purpose, and universal experience with regard to the human race and to domestic animals, agree in one point. They shew that *continued warmth renders the living system less capable of being called into strong, healthy, or pleasurable action.* Every muscle, steeped in a heated medium, loses of its contractility. Every nerve grows languid; and, when excited, acquires a disposition to throw the moving fibres, with which it is connected, into starts, shocks, twitchings, and convulsive movements of whatever denomination.

Those therefore who have not accustomed themselves to sit in warm rooms, should most carefully avoid making this unwholesome indulgence necessary to themselves. Others should, by degrees endeavour to render their comfort independent of it. For this purpose they should retreat out of their *stoves* at short and frequent intervals. They may easily, by a little attention, discover times, when the open air will be more grateful to the sensations, as well as more salutary to the habit. Those who complain of being most chilly in common, have their periods of dry heat: and then they will always find a cooler atmosphere pleasant and refreshing. One such period will often occur after dinner. And on that account, it were to be wished that we took our principal meal at an earlier hour; for we should be able afterwards to enjoy the benefit of the air, during a much longer portion of the year, by day-light. It would be better still, if breakfast were a more nutritious meal, and in the case of the puny, if it consisted of some preparation of milk or of animal substance, without tea or coffee. A short walk in the air, or pacing backwards and forwards in a room without a fire, will take off the unwholesome and strength-consuming after-dinner flush. But instead of availing themselves of the mildly cooling quality of the air, our genteel multitudes do every thing they can to add to the over-stimulating operation of dinner. After taking strong wine with their food, they sit in rooms rendered progressively warmer all afternoon by the presence of company, by the increase of fire, and, for more than half the year, by early closing the shutters and letting down the window-curtains. At a short interval, tea or coffee succeed—liquors, stimulating both by their inherent qualities, and by virtue of the temperature, at which they are often drank. And that nothing may be wanting to their pernicious effect, they are frequently taken in the very stew and squeeze of a fashionable mob. The season for sleep succeeds; and to crown the adventures of the evening, the bed-room is fastened close and made stifling by a fire. The robust are not so easily thrown out of the medium temperate state. The feeble, as I have said, are over-heated or chilled with nearly equal ease. And these, as they usually manage, rarely pass a few hours of sleep without feverishness and uneasy dreams; both which contribute to their finding themselves by far more spent and spiritless in the morning, than after their evening fit of forced excitement. Perhaps they drink tea before rising, and indulge in a morning, noon, or afternoon nap. This nap weakens much more than the greasest muscular exertion they would be capable of supporting for an equal time. For it is almost invariably disturbed, and attended by dry heat of the skin. The reason is plain. Almost every person is rendered more susceptible to the heat of the bed by a night's sleep. Much more those who are so liable to be heated by slighter causes. And the tea acts, if not so powerfully as to be an absolute sudorific, yet as a stimulant of the whole surface. The remedy for this is obvious at least, if in consideration of the power of habit and example, it cannot be said to be easy. It consists in deserting the heated rooms at least for *part* of the time they have been usually occupied—in abstaining from the strong wines of our tables—in forbearing to drink other hot and heating liquors—in renouncing the crowded resorts of fashion, and in extinguishing bed-room fires, except during severe weather, when the temperature should not be raised beyond fifty degrees *. When young people appear to be on the decline, and especially when they complain of unrefreshing sleep, they should be examined in the night, and be awakened without compunction, if found too warm. The bed-clothes should then be thrown off, or if the dry heat be considerable, it will be best to walk up and down the room in a

---

* A writer on regimen must frequently find it necessary to pause and review what he advises, lest any thing, too generally put, should mislead simplicity or arm malignity with a pretext. Inadvertent parents lodge in every street, and I am too much afraid I have once or twice in my life fallen upon the track of such as have the art of harassing their children to death, without subjecting themselves to legal animadversion, and indeed carrying towards the world a face of the utmost tenderness. To caution the one, and to prevent the other from making sinister use of my authority, I here expressly subjoin that the direction for extinguishing bed-room fires is not to be extended to the time when girls are dressing for public places in winter. I have known dangerous complaints take place apparently from this cause, in families where a parental prohibition issued against a fire on such occasions, though the voice and feelings of a daughter joined to call for it. When one considers what a frequent business it is at some places, as at Bath or at Clifton, to prepare for the ball, it is not too much to assert that one season's dressing in the cold, may undermine the firmest female constitution.

dress, which should be contrived for guarding the hands and feet from chill, while it suffered the trunk of the body to be freely ventilated. Those who have courage to persevere in these measures for three months, will often be rewarded with a return of some portion of their lost vitality. They will shiver less at one time, and flush less at another. They will be recruited, and not oppressed, by their food. Their ears will be less subject to ringing; they will start less at sudden noises; the general tenor of their feelings will cease to be comfortless; they will not catch cold from every shower that moistens the *surface* of their garments, and every breath of air that strikes a part usually covered; or from every slight insinuating damp that comes in contact with the feet. From being less debilitated by the perpetual stimulus of warm air, the *mucous* membrane that lines the nostrils and air-pipes, will be less liable to be affected by slight vicissitudes of temperature in the atmosphere: and we shall hear less of catarrhs.

There is a class of persons who lead a sedentary and a solitary life, to whom I particularly recommend these considerations. They have frequently a literary turn, and feel very little anxiety to acommodate their habits and appearance to the taste of fashionable ignorance and folly. Such persons may greatly benefit by some of the preceding directions; and that with scarce even a momentary sacrifice of feeling. They have only, for their hours of study, to contrive a dress which shall keep the extremities sufficiently warm in a temperature of about fifty degrees. The eyes of people, who read and write a great deal, not only come to feel painfully, but vision is at length impaired. And this happens not so much from the mere exercise of the organ, as from the degree of heat in which it is exercised. If a number of persons, who complain of that burning and stiffness of the eyes, to which the studious are so frequently subject, will make the experiment and compare notes, a very few evenings will satisfy them as to the result. The nervous head-ache, to which the same persons are sometimes liable, may possibly be prevented in the same way.

Exercise is another expedient for getting rid of the necessity of living in the unwholesome climate of a hot and close apartment. The feeblest and most delicate might devise some employment, which should enable them to support existence, for a short time, in a temperature below that to which they have been accustomed. These intervals may be successively prolonged: for the beginning only of the road back to health, as in other matters, is irksome. But every half hour, spent out of the carpeted, stuccoed, and stoved sitting-room, will contribute its mite towards the redemption of the constitution from oppressive languor and sickliness. The lathe I have already mentioned, and gardening. I have known these two resources produce the effect in a greatly debilitated habit. Many chemical operations are attended with sufficient muscular action to prevent chilliness. It is a great happiness for a child to have been taught the use of mechanic tools, as this will always render deliverance from indolent habits less difficult in subsequent years. Walking I cannot recommend for walking's sake. I think we actually find it of scarce perceptible service to health. But in quest of mineralogical or botanical, or indeed of any other objects, it will sufficiently stir the blood and chear the mind.

Exertion in a cool temperature will tend greatly to mitigate the propensity to catarrh (catching cold) which is the daily plague of many persons. But in resorting to cool air within doors for refreshment, when they are overcome, it is necessary for them to be extremely cautious, unless they do make some exertion at the time. If the door be opened, the stream of air flows against the feet, where the air of the room is warmer than the external atmosphere. In sitting near high windows, the head will be disproportionately cooled. In both cases the trunk of the body, which needs cooling applications most, and bears them best, will be little affected. A catarrh will be the general consequence; rheumatism sometimes. And the different parts will imperceptibly acquire the morbid habit of being in opposite states as to temperature and action at the same time, so that the feet shall be stone-cold, while the face and chest burn. And this will be an eternal source of teazing complaints. A disposition of this nature may be generally corrected. But its correction will require great judgment and experience in medical subjects, and cannot therefore properly enter into a popular tract.

The article of dress, as far as dress is a mere covering, requires very little discussion. Weakly people should put on clothes enough to keep themselves comfortably warm. When the animal powers are languid, it is indeed nearly impossible to keep the extremities warm. Whence, in Holland, the women have been obliged to have recourse to a particular apparatus for warming their feet. But for this purpose there is in reality only one good and wholesome expedient, *bodily action.* However there is no comparison between a load of clothes in cool rooms, and light dress in hot rooms, where these two circumstances only are to be taken into account. Those who follow the former usage will escape innumerable illnesses, to which the followers of the latter will be subject. In the one case, the mucous membrane will be enabled to resist the vicissitudes of the atmosphere. In the former it will be so tender, that very slight changes of atmospheric temperature will throw it into disorder. And not only so: but cold or moisture applied to any other part of the body, will be the cause of injury to that membrane. This will be the weakest part; and it is a common observation, that what tends to induce disease, wherever it be applied, will affect the weakest part. Metaphysicians, and indeed medical writers, express this observation in more learned terms. But it is sufficient to be sensible of its truth. It holds alike with respect to every organ. In the instance of the mucous membrane, the Dutch and English are two nations, where the experiment is daily repeated upon an immense scale. The Dutch are comparatively but little subject to harassing indispositions or serious diseases, affecting that membrane. But these are the national diseases of the English. The habits of the former tend to keep the membrane in a robust state; those of the latter to reduce it to the extreme of diseased sensibility. Some years ago, when employed upon an essay on consumption, I had the good fortune to procure on the subject of the Dutch and English, one of the most valuable contributions towards the history of the health of nations, that has ever been presented to the public. It comes from Dr. Cogan, a writer well known in the literary world. Dr. Cogan from a long residence as physician both in England and Holland for a great number of years, was better perhaps qualified in such a case to give information, than any other per-person in Europe. "I remember" says this observer, "when I had the pleasure of seeing "you, to have remarked that the Dutch, and "even the English, who had resided some "time in Holland, were forcibly struck with "the coughs (whether catarrhal or consump-"tive) so universally prevalent in this coun-"try (England) at almost every season of "the year. At church and at the theatre, "devotion and pleasure are always inter-"rupted, and sometimes totally destroyed

" (here) by incessant coughing and expec-
" toration, while in the largest assemblies in
" Holland, *instances of a similar kind are*
" *hardly known.* This very striking differ-
" ence I have been induced to ascribe to the
" contrast observable between the two coun-
" tries *in the construction of their habitations,*
" *and in the peculiarities of dress.*"—The doctor then goes on to describe the coolness of the Dutch apartments, and the warmth of the cloathing. A Dutch woman, says he, feels herself insufferably oppressed in an apartment we should deem moderately warm. And as all accounts represent the ladies of Holland as nervous or feeble, it seems to be a clear inference that a formidable portion of unhealthiness and mortality in England depends precisely upon the effect of our opposite customs in regard to clothes and lodging.

HOW TO ESCAPE THE COMMON DISEASES FROM VARIATION OF TEMPERATURE.

There are a number of common complaints, in most instances arising from the vicissitudes of temperature solely, though they are, strictly speaking, excitable by other causes also. These disorders, by frequent repetition, injure particular organs, and, after a time, totally destroy the harmonious movements of the system. When the mischief falls upon vital organs, and rises to a certain height, it of course proves fatal. No nation is infested with a greater number of these complaints than the English. We owe them to our variable atmosphere, acting, as already intimated, by favour of imprudent habits. The principal of these complaints is the *catarrh*, vulgarly denominated *a cold*. When the powers producing it operate with great intensity, in infants it becomes croup, or at least one kind of croup, and pleurisy or peripneumony in adults. These several maladies seem to differ but in gradation and in their seat, but agree perfectly as to the manner in which they are commonly brought on. Inflammatory rheumatism has the same origin. To occasion the one or the other, it is only necessary that this or that part shall be most exposed to the agents proper for exciting it.

The opinion prevalent among the faculty and the public, was not only erroneous concerning the production of these diseases, but directly led to the most dangerous management. Within these few years the mystery so long hidden, was unveiled by the sagacity of Dr. JOHN BROWN of Edinburgh, an author of powerful genius, whose merits and fate form a striking contrast with those of the *petits maîtres* of physic that figure as favorites of the great. The discovery deserves to be regarded as one of the most ingenious and happy combinations ever formed by the human mind, and in relation to these islands as perhaps eventually the most useful, recorded in the annals of medicine.

That complaint of the membranes of the head, windpipe, and chest, which properly deserves the name of *hot* or inflammatory catarrh, is not owing simply to cold. Except in a very few instances, where acrid fumes are inhaled, it never takes place but from the concurrent or successive action of cold and heat, or of stimuli, equivalent to heat. The ordinary process is the respiration, first of a cold, and then of a heated atmosphere. But after breathing the cold air, it is not perhaps absolutely necessary, even at first, that the immediately exciting stimulus should be warmer air. The reception of hot liquids into the stomach will probably be followed by the same consequences. It seems also certain that if the body be warmed by strong exercise, a catarrh will be produced by inhaling cold air, provided the increased action of the rest of the body keep up for some time, as will generally be the case. But after the repeated occurrence of this troublesome complaint, the first step in the process ceases to be indispensable. The liability of the mucous membrane to be affected, increases with every repetition. And from the rapid communication of the parts of the living system with each other, by which what comes in contact with any one will influence another, ever so remote, catarrhs are brought on by the most trivial seeming operation on a distant member. Thus leaving bare the arm for a few moments, if it have usually been covered with a glove, the slightest moisture finding its way to the feet, a quantity of rain not sufficient to wet a handkerchief through, shall produce an instantaneous sneezing, and a speedily following inflammation of the mucous membrane. This shall pass from spot to spot down the wind-pipe into the chest, and keep up an irritation to cough for many weeks. In this state the membrane being probably more than usually prone to inflame, causes still slighter, if possible, than the first, shall re-produce the whole series of symptoms for six months together. Variations of temperature too slight to be sensible to the nerves of the skin, shall give rise to all this continuance and severity of distress, when the membrane has once been rendered highly susceptible. So true is it that we may easily bring ourselves to such a pass, as not often to know *how we take cold.* Since the slightest cooling cause, succeeded by the slightest heating cause, or concurring with an exertion that scarce perceptibly increases the animal actions, is in very delicate people adequate to the effect, the order of events might for ever have remained undiscovered, if in others the phenomena had not been more distinct.

It is common to the whole exterior surface of the body to be thrown into a state of violent action, if it be first cooled, and then suddenly heated. Single parts are affected in the same way. The effect of the cold bath is one of the slightest general operations of this nature. The skin is rendered so susceptible by the application of the cold water, that a glow is brought on by the contact of air at so low a temperature, that perhaps it chilled but the minute before. Coming out of the cold air into a warm room, produces lively and continued warmth in the parts that have been exposed. A hand dipped in snow, and then held close to the fire, is stimulated into more considerable, and even painful action. It is now not an agreeable transient glow, but a burning or throbbing with an aching. The preparatory cold was more considerable. The exciting heat more intense. The affection of the hand therefore more violent in proportion. It is well known (and this is the extreme effect of the causes in question) that frozen limbs will inflame so as to mortify, if they be not very carefully kept from the contact of mediums, which are considerably above the freezing point. It is on this account that to be restored, they are rubbed, according to a practice long established in cold countries, with snow.

The membranes, which the air sweeps in respiration, appear still more liable to be thrown into an inflammatory state than the skin. When the nose has been chilled by frosty air, applying a heated handkerchief will bring on a sneezing, and sometimes that inflammation of the membrane lining the nostrils, and communicating cavities which we call a *cold in the head.* In medical language it is named *coryza.* But these are frivolous distinctions. By attention to the circumstances preceding catarrh, every doubt as to the manner of its taking place will be dissipated, if the mucous membrane of the observer be not already too diseased to admit of distinct perception of the successive phenomena. None of that stuffing in the head, or huskiness in the throat, which mark the beginning of this complaint when the person is of a hardy constitution, and not very liable to cartarrh, will be felt as long as he continues out of doors. But after seating himself by the fire, he will begin to feel universally on fire, and these sensations will be very distinctly perceived. And in addition to the feelings about the passages of the breath, feverish shiverings, attended by a quicker and stronger pulse, will succeed. And if the cold to which the system has been exposed was considerable, and the apartment be heated, and warm spirituous liquors be taken, a very severe catarrh will not fail to make its appearance, and perhaps a pleurisy in full form come on. The original feverish feelings, belonging to inflammatory catarrh, when at all severe, are very easily confounded with those at the beginning of the most dangerous eruptive fevers, being in fact of the same nature. I have, for example, known a patient, who had inadvertently caught the natural small pox, keep close by the fire-side all day long, and fill himself with warm liquids, under the idea of having taken cold. The consequence has been confluent small pox and death. And from the uniform result of experience, there is no doubt but the disease must have been excessively aggravated by the mismanagement that preceded the eruption.

When the supposition that *it is only a cold*, happens justly to apply to symptoms like these, such a method is of all others the one most calculated to render it as severe as possible. The warm air that is respired, and the the warm liquids that are drank, concur in carrying the inflammation to the greatest height. Cool treatment is just as necessary to prevent catarrh, as to mitigate the small-pox. The analogy of frozen limbs should be strictly followed. We cannot indeed rub the nostrils, and the continuation of the mucous membrane with snow or cold water. But we can do what amounts exactly to the same thing. We can keep an atmosphere, not much exceeding in temperature that by which these parts have been chilled, traversing them for any length of time. When the dryness in the nostrils, the huskiness in the throat, and the other feelings that denote the commencement of catarrh, have arisen from the chill of a distant part, the same expedient will be equally proper for prevention or mitigation. Even when there has been scarce any difference in the state of the air, by which the wind-pipe and its branches have been chilled, but in respect to motion, I have known a very considerable glow over the whole body, and corresponding sensations in the nostrils and throat come on, when a person has continued to walk about in the calm, or in an apartment without a fire. So considerable is the effect of a rapid current of cold air in carrying away their heat from the animal surfaces! But in this case, the external and internal glow both passed away without leaving a permanent increase of action, or fixed inflammation. The effect of continuing for a time in a low temperature, after a chill of the mucous membrane, may be materially assisted by drinking cold water. And where there is a strong propensity to catarrh, the food should be taken cold, and all warm and heating liquids be carefully avoided for the day. For I have sometimes known a disposition to become inflamed, remain for many hours after all action, strong enough to excite distinct feeling, had ceased through the extent of the mucous membrane.

In a variety of cases, mankind act upon associations formed by sounds, without further enquiry. What more natural therefore, than that he who has contracted a disorder called *a cold*, should expect to remove it by heat. The unquestionably just observation that cold has its share in the origin of this disorder, no doubt gave rise to the denomination, and contributed to make people inattentive to the succeeding heat. And after the inflammation has fully subsided, a warm temperature may have been felt useful in stimulating the weakened surface, and putting an end to the defluxion. In order to destroy the delusion, however produced, nothing can be more expedient than to banish the received term. I therefore risk the imputation of pedantry in using catarrh instead—leaving it to others to weigh the merits of this word against *rheum*, which is found in our older medical writers. It is in truth synonimous with catarrh.

Some authentic accidental occurrences, recorded within these few years, add considerably to the above evidence. If a low temperature alone produce the inflammatory catarrh, it should seem as if long exposure to it ought to aggravate that disorder, or at least to keep it up when formed. But severe cold has, on the contrary been found to effect a cure in a very short space of time. One of the most striking instances is minutely related by Dr. Hamilton of Ipswich, in a letter to the author *(West Country Contributions)*. A boy in Dr. Hamilton's service, with a considerable cough and febrile heat, played the truant one evening in February 1797, and passed the whole night, walking or sitting in the streets. The night was the coldest of that season;—the thermometer, at 7 o'clock the next morning, standing at 10 degrees below the freezing point.—"The boy's complaint, says Dr. Hamilton, was somewhat alarming; and I felt considerable uneasiness lest it should increase from his being first heated, as I presumed by play, and then suddenly cooled by the frost." "During the interrogation he underwent next morning, I observed he did not once cough, although in the same space of time, ever since the commencement of his catarrh, the cough would have often harassed him and interrupted his narrative."—"I watched him narrowly, as I apprehended a fever might be the consequence of his midnight ramble. But here I was agreeably disappointed; for, in place of fever, his catarrh was cured, his cough ceased, and never afterwards returned." The same physician cloathed a ragged old soldier, turned vagrant, in warm garments lined with flannel. Before putting on his new and entire garb, his bare body had been buffeted by rain, snow, and sleet. Next day he was very ill with catarrh, attended by fever; and the succeeding day his complaints increasing, he threw off his warm clothes, and betook himself again to his rags, swearing *his warm clothes had killed him.*—No doubt former experience had taught him that he could bear the severity of the elements in his rags with impunity. So that here, after a chill, the unusually accumulated warmth of the exterior surface of the body, seems to have had the same effect in bringing on catarrh, as warm air generally has on being respired. In an incipient catarrh, those who follow the example of Dr. Hamilton's boy, so far as to continue moving in the cold for some hours, will find it greatly reduced, if not taken off. But they must not afterwards heat themselves. They may sit at times, as well as walk, covering the extremities so as to prevent them from being chilled. The object is to apply a cool medium to the inflamed mucous membrane. It is probable that giving up a single night's sleep would be attended with the happiest effect in such cases. The heat of the bed-clothes, (like the warmer dress of the military vagrant above mentioned) and the change which sleep itself induces on the system, are unfavourable in inflammatory diseases. In the croup, in the scarlet fever, in analogous complaints, and probably in low contagious fever, it would, I believe, be much better to lay the patient upon stretched canvas, supported by tressels, instead of a bed.—But the childish demand among the opulent for toys and tinsel ornaments has heretofore so much and so mischievously engrossed the arts, that the essential accommodations of humanity are utterly neglected. But some time or other, (and may it be soon!) an ingenious mechanic will deserve nobly of mankind, by contriving a species of household furniture, sacred to the relief and preservation of the sick.

On minutely considering the production of catarrh, and the more serious maladies of the same parts, many remarks present themselves. But every attentive observer of his own feelings, once apprised of the true principles, will find out how to adapt his measures to his own circumstances. And to enter into individual peculiarities would be endless. I shall, therefore, content myself with singling out one or two cases, more important than the rest.

When there happens to be a triple succession of variations of temperature, there appears reason to believe, that the mischief will always be greater. I suppose, for instance, a young person, leaving an assembly-room in a perspiring glow. He is then exposed to a cold frosty night. And tremblingly alive to its rigour, he rushes to the fire-side, and drinks freely of hot wine and water. The case, indeed, may sometimes be simpler. It may be similar to immersion of the face in cold water, the rest of the body continuing strongly excited for some time afterwards, and a partial chill falling upon the air-pipes only. But if the renewed stimulation, subsequent to the influence of the cold, should be strong, the inflammation will, perhaps, run higher than in less complicated circumstances. For the effect of the chilling power will be much greater, if it act to a certain extent, than if the part had been in its ordinary state. Thus on sitting down, while we are warm with perspiration, in an apartment at other times comfortable, a sense of cold will soon come on. After hectic sweats, we always find the patient chilly; and in the same state from sudorific medicines, the continued contact of a cold medium is dangerous. It results from these facts, that there is a greater propensity to chill in the state of moist warmth, and, perhaps also, in a dry heat during health, where the application of the cold-producing cause is prolonged beyond the reduction of the system to its average temperature.—I say, *during health*, because in the early stage at least of fever, the diseased action generates animal heat for a considerable time nearly as fast as the most rigorous atmosphere in our latitudes can carry it off. Though from some observations it would appear that if the heat in fever be sunk below par by sudden application of cold water at an improper time, the shock will be dangerous or irrecoverable.—Should this distinction concerning a triple succession of exciting and depressing powers be found more subtle than just, it will still remain an indubitable fact that particularly violent catarrhs and pleurisies, and still worse complaints, follow exposure during the warmth of perspiration. The whole difference may indeed depend upon the particular susceptibility of parts to chills, when they have been excited to perspiratory action. A knowledge of the simple fact is sufficient to demonstrate the necessity of extreme caution in such a condition.

It is needless to enter largely into the consideration of rheumatism. Whatever measures are calculated for the prevention of catarrh, will also prevent rheumatism. And in the coldness of external parts, we have the advantage of applying friction; which should never be neglected when they have been considerably chilled. It is probable that the effect of friction on living substances does not agree with that of heat in any respect, except in throwing them into an action, by which warmth is generated. The accompanying animal motions I suspect to be different in some material points. Friction with bodies not previously heated and changed the moment they become so by this use, will not stir up the inflammation, which heat and heated liquids are so apt to do.

The consideration of the seasons is not indifferent to the enquirer concerning the prevention of catarrh. *Beware the ides of March!* said the soothsayer to JULIUS CÆSAR. *Beware the ides of March!* is a caution which every writer on regimen may with equal propriety address to the delicate, and to the superintendants of children. From the foregoing detail, it will without further explanation be perceived how easily catarrh may be occasioned by the contrast of the vernal sun and shade. Nor is it only the aguish nature of the season that threatens the weakly. By virtue of their peculiar properties, I apprehend the reigning winds of spring to be peculiarly fitted to produce catarrh, croup, and kindred disorders of the organs affected by the respired air. These north and north-east winds are not only cold, but extremely dry. By their dryness alone, they must act as powerful refrigerants on moist surfaces. For it is a fact well established by the chemists, that the conversion of moisture into vapour, greatly diminishes the warmth of the bodies from which the moisture is taken away. How very powerful the winds of spring are in producing this diminution of heat by means of evaporation, is proved by the following observations of the late accurate and ingenious Dr. James Hutton of Edinburgh. "I used to amuse myself in walking in the fields, by observing the temperature of the air with the thermometer, and trying its dryness by the evaporation of water. The method I pursued was this: I had a thermometer included within a glass tube hermetically sealed; this I held in a proper situation until it acquired the temperature of the atmosphere; and then I dipped it into a little water also cooled to the same temperature. I then exposed my thermometer with its glass-case thus wetted, to the evaporation of the atmosphere by holding the ball of the thermometer, or end of the tube, in which the ball was inclosed, towards the current of the air; and I examined how much the evaporation from that glass tube cooled the ball of the thermometer which was included. During the summer season, in the dryest weather that I could find, I never sunk the thermometer, in that manner, to the best of my remembrance, above two, three, or four degrees. But in a cold east wind in the spring, I once sunk it between nine and ten degrees. It was I believe about the month of March or April; the sky was cloudy above and no sunshine, and the wind was cold to the feeling, steady blowing, but not strong." Hence we perceive that the superior dryness of the air in March, makes moisture evaporate faster than the superior heat of summer air; and this is independently of all sunshine.

It is evident that the surfaces, along which the dry, cold air passes in breathing, must be affected in the same manner as the surface of the wetted glass tube; and these surfaces thus cooled, will be ready to be thrown into intense action by the rays of a powerful sun in a sheltered spot, or beside a brisk fire at home. An expedient, to be mentioned in the next section, will serve to diminish the bad effect of the coldness, and still more, that of the dryness of the spring-air.

Into the cure of catarrh this is not the place for entering. It is a subject not sufficiently explored. But from a variety of scattered facts, there is cause to believe a catarrh very speedily curable, if proper remedies be employed as soon as it is formed.

CATARRH NOT INFLAMMATORY.

After frequent returns of catarrh, the disease generally appears in a form considerably different from its first. The variations, though well worthy of attention, have been little studied by medical observers. An attempt to describe them does not belong to my present plan. But it is important to observe, that they, frequently at least, end in a complaint known by the name of *chronic catarrh*, or the *catarrh of old age*. The first name they derive from their long continuation; the second from their more peculiarly affecting persons, advanced in years, or (what is in effect the same thing) of broken constitution. From repeated injury by catarrh, the mucous membrane degenerates from its original nature and functions. The same thing happens to all the membranous surfaces of the body from repeated inflammation. For then, instead of being merely kept moist by vessels appropriated for this purpose, a superfluity more or less considerable is secreted, and finds an issue at some orifice, with more or less inconvenience to the person, labouring under this particular species of debility. Or if the excess of moisture be not continual, it will be occasioned by circumstances, not capable of producing any such effect in a healthy condition of the part. This is the case in the catarrh of debility. The different cavities along which the air passes in respiration, during the cold part of the year, pour out a quantity of troublesome phlegm, often as viscid as the white of an egg; and the patient (particularly on exposure to a low temperature) is harrassed with violent and constant fits of coughing. Such a disease, in a climate like ours, must be the daily and nightly plague of existence for one half of the year. And it usually abridges or helps to abridge life.

It is with this class of individuals pretty much as with infants, who bring into the world a weakness of the mucous membrane. No sooner do they inhale a single breath of cold air than their distresses commence. And frequently, indeed, on passing from the open air into an apartment somewhat warm, the most violent fits of loud, straining and convulsive cough will equally come on. The discharge of tough, ropy phlegm is accompanied by a struggle, which persons, doomed to the severest handicraft labour, might pity. While an oppressive sense of fullness of the chest alternates with the cough and occasions almost as much distress.

With tractability and minute attention on the part of the sufferers, these evils may be frequently relieved by medicine and by the use of the cold or cool bath. But they can be relieved only on these conditions. The following are the principal circumstances in which tractability and care are so indispensably necessary.

Proper cloathing must be worn. The propriety of the use of flannel next the skin is a question not well determined by our medical writers. Worn in the day, it has always appeared to me useful in chronic catarrh, except in our settled hot weather.—I doubt whether it should ever be habitually worn by night. Bed-clothes can always be laid on in sufficient quantity to prevent chilliness. On which account flannel seems useless, and in that state of increased sensibility of the skin, which sleep induces, it is more likely to be hurtful than at any other time, by the stimulating effect of its piles, as also by the warmth it keeps collected round the body.

As far as I have been able to judge, this harrassing malady is less frequent among labouring persons than among the idle. And yet when it is once formed, many sorts of labour are carried on under circumstances that must have a tendency to increase it. But that general glow and activity of the skin and also of the mucous membrane, which is kept up by muscular exertion are probably more than sufficient to counteract the operation of those unfavourable circumstances. It would, therefore, deserve to be recommended to rich persons, afflicted with this species of catarrh, to submit to such a degree of manual labour as is consistent with their strength.

But the most essential thing of all is, unhappily, the most difficult, or, at least, that to which it requires the greatest share of self-denial to submit.—The open air must be avoided when it is of that temperature, which provokes violent coughing and discharge of phlegm; for thus the disease is increased. It may be true that in consequence of the mutual sympathy of different parts of the body, the warmth of the skin must have some tendency to diminish the susceptibility of the mucous membrane, yet this will not be a sufficient protection, when the cold is of any severity. And it is impossible to line the passages from the mouth and nose to the remotest ramifications of the wind-pipe.

There, however, is one most simple expedient which I hav: been lately accustomed to recommend, and which certainly has allowed some invalids to range more at large with impunity. But I am not yet able to say whether it will answer in the most rigorous weather. It is merely an improvement upon the common practice of holding a handkerchief to the face on coming out of a hot room, and consists in a sufficient number of folds of gauze over the mouth and nose. The intention is not merely to warm the air, before it enters the respiratory passages, but likewise when the weather is dry as well as cold, to give it moisture. On looking into the accounts of those adventurers, who have at so much peril, and, in many cases, with so much damage to their constitution, lately climbed a variety of the most elevated mountains in Europe, I was much gratified to find the strongest indications of the efficacy of this measure. On the summits of very lofty mountains the atmosphere is known to be much drier as well as much colder than elsewhere. A short account of what travellers experience in those high regions will, in some measure, illustrate the manner in which a very dry, keen air acts upon living surfaces, when freely exposed to it, and clearly shew how very easily some of the most disagreeable effects may be avoided.

A letter of August 2, 1801, from Dr. Hoppe a physician, who dates at *Heiligenblut in the*

Upper Moellthale, contains some interesting medical observations, made upon the GROSS-GLOCKNER, one of the highest mountains in Europe. "The whole party, says he, who ascended to the summit, after a couple of hours were attacked with a superficial inflammation of the face. It was attended with considerable swelling and pain. The redness in some was more than in others. Nor did this attack disappear in less than two days, when the epidermis peeled off. An inflammation of the eyes was, in a few, more violent than any other symptom. In these the eye-lids were exceedingly swoln and turgid. The ball of the eye was also bloodshot. I myself could not go to sleep for pain of the eyes; but was obliged to get up and open my eyes. The rector of Drauthale however suffered most. He was not only unable to sleep, but spent the entire night in the open air trying to soothe his sufferings by pacing backwards and forwards. - - - Those of the party who had taken the precaution of guarding the face by crape, were perfectly exempt from these sufferings—a precaution to be recommended to every one who ascends snow mountains or glaciers. Here every body is acquainted with the effects above related; and the people who have to climb the Tauern in spring, and are obliged to walk for several hours together over the fields of snow, guard themselves in a different manner. They make a mass out of earth and water, and besmear the face with it, which renders the same service as the black gauze."

Former travellers, and especially M. de Saussure, had accurately noticed these singular phenomena. But without excluding the dryness of the air*, and indeed in a sub-

* Un troisieme effet de l'air des hautes montagnes, est de rougir et souvent meme d'excorier les parties decouvertes de la peau, celles du visage principalement. Cet effet depend *en partie* de la vivacite de la lumiere; car il est plus sensible lorsque le soleil brille; et quand on parcourt des montagnes, couvertes de neiges et de glaces. *L'air y entre cependant pour quelque chose*: *Voyages* § 561, published in 1779. In § 2061 he states an opinion somewhat different. La consideration de la secheresse de l'air, plus grande sur les montagnes que dans les plaines, s'etoit bien presentee a mon esprit; mais je trouvois l'effet beaucoup plus grand que la cause: d'autant plus que l'on epreuve souvent cet effet sur les montagnes a un degre de l'hygrometre, auquel on ne l'eprouve point dans les plaines.---Depuis que l'experience m'a appris que dans l'air rare des hautes montagnes la chaleur possede une force dessicative presque triple qu'elle a dans la plaine, j'ai compris que la chaleur animale, la chaleur interne de nos corps, agissant sur notre peau dans cet air rare, doit la reduire a un etat de secheresse extraordinaire. Et si les rayons du soleil directs ou reverberes par les neiges, viennent frapper cette peau dessechee et devenue par cela meme susceptible d'une plus

sequent explanation imputing the greatest part of the effect to the dryness of the skin, he lays a considerable stress upon the brightness of the light, which may have some, but probably no essential share in the phenomenon. And what share it has, I should ascribe to the heat of the face, occasioned by the light. And this heat, alternating or co-operating with the chills from the air, probably acts upon the skin as the same causes do in producing inflammatory catarrh. The words of Dr. Hoppe seem to imply that the people of the country, whenever they cross the cold snow-fields, are in the constant habit of besmearing the face, whether the weather be cloudy or bright. In our own dry winters, and particularly in a dry spring, every body finds that the skin is greatly affected, whether the day be dark or otherwise. I have often remarked the effect to be most considerable, when the sky has

---

grande chaleur, ces rayons exerceront sur elle une action beaucoup plus grande, et produiront la sensation de brulure, le hale, les gerçures, la bonfissure et les autres incommodites que l'on eprouve quand on ne couvre pas sa peau de maniere a la garantir, et de l'action du soleil et de cette excessive evaporation. *Ce meme dessechement produit aussi la grande alteration que l'on eprouve a ces hauteurs.* This last part was published in 1796.

been overcast, for then the east wind is most severe and parching. But what strikes me as particularly important in the relation of Dr. Hoppe, is the turgid, inflammatory, and bloodshot condition of the eye-lids and eye-balls when they were unprotected. Here we have mucous surfaces, sufficiently analogous to those along which the air passes in respiration. The very phenomena themselves, that are produced by a rigorous atmosphere when it traverses the wind-pipe and the whole extent of its numerous branches in persons subject to chronic catarrh, are here probably brought into view. Could the parts be inspected, we should observe a similar tumour, giving the sensation of fulness, and a similar redness from the effect of the cold air upon the circulation of the weak membrane; and must not the corresponding pain provoke violent fits of coughing? It is notorious that a discharge of mucus in greater or less profusion, always takes place from parts of this structure, in proportion as they have been more or less debilitated by previous causes. The difference therefore between wearing crape on our highest mountains, and exposing the face unguarded to the blasts, which those who scale them have to encounter, seems clearly to be an experiment for the direct benefit of invalids afflicted with chronic catarrh. I do not know whether all will be deterred by the dread of appearing singular from its application.—It is obvious that the construction of these *muzzles* ought to vary according to the case. Each individual will soon find how many folds of whatever material he may choose to employ, will communicate heat and moisture enough to the air he breathes, without injury to the freedom of respiration. Common veils will be generally insufficient; and they are not worn by men.

Persons, not considerably affected, need only wear such a guard on first going abroad, provided they are on foot, and the weather be not too sharp. Walking soon communicates to the mucous surfaces as well as to the skin a degree of activity, during which the functions of the chest are more easily and pleasantly performed; and in this situation a cool atmosphere received directly into the lungs will not produce inconvenience. Of course the muzzle may be discarded. It is obvious that the consumptive and the asthmatic will be enabled, by the same means, to indulge safely in air and exercise, at times when otherwise they must debar themselves the use of both.

In cases of chronic catarrh, it will be easily

understood that the bed-chamber should never be suffered to cool to that degree, at which the cough comes on. Fifty will probably be the lowest admissible temperature, and beyond sixty degrees it will seldom require to be raised.

ADVANCED YEARS.

As we approach to the term of second childhood, we require to be tended nearly with equal care, and to be cherished with more than equal warmth. During the coldest months there is regularly the greatest number of deaths among those aged above 60, and the fewest in the middle of summer, facts which do not appear from the bills of mortality to hold regularly of infants under two years (*Heberden's Observations*, Payne. p. 5152), though every nurse observes how much infants suffer in severe weather.

In a tract on consumptive and scrophulous complaints *, I have related a number of trials, in which invalids were kept for months in a temperature rather exceeding sixty de-

* *Observations on the Management of the Consumptive.* Longman and Rees.

grees. The result was the rescue of one person from the last stage of consumption *, and of more than one from the confirmed stage of that disease, when almost every variety of medicine had been tried in vain, beside unspeakable comfort where no cure was effected. In the course of these trials, I made observations, which emboldened me to promise sufferers from chronic catarrh compleat relief. Chronic rheumatism was also incidentally removed; and extension of a comfortable existence may undoubtedly be secured to the aged in the same way. I have shewn that these advantages lie compleatly within the reach of persons of very moderate fortune. Were the plan to be executed on a scale sufficiently large, nearly every enjoyment, of which the infirm are capable, might be brought within their reach. Should mankind ever cease to be content with slaughter, famine, and pestilence, as a recompense for their senseless admiration of political adventurers, and should the power of society ever be employed for the greatest good of the members, of which it consists, such a plan will certainly be carried into effect. Edifices, by their destina-

* Mrs. Finch, daughter to the celebrated Dr. Priestly.

tion putting to shame the monuments of ancient art, would rise under every inhospitable sky; and these CONSERVATORIES OF OLD AGE, scarcely subjecting their occupants to privation, would afford compleat shelter against the inclemencies and the dangers of the seasons.

END OF ESSAY V.

Printed by MILLS, St. Augustine's, Bristol.

ESSAYS

ON

THE MEANS

OF

AVOIDING

HABITUAL SICKLINESS,

AND

PREMATURE MORTALITY.

ESSAY SIXTH.

Printed by MILLS, St. Augustine's, Bristol.

Nullum ferè morborum chronicorum genus existit, quod a scrofulosâ causâ ortum traxisse non observatum fuerit aliquando.

There is scarce any species of chronic disease which has not, one time or other, been observed to derive its origin from a scrophulous cause.

ESSAY

OF

SCROPHULA.

TWO CLASSES OF DANGEROUS COMPLAINTS.

THERE are various disorders which compleat their course in a few days or in a few weeks. Of these some are accompanied by instantaneous disability, by violent commotions in the system, or by some other striking change. Hence they have been denominated *acute* by medical writers. It is but seldom indeed, when these disorders have once shewn themselves in their proper form, that there is any occasion to caution the patient, or those about him, against procrastination in seeking relief. When they rage in a number of families at the same time, an universal consternation pervades a city or a country. The municipal magistracy, if not the government itself, shall sometimes think it necessary to have recourse to public measures for preventing the progress of the danger. And if we advert to the conduct of the very individuals whose supreme delight, at other times, consists in hurling defiance at the medical art, we shall observe that their tone is of all others the most changed, when they find themselves in the gripe of an *acute* malady. There are none who more piteously invoke the aid of physicians, or more timorously obey their injunctions.

Disorders which follow an opposite progress are distinguished by the title of *chronic.* They are, doubtless, often equally formidable with those of the former class, whether we consider their issue, or certain periods of their progress. But when unattended at their commencement with severe pain or alarming reduction of the animal powers, (which is the case with some, most eminently deserving to be styled *chronic*), they excite no active interest in society at large. The persons affected, and those about them, remain insensible to their situation. Hence they are often suffered to run on to a dangerous or an irremediable length. And full security, or at most slight suspicion, is at once succeeded by despair. Though sometimes the security and the despair of the nearest relations of chronic invalids are separated by an interval of incredulity, than which nothing is apt more strongly to excite the indignation or the pity of the medical practitioner, according to the habitual or momentary state of his temper.

From the bills of mortality and the records of medicine, it is manifest that complaints of slow progress occasion as much destruction as those of an opposite nature. And if this be true, it will follow that they must, upon the whole, occasion much more distress. Hence it becomes the duty of writers on preventive medicine, to treat of the more formidable chronic disorders with particular care. There is nothing capable of effectually banishing false security, but a knowledge of the connection between first slight appearances, and the secret operations of fatal causes. A conviction of the understanding, produced by a knowledge of the whole course of slow disorders, will supply the want of that, which in more urgent cases presses upon sense and feeling with irresistible force, and urges all parties to grasp with eagerness at every kind of assistance.

PUBLIC IMPRESSION.

It is not that scrophula is a sound indifferent to the public ear. But the most important truth, however fully acknowledged, must continue as useless as if it were totally unknown, unless it be particularly applied. We daily hear our divines reproaching us with this species of neglect. Every human individual, they say, is apt to act and speak as if he thought every other subject to disease and death, but scarce ever speculates upon his own insecurity. In like manner, we find the whole world, lamenting the frequency of scrophula in this country. Its inveteracy and the variety of forms, in which it shews itself, are mentioned with terror. It is declared sometimes to make the vital organs its prey; and sometimes, with greater malignity in its method of attack, to derange the functions of the brain, and to bereave those, to whom it has descended, of their reason. To put a total stop to such calamities, the most humane persons profess themselves disposed to severe coercive measures. Nor is it uncommon to hear a law, prohibiting the scrophulous from marrying, proposed by those upon whose constitution or upon whose family the disorder is at the very instant committing unperceived ravages.

It is indeed undeniable that scrophula is exceedingly common in Great Britain. It is equally true that nothing can be imagined more disgusting to behold, or more miserable to endure, than the effects it sometimes produces. The object, however, begins to have a less gloomy side. For there are few diseases in which the triumph of art has been rendered more compleat. In its worst form, scrophula is to be removed by means, from which not the smallest collateral injury results. In general, its entire eradication from the habit is attended with little difficulty. I need scarce explain the conditions under which I make these declarations. It will be self-evident that there must be means and disposition for a proper course of regimen and medicines. The patient must not be just at the point of death; and the destruction or disorganization of certain vital parts must not have proceeded beyond a given degree; though in parts of too little importance to deserve to be termed *vital*, it is wonderful what damage can be repaired by proper remedies. Indeed, it is scarce in any instance possible, that this complaint should proceed to its incurable stage without extreme penury, gross neglect, or prejudice on the one hand, or unpardonable deficiency of skill on the other. The latter of these causes does not belong to my undertaking. But to obviate the mischiefs of penury and neglect, I think it necessary not only to treat singly of the appearances which scrophula usually puts on, but also to give the character of the temperaments which it most readily invades, and to consider some opinions which very commonly prevail, and which I look upon as so many dangerous errors—-Those who take up the present essay for the mere amusement of a vacant hour, will probably find me more than usually tedious. But numbers must feel a desire for the fullest information of which they are capable; and these will be better enabled to support a degree of minuteness by the prospect of the advantages, which I have given them to understand they may draw from it.

FIRST RUDE NOTION OF SCROPHULA.

The term scrophula is derived from the Latin word *scrofa*, which signifies a sow. The appearances, whatever they were, that gave rise to this denomination, occasioned the Greeks to follow the same analogy in their denomination: χοιρας being used both for the animal and the complaint. Various explanations, as is usual in such cases, are brought forward. According to some, the name was adopted because swine are liable to certain tumours, resembling those in scrophula, though the resemblance extends not farther than form, it having been discovered a few years ago that the *measles* in the hog race are produced by a species of taenia.—According to another derivation, the human neck is said to acquire from scrophulous tumours a degree of that grossness which that of the hog has from nature. Again, the great numbers, in which these tumours are often seen, is childishly supposed to bring to mind the number of young produced by a sow at one litter.

All these etymologies are founded upon the existence of external, and therefore tangible swellings; and to this we generally find that the notion of the disorder is limited among the unprofessional. But this is much too narrow an idea. It is true, and it is essential to understand, that the enlargement of a particular class of glands, called *lymphatic glands*, is one of the earliest and most frequent phenomena exhibited by the complaint. These glands are those which most nearly resemble the acorn, when out of its cup or calyx, from which glands universally derive their appellation. The lymphatic glands are largely distributed through the interior and over the surface of the body.

But as long as they remain in a healthy state, the most superficial do not present any thing elevated, knotty, or rough to the sight or the touch. But as soon as they become affected by scrophula, they immediately enlarge and become very sensible on pressure. At first, they are rather soft, and in some degree elastic. In process of time they grow harder, and swell in knots or clusters, till the formation of *matter* within them restores their softness. Their bulk varies from the size of a pea to that of a kidney-bean and hazle nut. They enlarge to the magnitude of a pigeon's egg, rarely to that of a pullet's. For though glandular lymphatic swellings often attain the size of a goose's egg, particularly under the arm-pit or in the groin, this happens from the enlargement of a whole cluster, and seldom of a single gland. And as to those large swellings which precede the disfiguring scars in the front of the neck under the chin, they take place in glands of a different description, when the habit is more seriously infected by scrophula.

One of the slightest of scrophulous affections, viz. when the swoln glands take the name of *growing kernels,* will give the best idea of the incipient stage of this complaint. Whoever examines this very common appearance, will find the single tumours moveable, soft, indolent or without pain, and producing no discoloration of the skin. In this state they will remain for a long time, even for a number of years. Afterwards they either disappear from accidental increase of robustness, or enlarge, coalesce with the contiguous skin or *common integuments*, become immoveable, occasion the skin to change from a pale to dark red hue, and give pain; all which circumstances denote progressive inflammation.

Even where they do not give pain on being handled, but are thick set, and somewhat enlarged about the neck, they produce a disagreeable sense of tension, when the head is turned round.

But it is to be observed that all tumours and indurations of the lymphatic glands, are by no means referable to the head of scrophula. I have seen the lymphatic glands of the groin enlarged from a seton in the side, and from an abscess in the buttock. The same thing is well known to happen from the venereal poison. A cancerous tumour will be attended with the same effect. And a temporary enlargement of the lymphatic glands near the inflamed parts, will often accompany catarrh, and, not very unfrequently, a bad tooth. In these, and other instances, the occasion is so evident, and the origin of the affection of the lymphatic glands so sensible, that there can be little danger of mistake. However, to enable those who have the charge of young persons, (for it is they who are peculiarly susceptible of scrophulous complaints), to judge when the habit may require correction from regimen or drugs, it will be necessary to enter more minutely into particulars.

SCROPHULOUS CONSTITUTION.

The great attention which the physicians of modern times, and particularly those of this country, have been called upon to pay to scrophula, has enabled them to fix, with tolerable precision, upon certain signs, indicating the habit most susceptible of this disease. It is particularly important that the public should be well acquainted with these signs, in order that a preventive regimen may be followed from the beginning, and that recourse may be had to proper remedies, where regimen alone proves insufficient, and the disease seems gaining ground.

The quality of the skin, and particularly that of the complexion, is one of the leading indications. The skin is fine and sleek; the complexion bright and ruddy. It is often such as, from first appearances, a mother would most desire for a daughter; and an unwary spectator consider as a mark of the most perfect health. This is so frequently the case, that some of those authors who have laboured so much in vain to give *definitions* of diseases, similar to those which occur in books of botanical nomenclature, have made a *fine complexion*, part of the character of scrophula. This part of the character is not merely objectionable, because the disease does not take place in all who have the assigned complexion. For causes to be mentioned hereafter will much more certainly occasion it, where this kind of complexion exists; and if a parent with such a complexion escapes, the progeny is often observed to become scrophulous.—The principal objection against this part of the definition is, that subjects as far as can be ascertained originally pale and sallow, are very often affected by scrophula in some of its worst forms.

Nevertheless, as great stress deserves to be laid upon this, as upon most single indications. There are many readers, to whom I hope I need not explain, that *beauty* is a thing perfectly arbitrary or conventional.

There is no possible cast of features or colour of complexion, to which persons of different age, country, and education, will not apply the epithets both *ugly* and *handsome*. It is true that when associations are more strongly formed, they cannot without much effort of the mind be broken even in persons of reflection, nor broken at all in those who do not reflect. One generation goes on to think and feel, as they have been taught by the preceding. In some points respecting the human person and countenance, we are at this day the disciples of Zeuxis and Praxiteles; the opinions, by whatever causes established in Greece, having descended to us by regular affiliation. But it seems far from impossible to alter these associations. It may indeed be effected by any individual of penetration and address, under whose eye the notions of young people are formed. And who has meditated upon the sources of good and evil in the world, profoundly enough to be able to conceive how much the sum of human happiness and virtue would be enlarged, *if the standard of health were rendered the standard of beauty also?* Towards this indeed every step, made by society at large in the knowledge of animal nature, must inevitably be an advance.

In many novels we meet with a whole exhibition of portraits. The favourite figures however must excite pity or aversion, instead of admiration, in every reader who happens to have a little the advantage over his author in point of information concerning the properties of the human frame.

Were an artist, after the fashion of antiquity, to select the most characteristic circumstances from a number of patients affected with the glandular variety of the complaint under consideration, he might compound a countenance, which the undiscerning should admire exceedingly, both for feature and complexion. It would certainly not fall short of the most perfect ideal beauty of a novelist, and exhibit every trait of the heroine, so far as the description of her charms has any thing in nature corresponding to it. The painting might be so framed, that the face only should in general be visible. But while it was under inspection, it might be contrived that the touch of a spring should bring into view the neck and one of the arms, crossed before the body, in which, with the strictest adherence to nature, might appear various tumours, some suppurating, some discharging matter, and others having left indelible scars, and for the present surrounded by an extensive area of a disagreeable dark red. Such a design might be called *a nimium-ne-crede-colori*, or *look-more-than-skin-deep*. If, on the first impression, it feel cruel to call up such images before the public eye, I hope, on second thoughts, that it will be allowed at least cruel kindness. I would appeal to any one, unfortunately in a condition to sit for such a picture, whether there is most cruelty on the part of those, who would hinder him by law from forming the most endearing of ties, or on the part of one, who gives warning, calculated for preventing the impediment from occurring; and who, if the tie have been formed, would fain persuade the party afflicted to adopt the measures, necessary for preventing the taint from being transmitted to other generations.—Sir Thomas More, by the example of his ideal commonwealth, recommends a previous personal search, lest the discovery of blemishes after marriage should create disgust. It would be better to take care betimes that certain blemishes should not arise before or after this change of state.

When the temperament verges in the smallest degree towards the formation of the actual disease, the face appears fuller or somewhat bloated. This appearance is more observable in some parts of the face than in others. It shews itself in the cheeks, and between the cheeks and ears. Both eye-lids are puffed, particularly the lower. The nose looks thicker than usual. But more remarkable and more constant than any of these indications, is the enlargement of the upper lip, which often reaches to the partition between the nostrils, and to the *wings* of the nose. The natural depression in the middle of the upper lip becomes now much more conspicuous. Many authors look upon this enlargement of the upper lip as the most certain single mark of a scrophulous tendency. It is true that it is not always present, particularly in grown persons, where the disease is characterized by a concurrence of other undoubted symptoms. It is true also that many individuals and families, entirely free from all taint of this nature, have a tumid and prominent upper lip. But these cases may be generally distinguished by the constancy and original existence of the peculiarity. In the scrophulous, it will be remembered not to have existed, and it will generally fluctuate in size from time to time. —One thing too must carefully be held in mind during the perusal of this and the subsequent tracts. In the present general deficiency of preparatory knowledge, no writer in his senses will think of enabling the unprofessional to distinguish diseases with certainty. It is sufficient to make it clearly understood when there is room for apprehension. Practitioners of medicine, as well as judges and juries, must often proceed upon circumstantial evidence. When the symptoms are of dangerous tendency, when the disorder they portend has infested the family, or is even endemial in the country, when nature is likely to do nothing for the patient, and when the physician is satisfied that he can conduct the treatment without injury, he ought to resort to the most probable remedies without scruple. Men must frequently act on other momentous occasions with the chances less in their favour.

This enlargement of the upper lip, of which I am speaking, is vulgarly considered too as a sign of worms. But nothing is more likely than for worms and the complaint in question to co-exist. Parents are always falling into mischievous errors upon the subject of worms. Worms very often infest children weakened by different maladies. When they are detected, they are considered as the prime or the sole object of regard. Domestic or advertised medicines, often of the most hazardous quality, are administered. These impair the constitution; or if they expel the worms without injury, the administrator of the remedy sits down content, and the main complaint goes on daily increasing, and infallibly avenges the neglect at some future period.

But not only does the upper lip become enlarged. The nostrils sometimes itch, grow excoriated and ulcerate, or are covered with a thick coating of crusts or scabs. This arises from a too copious discharge of mucus, the membrane, by which it is secreted, being very apt to be weak, or *relaxed* as it is termed, in constitutions of this nature.

Of the complexion, it still remains to be observed, that even in the class of scrophulous subjects, who have cheeks ever so florid, the enlarged upper lip, the nose, and the circumjacent skin are of a dead, pale, and unwholesome hue; by which contrast those who are accustomed to medical physiognomy will often recognize the complaint at first glance, and it ought to give others suspicion, if they will but avail themselves of their senses, and beware of a cheek like rouge, with a muzzle like white lead.

Before any positive symptoms of disease exist, and especially when the complexion is bright, the eyes are large, lucid, and such as to be esteemed highly beautiful. At this period too, the pupil is unusually dilated. In infants of the soundest habit, the pupil is larger than in adults, but on comparing such as from other circumstances appear to have a disposition to scrophula, the appearance will be found much more distinct, and to give the eye a darkness in the centre, which by the uninformed is esteemed as contributing to beauty. But from a variety of considerations, there is no doubt but it betokens constitutional debility, wherever it is habitual. Indeed when only occasional, it accompanies those states, in which the system is most remarkably deprived of its vigour. It is seen in fainting, in low fever, and in the weaknesses brought on by the most hurtful excesses of youth. No wonder that the fibres of the iris should lose their tone, in common with the rest of the living fibres.

A light, grey or blue eye, has been enumerated among the marks of scrophula. It certainly often occurs, as well as light hair, in this temperament. Those however who have wide opportunities of observation, will find that the other tokens frequently concur with black hair and black eyes.

The general appearance of the skin—(and not merely that of the face)—as also of those veins which run along the surface of the body, is particularly worthy of regard, insomuch that this one indication, where it shews itself in young people, with general feebleness or indisposition to exercise, ought to put their friends upon their guard. The skin is conspicuous for the superior smoothness and fineness of its texture. It exhibits appearances that give the idea of transparency. These are owing to the full, fine blue of the veins. In patients far advanced in scrophulous disorders, those vessels have actually looked to me as if filled with ink. And besides their colour, they are unusually turgid.

The older medical writers have laid great stress upon shortness of the neck, as well as on the shortness and narrowness of the forehead, the flatness of the temples, and the depth of the jaw-bones. But the very contrary to some of these circumstances for the most part obtains; as, for instance, in regard to the neck. The depth of the jaw is also produced by the actual existence of the disease, as manifested in the enlargement of the subjacent glands. The same is the case with regard to the neck, where the lymphatic and the other glands come to be swelled. Now as the lower parts of the face are dilated in size, the whole proportion of the features must evidently alter, and the forehead of course will look narrower and lower.

It is undoubted however that in the scrophulous temperament, the whole head, particularly its hinder part, does attain a larger size. This happens particularly to those who have been rickety in infancy. For between the one disease and the other, there exists probably no difference, except in the time of life, and the parts attacked. We well know that one of the most important early processes in the constitution, is to form the compound (perfectly understood by modern chemists, as being formed of lime and acids *) from which the bones derive their hardness. In many scrophulous constitutions this process is carried on too sluggishly, and the bones swell and become crooked in

---

* I have found by experiments made at the pneumatic institution, that when growing animals are put to breathe a larger portion of oxygen air, their bones are rendered wonderfully solid, and their growth checked—a fact, immediately applicable to certain scrophulous invalids, where the growth is too rapid. I mention the circumstance here out of place, lest the publication of the experiments should be delayed longer than could be wished, by engravings, intended for their illustration.

consequence. It is common to find those who have been rickety at one period, afflicted afterwards with the *evil*. And the children of mothers who have been rickety, will very frequently turn out scrophulous, without any disposition to that disease on the part of the father, and without the operation of any particularly unfavourable cause on the offspring. And in some families, part of the children shall be rickety and part scrophulous, so that not one among a great number of brothers and sisters, shall escape the one complaint or the other.

The intellectual superiority of children of the scrophulous temperament has long been noticed, and it is certainly found to exist in a large proportion of cases. That sensibility or openness to impression, which is one of the principal constituents of genius, has often been observed to accompany different diseases of debility. The compleat distinction of circumstances, and ascertainment of causes is one of the remotest benefits, to be expected from the knowledge of the human mind, when that shall attain the point of perfection, for which the labours of Hobbes, Hartley, Locke, and Abraham Tucker have prepared the way.

In some cases of scrophula, however, when

the disease has made considerable progress, the effect is entirely opposite. For there is a form of scrophula, in which the intellectual and other functions are equally impaired. Such cases almost universally belong to the class, marked by a symptom easy to be distinguished, and of which the very first appearance should be the signal for seeking instruction how to arrest the progress of the whole complaint, and for persevering not only till so much of it, as has actually taken place, shall be removed, but till the constitution has lost its scrophulous tendency. The symptom to which I allude is an enlargement and a hardness of the belly, especially towards evening. In the infants and young children of the poor, and in those who are ill-fed and generally ill-managed, this enlargement and hardness of the belly passes by gradations more or less quick into perfect atrophy. The character of the countenance here often differs totally from that, which is usually described as denoting the scrophulous temperament. The face is indeed bloated, and that in a still greater degree, but it is of a uniform sallow. The eyes are dead, and disposed to weep. The little invalids are heavy, peevish, gloomy, and without relish for the usual sports of their age. They have frequently that superiority of the understanding which belongs to the other variety of the disease. But in place of being lively, they are considerate beyond their years. So that, in pursuance of a language long adopted in medicine, the one may be called *sanguineous,* the other *phlegmatic* scrophula. When the symptoms are suffered to increase, and when the disease bears upon the mental faculties, not only the face, but the body and limbs are apt to be universally bloated. The hands and feet are habitually cold, and assume a purple hue. The bowels are so sluggish, that opening medicines are almost always necessary, and these to produce their effect must be of the sharpest kind. Every thing, in short, conspires to indicate torpor of body and mind. If there be any enlarged lymphatic or other glands about the neck and throat, (and this symptom is frequently absent), they partake of the predominating quality of the whole system. They are almost sure to be without pain. They enlarge slowly, or not at all, after once becoming perceptible; and, of course, it is exceedingly unusual for them to gather and break. That rickety tendency, which is denoted by a head of unusual size, and by large joints, is often seen in this species or variety of scrophula.

Opulent parents will never suffer such a complaint as this, to arise to a certain height unchecked. But it is here that they deceive themselves; and, after measures of small avail sit down content, while the disorder only changes its seat and form. The enlargement of the belly, and perhaps also of the upper lip, are considered as implying the presence of worms. Recourse is accordingly had to some standing family composition against these animals, or to the pill and lozenge most recently announced in the newspapers, under the sanction of reverend and titled names—Worms are expelled. The pill or lozenge, if it contain mercury, as it frequently will be found to do, may diminish for a time the scrophulous affection of the glands lodged in the belly. And the consequence of this may be, to favour the eruption of the disorder elsewhere. At least, it is commonly enough to be remarked that children, early affected with enlargement of the belly, and with the usual concomitant symptoms, are seized in the course of two or three years in more superficial parts, if by accident or design the first complaint should be lessened or removed, but no sufficient means adopted for clearing the constitution entirely.

SCROPHULOUS ATROPHY OF THE CHILDREN OF THE POOR, MORE PARTICULARLY CONSIDERED.

We have many proprietors of extensive estates, and of large manufactories, who interest themselves not only about the comfort, but about the health of their labourers, and of their labourers' families. In a time of sickness, these proprietors not only give away money and costly liquors to a considerable amount, but (what bespeaks infinitely more charity of heart), they do not grudge time and attendance. The gentlemen sometimes, and sometimes the ladies, regularly distribute remedies to a crowd of needy applicants. Their excellent intentions would however be crowned with far greater success, and by the same expence and trouble they would relieve a much more considerable number of objects, if they made it their aim to prevent, instead of curing, or even employing medical practitioners to cure, the disorders of their inferiors. And by this change of plan they would escape many of those disagreeable feelings that at present they owe to their kind offices. It will be sufficient, in the first place, that they make themselves acquainted with the marks of the most common chronic disorders, and afterwards that they inspect the families of the poor from time to time, with a view to the state of their health. It is by no means incumbent upon the superiors to ascertain the particular ailment with the accuracy, which ought always to precede the administration of remedies. A little practice and information will soon enable them to distinguish, among the various drooping human creatures that will catch their attention, such as a very moderate quantity of drugs, seasonably employed, will snatch from a miserable death, though they would not have the smallest chance of relief from the bounty of nature. The success of a mode of charity, so extensive and yet so prudently restricted, would create a new interest in life; and to the satisfaction of loading the memory of Howard with benedictions, would be added one much more solid, that of following this example.—When the opulent have not leisure or inclination, or do not suppose themselves qualified, to act as inspectors of those who stand to them in some subordinate relation, how easy to employ a professional deputy! This would not only be serving the indigent invalids in their neighbourhood. It would be also promoting the most useful part of medical science. Nothing is so much wanting as incentives to study the prognostics of slow diseases, and the means of prevention. Providence, for some inscrutable reason, in former ages almost always saw it fitting to disjoin wealth, and the inclination to use wealth to the best purposes. The means and the inclination we now frequently see united. But the knowledge is still for the most part wanting!

No portion of the afflicted poor would profit more by such a system of beneficence than the scrophulous. The peculiar *waste* or *atrophy* occasioned by this disease, when it affects the *mesenteric* glands, is not indeed by any means confined to the abodes of poverty. But it is infinitely more frequent there than elsewhere. And in consequence of the supineness, the ignorance, and the distress of the parents, it often proves fatal, though in all but its extreme stage, it is perfectly curable. No complaint bears more distinct characters. A very brief description of it, in addition to what has been said in the former part of this essay, will enable those who are willing to exert themselves in behalf of their helpless neighbours or dependants, to recognize it with the utmost certainty.

The belly of scrophulous young persons is almost always inclined to protuberance. But when that affection of the glands, lodged within the belly, or of the mesenteric glands, by which the protuberance is caused, has proceeded to the length of atrophy, this appearance becomes peculiarly striking. And not only a fulness, but a hardness will be felt about the navel. Sometimes, and at a certain period, hard inequalities will be sensible to the touch. The little sufferers (for it is always found to affect young children principally), will hardly be so stupid as not to have noticed those sensations, which in most cases are present, and which, where they are, will materially assist the judgment. One of these sensations is a weight about or below the navel. Another is frequent griping pain. There will exist an interior disorder, corresponding to the exterior. Sometimes it will be aversion to food; at others, an insatiable appetite, or the desire of gross, farinaceous aliment, or of substances not fit to be used as articles of diet. The bowels will seldom be regular; often too open; now and then, the contrary. The evacuations will have an unhealthy look, from being charged with slime or from being too light of colour, because the liver does not secrete bile of a proper tinging quality. The advance of the complaint (and occasionally its more early stage), is accompanied by a very offensive purging.

The emaciation of the limbs will form a most disfiguring contrast with the enlargement of the belly. The skin will be found to hang loose, and sometimes, if doubled, to go almost round the member to which it belongs. It will also be of a dingy or dark colour. The complexion of course will be of a sallow or deadly pale, except when the hectic flush prints its deceitful and ill-omened animation on the cheek.

The fever will soon declare itself by other signals. A general indisposition and lassitude will come on, especially in the evening. Shiverings will be felt; and to these an universal dry heat will succed, with a degree of general or partial moisture in the course of the night.

AFFECTION OF THE EYES.

With the complexional signs of scrophula, we shall often find conjoined a weeping and weakness, with inflammation of the eyes. If the lower eye-lid be turned down so that the interior surface shall come into view, it will appear more red than in the healthy; and vessels will be manifest, turgid with blood, on the white part of the eye-ball. This sort of inflammation is peculiarly conspicuous after waking in the morning. Frequently white spots will be found to occupy the transparent part of the cornea. These give alarm at first. But as they generally disappear altogether after a time, or speedily shift their situation, they are overlooked or forgotten by uninformed parents. Nor is it uncommon for short-sighted medical practitioners to tranquillize them by an assurance that *this is a thing which the child will not fail to overgrow.* It is indeed but seldom that these opake spots deprive children of sight. But other scrophulous complaints supervene; and unless the party affected has the uncommon good fortune to be thrown into a way of life, highly favourable to a delicate constitution, he will either be disfigured, mutilated, or destroyed, between childhood and maturity. This disorder of the eyes rarely occurs alone. When attended, as it usually is, by general weakness, the most unremitting endeavours should be used, for years together if necessary, to remove both the one and the other. If by chance there should be discovered, on examination, no want of general strength at the moment, it will be highly necessary to keep a sharp look-out for those more serious symptoms, which will hardly fail to break forth in due course of time.

CHILBLAINS.

Those who have opportunities of inspecting numerous troops of English children, will easily remark that the most scrophulous by habit and by descent, are most troubled with chilblains. In Scotland, where a different mode of dress in early life inures the feet to the vicissitudes of heat and cold, chilblains, I believe, are infinitely less known; and in Scotland of course, scrophulous ailments attack less hardy parts.

It is very commonly observed, that those who have chilblains when young, are after their disappearance destroyed by some complaint of the scrophulous class. Hence it is concluded that the cause producing chilblains has migrated from the feet (and it may be from the hands,) to some nobler organ, and that if the chilblains had continued to break out as before, no disorder more serious would have occurred. This opinion must be allowed to have some semblance of truth. But even in speculation, as thus generally enuntiated, it is, I apprehend, incorrect; and, in its influence on practice, it is probably as mischievous as the most arrant falsehood. Of many disorders it is unquestionable that while they are in force in one part, they do not invade any other. As long, for example, as the gouty inflammation keeps up in the ball of the great toe, the stomach will be safe. But again, it should be considered that as long as the morbid habit continues, vital organs must be liable to seizure. And by a slight change in his mode of life, the patient may unwittingly occasion the retrocession of the disease. Independently of any such change, and in spite of the most scrupulous care, the progressive developement or the unavoidable decay of the system may give rise to the same disaster.

In regard to chilblains, it is obvious that their occasional causes, merely in consequence of our usages, will operate with less force and frequency after the period of puberty. The feet and hands of men and women are not chilled in the same cruel manner, as we have found those of boys and girls to be. But it is not unlikely that natural and necessary alterations in the constitution would frequently call the scrophulous action from the extremities to the centre of the machine, even though the utmost pains were taken to secure an annual crop of chilblains. It is common enough to see a local scrophulous affection, far more severe than chilblains, superseded by others still more violent. In the lymphatic glands about the neck, it usually happens that as one dries up, another suppurates. When even ulcers affect the bones of the fingers, they shall heal, if that kind of seizure of the spine come on, which is followed by distortion. In a patient where I certainly knew that the salivary glands were diseased, because I could feel them and because there was present every mark of the scrophulous temperament, and where the strongest evidence existed of disease in the mesenteric glands also, I have seen the morbid action pass almost as suddenly from the neck to the belly, and from the belly to the neck, as it ever does from limb to limb in gout. When the belly became turgid, tense, and painful, the salivary glands sensibly subsided and grew easy, and *vice versâ*. These changes took place repeatedly in the space of a single day.

The true policy, therefore, is plainly still

the same. If we would avoid sudden death from gouty seizures in the stomach, let us use such medicines and pursue such a regimen as will clear the constitution, and prevent this terrible disease from taking possession either of the centre, or of the extremities. If we would avoid slow death from the destruction of vital organs, let us put an end to chilblains by agents that will produce a general improvement of the habit; and not absurdly lament that our heels cease to inflame and chap. For how should it be otherwise, when they are no longer affected by the proper causes of chilblains, considerable inequalities of temperature?

OBSERVATION TO PREVENT MISCONSTRUCTION.

It is proper that one thing should be clearly understood concerning several of the indications of scrophula. They have no necessary and general, but only a local and accidental connection with this disease. In a different climate, all, except the slow tumours of the glands, would betoken a tendency to complaints of a nature altogether different. And as to the tumours of glands, even though they should never advance beyond the state of *growing kernels*, they ought in strictness to be regarded as real scrophula. Thus in the West Indies, for example, enlargement of the pupil of the eye, and distension of the superficial veins would not, in one instance in twenty, be followed by any of those disorders, which in this country we consider as scrophulous, but by feebleness of the digestive organs, or by a state of excessive general languor. These signs therefore in reality do but indicate constitutional debility. And it is owing to the nature of our climate, to our diet, to propensities transmitted from parent to child, or to some other unexplored causes, that scrophula so frequently attacks people distinguished by these particularities. It is nevertheless obvious that an acquaintance with them is nearly as useful as if they were absolutely specific in their import.

The general reader will be at no loss to perceive how preposterously certain authors have judged in restricting some of these signs to a particular ailment. Thus, for example, a professor in an university, celebrated as a medical school, insists upon it that a puffing of the wings of the nose, of the upper lip, and of the parts adjacent, is to be considered as an infallible token of worms *. But neither this puffing, nor the dilatation of the pupil, have any fixed inseparable connection with the state of other parts. They are properly affections of the parts themselves. They may originate in local causes, and be quite confined to the spot. They may also be associated with an unhealthy condition of a distant organ. The proof is easy. The one or the other can be produced at pleasure by art. The dilatation of the pupil by introducing the preparation of a narcotic plant into the eye: the puffing, by cold applications to the features in question.

That the marks of the scrophulous temperament, as distinguished from the early and slight, but positive symptoms of the disease itself, are merely signs of general or local debility, is rendered more than probable by another fact. Women are more liable to this complaint than men. In childhood the evil appears to be pretty equally distributed between the sexes. But as they grow up, every circumstance in the mode of life, and all the differences in the bodily functions, are unfavourable to females. For

* Dr. Home's Clinical Experiments, § xx.

one boy we find at least four, and perhaps six girls, labouring under some scrophulous complaints, as distortion or curvature of the spine; and in no instance is the proportion reversed.

OF SOME OF THE GRAVER SCROPHULOUS COMPLAINTS.

These are foul ulcers from suppurated and broken glands, scrophulous consumption, complaints of various name about the bones, and particularly the joints, as *spina ventosa*, water on the brain, disease of the hip joint, lumbar or psoas abscess, white swelling, impaired vision or blindness, ear-ache, discharge of matter from the ear, and deafness. Indeed these affections of the ear have as good a right as most of the other circumstances enumerated above, to be considered as admonitions to guard against more serious attacks of scrophula hereafter.

Glandular ulcers are the most easy to be foreseen: at least where the external glands are concerned. The swelling in this case gives timely and sufficient notice.

Of consumption I shall treat in the subsequent essay.

Spina ventosa, or the *windy thorn*, is an

absurd Arabian denomination. The epithet *windy*, or *puffy*, is deduced from the aspect of the bony tumour, as if it were inflated with air; and to account for the other term, the pain is said to be sometimes such as would arise from a thorn, lodged within the diseased part. Modern authors have called the complaint *Pædarthrocace*, or the *child's joint-disease*. Whatever be the most appropriate name, the scrophulous affection of the bone varies as much in its progress as that of the glands. Sometimes there shall be no pain, but only a swelling, for years before the skin is broken and a running ulcer formed. In more rapid cases, the patient shall be sensible of an indistinct feeling in the part, which shall exhibit no preternatural appearance, nor give any pain on pressure. After some time however a tumour shall arise, and in the course of a few weeks a redness be evident. By degrees the red shall change to a livid hue. At last it shall break, and discharge a thin, wheyish, or ruddy liquid. The ulcer on examination is found to reach to the bone; and this proves rugged, unequal, and spungy on its surface. The different variations of this complaint are observed to affect the larger and smaller bones alike; the joints of the shoulder, thigh, and wrist, as well as those of the fingers and toes.

Water on the brain, or enlarged head, is one of the most palpable of diseases, often affecting children before birth, and indeed extinguishing life in the womb. It will probably disappear altogether, or in great measure, as soon as the scrophulous part of the public shall have knowledge and good sense enough to take the proper measures for cleansing the constitution of this taint. There are some families where the hope of progeny is continually defeated by this and other kindred diseases, the children being either still-born, or regularly perishing about a given age. In these cases, much may be expected from proper preventive measures applied to all parties. If the fault be on the side of the mother, she might pursue a course of scrophulous remedies during pregnancy, and of these, she would find the most effectual not only safe during that period, but beneficial for the future. The children should be put upon a regimen, suggested by a true knowledge of the animal œconomy, and not taken up on the faith of books on the disorders of infants, of which the authors, in spite of all their experience, too often partake of that scantiness of ideas, and that impotence of combination, which is implied in the phrase—*Such a person is no better than an old woman.*

Swellings from inflammation usually become red in a short time. But we have seen that it is otherwise in glandular swellings from the king's evil. It is also long before the scrophulous swellings of the joints grow discoloured. Hence a particular variety has been called *white swelling*, or *fungus of the joints.* This disorder attacks the larger joints in preference. The term is indeed almost confined to the knee. The tumours it causes are often soft at first and elastic, yielding to pressure, but not pitting. The pain, for the most part, is excessive at the onset; though sometimes the usual indolent character of the scrophulous class of tumours prevails here. The pain, however, when it does exist, is generally very circumscribed—it is sometimes unremitting, sometimes periodical: even when most excruciating, it is scarcely attended at first by any perceptible swelling. It encreases upon the slightest motion; and the patient being obliged to keep his limb in the same posture, in no long time loses the use of the joint. In the natural course of the malady, the distress lessens as the joint enlarges. And soon an increase of the ends of the bone is perceptible. The soft parts partaking of the disease, the tumour grows more elastic, smooth, and shining, and by degrees acquires somewhat of a red hue. The disorganization advances; abscesses form with variety of fistulas and sinuses. These break at last, and discharge the ill-conditioned, wheyish, and curdy matter, peculiar to scrophulous sores. The first ulcers will occasionally heal, but new ones do not fail to succeed. A decay or caries of the bone is easily detected at the bottom of the ulcer, and long splinters or exfoliations run out along with the matter. Hectic fever is formed. Of course the patient falls away, and at last perishes amid those sufferings, which diseases accompanied by hectic fever always induce. This is, it should be remembered, an excessively meagre representation of the natural course and termination of white swelling. Means heretofore in use undoubtedly sometimes arrest the complaint. But if it be suffered to proceed to a certain length, the precarious measure of amputation has been considered as the only resource.

In lumbar or psoas abscess a slow, indistinct and often unperceived inflammation arises in the inside of the spine and ribs among the muscles, corresponding to the inside of the sirloin of beef. The patient is at length surprised with a soft tumour in the groin, or low down in the thighs, towards the inner knee. On the discovery the surgeon is summoned; the family is shocked by the intelligence, that the appearance is occasioned by purulent matter, descending from a psoas abcess; the most approved expedients of the chirurgical art are employed, and usually without success; the symptomatic fever comes on; floods of matter continue to be discharged; the body wastes in proportion, and scenes of misery succeed, which neither pen nor tongue is adequate to describe.

MORAL.

I content myself with a very superficial extract from the history of secondary or confirmed scrophulous affections, as they are sometimes called. But it will be sufficient for the purpose, or a longer would be equally unavailing. Its intention is to deter parents from ever remaining content as long as they perceive a child of scrophulous temperament, suffering from local or general debility. There are I know medical practitioners in abundance, who when the immediate symptoms are slight, though they be ever so distinct in their character, will overlook the continual admonitions of experience, and encourage a dependance on accident.—Indolence creates an aversion to active and persevering measures. Pride disinclines people to listen to truths, which they suppose derogatory to themselves, and to their progeny. Nor is it difficult to perceive how readily in such circumstances parental tenderness will take part with parental pride. For these reasons, the destructive opinions of the shallow and the complaisant among the members of the medical profession so often prevail over the saving counsel of the sagacious and the sincere. Yet if reference were made to a set of umpires, free from all improper bias, it is easy to see what their award must be, when the signs of a tainted habit are undeniable and the system less vigorous than it ought to be for the season of life. Each as an impartial spectator would recommend the very thing, which as a partial parent and as one called upon to act in conformity to his sentence, he might decline. For whatever the future may be likely to bring forth, can any thing better be done for a young person than to render him robust? What though the process be tedious? and what though it should require to be resumed from year to year? Who grudges pains and perseverance, when the question is to reclaim a piece of land from the waste? Shall we hesitate to devote as many months to health as we are willing to devote of years to gain? And shall we account it a small thing to escape, when we are threatened with deformity, mutilation, and torture?

But besides timely remedies and proper regimen, it is in our power to take another security against secondary scrophulous complaints, rising to an irremediable height, before they are discovered. This consists in teaching children accurately to distinguish the parts of the body. Such information will lead them to observe many important changes, which as they take place slowly, are apt to proceed unobserved. It would also render them alive to sensations, that would otherwise escape attention. We have seen that the psoas abscess is one of the most formidable of scrophulous complaints. It often, as we have also seen, runs on unawares to a fatal length. Yet it seldom happens that its progress is not attended by feelings distinct enough; only they are not painful, or not severely painful, and so are disregarded. Now the measure I recommend would, I believe, infallibly occasion the disease to be detected at its commencement; and then, I have little doubt but it could be cured with great ease in the majority of cases. It is obvious that the utility of this measure would extend far beyond the range of scrophulous complaints. The ignorant of all denominations, and the poor and the young among the rest, we perpetually find unable to fix, with any tolerable accuracy, the seat of their maladies. Hence arises one great difficulty in their treatment.

But the labour of the parents would bring with it its own reward. In teaching they would learn. The persons of their children being subject to more frequent examination, it would be more difficult for disorders to lurk about them undetected. And the oftner they should examine, the more accurately would they be qualified to judge of every deviation from the healthy standard.

These considerations, I flatter myself, will reconcile some readers to that recommendation of the study of anatomy, which hitherto they may have considered as a ridiculous whim. If I have succeeded in giving them any tolerable idea of some, among the principal varieties of scrophula, they will not be slow to acknowledge, that it is a species of knowledge capable of universal and daily subservience to the most beneficial purposes. That the state even of superficial parts will be understood the more clearly for some anatomical knowledge, is what I need not take much pains to prove: and for judging with any certainty of the affections of the deep-seated organs, it is obvious that such knowledge is indispensable. The less, indeed, any one is dazzled with the vain glare and glitter of outward trappings, and the deeper insight into the various conditions of human existence he has acquired, the more plainly will he discern the inestimable advantage which must accrue to every family from physiological,—that is, applied anatomical—information in its heads. It is probably not too much to expect, that when any considerable portion of mankind shall be capable of appreciating the value of the different sciences, and of distinguishing the most worthy objects of desire, physiology will come to be considered as the domestic science *par excellence.* At present we are in general poorly qualified to choose among the things which it is in our power to attain. We deem it profane to look forward, except to the precise distance, and just in the direction, which we have been taught. Hence it holds at this day nearly as in the time of that sagacious satyrist of antiquity, who on a survey of the pursuits of his contemporaries, declares that few could distinguish substantial from specious blessings.

CAUSES FAVOURING OR PRODUCING SCROPHULA.

Of these some may be said to exist before the individual, some are casual and sudden, some require application for a long time before any sensible effect follows. I shall consider the two former sets of causes in the present section. The condition of the parents is frequently decisive for the progeny. The fruit of too young matches has been observed to be subject to scrophulous atrophy, and other scrophulous disorders. This is particularly said to occur in manufacturing places, where the intercourse of the sexes takes place early, and various circumstances co-operate with original weakness of stamina, in producing the disease. The same calamity befals the children of marriages, contracted at too advanced a period of life;—no matter whether it be the old age of years, or the old age of debauchery.

All the world is aware that scrophula runs in families. Many curious opinions have been advanced by minute observers upon this point. It is well ascertained that when the disorder has infected one generation, it shall often disappear in the succeeding, but come out again in the next, or even in that following. Hence we see that in infected families endeavours to eradicate the evil should never be intermitted, when its precursory signs are apparent. Many instances have occurred of numerous families, where part have been exempt from scrophula, and part miserable sufferers. This has led the ingenious to investigate, or tempted them to divine, the reason of the difference. According to some, the children, resembling the diseased or morbidly disposed parents, are those that become diseased. To ascertain the actual observance of any such law by nature, requires more extensive experience than commonly falls to the lot of one man; and it is always to be feared that the exceptions may be as numerous as the instances confirming the law. We often see one parent give feature, and another complexion; and sometimes one contributes this feature, and that another. I have seen various scrophulous patients strikingly like the untainted mother, while the father only had been a martyr to the complaint. So that the real laws of transmission seem enveloped in the same obscurity as every thing else connected with generation. Some have believed that the morbid disposition follows the sex; and they quote instances, where all the sons of a scrophulous father have become scrophulous, and where all the girls have escaped, though the latter have borne a strong resemblance to the father.

Those who are intent upon keeping their family sound, had therefore better lay no stress upon such observations. If the father or mother, or a more remote ancestor have been infected, the circumstance ought to be regarded as a strong additional motive for adhering to those precautions, which the tokens of the scrophulous temperament or a weakly habit may suggest.

Nothing is more notorious than the tendency of the small-pox to induce scrophulous disorders, where there has existed the smallest predisposition. Probably also the small-pox has affected the glandular part of the system sufficiently to bring on the disease, independently of all predisposition. Nor is there any other consideration which induces experienced medical men to set a higher value upon the mode of inoculation lately introduced; since the fever attending the cow-pox appears by no means strong enough to make a dangerous impression on any set of organs. For the glands are far from being the only part injured by the inoculated small-pox, even when mild in its symptoms. I have very lately been consulted for a young person of a thoroughly scrophulous race, in whom the small-pox, though nothing severe, left behind it a palsy of one entire side.

What is true of the small-pox holds in an inferior degree of measles. Low fever, scarlet fever, the malignant sore throat, excite glandular tumours. But these tumours often follow a different course from the scrophulous, when they occupy the same seat. They ought not therefore to be referred to the same head. Still by these febrile complaints, genuine scrophula, and scrophula of the most dangerous kind, is sometimes excited. It is is therefore to be wished that the plan, now pursuing with vigour in several places for stopping the progress of febrile infection, may become general. *Fever wards*, or *houses of recovery* for the poor, might be established in every town, on a scale proportioned to its size and opulence. Contagion cannot pass long from person to person among the needy, without occasionally alighting upon the opulent. And when it does, they are not only, in common with others, exposed to the danger of speedy dissolution, but also, under certain circumstances at least, to that which arises from a secondary superinduced malady. It is true that the fire of malignant fever is generally kindled among the poor. But when once kindled, it may justly be regarded as rising in a *gerbe.* Being seated therefore a few seats higher in the amphitheatre of life, does not secure any one against being reached by the fatal spray. There is scarce a town or city in the whole united dominions, where some possessor of useless opulence would not give the necessary sum to redeem the immediate or remote mischief, that has happened under his own roof for want of a public provision against the dissemination of febrile contagion.

HINTS TOWARDS BRINGING UP CHILDREN NOT TO BE SCROPHULOUS.

It is of most importance to make the public acquainted with the causes that are of slow operation. Thus every parent in affluent circumstances will perceive that he has them in great measure in his power. It will then remain for him to determine whether he shall set them to work, or in other words, whether he chooses to have his children affected with scrophula, or that they should grow up free from this calamity.

The diet most proper for encouraging the scrophulous disposition, is less in vogue now than formerly. People have at last suffered themselves to be persuaded to indulge weakly children with animal food, and its various preparations. But the dread of rendering the blood sharp, or of generating foul, acrimonious humours, still occasions some lots of children to do penance upon vegetable fare. Some of the most dreadful instances I have seen of the ravages of the disease, and of the transmitted disposition, have been, where the mothers were strict in this article. A few years ago, two healthy parents, who had subjected a very numerous family to a meagre regimen, boasted to me of their success in education. Since that time, a third of the children have been attacked by scrophulous ailments, without the intervention of any acute illness, except in one case. It would appear that similar misfortunes still more frequently originate from the same source on the Continent. "There are parents" says a modern well-informed foreign writer, "among our no-"blesse, who altogether withhold flesh-meat "from their children till their fifth, and "even their seventh year, feeding them "entirely on garden-stuff, wheaten bread, "biscuit and milk, under an idea that some "mighty advantage is to result to the poor "little creatures. To such parents, on their "boasting of the effects of this plan, I have "been able to point out a distinct protu-"berance of the belly, with other marks of "tendency to scrophula. Thus I have "wrought a conviction that the children "were not so healthy as was supposed. "Doubtless, in our temperate latitudes, a diet "merely vegetable is not suited to infants of "two years old and upwards."—With us, preparations of animal food are most suitable to some children, less than two months old. Within the first year, infants will suck a bone with great glee. Nor when weakly, do they digest bread properly without a mixture of animal matter. The permanent state of the constitution depends in many instances upon a well adapted diet during the two first years. And I do not understand how parents can depend upon the strict observance of good rules, unless they will enforce it by personal inspection.

Another practice, formerly general, is very much gone into disuse. Otherwise, as constitutions go now, it would be difficult to say who could escape scrophula. Should it be any where carried on at present, (and if so, it is on quite young infants), or should it be revived hereafter, let Sydenham unfold the consequences. " It is certain" says this celebrated observer, " that frequent purging " does much mischief in weakly habits, and " particularly at a tender age. — — — — " Tumours are produced in the belly; and " they increase in bulk the more, the oftner " the purging process is repeated. The whole " œconomy of the bowels is sometimes dis- " turbed by frequent purgatives. The me- " senteric glands acquire a preternatural " size, and these and similar complaints " pave the road to the grave." At this day, the observation may not be unworthy the notice of persons, ready on every slight pretext to recur to worm medicines.

No single cause contributes more to the production of scrophula than cold. A single severe chill has often been found sufficient to produce true scrophulous enlargement of the glands; and there is much reason to believe that psoas abscess itself has often, among other causes, originated from continuing too long, though but for once, in cold water; from damp sheets; from wet unchanged cloaths; and from a single exposure of some continuance to a rigourous atmosphere, particularly when the chill was directed to the back and loins. I have most certainly known abscesses, succeeded by a fistula, to take place in a scrophulous subject, in consequence of sitting down for a time on a wet seat; and it has gone exceedingly hard with the patient.

I have already spoken at large of protracted chills, and I have also considered the effect of mental depression upon the general health of young people. In what relation the latter cause, and those other causes with which it is usually associated, stand to scrophula, shall be stated in the words of an author, with whose sentiments on this point I have only become acquainted since the first part of this essay went to the press. Let his hypothetical phrases be excused, or rather let them be considered as awkward forms of expression. In recounting the occasional causes of scrophula, he says—" depressing affections of the " mind have the most pernicious influence " on the human body, particularly sadness " and grief, when long protracted. The ner- " vous power languishes; the parts lose their " tone; the vital movements fail; hence the " circulation of the humours becomes slow, " the functions of the abdominal viscera are "disordered, the fluids grow viscid, coction "and nutrition proceed imperfectly. That "such affections of the mind may give birth "to scrophula, appears as well from direct "experience, as from analogy." — — — — "They exert their most deleterious influence "upon children and infants. For it is on "young animals that nature has especially "bestowed vivacity of spirits and lightness "of heart, in order to promote their growth "and invigorate the system. Nevertheless, "a senseless mode of education * presses "upon the feelings of infants in a manner, "scarce conceivable. Image to yourself "little boys confined almost all day in a "school, forced to commit a tribe of words "and other things to memory, and kept in "constant dread by that apparatus of terri-"fying objects, which surrounds the master. "Can any thing be conceived more perni-"cious, and better calculated to ruin the "health of beings, so delicate and so tender?

---

* Severior sanæque rationi contraria educatio eo haud raro adigit infantes, ut moerore excrucientur, *acerbo sæpe magis, quam quis crederet*. Cogitemus pueros, qui per totum fere diem in scholis sedere coguntur, ad vocabula aliaque memoriæ imprimenda adstringuntur multiplici terriculamentorum apparatu a magistris in perpetuo quasi metu detinentur, &c. &c.

" What a variety of noxious powers combine
" their influence! an atmosphere vitiated by
" the multitude of those who breathe it—
" exercise of mind out of all proportion to
" the strength—fear, grief, sadness—defect
" of motion, which is so essentially neces-
" sary! There are many parents, who, with
" ill-advised kindness, place their little chil-
" dren the whole day long under the severe
" inspection of preceptors, to keep them from
" running about and coming to harm. Such
" abuses, we are convinced, must impress a
" character of melancholy on the minds of
" infants, and sow the seeds of numberless
" diseases in the body. Daily experience
" shews how efficacious they are in gene-
" rating the scrophulous diathesis."

Among several instances I could adduce to shew how justly the author insists upon the *extreme bitterness of grief*, experienced by children on first being put to school, and its bad effects, I shall mention two. A boy of fair average capacity had passed his time agreeably between exercise and learning. He had acquired a strong desire to learn Latin, and was put to a day-school. The unintelligible nonsense at the beginning of the accidence, however, immediately checked his ardour. Whole nights of the first week were spent in tears. By day he was gloomy and without appetite. His whole soul seemed to be absorbed by hatred of his tasks, and by fear lest he should not learn them against the appointed hour. The farther he went on, the more distressful were his sensations. While my opportunities of observing him lasted, he seemed to be in that despairing state, in which he might have said with Goldsmith's benighted traveller—

> Lo! here forlorn and lost I tread,
> With fainting steps and slow,
> Where wilds, immeasurably spread,
> Seem lengthening as I go.

In the other case a more reserved melancholy followed the inhuman initiation into humane letters *(literæ humaniores)*: and a protuberance of the abdomen, accompanied by other scrophulous signs, seemed heightened by the mental sufferings into incipient atrophy. For the season insured us against the operation of cold, and there was no deficiency of substantial nourishment.

Our English schools seem not at all behind those abroad. By one who did not know their professed purpose, both might be taken for nurseries for scrophula. How well they answer to the character, is evident in general from the immense number of victims to fatal scrophulous complaints among the youth in genteel life; and particularly from the prodigiously increasing frequency of crookedness or distortion in girls.

To dwell upon the other principal exciting causes of scrophula would be to no purpose. After what has been said, the following list will equally suffice for those, who may resolve to take the most effectual steps for obviating this evil, and for those who, after the fullest deliberation, imagine themselves obliged to determine that their sons and daughters must still take their chance.

An ascescent diet in infancy.
Abuse of purgatives.
Severe chills, or continued low temperature.
Of course, too light cloathing.
Habitual, or frequently renewed uneasiness of mind.
Want of fresh air and exercise.
Too much indulgence in bed, and in hot rooms.
Any other debilitating practices.

There are families, in which as there can be no doubt as to the imminence of the danger, there ought to be no hesitation as to the propriety of adopting a preventive plan, almost from the very moment of birth. Here if the mother be both feeble and scrophulous, it should be forbidden her to nurse. To the milk of a perfectly healthy foster-nurse, it will often be proper to join broth and diluted gravy. And should the history of the family, and the state of the little subject give ground for great alarm, a diet, in which milk forms but a very small proportion, will prove the most wholesome. Writers on infancy have not been insensible to the necessity of early recourse to preparations of animal food, where the constitution is feeble. But a superstitious regard to milk, derived from its real superiority in the healthy state, has hindered them from going far enough in those extreme cases, which I have at present in view. To a weakly child it is of no small consequence that it should interest its nurse. To this, digestible food will contribute more than may be supposed at first view, and without observation. An infant, tortured with perpetual acidity, will be perpetually torturing the ears of the attendants with its outcries. Hurtful methods of procuring an interval of peace will be pursued, or the nurse will desert her charge to breathe a few moments in quiet. I have known very young children become immediately placid on a change of food, and afford daily entertainment to all about them, by the expressions of their happy state of being.

The most sedulous attention should be bestowed upon cloathing and temperature. It is the most common part of the history of scrophulous invalids, that while they fluctuate between health and those terrible seizures, after which any moderate variation in the ordinary powers of nature can make little difference, threatening appearances shall arise in cold weather, and vanish in warm. Thus nothing is more familiar than to see glandular tumours about the neck come and go with the seasons. But it will not be sufficient in scrophulous habits to keep up a due warmth by mere covering. In children, able to use their limbs, the circulation should be kept brisk by frequent and gentle exercise; and they should never be suffered to languish on a bench for whole hours. At an earlier age, friction should be assiduously employed, as recommended in the preceding essay. On the approach of winter, and during an uncommonly rigorous season, it will be particularly expedient to be on one's guard against those relapses, which are so common in this complaint.

OF SOME MISTAKES IN OPINION AND PRACTICE.

Medical superstition is to be sought, not only in the peasant's hut, but in the city and in the palace; beside the toilette of the lady of first fashion, and in the cabinet of the philosopher—more or less disguised perhaps, but still the same in substance.

Doctor Chr. Aug. Struve Miscellan. I. 77.

The negative detriment from erroneous domestic opinions concerning disorders, is full as great as the positive. While the nugatory processes, to which they give occasion, are carrying on, the morbid disposition increases into disease, or a precarious changes into a desperate state. It may therefore be of service to enumerate some of the more common vulgar errors, that they may cease to be an impediment to the recovery of lost health, and to the improvement of bad constitutions.

It is frequently imagined that scrophulous tumours arise from tough, ropy humours, causing an obstruction in the tumified parts. The idea of viscidity of the humours must have arisen from some actual phenomenon, referred by the fancy to a place, in which it did not exist. The slimy or mucous stools in mesenteric atrophy are as likely to have given rise to this mistake, as any other scrophulous symptom. In other forms of the disease, this slime must have been supposed to occupy the glands, and by clogging the vessels, to cause them to swell, as a stream enlarges when a dam is thrown across it. But there is no one reason for believing that such viscid humour exists in scrophulous glands. Its being discharged in the manner abovementioned, is certainly no better a reason than the discharge of similar mucus, when a person sneezes, blows his nose, or has breathed a cold atmosphere for some time. The surface of the intestines is destined to secrete mucus or phlegm; and an excess of this humour at particular times, indicates nothing in regard to distant parts, to which this function does not belong. But if we reject the supposition of viscidity in the humours, because it is totally without evidence, we can go a step farther as to obstruction. For here the negative can be proved. In glandular enlargements, the vessels are so far from being impervious, or less permeable than in the healthy state, that their area enlarges, and new ones form to carry on the circulation in proportion to the growth of the gland beyond its natural size *. The case is exactly the same as with the growth of the whole body. There is an addition of mass and bulk; and the arteries and veins are found to become of larger diameter.

The exhibition therefore of those infusions and expressed juices (as that of cleavers or goose-grass, water-cress, scurvy-grass, sassafras tea), which are intended to attenuate the humours, and purify the blood, or to clear the passages, rests upon a wrong principle. Nor are objections to butter, as an article promoting obstructions, and relaxing the stomach and bowels better founded †. Its nutritive quality, on the contrary, renders it very fit for the scrophulous, in any quantity in which it is easily digestible.

Hard selenitic and calcareous waters have been given out by respectable observers for a cause of scrophula. M. de Luc, for in-

---

* This is a matter of direct experiment. Ipse expertus sum, says Dr. Soemmerring, per glandulas absorbentes *tumidas* expeditiorem esse mercurii viam quam per glandulas *solitæ magnitudinis*.

† Panis, butyro illitus, sæpius sumptus, proculdubio infarctus facilitat, cum butyrum insuper primas vias relaxet, nimiaque quantitate difficulter digeratur.

stance, (Lettres I. 17) remarks that where he has found incrusting or petrifying springs, there the people were scrophulous. In like manner, the water of melted snow has been held by many authors to be the cause of the bronchocele, or Derbyshire neck. But the property of waters of the first description, to deposit calcareous particles on the surface of the substances, over which they flow, combined with the hypothesis of obstruction, has obtained from various medical writers a sanction for this opinion, which it does not deserve *. The production of scrophula by cold and damp is supported by facts so pointed, that the greater number of cases in situations abounding with hard waters, may safely be referred to this head, though food of bad quality, and other evils attendant upon poverty, may claim their share also. BORDEU affirms, that in some valleys divided by a river, the part of the inhabitants exposed to the north are subject to glan-

* Quod vero assertum, licet ad strumas potissimum endemias pertineat, nullus tamen dubito tales aquas etiam diatheseos scrofulosæ evolutionem promovere malumque augere posse. - - Gottingæ scrofulæ frequentissimæ sunt; aquæ vero ibidem scaturientes calcareis particulis insigniter abundant.

dular swellings, while those on the other side are exempt, though both use the same food, and live in the same manner. The glandular swellings however in these instances seem to be the Derbyshire neck, or bronchocele, a disease confined to one particular gland low down in the neck, and nearly allied to scrophula, though persons affected by bronchocele have often no other scrophulous mark. Respecting the Derbyshire neck, accompanied, as it appears to me, with phænomena, in these cases, manifestly scrophulous, there is an authentic observation which deserves to be given entire, as it distinctly shews how glandular swellings may be produced in a short time by natural agents, operating constantly and with great intensity, though in general they take place slowly and imperceptibly, in consequence of the more gradual application of their exciting causes. " On occasion of tumours in the " neck, very prevalent in our military school," says the author of the observation, Dr. Lange of Cronstadt in Siebenburgen, " the com" manding officer desired me to examine " the children —— I actually found thirty" six among forty-eight affected. More" over, of seven adults in the house, two " had the same complaint. All looked very "cachectic, pale, and bloated. Many had "weeping eyes; the appetite was good in "all. The nature of the tumours was dif-"ferent in the different patients. In some "there was but one. Others had two or "three distinct. In others, several had con-"creted into a single cluster, which some-"times had its seat in the middle, and "sometimes on one side of the neck. Some "were of the size of a hazle nut, others of "that of a hen's egg. Some were spungy, "others fleshy, others cartilaginous to the "touch. On other parts of the body, even "in the belly, there was no hardness or en-"largement to be observed. The affection "had been noticed in the school for the "first time since the last winter. The di-"rector shewed me a child that had come "into the house in perfect health but three "weeks before, and he had already a tumour "of the size of a walnut."

"The officers seemed willing to seek the "cause of this complaint in the water, which "the children used for drinking. The water, "however, appeared to me perfectly inno-"cent, because on a chemical examina-"tion, it turned out the purest water in "Cronstadt; and also because the whole "neighbourhood, where no such complaint

" existed, was supplied from the very same " well. The cause, therefore, must have been " confined to the house itself. I asked if the " children were made to wear their stock " very tight? or if their scald heads were " treated with hazardous external means? " Both questions were answered in the ne- " gative. After the strictest enquiry, I con- " cluded that the cause of this general sei- " zure was twofold."

" The first cause appeared to me to be the " low, confined, damp, and crowded apart- " ments. The whole forty-eight children " lived in four small apartments on the " ground floor. They mostly slept two in " a bed; and the building was close by the " foot of a hill, planted with fruit-trees. " Moreover, some of the rooms had been " stabling a few years before, in consequence " of which every thing on the ground was " mouldy."

" The second leading cause appeared to " me to be the frequent use of leguminous " fruits, and of other crude food. That " these were the most probable causes of the " malady, I was farther led to conclude from " two soldiers' children, in whose cases both " causes were very striking, and of whom " both were similarly affected. I advised "therefore, above all other things, to provide "the children with a healthy and spacious "habitation, and to put them upon light "vegetable diet. But as it was not so easy "to provide fresh quarters at a moment's "warning, I proposed in the mean time to "open the windows, to fumigate the apart- "ments with juniper-berries, to omit sprink- "ling the floors when the rooms were swept "out, to furnish them with vent-stoves, to "make air-holes opposite the windows, and "to give the children some wine daily."

By help of these measures, which in the main appear judicious, though some are of very questionable propriety, and by burnt spunge, mercurial laxatives, by some other medicines, and an eventual change of dwelling, "all the children" says Dr. Lange, "soon "regained their natural colour, and in a few "weeks not a glandular tumour was to be "found among them." After the perusal of such very striking facts, the friends of young people will need no farther inducement to examine whether there be any connection between their glandular ailments, and cold from moisture. I have known instances where these ailments appeared certainly to depend on the damp of a wall in the bed-room of a school, upon the same condition of a miserable office where an attorney's clerk has been confined to his desk for a large portion of the day, and of ill-lighted, and ill-ventilated merchants' and bankers' compting-houses.

Concerning the treatment of scrophula, when it affects superficial parts, two opposite incorrect opinions seem to have been entertained. The one, that it is dangerous to disperse tumours or to heal ulcers, lest the scrophulous acrimony should migrate to more important organs, and excite fatal complaints: the second, that nothing further need be done than remove these local affections.

I have already noticed the great chance there is of giving mercury to the scrophulous, unawares, in the form of worm-medicines of secret composition. Mercury is known in very small quantity to bring on that state of the system, in which slight exposure will occasion mischief. Sometimes the injury is sudden and marked; more frequently, it is slighter and unnoticed. In all its degrees, it is more likely to fall to the share of the scrophulous, than of others. None suffering so much from inflammatory affections on the one hand, or from the severity of the elements on the other. And we know that medicines of the above description are usually administered at random, without sufficient probability of their irregular action being observed or restrained, or of the necessary precautions being enforced. A few years ago, I was consulted by letter for the daughter of the keeper of a turnpike gate in ——shire. A quack medicine had been given her, which from its effects must have contained mercury. I could procure none for examination. But the surgeon who had been called in immediately on the seizure, ascertained that a degree of salivation had been excited. In this state she went out one bitter night to open the gate, and was detained by some embarrassment for several minutes in the cold. She felt the chill strike through her, and soon afterwards an ascites, or dropsy of the belly, came on, and proved fatal. The girl had glandular tumours about her neck before taking the medicine, and was of a scrophulous complexion.

Leaving imprudences like these out of the question, I am exceedingly doubtful if the removal of scrophulous affections in one part, will ever excite them in another. It is true indeed that dependance ought never to be placed upon this partial method. In various places, we meet with ignorant country practitioners, who profess to cure the scrophulous ulcers of the common people; and they often succeed for a time. I have ascertained in three different parts of England, that they use plaisters with calomel for this purpose. But though the ulcer heals, another almost always breaks out. Whence it clearly appears to be in general absurd to think of curing scrophula, without means, that affect the whole constitution. But there is no good reason to believe that mischief arises from this imperfect practice, when the substances employed do not get into the habit. A compleat collection of facts would, I am inclined to suppose, tend to shew that to subdue a local affection, is to lessen the whole disease. Sometimes indeed no more of it is seen after the healing of an ulcer by local applications merely.

There is another reason for doubting whether the more dangerous forms of the disease are owing to topical remedies, even when they follow the use of them. Nothing is more familiar to medical practitioners than to see a large joint, or an internal part, attacked at the very time of the spontaneous subsidence of swoln glands, or of the spontaneous drying up of an ulcer, or some time afterwards. Nay, one of the more serious scrophulous ailments shall begin and hold on its course, while a slighter one of earlier origin is proceeding with equal pace. A number of instances have occurred among the crowd of scrophulous invalids, which have offered themselves to my observation; and from these I shall select a case, that will abundantly illustrate the position.

April 14, 1802.—Susanna L. aged 33, of a dull, muddy complexion, with light hair and eyes, No. 4, Cart-lane, Temple, Bristol, miscarried six months since. On coming down stairs, she took a most violent catarrh. Soon afterwards she became aware of a tumour, about the size of the end of her little finger, in the left submaxillary gland. It is at this time as large as a child's fist, has gathered, and appears on the point of breaking.

Six weeks ago, she was sensible of acute pain in her knee, and about a week since perceived that the joint was enlarged. This enlargement has all the characters of a white swelling.

On being questioned respecting the preceding state of her health, she said that at 16, she had an enlargement of her knee. This, she was told, was a white swelling. She remembers perfectly that it gave her much more pain, and was much more swelled than at present. It was then dispersed, as she supposes, by a liniment, though the time of life may be suspected to have chiefly contributed to her recovery. Now had the submaxillary gland been reduced in this case by external applications, how easily might any one have been misled to ascribe the present white swelling to that mode of treatment?

No case can speak more strongly in behalf of the necessity of carefully watching persons of a scrophulous habit, when weakened by any cause whatever. And I was glad to have such a confirmation of the important lesson, which I had been inculcating in a part of the present essay, printed before the case of Susanna L. fell under my observation. Here we have an example of a new attack at the interval of fifteen years, and at the age of thirty-three, in a person by no means subject to the rigours of penury.

The two following instances may further serve to shew that, on occasion of catarrh somewhat more severe than usual, equal vigilance is necessary, even in the more robust among scrophulous men. For the feverishness, attending catarrh, gives rise to genuine scrophula, as well as the fever, attending small-pox and other complaints formerly mentioned.

Samuel L. aged 24, with dark hair, grey eyes, and fair skin, pitted with the small-pox, husband of the preceding Susanna L. said he had caught a very bad cold, shortly before his wife miscarried; that in about three weeks afterwards he was sensible of a swelling under his jaw; and that observing it to go on encreasing, he applied, when it had reached the size of a pullet's egg, to a public charity, and was told *it would never come to hurt him.* He says his strength is considerably reduced; and his general appearance bespeaks indisposition. The swelling which is in the right submaxillary gland has suppurated, and now discharges matter. Near it are several enlarged lymphatic glands. One in the axilla is proceeding fast to suppuration, and there are others hard and swelled near the collar-bone.

William D——, Tyler's-Court, Lewin's-Mead, Bristol, aged 29, with dark hair and grey eyes, (which two circumstances often concur in scrophula), skin not sensibly smooth or fair, and of a robust habit—six years ago caught a very violent cold, and has never since been free from cough and mucous expectoration in cold weather. In about three months after the commencement of his catarrh, the right submaxillary gland began to swell. After a time it broke, leaving a scar of the usual appearance. In about a twelvemonth from the first seizure by the catarrh, a hardness was sensible in the submaxillary gland of the other side. This continued nearly in the same state for above four years; nor did it increase much till some time in the course of the winter just past. It broke only within these five days. The patient seems inclined to ascribe the last gathering also to another bad cold. His history clearly shews how unsafe it is to rest satisfied with any distinct scrophulous complaint, though it may subsist in the most perfect quiescence. Such a complaint ought to be considered as an enemy in ambush, ready to make his assault as soon as an opportunity offers: and to make it with more effect, as the person he attacks is less robust.

Should the proposal at the end of the present essay towards more effectual instruction in the art of preserving health be thought worthy of acceptance, it will be in my power, for some time to come, to bring forward these three instructive patients to those, who may wish to hear their story from their own lips.

The two latter declared that their relations, so far as they knew, had never been affected with any complaints of a like nature. Little stress indeed can be laid upon the testimony of the poor in a case of this kind. But in families, too well informed to be ignorant of such an event, and too candid to suppress any information, which the physician may require, I have many times met with the same reason for believing that *scrophula is often not hereditary*. From the nature of the complaint indeed, I think we ought to be assured that it may originate in any individual without the least hereditary predisposition. One might, without much risk of failure, if such trials were allowable, engage to produce it by art. Such trials indeed have been unintentionally made with a result as decisive as in experiments conducted according to the most deliberate plan. Men and quadrupeds transported from their native hot climates into our colder latitudes, become continually scrophulous. Numerous examples of this fact have been afforded by negroes, and by apes of different species. These observations combined with the information to be obtained in families, leave no room for doubting that Europeans are perpetually rendered scrophulous as well by privation of warmth and of other necessaries, as by causes of a different nature.

Another prevailing error of considerable importance, as it tends to the destruction of many constitutions, and to the encouragement of fraud, is the confounding of certain affections of the skin with scrophula. The news-papers are perpetually offering a cure for *scurvy in the blood*, and for the king's evil. Now the eruptions, vulgarly called *scorbutic* have no affinity to *scrophula;* and it is either the similarity of these two words in sound, or the occasional injury to the skin by the scrophulous swelling and suppuration of superficial glands, that keeps up this miserable delusion. So easily do property and life become the prey of the vilest and most clumsy impostures, when man is left in ignorance concerning himself, so that his hopes and fears, relative to this dearest of all objects, are subject to no sort of controul from reason!

About half a century ago, Dr. Russell published his celebrated dissertation on the use of sea-water in glandular atrophy, and in other diseases of the glands. The testimonies of other respectable practitioners have since been added; and a large part of the public has learned to look towards the sea as a resource, to which they may fly in case of need without further enquiry.

The effects of sea-water and of the sea-air seem to be confounded in many minds.

And it is for the benefit of the latter that, in the maturity and towards the decline of summer, the metropolis pours out part of its crowds to the Kentish coast. That the moral and physical effects of a journey and of fresh scenery will do much towards the restoration of certain invalids is a fact daily experienced in the practice of physic. But in scrophulous debility, or under any serious scrophulous ailment, no expectation of this kind ought for a moment to be entertained from mere removal either to the sea-coast or into the interior of the country. It is to be lamented that the rich should adopt such false measures, and waste a portion of time, that perhaps cannot be retrieved. It is more to be lamented that those, who are not rich, should throw away, after the same hopeless chance, that money which they can ill spare, and of which a portion, expended upon proper means, would certainly accomplish the object they have at heart.

Of *sea-water*, after the value set upon it by such a man as Mr. Hunter, I dare not speak in a tone of the same disparagement. Invalids who have used it have however often come under my inspection without exhibiting proof of its virtue. And experience assures me that there is an infinitely better method of treatment than any in which sea-water is concerned.

Nor should it be concealed from those invalids or from those friends of invalids, who may be disposed to take upon them the physician's office, that the internal use of sea-water—much more its use as a cold bath—has been manifestly followed by increase of scrophula and even by death. In the *Philosophical Transactions*, vol. lv. we find an account of a young lady of a thin, delicate constitution, very weak and sickly when a child, who "being incommoded now and "then with an inflammation and swelling of "the upper lip, which was thought strumous" (scrophulous), drank about a pint of water every morning for ten successive days. On a sudden, she was seized with a violent hæmorrhage, and on examination was found to be "perpetually spitting blood from the gums." - - "She at length expired choaked "with her own blood." Her blood before drinking the sea-water was in a pretty good state. But some taken in her last sickness was mere putrid, dissolved gore.—On this fact, the late celebrated Dr. Huxham remarks that "with the thin, tender, and hectical, "a course of sea-water seldom agrees." And among the thin, tender and hectical, are not most scrophulous members of affluent families comprehended? When cases, such as this, occur, we can by no means be certain that they will always be recorded. Nor is it necessary that there should occur many cases, so violent in their symptoms, and so rapid in their progress, to deter from the unadvised adoption of a plan, which in the feebler class of scrophulous subjects, many physicians will be disposed to join Dr. Huxham in considering as worse than inefficient. Parents, particularly observant of their children, have assured me that they have found those, sent to the sea for cold bathing against scrophula, to grow faster during this course, and for some time afterwards. At the same time, they were evidently reduced in strength. Nevertheless, in compliance with medical authority, the plan was continued for several seasons, though the accompanying debility and the accelerated increase of stature, a process already too rapid, failed not to raise doubts as to its propriety. These doubts were too often confirmed in the sequel.

It has been observed that scrophulous affections sometimes terminate favourably as to the general health, by sudden suppuration of the glands about the jaws. Some authors have fabricated whole theories out of this phenomenon.* And if, by any fortunate concurrence of circumstances, the cessation of growth should be succeeded by vigour, the indolent tumours will inflame, gather, break, and disappear for ever. Some of the strongest persons I know are marked by scrophulous scars under the jaw. They have been men, of active habits, using exercise in the free air, at no time subject to exhausting labour, always enjoying plenty, and exempt from long-continued mental suffering. Their mode of life was their cure.

A directly contrary crisis is much more common. It appears in the form of some one of the graver scrophulous diseases. The persons subject to it are those, who in their early years have been cruelly debarred from that enjoyment of the bodily faculties, which, by the laws of our nature, is destined to make youth happy, manhood healthy, and old age comfortable. The marks of this treatment will be apparent enough upon the person; and let these be the criterion for judging, by whom the unfavourable termina-

* Il semble qu'il y a, dans ce temps de crise, un mouvement interieur qui pousse, du centre à la circonference, une chaleur salutaire qui ranime les tumeurs indolentes, les conduit promptement à la suppuration, et procure des cicatrices solides.

tion of scrophula is to be expected and guarded against!

When the spine has grown awry, a machine for supporting the weight of the head is most strongly to be recommended. But let no one trust to machinery alone. It has no power to correct the vitiated constitution. Cases are for ever occurring, where the patient goes on with a steady pace to destruction, while the machine is perfectly answering its purpose. Courses of sea-bathing and of water-drinking are usually limited to some six or eight weeks. Unhappily for invalids, this limitation is transferred to the treatment of scrophula. But nature refuses to be circumscribed within such bounds. Concerning the practicability of a radical cure, I refer to what I have already said. But the means must often be continued for months; and it will be sometimes proper that they be omitted for a season and resumed. Whenever the enemy threatens to collect his scattered forces, we should constantly take the field, and put a stop to his operations.

Miss — —, for whom I have lately been consulted, had a white swelling of her knee, at five years old. This was perfectly cured by long application of blisters. However, proper medicines were either not given or not continued long enough to remove the constitutional affection. She therefore continued weak and unhealthy; and of late, at the age of 9, a formidable affection of the glands about the jaws has been observed. This is one proof, among hundreds, of the insufficiency of the most successful local treatment, and of the necessity of persevering till the scrophulous debility is entirely removed.

I feel that these conditions are not altogether pleasant. It were to be wished that the business could be dispatched in one short campaign. It were indeed to be wished that a set of tumefied glands could be reduced, as an overloaded stomach can be relieved, at one operation. Or rather it were to be wished that the end could be attained, while patient and practitioner are as much at their ease as Owen Glendower was, when he—

Called up spirits for the vasty deep.

But as it is always probable that the constitution will be retrieved, and as the patients are generally young, medicine must be allowed to be a less evil than the disease. That the one is an evil as well as the other, I have no inclination to deny. But the great concern of prudence lies in the comparison of evils; and to understand in what proportion these two stand to one another, I advise to acquire as accurate an acquaintance with scrophula as possible, even from actual observation of its worst forms. Let the master of a suspicious family frame to himself such pictures as the one sketched in an early part of this essay. In nature, he may easily discover the component parts of pictures more frightful. Does the reader revolt at so painful an employment of the thoughts? Does he not know that, in this world, disagreeable things must be either thought of beforehand, or be endured? And let him, if he can, assign a more powerful incentive towards the early use and patient continuance of the measures, necessary to prevent the realization of such pictures. I would have a person, threatened with such a domestic disaster, put himself first in the place of a scrophulous patient, whose flesh is melting down into matter, amid alternating chills and heats, and perspirations, and languors, and pangs; and then in the place of a parent, bending over so heart-breaking a spectacle. Nor would I have him rest even here. Let him once more dramatise these scenes in idea, according to the best powers of his imagination, making himself and his own children, the actors and the sufferers. His affection may not perhaps require the spur. But by exercises such as these, it will cease to be blind without ceasing to be ardent. I defy any one, who will submit to them for a time, to follow the general parental practice, which consists in devotion to the petty interests of the offspring, and in neglect of its great interests. Would it, after this, be possible that such or such an addition to the stock of a child's wealth should ever stand in competition with such or such an addition to its stock of health?

We know what a sublime function Aristotle assigned to the antient tragedy. But there is a far more effectual purifier of the soul from unwise purposes and mean desires than fictitious distress—a purifier, which in that first, faint dawn of the science of human nature, it could not be expected that the eye even of an Aristotle should discern. It is the *pity and terror*, raised by a proper contemplation of the actual, home-bred miseries of life.

Nor let it be supposed that the feelings, which will arise from such contemplations, must be pure and uncompensated pain. On the contrary, if they be followed up by that energy of conduct, which they may be expected to inspire, they will prove a momentary sacrifice of that species, which brings the amplest eventual reward.—Here is a lot of young people, doomed by descent, by constitution, by the usages of families to what is most loathsome and destructive in the accidents, to which humanity is liable. It is a fate, from which they would not have been rescued by the strictest observance of the rules, laid down even by some writers of the first authority on the management of children. But those, who had to form a scheme of life for these children, were clear-sighted enough to fix upon that object, to which the extemporaneous decision of common sense and the most careful consideration agree in assigning the preference above all others united. The issue corresponds to the wisdom of the choice. Then it is that the idea of past danger rises only to render the sense of present security more grateful, and the pleasure, springing from the difficult and well-accomplished task of education, is enhanced by remembrance of the pain, in which it was begun.

What piece upon the stage, with the help of the strictest poetical justice, can fill the mind with equal serenity? Or what emotions, excited by the drama, can be compared for effect in rectifying the moral sentiments to those, by which so happy a family denouement was prepared?

END OF ESSAY SIXTH.

ART OF

PRESERVING HEALTH.

It has always appeared to me extremely feasible to give PRACTICAL instruction, provided one could find persons really in earnest to become proficients, in this art. Five years ago, when employed in drawing up an address to be delivered before a promiscuous audience, assembled to receive some elementary information, respecting the structure of the human body, I ventured to deliver such an opinion. Nor will it be easy to find a fitter opportunity for reviving and enforcing it than at the close of a series of considerations on scrophulous diseases. The terms I used were the following — "*I do not see why popular*
"CLINICAL *lectures should not be joined to*
"*popular anatomical lectures. They should*
"*doubtless differ widely both in plan and exe-*
"*cution from clinical lectures for medical*
"*pupils. It must be their aim to make fully*

"*sensible the mischiefs arising from systematic*
"*irregularity; from injudicious management*
"*after exposure to the inclemencies of the*
"*weather, and from the other innumerable*
"*ordinary errors of individual conduct. They*
"*must explain the origin and conduct, much*
"*more minutely than the treatment of diseases.*
"*They may be conveniently undertaken where-*
"*ever there exists an infirmary.*"

It may not be unnecessary, for the sake of some readers, to observe that *clinical* lectures are instructions grounded upon the cases of patients, actually submitted to the inspection of the pupils. Their introduction is the greatest improvement made in the schools of physic. What will confer greater skill in the cure of disorders, must be equally efficacious as to their prevention. And in this, not the members of a particular profession only, but all mankind have an active interest.

The avidity, with which people read and converse about the means, continually proposed for influencing the fate of the rising generation, might lead one to believe that they would be equally eager to qualify themselves for promoting its happiness in the most effectual manner. And after the necessary allowance for the difference between talking and doing, it might assuredly be expected that the prospect of advantage to their children would induce numbers of parents to embrace a plan, demonstrably conducive to this end, and wholly free from danger, disgust, or inconvenience.

To a medical practitioner, who has constant intercourse and considerable influence among the invalid poor, it must be easy to form such a plan. He would only have to select such patients as would, without reluctance, and could, without impropriety, be presented to a number of spectators, before whom he should question them as to the origin and nature of their complaint. To do away every shadow of indelicacy of conduct towards the indigent sufferers themselves, he should, before their appearance, prepare his audience for observing the most significant peculiarities in the complexion, features, and person. But these points of management will easily occur to every reader's imagination. After the invalids had retired, the connection of cause and effect in the case should be pointed out. Explanations from anatomy, chemistry, and other sciences should be introduced, and the effect of the medicinal and dietetical treatment be shewn by producing the party, subjected to it, from time to time. It should be the principal object to illustrate the obscure origin, and to detect the insidious course, of those chronic diseases of the country, which infest all conditions alike. It would be impossible that there should not offer an immense multitude of facts extremely curious, if considered only as belonging to the natural history of the human animal, and acquiring a tenfold interest from a perception of their applicability.

The author has it most amply in his power, as will appear from the annexed list, to execute the scheme, of which he can here exhibit only a very imperfect sketch. The simplicity and the success in treating the worst and most common chronic diseases at the Pneumatic Institution is such, and the fact has become so well known in Bristol and its environs, that poor invalids are resorting to it in rapidly increasing numbers. And among these, the choice would be very much unrestrained. All the forms of scrophula would be seen, and it would appear in what a large proportion of instances the disorder could be removed, even under the disadvantage of too sparing and meagre a diet. The power of medicine in arresting the progress of consumption in its early stage would be witnessed. And it would be shewn with what perfect certainty indigestion, hypochondriasis, hysteria, and a kindred and very harrassing tribe of complaints can be removed by means, first brought into use at this establishment for medical research. Similar success would be witnessed in cutaneous eruptions. Many instructive particulars, relative to the effects of groundless fear, of vulgar prejudices, and ill-adapted dress, have come before us, and the same will not fail to occur hereafter.

There is one lesson, which I have been taught by my own experience and that of others. He, who would render mankind any great service, should qualify himself by as long a preparation of patience, as the Pythagoreans were enjoined of silence. Otherwise, what can the seed of his good intentions bring forth to him, but the bitter fruits of fretfulness and misanthropy? I shall therefore, without chagrin, see a project, which I think capable of producing boundless benefit to the human race, rise from the humblest beginnings. And if there be within reach of me only as many parents, disposed to learn from the book of nature how to guard their progeny against ill health, as the angel required of righteous to save Sodom and Gommorrah, I shall not disdain to give them the best assistance in my power. I shall impose upon them no heavy burden of trouble or of expence; I shall only require them to wait upon me at the Pneumatic Institution at my own seasons, and to bestow upon the selected invalids such small gratuities as may compensate for their loss of time, and render them punctual in their extra attendance. A few mothers of growing families I should be glad to find accepting this invitation. I would pledge all I am worth, that those, who can overcome their first nameless and unfounded apprehensions, shall not only not meet with any thing to give them the slightest feeling of repugnance, but that they shall be gratified in a high degree, as well as instructed.

Number and Kinds of Cases of Patients, who have applied at the Medical Pneumatic Institution, from January 1 to April 18, 1802—registered by J. KING, Surgeon.

	No. of Cases.
Fevers, intermittent	13
—— continued	7
Dysentery and Diarrhœa	8
Acute Rheumatism	4
Inflammatory Sore Throat	2
Catarrh chronic and acute	38
Hooping cough	36
Shortness of breath of different kinds	22
Blood spitting	4
Pulmonary Consumption	52
Chronic Rheumatism	39
Scrophulous affection of the	
- - - - - - eyes 11	
- - - - of the ears 8	
- - - - swellings 25	
- - - - - - - abscess 7	
- affections of bones and joints 17	
- mesenteric atrophy 21	
Scrophulous in all	89
Rickets	3
Indigestion	32
—— bilious	17
Pain in the stomach	2
Hypochondria	8
Hysterical, and other diseases peculiar to women	98

	No. of Cases.
Headache	9
Giddiness	7
Palsy	9
Epilepsy	6
Catalepsy	1
Bleeding at the nose	2
Urinary and calculous complaints	10
Dropsies	12
Piles	3
Fistula	2
Cancerous	3
Divers cutaneous eruptions	41
besides ditto from poison	1
Itch	4
Scald head	3
Venereal	10
Irregular action of the heart	2
Diseases of the eyes—external	6
ditto——internal	6
Deafness	9
Worms	12
Abscess and tumours	5
Sprains and contusions	7
Sore legs	12
Gout	3
Inoculated for the Cow-Pock	19
Total	678

Printed by MILLS, St. Augustine's, Bristol.

ESSAYS

ON

THE MEANS

OF

AVOIDING

HABITUAL SICKLINESS,

AND

PREMATURE MORTALITY.

ESSAY SEVENTH.

Printed by MILLS, St. Augustine's, Bristol.

On verra toujours beaucoup de pulmo- niques parmi nous, tant que nos modes tendront à l'affoiblissement de la poitrine et que peu de personnes auront l'esprit assez fort pour se mettre au dessus de l'usage.

BOURRU *des bains*.

ESSAY

ON

CONSUMPTION.

IN a series of tracts on the means of avoiding *habitual sickliness* and *premature mortality*, I cannot expect that it will be allowed me to decline all consideration of the principal source of both calamities in the countries which we inhabit. Otherwise I might content myself with referring to what I have already written for popular use *. To a physician, however, whose fortune it is to be consulted by numerous invalids of all ranks, new illustrations of the nature and origin of so destructive and common a malady will be offering themselves from time to time. Hence he will continually discover cause for new admonitions, and meet with examples that more severely put to the test the propriety of those formerly given. Nor might it be without its use to draw into a narrower compass the materials of a more extensive treatise,

* *Essay on Consumption*, 8vo. Longman and Rees.

even if later reading, reflection, and experience, had not placed the author far above the necessity of confining himself to an abridgment of his own work.

When the danger is general, admonitory writings cannot easily be too much multiplied and varied. Consumption may be regarded as a vast pit-fall, situated on the high road of life, which we have not sense enough of our common interest to agree to fill up, or fence round. Heedless fathers and mothers are for ever guiding their sons and daughters directly into it. Nor is it often that those who obtain, and who in other respects deserve, most credit for a diligent discharge of the parental office, will be at the smallest pains to conduct them safe over this treacherous ground. It is therefore the more incumbent upon those, who are acquainted with the margin of the precipice, to plant upon it marks, striking enough to catch every eye. And how happy if bands of volunteer watchmen would submit to the generous service of filling all the approaches with their warning cries!

MORTALITY FROM CONSUMPTION.

Should it be asked, *in what manner is the existence of every person, born and continuing to reside in the British islands, most likely to terminate?* the following table will supply the answer. The whole number of deaths for several years together, at two different periods, is taken from the bills of mortality; and the four largest items subjoined. It will be seen what a proportion consumption bears to the whole, and how regularly the deaths under it increase, with the increase of the total number of the burials.

Years.	Buried.	Stated to have died of Consumption.	Of Convulsions.†	Fevers.	Small-Pox.
1790	18,038	4,852	4,003	2,185	1,617
1791	18,760	5,090	4,386	2,013	1,747
1792	20,213	5,255	4,646	2,236	1,568
1793	21,749	5,474	4,783	2,426	2,382
1794	19,241	4,781	4,368	1,935	1,913
1795	21,179	5,733	4,758	1,947	1,040
1796	19,288	4,265	3,798	1,547	3,548
1797	17,014	4,776	3,804	1,526	512
1798	18,155	4,533	3,663	1,754	2,237
1799	23,068	5,721	4,512	2,712	2,400

Under one head, those that die of mesenteric atrophy, and of certain other species of decay, are thought to be registered along with the victims of pulmonary consumption. But in the same bills we find a number, seldom

† Convulsions is a name that stands for a variety of disorders, totally unlike each other. Children commonly die convulsed whatever may have been their complaint; and almost all such deaths are referred to this head.

greatly short of a thousand, under the titles asthma, phthisic, and cough. And these, if the cases were accurately sorted, would, I imagine, be found to go a considerable way towards replacing the sum it would be necessary to deduct, on account of inaccuracy, from the decay induced by genuine phthisical affection of the lungs. Of 1654 buried in one parish in Bristol—683 stand under the title of consumption or decline: and from a register at Shrewsbury it will be presently seen that for ten years, above one death in four is appropriated to consumption.

At Chester, in the years 1774 and 1775, particular pains were taken to obtain correct lists of mortality. Strict injunctions were given that no disorder, unless attended by a cough, should be called consumption. The faculty, in general, sent written certificates, or gave verbal information, to the clerks of each parish, of what disorder the patients died. And what was the result? Did these precautions afford any reason for supposing that the London bills of mortality give too formidable an idea of the havoc, made among the human species by consumption? So far from it, that in 1774 more people were found to die of this disorder, in the healthful city of Chester, between the ages of 10 and 50, than of all other diseases put together. And in 1775 more died between the ages of 15 and 50. (*Philosophical Transactions*, LXIV. & LXV.) The numbers will be given below.

It were greatly to be wished that it were by any means practicable to carry on the same enquiry from house to house, if not through the whole kingdom, yet in several cities, towns, and country parishes; so that an universal conclusion might be drawn. We might then have something, like a precise notion of the proportion of families, which this disorder puts into mourning, and of the average proportion of seizures to the whole number in each family. I feel that at first there would arise in innumerable bosoms a passionate determination to oppose such an enumeration. To some inconsiderate persons, especially when they are not personally affected, it is the very worst part of the worst evils, that they themselves should be made to fix their attention upon them, or that others should come to learn their existence. But it is not the part of the firm and the prudent "to hide from themselves their "state." They will bear to contemplate it, as well knowing that this is the first step towards its amendment; and the tender father of a family might be disarmed of his indignation, if he were requested to consider that it is not so much his own being and happiness that are involved in what he may, at first hearing, desire to conceal from himself and from the humane enquirer, but the being and happiness of others, over whom though he has great positive and almost uncircumscribed negative power, he would not deliberately stretch it to the hazard of their destruction.

SICKLINESS from the Consumptive Disposition, where the Disease itself never forms.

But neither the preceding statements, nor those to be produced below, give any thing like an adequate idea of the misery, connected with consumption. For I apprehend that, in consumptive families, numerous enough to afford field for observation, another phænomenon, not less melancholy, and not less demonstrative of the necessity of preventive measures, would very generally shew itself. The individuals, spared by consumption, would be found a prey to other complaints. They would hold their station among the living, not by the tenure of good health, but merely by favour of one bad disorder in possession keeping out a worse. Some months ago, I had under my care a patient of great talents and observation, who looked forward to that fate, which consumption so often brings upon females in their prime of life, and which in this instance could not be averted. When I meet with invalids, observant of themselves, and disposed to communicate, it has long been my custom to request a particular narrative, in writing, of all the circumstances, which they suppose to have any relation to their complaint. It is the only way, I believe, to get at the knowledge of the original causes of disorders, when they lie any way remote. This patient not only drew up the following account with great readiness, but begged I would use it for the benefit of persons, threatened with similar danger, if I thought it likely to answer so good an end. With some slight variations, which however make not the least difference for the better or the worse in regard to health, I believe it will apply in most of the instances, now become more frequent and likely to increase, where a plan is adopted for bringing up children, that has the credit of being extremely rational, and which is certainly, in many respects, preferable to the common routine, though still miserably deficient.

" Our family was large. There were twelve " of us in all, eight sisters and four brothers. " I do not reckon three others, that died in " their infancy. Our ancestors were said to " have lived to a good old age. My grand-" father was a great martyr to the gout. " Otherwise, no hereditary disorder had ap-" peared on either side. My father was of " a studious turn. He had always paid par-" ticular attention to the subject of educa-" tion. He was ambitious that his sons " should shine; and his daughters he wished " to have more information than in general " falls to the share of women. It lay near " his heart that his domestic arrangements " should be admired in the neighbourhood. " He accomplished his purpose *to all appear-" ance*, for it is by no means clear that the " compliments he received were always quite " sincere, though he was too full of his own " anxiety to perceive the circumstances that " often gave me suspicion. Alas! had we " been thoroughly known, we should have " been found to deserve pity, and not admi-" ration!"

" Do not suppose we were tortured into " shape, or fixed down beside the harpsichord " for half the day. Nothing of this. We " were suffered to pursue the accomplish-" ments of music and dancing only so far

" as not to appear awkward in company, or " quite ignorant of the subjects of common " chat. For the rest, pains enough were taken " to secure us against

A youth of frolicks, an old age of cards.

" Nothing was held cheaper with us than " the things, at which girls are set to work " so hard, as soon as they begin to be *Misses*, " and which they find so useless as soon as " they are come to be *Mistresses*. We were " scarce ever assembled in full circle, without " hearing these things made the theme of " ridicule. My mother had too good rea- " son for comparing what she had suffered " in the acquisition of accomplishments at " school, and their insignificance afterwards. " But what with us was substituted for ac- " complishments, I can most conscientiously " attest to have been scarcely more conducive " to my own happiness. You shall imme- " diately hear whether my sisters could attest " the same. Vanity was still the ruling " principle. It was vanity indeed of loftier " pretensions, and a more solemn aspect; " but still it was vanity. For instruction, it " seems, may be made matter of vanity, as " much as dress. Of this our feelings the " more forcibly convinced us, the older we " grew."

"The graver part of light reading was "our amusement, and the lighter parts of "science our study. The choicest of the new "publications for children and young people "were procured, with unfailing diligence, for "our use. They often entertained us, and we "were certainly indebted to them for much "improvement. But ah! how possible that "parents may avail themselves to the utmost "of these helps, and their children grow up "to be the veriest of wretches in themselves, "however they may shew in competition. "Our example shews that something else is "wanting. This *something* failing, all may "as well fail. There was not the least "idea of laying our appetite under any "restraint. But health was never a constant, "and seldom an occasional, concern. How "clearly do I perceive this deficiency since "my conversations with you! Different little "domestic incidents occasioned us now and "then to suffer from too long fasting. Our "sufferings consisted in a certain inward "sinking. We eat not so much from relish "for food, as to get rid of this insupportable "sinking. We were unacquainted with the "pains of hunger; but then we knew as "little of the pleasures its gratification "brings. At dinner, we collected with hea-"viness in our looks, and I know not with

" what sense of disrelish in our palate, and " of mawkishness in the spot, where the " heart-burn is felt at other times. I often " thought that a spectator, used to a party " with a proper appetite, would suppose we " had just taken, or were about to take, " physic. It is certain we regularly sate " down as if it had been to a task. Our saun- " tering walks were in the same listless tone. " We passed through a few fields. We set " out languid, and came home again without " feeling refreshed. Except in hot weather— " and then we were oppressed—our state " was uncomfortable from chill—our feet " generally stone-cold. But in sharp frosts " —oh! I want words to express the mi- " sery of that cramped, contracted feeling, " which then seemed to clasp us all over. " The whole day we had the goose-skin fast " upon us."

" We all fell off on the *change of consti-* " *tution.* To our surprize, we found our- " selves without our former childish spright- " liness. I, for my part, often looked round " to try if I could discover which way mine " had flown, and hoped it would return after " a time. But it visited us no more. There " was something in the head or the stomach, " or in *the nerves* I fancy, for we felt down-

" cast and cheerless, that kept us uneasy for " the moment; and we had no relief to ex- " pect in the regular changes of our day. " But I fear I shall not be able to make this " state of nameless discomfort well under- " stood. And if I could, it would prove as " tedious in description as it was to endure "

" At 17, my eldest sister began to look " thin and sallow. She had no spirit to do " any thing, though formerly very indus- " trious, and would be continually brooding " over the fire. She was told she was lazy, " and that if she would but exert herfelf, " she would be very well again. She did " exert herself, and grew worse and worse, " till it was necessary to call in a physician; " and he declared her lungs to be affected. " She died of a decline in about half a year. " Another died in the same way, after hav- " ing had the measles, which she never " properly got over, at least never regained " her strength. A third was seized with a " short, teazing cough in winter; which was " thought nothing of. But it got worse in " the spring, which turned out very cold, " and she quickly fell into a decline. I can " evidently trace back my own illness for " more than five years. I have been harrassed " by shortness of breath for all that time.

"There have been shooting pains too in my "chest; and I have been very subject to "flying shivers, and at times greatly flushed, "with most disagreeable dry heats in the "palms of my hands, and throbbings in my "neck and temples."

"In a *certain respect* we were scarce ever "proper, though otherwise we were unwell "in opposite ways. My sister, who has "applied most closely, is much plagued "with head-aches, which return more and "more frequently, and sometimes almost "distract her with the pain. The one you "have seen with me is as much troubled by "weakness of her stomach. Her digestion "is miserable; she is quite oppressed with "wind; and the loud noise it perpetually "makes, occasions her the greatest distress."

"My next younger sister was early sub-"ject to momentary faintings. Afterwards, "as it is believed in consequence of the "shock from my father's unexpected death, "a variety of nervous symptoms came on, "and the health has been entirely deranged "ever since. No remedy, that medical skill "could point out, has been omitted; but "every symptom has increased in violence, "occasioning the most agonising cramps "and spasms, which alternately invade the

" whole frame. It is quite impossible for " me to describe properly the whole of this " distressing case, the cramps and spasms " often coming to such a height as to cause " a deprivation of the use of the limbs, and " to render confinement to bed necessary for " two whole years."

" I have two sisters married. One has no " family. The other has had three children " in little more than four years. Neither " can be boasted of. But the latter has been " in a terrible state ever since her first con- " finement. She is said to be threatened with " a nervous atrophy. My brothers can none " of them be called strong. Two have " had no particular complaint. One is ex- " cessively subject to take cold. Apprehen- " sions are entertained for the fourth, who " grows fast, is narrow-chested, and has " bleedings at the nose very often."

" Since I became alarmed about myself, " I have done what I understand patients " are more apt to do in other complaints, " and what you medical gentlemen all set " your faces against. I have read different " treatises on consumption, and have been " particularly struck by the observations I " have met with, relative to its origin. These " observations unfolded to me the melan-

"choly secret of my own history. Must
"not the house-lamb always turn out a
"helpless, shivering, short-lived creature, in
"comparison with the field-lamb, though
"the utmost pains be taken in training
"him? I feel some satisfaction from the
"idea that my parents did not live to see
"the compleat issue of all their cares, though
"they might not have been aware that the
"miserable situation of a surviving child
"was the genuine issue of those cares. They
"seemed not to be sensible of it in the two
"they lived to lose. Would to heaven that
"other parents could be made sensible in
"time, when they are incurring the same
"danger! Nimble fingers and nimble heels are
"very far from being of the same value with
"a well-furnished head and a well-disposed
"heart. But what avails the one more than
"the other, in guarding against the horrors
"I have seen and felt? Nothing surely—or
"does your experience entitle you to object
"to this inference of

LOUISA ———?"

If a plan, formed upon mature deliberation and with some judgment, be liable to be thus cruelly defeated in consequence of one capital oversight, what can we expect in the total absence of all sort of consideration? In fact, we shall find that young people, confined to frivolous pursuits, grow up to be so many *stocks*, on which consumption, or some other complaint of debility, does not fail to engraft itself.

HAS CONSUMPTION INCREASED IN FREQUENCY?

La Phthisie pulmonaire est un des plus grands fleaux de l'humanite *par sa nature* et ses resultats. - - - Elle est très-commune, et chaque jour elle semble le devenir davantage.

J. B. F. BAUMES, 1795.

Phthisis recentiori ævo certo certius quam olim frequentior.

C. G. T. KORTUM, 1789.

An enumeration or census of phthisical families * would enable future generations

* Such an idea will perhaps be thought less absurd, if it be considered that the thing has actually been realized upon an inconsiderable scale. The author carried on an enumeration of this kind through a small town and two parishes, to discover whether the mortality among poor children was not in proportion to the number of children in a family. A parish in Chester was numbered for medical purposes. Dr. CAMPER, the late great anatomist of Holland, numbered the inhabitants of Franeker to

to determine whether the disorder be on the increase or the decline. This is by no means however, for the present, a question of idle curiosity. For there can scarce be imagined a more powerful incentive to every measure of prevention, than the knowledge of a fatal disorder having become more frequent by regular progression. As the children of consumptive parents are peculiarly liable to consumption, if it have once become more ge-

---

ascertain the proportion of the lame. "In the city where "I live," says he, there are 2,775 inhabitants, among "whom 96 are lame. Deducting 104 infants, not yet able "to walk, there remain 2,671, which number divided by "96, gives 28. Therefore there is one lame in 28."—*Haarlem Verhandelingen*, vii *Deels* 11 *Stuk. bl.* 451. There are many examples of a similar kind, though not subservient to the purposes of the exact medical enquirer, dispersed through the transactions of societies.

I believe that care to suppress the knowledge of the existence of complaints in families is often fatal to the weakly. The parents do not attend to the early stages, and for this reason, as they ought. Advice is tardily sought. The delicacy that would prevent an enumeration conceals nothing; because a general statement in numbers would disclose nothing with regard to individuals. The particular facts as known to friends and neighbours, are useless. A general statement might be of the highest utility. And what would strange readers and foreigners learn of A. and B. from numbers, expressing the proportion of consumptive families, to the whole families in Great Britain?

neral, it must go on spreading at an accelerated rate, unless it be checked by some new power. And does it not feel still more inhuman to withhold our endeavours, when the evil threatens to affect unborn generations, more and more widely, in indefinite succession, than if its channel and its course were limited? The narration, with which the preceding section concludes, gives an example of a very common calamity. On enquiry it will generally be found that *families, infected by consumption, are in other respects more sickly, or less acquainted with the enjoyments that flow from a vigorous habit of body, than families equal in number and other circumstances, but exempt from this disorder.* This will hold almost without exception, where consumption attacks part of the offspring of consumptive parents. The cause too, whatever it be, which renders us liable to this kind of malady, must have the same character as the effect. It must have gone on increasing or diminishing in proportion. And by attending to this, we shall be in some measure assisted in detecting it, than which we can have no concern more important.

It is certainly a very prevailing opinion among medical writers, that consumption is more frequent than formerly. But if we enquire what may be the reasons for this opinion, we shall not find them altogether satisfactory. Our predecessors have not left us any certain data for the comparison. No practitioner of medicine has it in his power to compare several generations. It falls to the lot of few to have very extensive experience of more than one generation. The experience of a practitioner may be partial even on the spot where he resides. He may be consulted by particular orders chiefly, or but for particular disorders. Much less may his observations apply to a whole country. My reason for acceding to the opinion of authors is this. I perpetually find consumption attacking new families. Nor does it appear to cease equally often in old ones, unless by rendering them extinct. If this be a fact generally observed, it must be allowed to have great weight in deciding the question.

What can be learned from the bills of mortality, perfectly confirms the same unpleasant conclusion. The confirmation is the stronger, because the conclusion appears to have been previously drawn from experience in the gross. I do not know, indeed, that any one had been at pains to examine the bills of mortality with this view, till Dr. Heberden, who, in the work formerly quoted, has presented us with the following statement. It was deduced, he tells us, in a coarse manner, from an average of about ten years. Its purpose is to exhibit a general comparative view of the mortality, occasioned by different causes, at three periods: namely at the beginning of the eighteenth century, at the middle, and at the close. Care was taken to select, from the bills belonging to each period, those years in which the total amount of deaths was nearly alike, viz. 21,000, a part of which are only introduced here.

	Beginning.	Middle.	End.
Abortive and dead-born	600	5,070	750
Bowel complaints -	1,100	135	20
Consumption - -	3,000	4,000	5,000
Dropsy - - -	850	900	900
Evil - - - -	70	15	8
Fever - - -	3,000	3,000	2,000
Gout - - -	26	40	66
Palsy and Apoplexy -	157	280	300
Ricketts - - -	380	11	1
Small-pox - -	1,600	2,000	2,000

The victims of consumption, therefore, if the same order of things continues, will probably soon be double of what they were a century ago. The two extracts from the register of the parish of Holy-cross in Salop, inserted in the Philosophical Transactions, would indicate a much more rapid rate of progression, unless the second ten years were peculiarly productive of phthisical complaints.

For the four articles before selected from the bills of mortality, they give as follows:

From Michaelmas 1750, to Michaelmas 1760.

Buried in all.	Consumption.	Convulsions.	Fever.	Small-Pox.
365	47	9	39	33

From Michaelmas 1760, to Michaelmas 1770.

Buried in all.	Consumption.	Convulsions.	Fever.	Small-Pox.
365	101	31	35	46

It is in this instance as in the London bills. And as far as the documents we possess enable us to judge, wherever the total of deaths appears remarkably swelled, a correspondent augmentation is found under the article consumption, which by no means holds of other articles so regularly. Nor is it the proportion of the consumptive only at different periods that has the strongest claim to our regard, but the proportion taken along with the absolute numbers. If from 3,000, 4,000, and 5,000, which are the sums assigned to the beginning, middle, and end of the century just closed, the cyphers were cut off, the addition of two-fifths to the mortality would be scarce more a matter of general concern, that the fatal accidents that have followed the prick or the scratch of a pin.

PERSONS MORE OR LESS LIABLE TO SEIZURE.

This great mass of disease is however far from being equally distributed among the members of society, though no age or station can flatter themselves with compleat security. Some fall into consumption accidentally; others from original disposition. The first class comprehends those who would never have been consumptive, but for the nature of the occupation, to which they have been put, or for some accidental occurrence. They may be therefore said to have been forced into the disease. The second class comprehends those who can have little hopes of escaping without particular pains, and therefore may be considered as naturally entitled to consumption. A brief comparative account of these two classes will pretty clearly lay open to us the course that nature follows in producing this dangerous disorganization of a viscus, so essential to life as we know the lungs to be. This being once understood, the manner of preserving them entire will be more easily understood.

We find by constant experience that there are certain chemical and mechanical irritating substances, which if applied to the parts of the body, disturb them in the exercise of their functions, or induce an incapacity to carry on those functions. Thus it has long been understood, and lately brought into general notice, that there is something in the atmosphere or soil of Egypt, which brings on an inflammation of the eye, and by a continued or more intense operation, occasions total blindness. I have heard it ingeniously conjectured, that this effect is produced by saline particles, which rise in the dust of the country, and violently stimulate the interior of the eye-lids, and the eye-ball. The idea was suggested, and is strongly corroborated, by the experience of Dr. Pallas, who describes himself as suffering exactly as the French and English soldiers in Egypt, when he was traversing the similarly impregnated soil of certain districts in the Russian empire, though mere flying sand did not materially incommode him. It is common enough at home to see dimness of sight, or total loss of vision, occasioned by quick-lime, carried by the wind into the eye. Certain fumes, as I shall presently have occasion to shew, will inflame the mucous membrane of the nostrils. The less exposed continuation of the same membrane, after it has descended into the chest, is not so sensible indeed. But although rough substances, finely divided and drawn in with the breath, produce no immediate *pain*, they lodge in the moisture, and when constantly inhaled, excite an inflammation, which injures the texture of the membrane, and spreading into the substance of the lungs, produces a train of symptoms that always end fatally. The most striking example of this species of injury is afforded by one of the processes in the needle manufactory. It is that of *dry* grinding, by which the needles are pointed. The persons employed in this labour are universally, in a short time, affected with the symptoms of approaching pulmonary consumption. They go on coughing till they either spit blood, or a thick substance, having the appearance of matter. They decline in flesh and strength, and scarce ever survive to the fortieth year*.

In behalf of children, bound apprentices to manufacturers of needles, it is stipulated that they shall not be employed in this pernicious branch of the business. Indeed its effects are so notorious, that the manufacturers are said

* Memoirs of the Medical Society, V. 90.

to find it difficult to engage hands for pointing the needles. The lure of high wages is absolutely necessary. These enable the miserable victims to drown their sufferings and the sense of their danger, in intoxicating liquors.

There can be no doubt whatever as to the cause of the peculiar species of mortality among the needle pointers. Particles, worn off from the steel and the grinding stone by their mutual attrition, fly about the workshop. They are drawn in during respiration, and accumulate in the air-cells of the lungs. There they cause the substance, known to physiologists by the name of coagulable lymph, to be thrown out by the secreting vessels. By this coagulable lymph, and by the thickened mucus, the particles are cemented into small balls. In the early stage of the disease, mucus only, or mucus mixed with blood, is expectorated. But the expectoration changes by degrees, as in other consumptive cases. Purulent matter comes up, and it is very often mixed with the concreted metallic and earthy powder.

There does not seem to be any thing in the latter stages of the disease, by which it is distinguished from consumption, induced by other causes, though indeed practitioners of medicine might require a fuller description of the state of the lungs after death, than is to be found in any account with which I am acquainted.

The workmen are said to be fully apprized of the end that awaits them, and in despair to neglect all means of relief. Without doubt, the feeling reader will be shocked at the infatuation of these poor wretches, and still more, when he is told that no attempt appears to have been made to preserve them by a guard before the mouth and nostrils. But the self-neglect of the workmen, and the unconcern of the masters, are not phænomena confined to the needle manufactory. We shall find them universal, wherever there is danger of pulmonary consumption. No wonder, therefore, they should occur amid groups of the most abject of human beings, under the total absence of mutual affection, where one party is solicited by a temptation, which ignorance can seldom resist, and the other, rendered insensible to all the motives of humanity, by desire of gain.

An immense list of artisans of different name, whose labours are carried on amid the floating particles of earth and metals, might be subjoined to the needle-grinders. In France, the pulmonary affection of the workmen in plaister of Paris and marble, has acquired a peculiar name. It is called *maladie de grés*, or *maladie de St. Roch.* It seems to vary in no material respect from that which is so general in this country. Its course is the same. After death, the substance of the lungs is found indurated and full of tubercles. There are adhesions of the lungs to the membrane lining the ribs, and often one entire lobe of the lungs is destroyed by suppuration.

The softer dust, that arises during the fabrication of various animal and vegetable substances, is equally capable of bringing on a fatal irritation of the membrane, investing the air-pipes. Flax-dressers may be quoted as an instance. But all those labourers, who live during a great part of their time in an atmosphere, charged with floating animal or vegetable filaments, and even with dust from animal or vegetable substances, are subject to serious complaints of the chest, either of the acute or chronic kind. Consumption is the reigning malady. Nor is even the mild influence of the climate of the South of France sufficient to counteract this sort of irritation. The girls who come fresh from the high and airy district of the Cevennes, to perform the various operations in the silk manufactory at Nimes, from the preparation of the cocoons to the spinning of the silk, suffer like our flax-dressers. Even in a few days, they are described as losing their bloom and all their vivacity. They are seized with a dry cough, by which they soon come to be continually harrassed. They complain of oppressive pain in the chest. Fever often succeeds, and they die consumptive.

On referring to various articles in the statistical reports for Scotland, it will be seen that this is very much the history of the girls who employ themselves so much in spinning in that country. All these are the effects of disorder in the functions of the lungs. It is more than probably the office of this organ to dye the white globules of the chyle into the red particles of the blood. This process being interrupted, no wonder the roses of the cheek should wither. This, though a very efficient one, is indeed not the only cause of that languid air which interests the spectator so much for young people, when their health begins to decline. For if the organs, that are destined to prepare the chyle, do not send it to the chest in a proper state, it cannot be converted into good blood. Even Mr. Boul-

TON's mint cannot stamp the unmalleable metals into beautiful pieces of coin.

The readers of these essays may think the preceding facts, foreign to their own situation. But they are, in truth, as will immediately appear, essential to the formation of clear ideas on the present subject; and it is only on opinions founded upon such ideas, that we can depend for any permanent regulation of conduct. Much more than half the ravages of consumption depend upon the general want of such opinions.

In passing under review the different orders of phthisical mechanics, it would not be impossible to fix upon a groupe, which should connect those who suffer from hard and coarse powders, with the case of invalids so delicate, that they almost faint, if by chance they enter a room newly swept, and scarce inspire a particle of dust, except what the summer winds brush from the surface of the earth. Among the variety of workpeople, employed in the preparation of materials for the ornaments of dress, there would, I imagine, be found a number, whom confinement and much sitting concur in disposing to consumption, though the disorder is immediately provoked by irritating particles, deposited upon the mucous membrane. Perhaps, were either cause removed, many would be preserved who now perish. Without the dust, which they so frequently breathe, even their feeble lungs would continue sound; and persons of stronger chest would breathe the soft floating particles, of which this dust consists, with impunity. It is certain that the phthisical race of carpet-manufacturers, taylors, lace-weavers *, and others employed in a like manner, must draw into the chest a considerable quantity of linen, woollen, silk, and cotton fibres; and why should not these occasionally produce upon their peculiarly susceptible and delicate mucous membrane, the effect which similar substances produce on such as breathe them more freely and more constantly?

---

* 'A Arras, les dentillieres, dont le metier exige qu'elles soient toujours courbees, se plaignent ordinairement de la poitrine et de l'estomac parce que ces parties essentielles se trouvent dans une gene continuelle. Ce metier est si pernicieux pour la santé de ce sexe delicat, qu'on voit la plupart des jeunes filles, arrivees a l'age de l'enjouement et des plaisirs, le passer sans gaiete, et perdre en peu de temps, la fraicheur de leur teint, l'eclat de leur coloris, et la vigueur de leur temperament; souvent etre attaquees d'une toux seche presque continuelle; se plaindre des douleurs sur le sternum et le long des cotes; etre vivement oppressees, manquer d'appetit et tomber enfin dans la phthisie.

RETZ.

The opulent sufferers from consumption, and the artists who spend their day in warm, close rooms, free from floating particles of any sort, belong to one and the same class. All such artists are peculiarly subject to consumption. The confinement alone is sufficient to generate the propensity, as we find in shoemakers, without a particularly warm temperature. Confinement is sufficient to induce consumption in the ox kind, although by no means predisposed to the disorder. But with heat it acquires a double efficacy. Thus LEPECQ DE LA CLOTURE, a physician of Rouen, who has particularly attended to the paper-makers there, informs us that these artisans, (who begin to work about two in the morning,) have a flabby skin, a sallow complexion, and every appearance of feebleness. They are exceedingly liable to complaints of the chest, and very commonly die of pulmonary consumption. The atmosphere they breathe is constantly full of warm vapours. Mr. Carlisle, surgeon in London, in a letter to the author, relates of the gilders in that city, who work in heated rooms, and are often induced to expose themselves to damp and cold, that not less than six out of seven are said to die consumptive in their apprenticeship.

If we raise our view from the lower to the higher classes, we shall here perceive that it is upon the lilies of the land, that neither toil nor spin, that the blight of consumption principally falls. Were a table to be drawn out, exhibiting the result of an accurate and general enquiry, the females, who live without labour, would occupy a station, not exceedingly remote from the manufacturers, destroyed by breathing flint and steel *. I have very little doubt but they would come next to those, who have been described as suffering from gross and palpable irritants of the smoother and softer kind.

To understand the gradation of cases from the wretched needle-pointers to the pulmonic patients, whose apartments are filled by an atmosphere where scarce a moat is discernible in the sun-beams, it is necessary to

* A physician of Vienna considers the dust of the roads, as a cause of consumption. "This holds especially of "those great cities, which are surrounded by distant sub- "urbs, to which unpaved roads lead. Working men are "often obliged to travel from the suburbs to the city once "or twice in a day. - - - It is particularly bad when they "are forced by hurry to walk fast, or when they do so to "get sooner out of the cloud of dust. For by their deep "and frequent respiration, they must inhale much more "dust than in going gently."

remember that consumption may be induced in two different ways. The first is by the constant operation of violent causes, where there is no more predisposition than the structure of the organs concerned necessarily brings with it. The second is by the operation of slight causes, where there exists a strong predisposition. Here the habit compensates for the inferiority of power in the immediate agents. Nor does it seem essential that there should be received hard angular or sharp particles into the nostrils and air-pipes, in order to induce any common disease, of which the air-passages are susceptible. I am acquainted with a person, in whom a pinch of ordinary snuff will induce sneezing in the first place, as he is not an habitual snuff-taker, and afterwards a genuine catarrh. His first account of the phænomenon was, that *a pinch of snuff always opened his head.* His conception seemed to be that there was a reservoir of mucus somewhere above the nose, which the operation of snuff occasioned to be in part discharged. I did not comprehend the circumstance clearly, till after some explanation. But this is somewhat singular. There are acrid substances which will immediately cause violent sneezing in every person, and leave a catarrhal affection behind in many. Nay, the oxygenated muriatic acid, which cannot be supposed to act by any angular points, will produce long-continued sneezing, and in some individuals a perfect artificial catarrh. M. Guyton-Morveau, on smelling to a phial of what he terms his *extemporaneous oxygenated muriatic acid*, was seized with " a sneezing, which held him constantly for several hours."—(*Moyens de desinfecter l'Air*, p. 135.) In M. Vauquelin and others, the effects of this active gas were perfectly analogous to a common cold. There was the sneezing, with which the latter frequently begins; the same disagreeable feelings about the head, and afterwards a similar copious running of clear liquid from the nose. During this defluxion, two ounces of mucus have been discharged in half an hour *! The subsequent

---

* Un resserrement et une gêne insupportable dans les sinus frontales et dans ces arriere-fosses nasales; l'eternuement suit de pres cette premiere action; bientot il s'etablit un ecoulement d'une liqueur limpide comme un chrystal. Les eternuemens sont quelquefois si promptement repetés qu'une sueur abondante couvre tout le corps. - - - M. Vauquelin craignoit l'hæmoptysie. Après que les symptômes les plus violens - - - sont calmès, il reste encore pendant plusieurs heures un resserrement, une espece de roideur insupportable dans toutes les parties, qui ont ressenti l'ac-

roughness and soreness were no less observed, and there came on the same impossibility of breathing through the nostrils from the thickening of the membrane, and afterwards the well-known, dense yellowish green discharge.—When the gas gets down into the chest, or when the affection spreads by degrees from the nose into the bronchia, a regular catarrh, with the symptoms proper to a cold in the chest, supervene. A huskiness or dry heat is first felt. A cough of several days' duration follows; then a hoarseness; the person loses his taste, grows feverish, and has that confusion or embarrass-

---

tion de l'acide muriatique oxygene. Lorsque l'ecoulement s'arrete entierement, les fosses et les sinus du nez s'embarrassent. Ils ne permettent plus le passage de l'air pour la respiration et l'on se trouve fortement enchiffrenè. L'humeur s'epaissit - - - - Elle se detache par masses considerables tres epaisses d'une couleur jaune verdâtre. Si la vapeur de l'acide muriatique oxygene a passe par la tracheartere, ou si son effet s'est porté de proche en proche jusqu'a cet organe, il en resulte un rhume de poitrine, qui a ses periodes regleés et constantes. L'on sent dans la poitrine une chaleur acre, la toux dure pendant plusieurs jours, la voix devient rauque, l'appetit diminue, et les alimens semblent ne point avoir de saveur. Enfin il y a souvent une fievre assez forte, et un mal de tete sourd, qui brouille les idees et met l'individu, que l'eprouve, dans une position desagreable pendant plusieurs jours.

ment of head, which is well known to sufferers from frequent catarrh.

But not only have the symptoms of catarrh been produced by chemical fumes, exactly as by a succession of atmospherical cold and heat, but a genuine consumption has been most distinctly observed to be brought on by these fumes. It is related by Dr. Fourcroy of an apothecary at Argentan that, finding his elaboratory too small for the process for making liver of antimony, he placed the materials in a vessel in his garden. After he had set them on fire, a gust of wind blew some of the smoke into his face, He was seized with a convulsive cough, which lasted for several months. The cough diminished in violence, by degrees. A slow fever, however, came on; the patient fell off in flesh, and finally died of a confirmed pulmonary consumption. It was not indeed till the fifth year after the original injury to the lungs. The distance of time, however, is far from being any objection to the supposition, that the fatal malady was occasioned by the volatile, pungent, sulphureous fumes. The patients' chest and general health appear to have been constantly affected from the very moment. And though this had not been the case, but intervals of apparent freedom had occurred, the conclusion might nevertheless have been just. Analogous effects are every day observed from the vicissitudes of the elements; and on account of the very great distance of time, at which the disorder becomes incurable and the fatal event happens, they deserve more particularly to be pointed out for the benefit of a numerous class of consumptive invalids. The following is one, out of a variety of instances, which I have seen. A gentleman of North-Wales of a healthy race, robust habit, and good conformation, at the age of 37, caught a violent catarrh in winter, and from perpetual renewals, laboured under severe cough with expectoration for several months. His health, in other respects, continued good. He had no hectic fever. He did not decline considerably in flesh or strength. At the setting in of summer his cough left him, and he became perfectly well. The next winter he caught cold again, and suffered as before; and once more recovered in the summer. This happened for six following years. Only he found that, every successive winter, he more readily contracted and renewed his complaint. The seventh winter he became unwell as before. But in the warm weather he did not lose his cough.

On the contrary, he grew weak, thin, and feverish. His lungs ulcerated and he died, after a terrible struggle, of confirmed consumption.

That inflammatory affections of the mucous membrane are produced by changes of temperature in the air, and that damp, when applied to the body, does, as a cause of cold, contribute to the same effect, are circumstances, about which all the world are perfectly agreed. Nor am I now anxious concerning the mode of operation. The preceding facts are adduced in proof of a strict analogy between stimulating powders, exhalations of a peculiar kind, and atmospherical changes, in occasioning superficial disease in a particular membrane. We have seen that sharp particles and filaments, abraded from a variety of solid substances, occasion a fatal disease, which must begin on a part of the surface of the same membrane, and spread into the substance of the organ, to which that membrane is attached. We have seen, in one striking instance, that acrimonious vapours, which cannot stimulate by any points or edges, act in the same manner. And were more instances wanting, we might collect them in abundance from the elaboratories of the chemist, and from particular manufactories. It is a matter of universal observation, that consumption of the lungs is equally excited among the civilized inhabitants of various climates, by the vicissitudes of the weather. A *superficial* inflammation first arises, and on frequent repetition, or from some other circumstance, the mischief penetrates deeper, and renders the lungs unfit to contribute their share to the maintenance of life. That catarrhal coughs, acquired in the usual manner, do terminate in this manner, is denied by scarce any medical writer. Respecting the frequency of the event, indeed, there is considerable disagreement. But the present is not the moment for discussing that point. It is sufficient if the existence of the fact be allowed.

If a rigourous and variable climate be capable of inducing this fatal complaint, on whom must it principally exert its baneful influence? Unquestionably upon those, who by descent, by defective original conformation, and by peculiar habits of life, have a more than common tenderness of the parts, upon which the atmosphere acts. Those who inherit a title to consumption, should take the fate of their immediate or remoter ancestors as a fair and full warning. In cases where the death of the parent has happened during the infancy of children, would it not be kinder to put them on their guard, than to attempt to conceal from them their danger? Who, on finding himself entangled in the snares of a relentless enemy, would not charge with cruelty the friends that foresaw the disaster likely to befal him, and forbore to make the disclosure?

The principal defect of conformation consists in too great narrowness of the chest, which occasions the shoulder-blades to stand prominent, like wings just raised from the body, not yet unfolded, though about to expand for flight. A long neck seems frequently to accompany this structure, as if the narrow chest had been lengthened or tapered out into neck, instead of being expanded into due breadth. The tokens of the scrophulous constitution are also very commonly observed to precede consumption. These (to enumerate them once more), are a sleek, fair skin, with turgid blue veins. I have already, however, explained that they are to be considered as signs of debility, rather than of any specific disease. Hence they will precede scrophula and consumption, or not, according to the climate, and the circumstances in whieh the individual is placed. As signs of debility; the physician must not be complaisant enough to regard these as so many personal advantages, whatever may be the case with the poet, the painter, and the lover. In this country, they must be looked upon as foreboding consumption. The two opposite extremes of light and dark hair and eyes, are often observed in the consumptive. But if the chest is well-formed, and the body not slightly built, I look upon these complexional characters as very uncertain in their signification. The want of a free sweep in the ribs, and the consequent narrowness of the thorax, I regard as the most important of all personal indications. The bad condition of the bony case or cover, betokens a bad condition of the soft parts contained within.—Habitual bleedings at the nose in youth frequently accompany a narrow chest and long neck, and are justly numbered among the tokens of future danger to the lungs.

Nothing in life is more familiar, than to find the effect of descent and conformation, seconded by habit. To be sensible of this, only call to mind the detail of a lady's day. I do not mean merely that of a *fine* lady: but of the innumerable tribe of females in different circumstances and situations, who by right or courtesy, enjoy the title of *lady*, in proof of their not being obliged to work hard for their subsistence. You will find them all just like so many divinities of Epicurus—not indeed basking upon clouds in the mild empyreal warmth of heaven, but fixed almost as immoveably upon well-cushioned chairs and sofas, in hot, close apartments. And if they resemble those of Epicurus in indolence, they will not be behind those of Homer in the tenuity of their diet.

> Ου γαρ σιτον εδουσι, ου πινουσ' αιθοπα οινον·
> Τουνεκ' αναιμονες εισι——και αθανατοι καλεονται.

They scarce partake of the coarse fare of mortals, and therefore they are without blood.—And are named immortals, adds the poet. But here the case is reversed. Our ladies, unable to receive and digest a proper quantity of aliment, are apt to fall farther short of immortality than their equals in society. They are scarce alive, while they continue to breathe: and, in numberless instances, a fiend, infinitely more cruel than the night-mare, makes a lodgment in the bosom; and finishes by strangling them, after he has oppressed them long. Effeminate men, of course, become the prey of the same malignant imp. But *men*, at least *young men*, cannot easily devote themselves so entirely to indolence, as women have the misfortune to be devoted by our usages. They indeed make it up in a degree, but not compleatly, by wantonly braving the agents of destruction.

It is universally observed that the persons, who render themselves most tender by their mode of life, are most susceptible of the immediate disagreeable effects of a medium, much colder than common. They are also more susceptible of the variety of injuries, more or less permanent, which such a medium has its share in occasioning. The application of the fact to consumption in particular, as a disease that may begin from irritation of the surface of the mucous membrane of the chest, is attended with no difficulty. I have explained in the fifth essay, how damp and cold, applied to any part of the skin, inflame the mucous membrane by association, and bring on catarrh. As consumption is, at times, the unquestioned progeny of catarrh, the tender will be more liable to it from this and from all those other causes, by which the mucous membrane is affected, independently of the immediate application of hot and cold air in succession.

PHTHISICAL DISORGANISATION OF THE LUNGS.

The examination of the bodies of the dead has greatly contributed to elucidate the destructive operation of hard, rough particles upon the lungs of the more robust, and of slighter powers upon those of the weak. In a great majority, it would appear, of the invalids who perish by consumption, the substance of this organ is found every where studded with roundish bodies, denominated *tubercles*, from their resemblance in shape to certain roots (*tubera*.) They are dense, light-coloured bodies, at first not larger than the heads of small pins. They are frequently found of the size of garden peas, and, as it is supposed, by the consolidation of a whole cluster they come to form masses very much larger. In the inside of the more advanced tubercles, is found matter, inclosed in the yet entire circumference of the tubercle, as in a bag. This matter by degrees finds its way into the air-pipes, and is what we see expectorated in confirmed consumption. The matter is sometimes of a peculiar consistence, being grumous or curdy, as in scrophulous sores. It is frequently without a mixture of more solid parts, and such as issues from a common sore.

Sometimes a substance, resembling tubercles, is found to occupy a considerable space of the lungs, but having no defined figure. It has a degree of tenacity like a scrophulous gland, just beginning to suppurate.

Tubercles are not peculiar to the lungs. They are very frequently formed in the liver, and often in the uterus, and sometimes, though very rarely, in the spleen and the kidney.

The origin of tubercles, like that of the other products of the animal organization, is obscure. The physiologist cannot say by accession of what materials these concretions are formed, as the chemist can explain the composition of a neutral salt. But the first, which is the most interesting step, in their formation, can be assigned with a high degree of probability. And by making a proper use of this fact, all that perplexity, which has hitherto so universally prevailed concerning the connection between tubercular consumption and scrophula, may be removed. The explanation belongs to this place, because it will be productive of some advantage to preventive medicine, though it may not, at the present moment, forward us a step in the cure of confirmed consumption.

Dr. Desgenettes, from the dissection of a stone-cutter, in whose lungs he found a number of stony concretions, was led to believe that these concretions were formed by particles of calcareous earth, taken up by the absorbents from the surface of the mucous membrane. In one lobe of the lungs, he says he met with lines, consisting of little stony grains, included in a peculiar membrane. One of these lines, he remarks, made up of three such grains, and near four lines in length, exhibited the form of a lymphatic vessel. Hence he thinks he has a right to conclude that fine calcareous particles are very often absorbed; that they are thrown out into the cellular substance, where they form calculi, and from the destruction of parts they soon occasion, they afterwards get into the air-pipes, and are expectorated.

Whatever may be the fidelity of this observation, to which a chemical examination of these pulmonary concretions is wanting, it is very certain that tubercles may be formed by introducing foreign matter into the lungs, and that a true consumption will regularly follow the formation of these tubercles. Dr. Haighton, one of our most ingenious physiologists, ascertained this very curious and important fact. He injected running quicksilver into the crural vein of a dog. The fluid metal, being circulated along with the blood, found its way to every part of the body. The animal did not seem to be disagreeably affected during the first day. It then became feverish, and afterwards laboured under difficulty of breathing. A cough succeeded. These complaints went on increasing till the death of the dog. In the lungs were found tubercles, of which many contained matter. That these tubercles had been produced by the injected mercury, was demonstrated by cutting into their substance, from which it appeared that each contained a particle of the metal. This experiment appeared to me of too much consequence to remain unconfirmed by repetition. I accordingly caused two puppies to be procured. Into the crural vein of each, at the age of about six weeks, running mercury was injected. The animals appeared not to suffer in the least for nine days. On the tenth, one was killed, and, on examination of the lungs, tubercles were discovered in considerable number, each of which, as in Dr. Haighton's experiment, contained a globule of the mercury. A few tubercles at the edges of the lungs, though not equalling a garden pea in size, contained purulent matter. On this, which is the fourteenth day, the remaining animal appears sluggish and indisposed. The pulse is quicker than in another of the same litter. The difference of time, at which the animals appeared to be affected, may have depended upon their difference of age, if that employed for the original experiment was full grown, and if we may suppose that the young of the dog kind are less susceptible of this particular sort of pulmonary disease in the same proportion as infants appear to be, in comparison with the adults of the human species.

The probable influence from these facts is, that tubercles originate from a small particle, in most cases thrown out by secreting vessels during a particular kind or degree of inflammatory action, and that this by its stimulating quality maintains the very action, by which it was thrown out, in the same manner as the globule of quicksilver does, till the new product of disease is sufficiently advanced to form matter within its substance, and to burst into the ramifications of the bronchiæ.

Nothing is more common, in certain deviations from the healthy state, than for the vessels destined to throw out liquids for the moistening of the surfaces or of the interstitial cells of the body, to secrete substances, having different degrees of tenacity, even to a stony hardness. Calculi are occasionally formed in the lungs in place of tubercles, and sometimes in the same subject and at the same time with tubercles. In a Wallachian horse, Dr. Camper found the lungs full of small, round stones, of which some had occasioned little abscesses.

Inflammation, excited in one spot, often diffuses itself gradually into the contiguous parts. This is one of the most common facts in pathology. It is no wonder then, that when the mucous membrane, investing the air-pipes or ramifications of the bronchiæ, is inflamed, the invested parts should become inflamed too. And then it is that the formation of tubercles *often* begins.

But this event depends probably upon the degree of superficial inflammation, and certainly, under ordinary circumstances, where irritating powders are not concerned, it depends upon the habit of the party suffering from inflammation. In what kind of habits tubercles are most likely to be the result of inflammation, we have no means of knowing but by reference to the persons of the consumptive. This, if we leave descent from consumptive parents out of the question, will lead us to the feeble and the delicate in general. Among the feeble and delicate, we shall find those of scrophulous constitution most subject to consumption.

In the natural history of tubercles, no circumstance is more singular than that they should not unfrequently exist in the lungs without injury, and even without inconvenience. For they have been found in the lungs of malefactors, executed in full health, and of persons, who have had no pulmonary symptom.

I conceive this fact in the following manner. Severe or repeated catarrhs, or else hard particles, occasionally, not habitually inhaled, shall produce the tubercular inflammation in persons little disposed to it by the nature of their constitution. The inflammation shall subside, and there being as little disposition to supply the materials, by which they are enlarged, as to form the central portion, they remain quiescent: and if death from some other cause take place before the mucous membrane again suffers an equal degree of irritation, neither the lungs nor the constitution will sustain any sensible injury. But in case of new and severe catarrhs, the tubercles will be enlarged by the inflammation of the adjacent substance of the lungs, and matured into abscesses, which when arrived at a certain size, discharge their contents into the air-pipes, and occasion consumption. In disposition to form tubercles, in the tendency of those tubercles when formed to enlarge, inflame, and suppurate, and to affect the constitution, when suppurated, the difference between different individuals appears immense. Dr. Camper says that the horse, whose lungs were full of round calculi, some of which had produced abscesses, had no cough nor any difficulty of breathing, even when put out at full speed.

RELATION OF SCROPHULA TO CONSUMPTION.

It is observed that consumption is particularly apt to rage in scrophulous families. It is also observed that many persons fall into consumption, whose ancestors have been exempt from scrophula, and who themselves have had neither any mark of the scrophulous constitution, nor any positive symptoms of the disorder. These two observations medical philosophers have not been able to reconcile *. Some have prudently contented themselves with addressing the difficulty to future reasoners. Others have used their pen, as Alexander did his sword, to cut the knots they could not untie. They have assumed that the existence of consumption—of tubercular consumption at least—was proof sufficient of the existence of scrophula, though all other evidence should be wanting.

A degree of resemblance in form between tubercles and lymphatic glands, occasioned these bodies to be confounded, at a time when anatomical knowledge was less accurate; and had tubercles proved lymphatic glands in a state of disease, the conclusion that tubercular consumption is always a scrophulous disease, would have been inevitable. But it is now perfectly understood that the substance of the lungs is destitute of glands of this description; and from all the facts it appears that tubercles are not pre-existing organs, changed by disease, but to all intents and purposes new concretions, like gall-stones and the calculi of the kidney or the bladder. The formation of tubercles in the manufacturers formerly enu-

* A treatise on Consumption, by Dr. Read, contains some of the best ideas we yet have on this difficulty.

merated, many of whom are of a temperament as opposite as possible to the scrophulous, is a sufficient refutation of an opinion which rests upon a foundation little firmer than the following argument. *Because many scrophulous persons become consumptive, therefore all consumptive persons are scrophulous.* It does not appear that the least difference has been observed between the tubercles, occasioned by the inhalation of foreign bodies, or by repeated catarrhs in the healthiest persons, and those of the most phthisical constitution. And when tubercular consumption can be excited in any dog, taken at random, it must be in vain to think of a necessary connection between scrophula and this species of consumption. Before these facts were discovered, to say that tubercles alone were sufficient evidence of scrophula was at best begging the question. But now the negative is demonstrated.

It remains an undoubted truth, that the slightest exciting cause is more apt to produce consumption in the scrophulous than in others. And independently of external causes, those who labour under the glandular variety of scrophula, are exposed to a peculiar kind of danger. Enlarged glands, from the unexplained, but unquestionable influence, which living parts exert upon one another, dispose to the formation of tubercles. This process in the lungs is much more apt to be excited about puberty and afterwards, than in earlier life. The progress of disease toward the chest may be observed upwards and downwards; from the mesenteric glands in one direction, and from the lymphatic glands that dip under the clavicle in another. It sometimes, but less frequently appears, that the glands become scrophulous from the agency of advancing tubercles. I have several times known lymphatic glands on the outside of the chest begin to enlarge, after all the symptoms of tubercles had, for some time, made their appearance. And so exact is the sympathy between lymphatic glands and tubercles in some scrophulous subjects that, after death, I have seen both filled partly with purulent and partly with calculous matter. The rules of caution, deducible from the preceding disquisition are equally simple and important.—*No person can flatter himself with being entirely exempt from consumption. No one should therefore wantonly expose himself to the exciting causes. The most perfect freedom from the scrophulous taint is a very small ground for confidence, where there is any other reason for apprehen-*

sion. But the scrophulous ought to be particularly on their guard.

Catarrhs, it should be observed, are particularly apt to lay the foundation for consumption in scrophulous habits. For in these the substance of the lungs is peculiarly apt to run into that sort of inflammation which produces tubercles. It is the observation of a physician, whom Tissot terms very able, *that colds cut off more invalids than the plague* *. From common observation, and the titles of many popular tracts, it would appear that the idea of consumption being frequently produced from catarrh is very general. But at present we are not acquainted with the proportion among these sufferers, that are not scrophulous. A majority of the consumptive patients I have seen above thirty, have been free from glandular affection, and all propensity to it. It is exceedingly common to find feeble young people attacked in spring with an inflammation of the chest, more severe than an ordinary catarrh, and approaching to what is commonly called *pleurisy.* This complaint, unless carefully treated, ends in consumption, though the constitution be free from all suspicion of taint. In some

---

* *Les rhumes emportent plus de malades que la peste.*

places, scrophulous affections are thought to be less frequent, without the least apparent diminution of consumption. For example, "the scrophula has been suspected as the "most general cause of consumptions; but "this disease is less common in Chester than "in most other places. - - - Very strong evi-"dence could be adduced that the most "frequent and most fatal cause of consump-"tions, in this town and neighbourhood, is "catching cold, which occasions a cough, "that is entirely neglected for many weeks "or months, and is never supposed to be "dangerous, till it becomes desperate."—(*Haygarth*, *Phil. Trans.* LXIV. & LXV.) The numbers given are as follow:

Buried in all.	Consumption.	Convulsions.	Fever.	Small-Pox.
In 1774—379	62	76	23	16
In 1775—352	73	69	34	1

In the former of those years the number 26 stands against asthma, and in the latter 16. It is not impossible but some cases of consumption may have been referred to asthma. And of the term *convulsions*, it seems that at Chester the nurses are full as prodigal as in London. For "when an in-"fant is seized with fever, or almost any "other disorder, if there be any distortion "of features, expressing pain, or any irre-

"gular motion of the eyes, nurses always "denominate the disease, *inward convulsions.*" The disorders of children were left to be named by women, medical practitioners being more rarely consulted in their cases.

Not to suffer a catarrh to go on without medicine above two or three weeks, is a common piece of advice. But it is safer not to suffer it to form, and if it have formed, and be at all violent, we should proceed to its removal without delay. We have a remedy in the fox-glove alone, which succeeds at once in the great majority of *colds.* We have powerful auxiliaries, when they are required: so that consumption from catarrh, measles, small-pox, and hooping-cough, (of which latter the present season threatens to be productive) ought to be an event almost unknown. The same may be said of chronic catarrh, of asthma, and of the various other evil consequences of a *neglected cold.*

If those, who cannot take sufficiency of sustenance for want of exertion, and who cannot exert themselves properly for want of sustenance—to whom hunger, fatigue, and sleep are never sources of pleasure or refreshment—if such persons are most subject to consumption, it will be easy to conjecture who are the most exempt. And circumstances or exposure being alike, where the animal powers are least suffered to dwindle from disuse, there the organs of the chest will be found to remain freest from ulceration, and the constitution from the terrible effects of such ulceration. To enumerate those who enjoy this privilege, would be to give a list of the professions, callings, and occupations that are *carried on in the free air, and that fully supply the means of answering the calls of appetite.* Butchers and fishwives supply curious examples of this exemption, as may be seen in my larger essay on consumption. Whoever will fix upon that part of the inhabitants of Great Britain, in whom these two conditions unite, will find that he has selected the least phthisical from the whole mass; and though the excesses of these lead to consequences sufficiently deplorable, they escape the extreme misery, attending the gradual destruction of the organ, by which we breathe.

I do not speak of the origin of consumption in high inflammatory pleurisy, when blood-letting is neglected, nor of the effects of blows on the side, or of severe courses of mercury and the like. In such cases prevention lies with the medical practitioner; nor is there any interval for domestic measures of safety. I pass over the cases of slow pulmonary ulcers, towards the formation of which tubercles are not the first step. Such a disquisition I consider as needlessly subtle for the readers of these essays. I shall only add a caution not to let chronic catarrh degenerate into consumption. This caution more particularly respects elderly people.

VOCAL EFFORTS.

A considerable effort in singing or speaking will often bring on a spitting of blood, even when the chest is not originally weak. The celebrated composer GRETRY, brought on a pulmonary hæmorrhage by singing; and he has given to those who may meet with the same accident some advice, which the British physician will hardly confirm *. The

* *Memoires*, Paris 1789. He objects totally to bleeding, which perhaps is solely or principally adviseable in this kind of hæmorrhage from violence. You are to lie on the back two days and two nights without stirring or speaking —not bad counsel in such a case.—But he says that the free expectoration of coagulated blood is a proof that the cicatrix has been formed.—Alas! the closing of a ruptured vessel is no security against abscess or tubercles, without stronger measures than he advises; and the danger of perishing from loss of blood is very small. Few attacked

injury in these cases appears to arise in two ways. A quantity of blood is driven by an effort or a blow into the substance of the lungs, and there occasions the formation of a large abscess or imposthume, or else coagulable lymph is effused, inflammation commences in a number of points, and a formation of tubercles, with all their bad consequences, takes place. The following very distinct case, recorded by a foreign observer, shews how the abscesses are produced. A labourer's infant, about seven months old, was put into the hands of some neighbours, by its mother, who was obliged to be from home almost all day. The child fell into the most violent fits of crying, and its mother, on her return, found it quite altered. The face was almost black, and it was still fetching deep sobs. The terrible agitations which it had undergone on being first forsaken, had subsided. But it appeared to be oppressed and coughed much, which was

---

with spitting of blood die in this way; and few even sustain serious injury from the quantity of blood lost.—If your chest is heated, which you will know by a short, dry cough, proceeds the musician, take syrup of vinegar in water—if chilled, a glass of wine after dinner.—Excess of refrigerants once brought on my spitting of blood for five days.

not the case in the morning. So it was supposed to have taken cold. But after a little time, it was observed to cry and to suffer in being laid on the right. Two months afterwards, febrile exacerbations came on and went off from time to time. The cough and the oppression increased, progressive emaciation took place, and now the little invalid had a burning skin every evening after slight chills. The cheeks were of the most beautiful red, the palms of the hands, and the soles of the feet, particularly hot. As the symptoms grew more violent, the cough was attended with a sound as of something to expectorate. Diarrhœa succeeded; the loss of flesh was such, that all the windings of the intestines are said to have been apparent through the skin of the belly. The child was suckled, and its appetite continued good to the last gasp. It died after eleven months of suffering. On opening the body, a large abscess was found in the left lobe of the lungs; the right lobe was sound.—There cannot be an example of the effect of inordinate exertion, more distinctly marked as to time and circumstances. It was supposed that an effusion of blood had taken place into the substance of that side of the lungs, where the disorder was discovered after death.

Some effusion there probably was, as the affection of the chest never went entirely off from the hour it commenced.

The existence of tubercles where consumption has begun in spitting of blood, is a fact of perpetual occurrence. But whether blood, thrown out into the cellular substance of the lungs, as well as into the air-pipes, directly promotes tubercles, or whether they arise from some concomitant irritation, the time at which the short dry cough, and other signs of their existence are observed, often coincides sufficiently with their production by quicksilver, thrown into the blood-vessels.

After these explanations it will be readily enough conceived how singing and similar exercises will excite an inflammatory action, which shall terminate in consumption; and this independently of all spitting of blood. I have several times been consulted for invalids, in whom a pain of the side had come on while they were learning to sing, and in whom it disappeared on relinquishing the practice. It is sometimes thought that efforts of the voice strengthen the chest. But to escape hazard, they should be begun with great moderation, and increased with great caution. Fatigue should always be avoided; nor can there be much chance of benefit from bringing a single part now and then into play, under the continuance of a life that debilitates the rest of the frame.

CORRECTION OF THE PHTHISICAL TEMPERAMENT.

The difficulty here is to create a resolution, not to give proper directions. If a compassionate spectator were to find a cluster of human beings, employed at one of the abovementioned deadly occupations, how would their hard dry cough, their flying feverish heats, and shooting pains in the chest, alarm him for their fate; and it is easy to conceive with what warmth he would urge them to fly to the medical art for relief, while it was yet time, and afterwards to turn their hand to some safer labour. The monitor may be one of those who equally needs admonition against similar danger. And who does not wish that the party admonished could but put into forcible language the power of choice, which a certain degree of opulence bestows, the superiority to temptation, which a knowledge of consequences ought to confer, and all those other circumstances which render blindness to situation, and perseverance in destructive habits, so much more inexcusable in a person of education? How much less gross must the infatuation of the mechanic appear on a fair comparison? But alas! can one hope that the clearest prospect of their common end would move the one to quit his murderous habits, any more than the other his murderous trade!

To those who will listen to counsel, I would recommend it to begin their measures for preventing consumption from the very cradle. To the puny progeny of a puny, consumptively disposed mother, I would forbid the mother's breast. Without supposing that the seeds of disease could be conveyed by suckling, I should prefer the milk even of a healthy cow to that, secreted by a gland, which probably partakes of the constitutional feebleness. With cow's milk, if a wet nurse could not be procured to my satisfaction, I would alternate diluted gravy to as great extent as observation of the infant's progress should teach me was suitable. I would join bathing at such a temperature as the countenance and habitudes should indicate. I would employ gentle friction of the extremities, and even of the chest and stomach, that the system might be cherished with heat of its own producing, and animated by the vigour of its own actions. And, in all respects I would bring into practice the principles, already laid down for the benefit of weakly constitutions. Chills, confinement, sedentary occupations, pressure from dress, and employments that throw the body into constrained, uniform attitudes, I would carefully shun.

The dumb-bell I consider as a miserable substitute for running, digging, working with tools, and those other exertions that engage the mind, while they put the limbs into motion. Curious sports, such as archery, have not a permanent interest. They do not incorporate into a system of actions for life. *Task-exercises,* in my opinion, hold the same proportion to health, as the castigations of penitents to piety or virtue. Neither have they, at the moment, that salutary effect, which employment, connected with interesting ideas, always has within certain bounds, and when it fills a proper, that is, a large portion of a child's time. It has been my lot to see many young ladies, who have wielded the rope with great assiduity, but I have seen no one of much delicacy of habit, who has skipped herself into good health. Sometimes skipping has appeared to accelerate curvature of the spine, and to increase that feebleness, which is the forerunner of consumption. It is a great mistake to suppose that occasional efforts repair any part of the mischief which habitual indolence produces. A girl, who does not move enough to keep herself warm in common, but heats herself every now and then by dancing, will, I believe, be weaker in the long run, than if she never started out of her indolence. The more reduced the powers, the more cautiously does exertion require to be apportioned to them; and this rule holds through all the gradations, from the convalescent in fever that cannot bear all at once to be raised upright in his bed, to the phthisical and the nervous, who are fluttered with palpitations, dissolved in languor, or teazed with head-aches, after every movement a little more brisk than usual.

IMMINENT DANGER.

If thousands and tens of thousands, now fast approaching to maturity, have been brought up with entire disregard to their constitution, it may be collected from the preceding statements what numbers are, at this moment, on the eve of being seized with fatal affections of the chest. Of how much importance then is it that the chiefs of families should be made acquainted with the appearances, most probably denoting the commencement of that sort of pulmonary disorganization, which in a few months or a few years will be followed by so highly dangerous a disorder!

The aptitude to consumption is become so great in modern times, if it was not always so, that we daily see complaints apparently the most remote, terminating in this. In delicate habits, therefore, almost any indistinct indisposition, after the period of childhood, especially if it continue for any length of time, ought to be suspected. Should it be accompanied by falling off in flesh, a situation is indicated, which will not admit of delay. We frequently observe, in fact, (and I shall quote a clear instance below), that consumption can be clearly traced back to such beginnings: and we may perceive by the light of physiology, faint as it yet is, that these consequences very naturally result from an obstruction to the action of the atmosphere on the blood. That fluid will probably be impoverished; and there will be less substance added to solids, and what is added, may be of a less healthy quality.

These and other circumstances, which occur in a variety of instances and call for medical interposition, are faithfully copied from nature in the following letter of consultation.

" I am desirous of your sentiments upon my

" daughter's case, which possibly requires
" nothing more at present than attention
" to prevent those consequences, which may
" arise from her spare habit and delicate
" constitution. She is 14, and not only thin,
" but narrow-chested. Her spirits have al-
" ways been unusually good. She has had
" no complaints but the usual ones of child-
" hood. She has had little appetite for animal
" food, always preferring bread, vegetables,
" and fruit. In warm weather this prefe-
" rence is particularly marked. Finding her
" greatly relaxed last summer, I had her to
" use first a cold, and afterwards a temperate
" bath. Within the last six months, she has
" lost much of her strength and spirits, with-
" out any specific disease. She complains of
" extreme lassitude. The least exertion of
" body or mind is disagreeable to her. She
" complains also of sickness upon first rising
" in a morning, and at no other time. Of
" late she has found a sudden flushing of
" heat come over her, and she sometimes
" feels a slight oppression upon her breath,
" but not often. Of her pulses I can judge
" no farther than that they are always
" quicker, and often much more so, than in
" either of her brothers. Within the last
" six weeks she has rode a good deal, and
" the exercise seems to agree with her, though

"she feels fatigued after it; but on lying "down a quarter of an hour, the fatigue "goes off. Her general habit of body is "such as appears to require opening medi-"cines, and at particular times she con-"tinues unwell too long, and is much ex-"hausted."

Succeeding chills, a burning skin, a flushed countenance, and quicker respiration, ought to alarm the unprofessional. Nor will it be difficult for any one to ascertain the difference in the frequency of the pulse at different times of the day, and after such an exertion as walking up one or two flights of stairs. It should also be enquired, whether the surface of the body be in a burning or a moist state on awaking in the morning. These symptoms do not indeed demonstrate the lungs to be in a bad state. But in young men, that there are tubercles, and those tending to suppuration, is infinitely the most probable supposition. In young women, under the absence of any complaint peculiar to the sex, it is also the most probable.

It shall often happen that no expectoration but of a little frothy mucus, which is absolutely forced from the membrane by coughing, shall take place till the lungs become really ulcerated. But sometimes there shall be a pretty free expectoration of phlegm for some time before. It shall take place principally in the morning; and this ought to add to the alarm, when there are other causes to excite it. Indeed, habitual excess of phlegm alone, about the phthisical age, calls for attention, and whoever sets about corrective measures, does no more than what prudence requires. The great danger which it is intended to obviate is highly probable; the present evil, though seemingly slight, very certain.

When there are shooting pains, or uneasiness in the chest of a subject, in any other respect suspicious, and when these feelings depend on no other manifest disorder, *the bill ought to be found* without scruple, and the case remitted to a person or persons, competent to decide upon the whole of the evidence. When there is a cough, with any degree of feverishness, and a difficulty of lying on one side, there can scarce remain a question or a doubt. Consumption may be considered as fairly formed, and the object will be to hinder its proceeding to the second stage, when it is characterized by purulent expectoration and regular hectic fever, in addition to the preceding symptoms.

It is true, and it might be interpreted as want of candour, to conceal, that once in a great number of times, the most intelligent physicians, after full and repeated examination, shall confidently consider an affection as genuine pulmonary consumption, in which examination after death * shall prove the

---

" * It is not long since I was present at the dissection " of a certain person of quality, who was pronounced con- " sumptive by various experienced physicians. They of " course expected to find the lungs ulcerated; and they " supposed this to be the case also with the liver. What " was their astonishment then on finding both the lungs " and liver in the best condition? A fleshy excrescence, " however, of vast size, was found attached to the intes- " tines, by which these and the other viscera of the lower " belly were violently compressed. - - - - It was recollected " that the patient had had a bad fall twenty years before, " since which he had complained of pain in his right side. " - - - Hence we see how necessary it is to pay attention to " every symptom and to the cause preceding it, if we would " judge with accuracy of complaints, and treat them pro- " perly. For we often find disorders, wearing the appear- " ance of true consumption, with cough, pain of the chest, " difficulty of breathing, and hectic heats, where the lungs " are perfectly sound." (Marx, v. d. Schwind-lungensucht ss. 89.) - - - " A mason, as he was at work, was obliged " for some instants to sustain a very large stone, which " would have carried him along with it in its fall. Though " speedily relieved, he had immediate spitting of blood, " with pricking pain in the middle of the spine. Bleeding " and other means restored him in all respects, except that " there remained an obscure pain in the same spot. But " his strength gradually declined; he lost his appetite; he " had continual cough, either dry or with expectoration; " he fell off in flesh, and after complaining all the time of " painful spasms in the chest with expectoration of hectic lungs to be untainted. But these deceitful cases are generally fatal, if neglected, and therefore do not less require the earliest interposition of the medical art; and probably they would sometimes be cured by that general and local treatment, which is most efficacious in consumption of the lungs. The use of such cases out of the profession is to deter the friends of invalids from taking upon themselves to decide ultimately in an affair so dangerous and obscure, as to perplex and baffle the most sagacious and experienced of the faculty.

FALSE OPINIONS GENERATING FATAL DELAYS AND ERRORS.

In an *almanack of health*, published abroad last year *, I find this given as a common

---

" fever, he died nine months after the accident with genuine " signs of consumption; It was natural to expect ulceration of the lungs, as spitting of blood had preceded; but " no such thing. The lungs were quite sound. But the " interior muscles of the spine were almost destroyed by " ulceration." (*Baumes de la phthisie pulmonaire.*) This is a case that would probably have been cured by caustics and digitalis—and of course have passed for cured consumption.—And would it not have proved as much in favour of the treatment as if it had?

* Gesundheitsalmanack. 1801.

sort of dialogue between the physician and his consumptive patient:

Physician. Do you cough a great deal?

Patient (coughing). Oh no—by no means.

Physician. Have you spit any blood?

Patient. No.

Physician (looking him hard in the face). Have you really not?

Patient. Yes, indeed; a trifle—but it is some time since, and it did not come from the lungs.

Physician. Do you sweat in the night?

Patient. Why, I do sweat; but then it is because I lie under a heap of cloaths.

Physician. It is well if you have not something the matter with the chest?

Patient. God forbid! but it can be no such thing. My chest is sound (*striking it*). Observe, I feel nothing from that. But my stomach indeed may be relaxed. This plaguy wind——

It is commonly observed with some surprize, that persons affected with so unmerciful a malady as consumption, are insensible to their danger; and not only so, but when it is confirmed, and their last hour approaches, they cherish the expectation of recovery, and amuse themselves with plans to be executed at some distant day. This is sometimes represented as if the disorder itself generated the chearful feeling of hope. But any person, acquainted with the manner, in which our thoughts and sensations are formed into trains, would be able to perceive, from the history of its symptoms, that consumption must be a *flattering* complaint. No two things can be more different than the impression upon an impartial spectator from the multitude of deaths, exhibited by the bills of mortality, and that which is made upon the patient by the actual presence of the first symptoms of the complaint. There are several circumstances that render invalids blind to their danger during the early stage, and which indeed, except in very few instances*, preclude, even to the very close, any

* I have not met with above half a dozen instances, where consumptive patients had renounced all hope of escape, before the tragedy drew near its termination.. In none of these did the steady prospect of dissolution produce the smallest lowness of spirits or self-bewailing. But this very serenity of mind rendered the cases more affecting. In one, a young lady of Norwich, aged 17, the flower and favourite of a numerous family, requested her friends to retire that she might have an opportunity to ask me how long it was probable she could survive. On hearing my answer, she very calmly regretted that her period of suffering was likely to be longer than she had hoped.

The late Lady John R—— told me with the utmost tranquillity that she knew herself to be in a confirmed consumption, and that she had no hope from medicine, steady and unvarying sense of their danger. In the first place, the deviation from health is commonly gradual. Indeed in many cases there cannot be properly said to be a deviation from health. It is only a transition from a valetudinary state into a serious disease; whence the habits have been already moulded to suffering. The most healthy women know something of the feelings, accompanying indisposition. They therefore exhibit more patience under sickness than men. But phthisical women shall be for years in a state of probation for the scenes of wretchedness, brought on by this complaint. Hence it is less likely to give alarm, while forming.—The morbid process is seldom attended with acute pain; and sometimes no pain at all is felt in the chest, or none till towards the last. But what principally diverts attention from the growing evil is, that it cannot be seen or felt. Were pulmonary tubercles as much exposed to observation as schirrus of the breast, consumption would be nearly as much an object of apprehension to those who are falling into it, as cancer.—Then we must remember that something has

---

"Nor," added she, "do I regret my fate, but for the "sake of these"—pointing to her children.——Life does not often present more moving scenes.

got abroad concerning the difficulty of discriminating pulmonary consumption, which is probably conceived to be greater and more frequent than it really proves. But almost every one will interpret the chance in his own favour; and even medical practitioners, influenced by some of these causes of deception, as by the multiplied and fruitless, and perhaps unimportant attempts to find a criterion for purulent matter, will shut their eyes to the fullest evidence, when it is presented in their own person.

It is a general opinion that, in this deplorable disease, ignorance is bliss, and that the knowledge of the real danger is more to be deprecated than any fluctuation of hope and undefined fear—for decay and suffering generally give birth to apprehensions, though the terrible idea of consumption should never arise. But whatever we may look upon as most desirable with regard to invalids, there can be no reason why those about them should be lulled into neglect by the same delusions. Let it not therefore by them be considered as any argument of security that young persons do not complain; that they can seemingly carry on their accustomed occupations as well or nearly as well as ever; or that the appetite has not fallen off. It is sometimes impossible to make it understood that those may be in most imminent danger, who can go about and take food. The idea of serious illness one finds inseparably united in some minds, with those of entire helplessness and a sick bed. The condition of the lungs, however, is so far independent of that of the digestive organs, that there are instances where the consumptive are seen to eat to the day of their death as plentifully as in their best health. Nor is there any organ of the body but what may have its structure irreparably injured by a process, which for a long time shall give no alarm.

With a narrow chest, one of the most suspicious symptoms is habitual quickness of respiration, when this is ascertained not to depend on any female complaint. It is more suspicious still, if it increase with flushing and feverishness in the latter part of the day, and on every slight exertion. Yet this symptom will come and go for a number of times, before it is rendered permanent by extensive disorganization. Divers moral and physical causes will produce this variation. From what has been already said, it may be easily understood that it is not so much the existence of tubercles of moderate size, as their being in an increasing state, and irritating the contiguous substance of the lungs, that produces the distress of incipient consumption. Respiration would go on without much inconvenience, if the tubercles, such as they generally are in the early stage, only precluded the entrance of a quantity of air, corresponding to their own bulk, into the lungs. How much difficulty of breathing depends on irritation, and how little on the mere bulk of an obstructing cause, is proved by appearances, that often enough occur to medical men. Hard substances for example of inconsiderable size have been drawn into the windpipe, and have occasioned an inflammation of the superior part of the lungs, so violent as to end in an abscess. From the effects of a small fragment of soft food, which has gone the wrong way, it may be conceived what agonies would follow so terrible an accident. The labour of the breast has in reality been such as to throw the whole frame into convulsions *.

* A maid-servant of W. Martin's, Esq. of Shadwell, attempting to speak hastily in the act of deglutition, drew some of the crusty part of a piece of bread and butter into the wind-pipe. In a few hours, she was deprived of her senses and of her speech, and lay in strong convulsions, fetching her breath very short, so that it was expected she would expire that night. After some time the violent cough left her, the extraneous body appearing to have made its way into one lobe of the lungs. "Notwithstanding the great agonies she was in while awake, the second

Yet when the strength has sunk under these furious agitations, and an interval of sleep has taken place, the function of respiration has been performed with as much tranquillity as in the most perfect health. It will therefore hardly appear surprizing that affections of the mind, changes of food, another complaint, a journey, and personal incidents of which no account can be taken, should procure a temporary respite. The same thing will happen from the same causes with regard to the feverishness, excited by tubercles in their progress towards suppuration. It will often subside and often return, before it is matured into that fatal and unrelenting hectic, which leaves the patient no single day or night unmolested.

We have seen that their inactivity produces greater frequency of disposition to consumption among females. From the greater uniformity of their life, they experience fewer escapes or accidental cures. For, considering the multiplicity of facts, by which the conclusion is supported, there can scarce be any doubt but the very causes, which suspend

---

night after the accident she fell asleep, while insensible, slept sound for several hours, and while in that state fetched her breath quite easy, as she did every night after, when asleep, with a serene pulse, but always awoke in exquisite pain and in great agitations." On the eleventh day an imposthume which had been formed broke, and a piece of crust about the size of a filberd was discharged. The account leaves the patient recovering, but weak, and with a hollow cough, "as if the lungs were impaired." It seems never to have been stated whether she fell into consumption or not.

consumption at the period of its first formation, will in some constitutions and under some circumstances, put an entire stop to it. If those incidents therefore, to which the male sex is more particularly exposed, are occasionally fatal by hurting the chest, they full as often perhaps counteract the constitutional or hereditary tendency to consumption.

Among the instances that favour this opinion, I reckon those, where it appears that disorganization of the lungs and even tubercles in a state, by no means quiescent, may subsist for a long series of years without occasioning death, though they do occasion death at last. Facts of this class teach two important lessons; the first, to lay hold of the opportunity, afforded by slow progress of tubercles, or its occasional suspension; the second, that we are by no means to conclude that calamity will never arrive, because it advances with a heavy pace.

By way of illustration, I select the following narrative in preference to a case from my own experience. It is given as a simple, unapplied occurrence, by an author of good credit, already quoted. "A nobleman of "remarkably stout constitution, and choleric temperament, was unfortunate enough, "at the age of 19, to have a cannon ball "pass close by his chest. He instantly sunk "to the ground, and had a violent hæmorrhage from his lungs. He got speedy assistance, however, but from that time forwards, he had frequent and violent returns "of the hæmorrhage. He was troubled with "constant cough and purulent expectora-

"tion. The febrile exacerbations were some-"times mild, sometimes more severe, and "the purulent matter discharged varied in "quantity. But the difficulty of breathing "was incessant. He moreover often com-"plained of distress on the right side beneath "the short ribs, and after eating, he had a "sense of oppression in the region of the "stomach. Nevertheless, except at the time "of fever he had a good appetite, and the "bowels were regular.

"He made seven campaigns in the quality "of colonel, and consequently was subject "to violent exertion. Afterwards he per-"formed variety of journeys, being from "first to last of an active turn. From his "rank, he was under the necessity of fre-"quenting the tables of the great, where he "would make a good meal of substantial "food—always abiding firmly by the pur-"pose he had adopted immediately upon "the accident—never to touch wine or any "heating liquor. Happily for his family he "attained his 55th year."—The author goes on to relate, that not long before his death he was injudiciously treated for a catarrhal fever with Peruvian balsam and with the bark, by which his hæmorrhage and shortness of breath were aggravated.

On opening the body, "both lobes of the "lungs were found more than usually large. "They had the appearance of the liver, being "of a blue red colour. There were innu-"merable tubercles dispersed through their "substance, which tumours had suppurated, "and formed open abscesses even to the

" very edges of the organ."—The author then mentions that the liver was of most uncommon size, and concludes thus—" an " example without example! to exist 35 " years with lungs in such a state. I can " ascribe it to nothing but the wonderful " robustness of the constitution, and to absti- " nence from all fermented liquors." (*Marx ubi supra.*)—Should it be asked how we can be sure that the lungs had been in the same diseased condition for so long a time, it must be confessed that there can be no certainty of a point, not ascertainable by any of the senses. But the characteristic symptoms of tubercles existed from first to last. There was no change of symptoms but in degree, and the original injury and the hæmorrhages were likely to produce the very effects discovered after death.

Where consumption has not occurred in any preceding generation, it is natural that incredulity should be most obstinate. I do not know that I can do any thing more effectual towards banishing this incredulity, when it resists all probabilities united, than to transcribe a letter, which passed upon a real occasion. The occasion is certainly one of very common occurrence.—" I al- " ways wish to give the friends of inva- " lids the fullest information in my power. " I shall therefore reply minutely to your " general letter of enquiry. I cannot pos- " sibly convey to you my opinion along " with the degree of conviction, attached " to it in my own mind, but it is not dif- " ficult to lay before you the principal " grounds on which that opinion rests.

" I can easily perceive that in spite of the
" ravages, made *for the first time* by con-
" sumption in the family of one of your
" nearest neighbours and most intimate
" friends, you are more than half inclined
" to think I over-rate your daughter's
" danger. This hardness of belief is com-
" mon; and it is certainly a principal reason
" why the annual deaths from consumption
" in Great Britain amount nearly to the
" sum they do. The exemption of your
" family hitherto is likely to weigh with
" you against a truth so unwelcome. In a
" case where the appearances could be plau-
" sibly referred to any other source, it would
" weigh with every body. But the positive
" assurance we have, that this is a most pre-
" valent disease among delicate females, and
" the high probability that some universal
" influence increases its frequency, where
" already known, and introduces it into new
" houses, overbalances this consideration.

" Then is it not evident, that Miss ——, by
" the conformation of the chest, by her com-
" plexion, by the colour of her eyes, the
" enlargement of the pupil, and the general
" feebleness of her habit belongs to a class,
" marked by nature as consumptive? And if
" we are to lay no stress on remoter points,
" however important, what shall we say to
" positive, recent symptoms? Have we not
" the early indications of pulmonary tu-
" bercles as clear as indications can exist in
" a slow and deep-seated disorder?—Subse-
" quent, as I am given to understand, to
" an inflammation of the chest in spring, a

" short cough came on, and it has continued " now for seven weeks. She has quick dart- " ing pains in various parts of the chest. " She has the sort of flying chills and heats " that often precede ulceration of the lungs. " Her breath is generally quick. She cannot " walk up a flight of stairs without rendering " it more so, and without increasing the " pulse to 120 or 150 strokes in a minute. " The bronchial membrane is relaxed. This " appears from the constant expectoration " of phlegm; and another weakness shews " that another membrane is in no better " state; not to speak of too much disorder " at *particular* times.

" From all the accounts, too, she has fallen " away, and rather rapidly regained flesh " within these twelve months—a very com- " mon occurrence on the eve of consump- " tion, when people move about much from " place to place, and sometimes happening " from no assignable cause.

" I really know no other complaint, to " which all these symptoms can be referred: " decided and constant as they are, will you " cherish the dangerous persuasion that they " belong to no denomination of disease, but " will vanish of themselves as they came? " In less advanced cases, I have no doubt " but journeys and accidents one cannot " specify, do sometimes procure a respite, " and even accomplish a permanent cure:— " an observation, which will give you to " understand in part, why boys are better " off in regard to consumption than girls.— " When this sort of spontaneous cure takes

" place, people are deceived into the belief " that no just reason for alarm ever existed " in the individual case; and that there can " exist none in similar cases. You are well " acquainted with F. S——. That young man " was thought by more than one medical prac- " titioner to be on the brink of a decline. " His business led him at that crisis to un- " dertake a long voyage, and he has been " since continually in motion. So far he " has struggled on, though with difficulty. " But the girls had not the same good for- " tune. Yet the symptoms at first were just " the same in both; and the fate of one of " their parents was equally alarming for " both.

" I doubt whether all this will be con- " vincing. However I give the opinion after " much observation of the complaint, and " much attention to the party, whose health " I consider as at hazard. To mistake real " warnings is not easily excusable in a phy- " sician. What I would next be cautious " to avoid is the raising of groundless ap- " prehensions. And next to being right in " my judgment, I should wish the friends " of my patients to think me so."— — —

But when continued indisposition and manifest decay have produced a tardy conviction, what is more common than the adoption of frivolous or hurtful measures by the managers of invalids? Of two opinions from different medical practitioners, they will often follow the more inefficient. They cannot persuade themselves that what is so trifling in shew, can be so formidable in reality.

"Something however may be necessary to "be done." So juleps, emulsions, and saline draughts under the appellation of *gentle alteratives*, are poured down without mercy. The complaint holds on its course and terminates as it always does, when the most active remedies are put off beyond a certain season. Those however under whose sanction a life has been thus trifled away, force themselves to believe that every thing has been done of which circumstances admitted.

At other times the interposition of a gossip shall procure a trial for some decoction of figs and raisins, for some compound of sweet oil, acid and honey or sugar-candy, for some soup or jelly prepared of materials less palatable and not more nutritive than ordinary. In a catarrh, which would go off by itself after a certain period, such preparations may twenty times have appeared to the ignorant to effect a cure; and I grant they might really have been of transient use in lubricating the top of the throat, and easing a fit of coughing. But the idea of their having power to remove a cough, occasioned by beginning alteration in the substance of the lungs, can never be entertained for a moment, by any one who has an idea of the real condition of that organ. And this is a case in which a very small portion of anatomical information would be a safeguard against very gross and very fatal, but very common errors and impositions. To have considerately inspected a pair of lungs, beset with tubercles, and to have heard at the same time such an explanation as will place the concomitant phænomena among a man's current ideas, may be the means of preserving the constitution of several children. And where there is a horror of preparations from the dead human body, we now see how easy it will be to obtain a full illustration of the nature of this exterminating disease from the bodies of brute animals.

At other times, entertaining just suspicions, but for want of accurate information, flattering themselves that the internal cause corresponds to the slightness of the external signs, parents shall take the cure of incipient consumption into their own hands. For this purpose, they will often convey their invalid to a watering-place, and set him upon a course of water-drinking on their own judgment. Not infrequently they repair to the sea-side, though there is no reason to suppose that station beneficial in any circumstance of pulmonary consumption; and some medical invalids have reported it injurious to themselves. It is well, if they do not venture upon the hazardous prescription of bathing in the open sea, or in cold fresh water.

That cold mediums may be applied with advantage in all the stages of consumption, the reader will easily collect from the observations on this subject in Essay fifth. But some discrimination is requisite to adjust the interest of the system with regard to heat and cold. And it is very seldom indeed that a physician, in possession of all our present stock of information, would venture to advise immersion in cold water. As a promiscuous measure, without regard to states and seasons, it cannot be too strongly condemned. In an instant, cold bathing will accelerate the progress of tubercles—at least, it will aggravate every phthisical symptom—more than the other ordinary causes in weeks and months. Nay, I will venture to say, that it will fix the disease where fluctuating, recall it where departing, and ensure it, where uncertain. A person who has been often a witness to such occurrences, must despair of being able to find admonitory expressions sufficiently strong; while he, who has never witnessed such occurrences, will hardly conceive that applying water at the temperature of our springs to the skin, is very often as surely, though not so quickly, destructive as giving a draught of laurel-water would be. There are analogous cases, in which the effects of the same cause are more palpable. And by assisting the reader's conception, these analogies may induce him the more readily to believe that in particular states, one, two, or three plunges in the cold bath are capable of disorganizing any of the viscera in the chest. Thus the late Dr. R. Pulteney of Blandford, whose authority is a guarantee quite sufficient for such a fact, relates of T. C. that he was ricketty in his infancy, and very weakly for several years after; that in the winter of 1759 he had pleuritic symptoms; that a rheumatic fever the next summer left him afflicted with chronic rheumatism; and that he was advised to go into the cold bath. "He did so: but upon coming out again, "felt such an increased load, fainting, and "anxiety about the præcordia, that he thought "he should hardly have recovered the shock "it gave him. Nevertheless he ventured in "again a day or two after, but experienced "the former symptoms in an aggravated "degree, and from this time dated the dis-"order which terminated his life!" A palpitation of the heart, to which he had been before subject, increased greatly; he became weaker, breathed short on the least exertion; sometimes could not sleep above an hour or two in the night; commonly awaked with a sense of suffocation, and at the end of a year and a half was actually suffocated by a discharge of blood from his lungs. The heart was found of an enormous size; of a very pale colour; loose and flaccid in its texture, and with the sides very thin.—The great increase of this man's disorder on using the cold bath only twice, does not indeed prove any thing with regard to the consumptive. It only demonstrates the formidable power of the agent in a constitution nearly allied to the phthisical, and where some mischief pre-existed in the chest.

In the life of J. H. Beattie by his father, the well known author of the Minstrel, we have strong indications of the tendency of cold mediums, incautiously applied, to occasion or accelerate the destruction of the lungs. Indeed the preceding opinions concerning the JUVANTIA and LÆDENTIA—the things that help and those that hurt, are singularly corroborated by the whole narrative. The coincidence must give the greater satisfaction as the narrative was not drawn up for any medical purpose, and as the opinions, in another form, were given to the public first.

Young BEATTIE, we are told, was born with "a very delicate constitution." His head-aches persecuted him so that he was unable to make the proficiency he desired in drawing. About the precarious age of 15, as it would seem,* "he laboured for several "weeks in endeavouring to learn to swim -- "but it appeared at last, *that the chillness of "the sea-water did him more harm than good.*"

The compositions and projects he left behind, shew him to have read and meditated with dreadful assiduity. At 20, he was able to deliver lectures, as professor of moral philosophy. Shortly before, he was advised to "apply to music as an innocent and useful "recreation after the fatigue of study!" He not only learned to perform with good expression both on the organ and the violin, but went deep into the theory of the art! He occupied himself with incessant diligence in literary accomplishments. He wrote English and Latin verse, and did not altogether neglect composition in Greek. No wonder that, in a feeble habit, and with such intemperance of application, he should be subject to occasional indispositions. Change of air and scene, however, deluded him from time to time with apparent recoveries. "The air "of Peterhead several times restored his "health, when it was declining in a manner "similar" to that, which preceded his last illness—

About a year before his death, he was sud-

---

* Minstrel and other Tracts, by Dr. Beattie, and by J. H. Beattie. *Dilly*—1799.

denly seized with a violent fever fit. Ever afterwards he was molested by night-sweats, and sensibly fell off. For a time he laboured under that obscure kind of indisposition, which, in the puny and narrow chested, I have marked as peculiarly suspicious. Here, the whole tenour of the narrative shews that it arose from tubercles in an active state. In spite, however, of that tenderness of form, which the portrait exhibits, and in spite of the consideration that at such an age, and in such subjects, consumption is the reigning malady, the case was not considered as consumptive. It was called a *nervous atrophy*. But a cough at last made its appearance, and in a few months he died at the age of 22.

Rowing, fishing, shooting, and even archery are numbered among his active pursuits. These are however of such a nature, and his passive pursuits were so various, that he was not likely to recur to them but by fits, and then he would probably be tempted to over-exertion. It is remarkable that once, when his health was impaired, he fell upon a continued and gentle exercise. From the effect we may, in some measure, judge of the advantage of such occupations, when they are begun early enough, and fill up a sufficient portion of life. Building an organ for himself " occupied him for five months, and the " amusement he derived from the work, and " the musick with which it afterwards sup-" plied him, had the happiest effect on his " constitution."

The friends of this remarkable young man will not read the foregoing analysis unmoved, should it ever fall into their hands. It may, perhaps, excite an emotion of ill-will towards the author, as if he had disturbed the ashes of the dead, or sported with the feelings of the living. But the impartial will, I hope, consider it as more than a right to use the information, that is laid before the public, for the public benefit. To mark the rocks on which former adventurers have struck, though it recal past calamities, can hardly be a cruel office, even if we did not perceive the numbers to increase that ply in the same fatal track.

MAXIMS, *to be hung up in the heart of every father of a family.*

1. If either side have been consumptive, use the earliest precautions, and do not relax, under peril of attending your daughter in her shroud, instead of her wedding garments.

2. Should ſhe escape, see if she owe not her preservation to some other standing disorder.

3. Your son's chance will be better, by how much he is more robust, and the less he is exposed to hardships without the most gradual seasoning.

4. Though consumption have not been on either side, the chance, without an anti-phthisical regimen, is still bad. Two or three colds upon colds in winter, or a cutting blaſt in spring will do the business; and in the mean time, there shall be wretched health almost to a certainty.

5. Set not your heart upon accomplishments, elegant or literary. Book-learning should be the least concern of the delicately constituted. Living instruction turns out its pupils not only stouter but abler.

6. When a son or daughter droops between 14 and 34, suspect that a secret enemy is sapping the lungs.

7. When those, who *must* be ignorant of the essential difference between a common cold and consumption, boast of their cures, hear but heed them not. Ask this question of your common sense—*what experience can instruct such pretenders?*

8. A little cough may be the sign of a great disease. Beware then how you play the doctor's part. Would you consent to be turned blindfold into an apothecary's shop, and give your child the first drug you may lay your hands on?

9. It is wise to check a bad cold the first week; but much wiser the first four and twenty hours.

9. After the small-pox, hooping-cough, scarlet-fever, and measles, watch your young convalescent close. If he bark but once, fear lest there be a murderer within.

10. Though dislodged, expect him again; he now knows the way.

11. You think perhaps a single course of medicine *ought* to be effectual, and that once cured is cured for good. But nature, be assured, will not be regulated by your fantasies; you have probably been acting in defiance of her for years, and then you may think yourself happy to compound with her on her own terms.

12. The less consumptive any one is rendered in the rearing, the greater chance of recovery, if he becomes so.

13. When consumption is hanging about a girl, the distance between the marriage-bed and the grave is usually short with her. The husband, if he do not become a widower soon after the birth of the first child, may reckon upon a perpetual ailing wife.

IS CONSUMPTION CATCHING?

YES, answer the writers of the South with great consent, though not unanimously. The writers of the North say, NO. To our comfort, the former are either dupes to the phantoms of their own imagination, or their facts do not apply to us. We are certain that furniture and apparel, more bulky than bolt-ropes and gloves,* and certainly not less

* I have heard Dr. Luzuriaga, in whose veracity I have entire confidence, relate the following fact. He has since published it, and it has been quoted as one of those most strongly in point. A Nun died at Bilboa of consumption. The furniture was burned; the walls, cieling, and door washed, the floor taken up and sand laid in its place. The cell was then tenanted by another nun in perfect health and of an excellent constitution. In two months, she began to decline; her flesh wasted, the chest was attacked, the cough became worse and worse—in short she died consump-

likely to foster the matter of phthisical contagion, has no such effect with us. In any country is not accidental succession more probable?

What alone gives a colour to the idea here, is that a husband or a wife, no way predisposed, shall sometimes speedily follow the fate of the partner for life. But an experienced physician shall not see the fact above thrice in his life. Other probable causes, too, sometimes occur. And as there is no absolute exemption, must not such coincidencies sometimes happen from the very frequency of the disorder?

An especial reason may be assigned in the case of married people.—The physician is doomed to witness the most humiliating scenes in this tragi-comedy of life. But he is in return the privileged spectator of the most ennobling—where the fortitude of those who suffer generously contends with the devotion of those who soothe. But what may most console us for the base and selfish alloy

---

tive in eight months after taking possession of the cell. General purification as before. It was supposed that the disease was hereditary in the last case. A third healthy nun succeeded and died in the course of the year. New examinations took place, and it appeared that the cord near the bed, attached to the dropping bolt of the door, had not been removed. This, it is said, was impregnated with the sweat of the patients, and with other noxious exhalations. It was removed; new furniture introduced, and a fourth has lived in the same cell for five years.—In Italy there is the story of a beautiful pair of gloves worn by a first wife, and alone of all her appurtenances not destroyed upon her dying consumptive. Long after, a second wife spies the gloves, wears them, and perishes like their former owner.

in our nature is the affection we find subsisting between persons that have been long united. Where neither dislike nor indifference has followed intimacy, this sentiment, which in ordinary situations retires from view, bursts forth, in the hour of danger, strong and undisguised, as it shewed itself in ages, when the sincere expression of the feelings stood in the place of that circumspect and disciplined demeanour, which looks round among the by-standers, before it dares listen to the voice within.

Every one sees how much self-neglect is attached to affection of this degree. The consequences of self-neglect can as readily be conceived. What more in order than the strongest constitution should be enfeebled by the impression of cold, received under the most unfavourable circumstances, and which must be intense indeed, before it can be sensible to an occupied mind. What more likely than that the firmest chest should be shattered by the incessant aggravation of catarrh? Sorrow willingly joins exposure and fatigue. So that, without contagion, these allies will find no more difficulty in their operations than associated armies before an otherwise impregnable fortress, when its walls have been rent by an earthquake to their foundation.

The succession of the disease, in persons not predisposed, is too rare to admit of the supposition of contagion. In the general hospital of Vienna, Dr. SCHMIDT, one of the physicians, tells us that there died in three years, from 1797 to 1799 inclusive, 2,735 patients of consumption, without the physicians or nurses being ever infected.

PROSPECT.

Who shall undertake to tell the fortune of the coming generations? and satisfy us whether we are still to hear of the domestic horrors of consumption being repeated, among us, eighty or a hundred thousand times every year? A lot of young men and maidens from the first families in Athens were annually sent to glut the monster of the Cretan labyrinth. But the sentiments of the people rose against the horrid tribute, though it consisted but of seven of a sex, and it ceased.

Without being a Theseus, one might engage at once to reduce the tribute of lives we render to consumption. It must first, however, be generally believed with SYDENHAM, that our *chronic maladies are of our own creating.* An universal interest too must be taken in the condition of those about us, not less lively than the desire we see so prevalent of dazzling strangers by the splendour of an equipage, or by the lights of the understanding. Our taste must be corrected. We must all learn not to bear to have every thing about us—cloaths, tables, chairs, pictures, statues—all exquisite in their kind—except our progeny. We must be shocked at seeing, under the same roof, perfect representations of the human figure in contrast with the most wretched realities. This we must feel to be, as it actually is, in a more barbarous style than the Prince of Palagonia's groupes of marble monsters. Parents, in short, must find and seek honour in the healthiness of their children.

I suppose this time will come, because I think myself able to point out its most certain sign.—Wise and worthy men have fixed upon the state of intercourse between the sexes, as the surest criterion of the happiness and virtue of a community.* In spite of the celebrated forgeries of Macpherson, we must indeed still regard the importance of women as a proof of civilization. But it is more useful to hold the principle in view than to rest upon an example, the happy choice of which is extremely questionable.

The principle of civilization is no other than the spirit of equity. It is the abolition of the *jus fortioris;* the resignation of those advantages which the stronger owes merely to his strength.—Readiness to make a present sacrifice for a distant good is but the same principle in another form. It is merely the spirit of equity, which always first emanates abroad, reflected.

We may, then, set down justice to the weaker

---

* "In appreciating the happiness and virtue of a community, there is perhaps no single criterion, on which so much dependance may be placed as the state of intercourse between the sexes. Where this displays ardour of attachment, the character and the influence of women rise in society; our imperfect nature mounts on the scale of moral excellence, &c."—*Life of Burns, preface p.* 18.

sex as a happy commencement, but by no means as the completion of civilization. It is only the first infringement of the authority of brute violence.—Where the greatest physical and moral means of enforcing claims exist, there will the claims be first admitted. In their intelligence, stature, tenderness, services and confidential intercourse, women find these greatest means. A community therefore may be much civilized in regard to women, while it remains barbarous with regard to two other classes of human beings. In declining Rome, women were of importance, and could torture the poor slaves for their amusement, while the condition of children, except by courtesy, was little inferior to that of slaves. The history of modern cultivated states shews the abolition of vassalage and slavery at home, while the ancient sternness towards children continued in doctrine, and was often exhibited in practice. There is a sculpture, from which it has been collected that Sully's children stood while he dined, and I have been well informed that a British nobleman, not long since deceased, never allowed his sons or daughters to sit down in his presence unbidden. How remarkable that about the same period of the last century, the cause of colonial slaves and of children should have been pleaded with all the fire of eloquence and force of truth!

The scheme to which these ideas belong, is easily compleated. Pleas in behalf of weakness must be nearly alike. To admit one is a step towards admitting all. It is the honourable distinction and most striking character of the present age, that sentiments of perfect equity are extending themselves to the most helpless class. We every where meet with parents renouncing all capricious exertion of authority, and careful to weigh their children's welfare in the scales of justice against their own inclinations. By this truest criterion, we may safely decide that civilization has advanced. And our contemporaries must obtain as individuals that applause, which will less readily be granted them in their collective capacity.

It follows that consumption will not be suffered to continue its present ravages. For nothing further is required to stop them, than to carry the disposition, already formed in favour of young people, fully into act. It is manifestly a disposition that cannot remain at rest, while its objects are so cruelly suffering. Nor will it be content with doing for them what is *good*, but must go on, by its own progressive spirit, to what is *best*. And what is this but the removal of such causes of unhappiness as are inherent and personal, not transient and depending on the pressure of external circumstances? Some few families I know, where the most determined steps have been taken, from the very cradle, to resist the "GIANT-MALADY" of our island. There the human plantation flourishes like scions, removed from shade and rubble to a rich and sunny soil. I look upon it that such families have nearly as much the advance upon many respectable authors on éducation, as these authors themselves have upon the monkish institutors of our school-discipline. As soon as such examples shall become a little more common, it will appear manifest that the devastations of consumption proceed from domestic mismanagement, and not from the inalterable dispositions of nature.

END OF ESSAY VII.

Printed by MILLS, St. Augustine's, Bristol.

ESSAYS

ON

THE MEANS

OF

AVOIDING

HABITUAL SICKLINESS,

AND

PREMATURE MORTALITY.

ESSAY EIGHTH.

Printed by MILLS, St. Augustine's, Bristol.

Quanta morborum hinc cohors, si vires digerentes non bene se habeant? --- PRIMÆ VIÆ magis quam aliæ machinæ nostræ partes morbificarum causarum influxui exposita sunt: nec non ob organorum, ex quibus constant actionumque particularium ab iis perficiendarum varietatem innumeris obnoxiæ sunt a sanitate deflexionibus. Quis, rogo, plus morbis obnoxius est? au pauper agricola qui äëris vicissitudinibus absque ulla fere cautione corpus exponit, quique quod ad cæteras res sic dictas *non naturales,* imo naturales, attinet, multo incautius vivit quam dives civis? An verò hic, qui licet multo solertius evitet noxium plurimarum causarum occasionalium influxum; tamen multo frequentius morbis laborat, cum in uno tantum peccet, quod scilicet vitæ suæ genere primarum viarum actionem debilitet?

What a host of complaints, if the digestive faculties are out of order? --- The PRIMÆ VIÆ are more exposed to morbific influences than other parts. On account of the variety of organs, and of the offices to be performed, they are liable to numberless deflexions from health. Which, I pray, is most liable to disease --- the peasant exposed, without precaution, to the vicissitudes of the air, and inattentive to the *non naturals* as they are called and to the *naturals* also, or the rich citizen? Does not the latter much more frequently suffer from indisposition, because by his mode of living he debilitates the action of the *primæ viæ*?

WEDEKIND.

ESSAY

ON

THE PRESERVATION

OF

THE PHYSICAL POWER

OF

ENJOYMENT;

WITH REMARKS ON

FOOD AND DIGESTION.

THE SUBJECT STATED.

HITHERTO I have been principally occupied in considering the first stages of life. By resolute adherence to the simple practices inculcated in the foregoing series of Essays, we shall no longer see those members of society, who seem to have the fairest title to a happy and protracted existence, mowed down by whole ranks every year. The agents which slowly and silently loosen the ties, and open the seams, of the constitution, leaving it a wreck to the mercy of the elements, will be counteracted in their operation. A few scattered individuals here and there may perish from inevitable accidents. But the bulk of our youth will be snatched from that fate, which at present they so frequently experience. I proceed, in the next place, to consider how they may be delivered into their own keeping with *the faculty of pleasurable sensation* unimpaired. Nor is this the affair only of youth. For in the progress of existence, what shall we look upon as the first concern of all, but more particularly of those privileged classes, who, by favour of Fortune, find perpetual admission to the feast of Nature?—It is surely to know how they may maintain this faculty in its most perfect state of freshness and vigour.

To beings, endowed with feeling, it is essential to strive to draw all possible gratification from the things, by which they are surrounded. The attempt to keep this propensity under forced restraint has always proved as abortive as it is unjust and injudicious. In the great majority it will not fail to break out occasionally with blind and destructive fury, or wear out mind and body, already enfeebled from continual privation, by the efforts which will be necessary to resist perpetual solicitation. In the state of moral improvement, however, to which this country has advanced, mortification is rarely assumed or imposed as a duty. Our danger arises from untutored appetite. To indulge in such a way, that desire may constantly spring up anew to administer the occasion of fresh indulgence, constitutes the *virtue of temperance*, which, according to my apprehension, can be no otherwise considered than as synonymous with the *art of enjoyment*. In endeavouring to illustrate such of the fundamental principles of this art, as are entirely of medical resort, I should be happy if I were able to assume a tone the very reverse of that which we sometimes hear from our moral and medical advisers. For in addressing individuals, full of relish, hope and animation, and who have the world at their feet, they either require of them to renounce their advantages; or in pointing out to them how they may avail themselves to the utmost of these advantages, they speak as if they were imposing the most rigorous penances. What success must attend the proposers of self-denying measures, when they have no authority to enforce them, and when they do not thoroughly explain the grounds on which they are adviseable, cannot be difficult to conceive.

Of the ORGANS that contribute in the first instance, to OUR NOURISHMENT.

The gratification most universally coveted, most frequently returning, and that qualifies us best for the enjoyment of the most refined pleasures and for the performance of the most important duties, is connected with these organs. "For physic, metaphysic," as a lady wrote to her sick son, "all depend "upon the inspiration of roast beef. If you "would do well, you must eat and digest "like a ploughman; nay, if you would walk "well, think well, write well, &c." The exterior of the human body is frequently contemplated with the highest delight: and but seldom, in its healthy state, with disgust. Familiarity with the internal parts, the exercise of the arts that disclose their structure and the study of their uses soon occasion them also to be viewed with deep interest, though as mens' minds are usually disposed by our customs, it requires an exertion to overcome the repugnance, which arises on observing them attentively for the first time.

The stomach in man and in animals resembling man, is far from recommending itself by any elegance of appearance. It will, on the contrary generally be reputed an unsightly, membranous pouch. But the delicacy of its texture, and the consideration of its extraordinary powers, though it be, at first sight, so unpromising, may create a salutary reluctance to hazard any practice, by which it may be injured; even though the integrity of the whole machine did not depend upon the sound state of this organ.

MECHANICAL FUNCTIONS of the STOMACH.

I call the functions, of which I am going to say a few words, *mechanical*, merely because they affect the whole bulk of the stomach. I shall not digress into the consideration of the almost incredible, but perfectly authenticated strength of the gizzard, by which animals possessing this very powerful machine, grind down globules of glass and sharp steel instruments, without receiving any hurt themselves. It would be beside my purpose to speak of any thing but the contractility of the thin and flexible human stomach, by means of which it constantly adapts itself with accuracy to its contents. The stomach is never flaccid; its sides do not, at any time, fall flat upon each other, as those of a bag of no greater thickness, would instantly do, after the contents were discharged. If distended only to half its capacity, the inner surface is closely applied to the materials which it encloses. Nay, when quite empty, it purses itself up so completely as to leave no vacuity within. The difference in its dimensions will be easily imagined, when it is known that various physiological authors have estimated the quantity, which this viscus is able to hold at once, at upwards of ten pints. It is sufficient however if we suppose that it may sometimes receive two quarts, and sometimes be filled by less than an ounce, and if we remember that its sides, when separated by the introduction of aliment, approach again more or less closely by virtue of their inherent contractile power several times each day through the years of a long life.

Its living energy, when considered as a whole, is most apparent in its tendency to contract readily upon the remainder of the food, as part passes into the intestines, till nothing is left but that small quantity of fluid, which issues perpetually, in greater or smaller quantity, from its own sides.

The disagreeable sensation we experience, when we say that *something lies heavy on the stomach*, may, in some cases, (as where it follows the use of certain liquids,) depend on the diminished contractility of its muscular fibres, though it oftener depends as much on another kind of weakness, which will be considered below. But the internal sinking, so often the subject of bitter complaint among the feeble, if they remain without food but a few minutes too long, depends perhaps altogether upon this diminution of contractility. The more vigorous the fibre, the more easily does it reduce itself to the dimensions prescribed by the healthy state of the organization. The more frequently also will it repeat this motion without exhaustion, and the longer continue the effort of contraction. But the enfeebled organ experiences greater and greater difficulty, as it retires into a smaller compass; and the utmost degree of retraction is accompanied by a most distressing effort, which distension from the introduction of a little food, immediately relieves. The sensation, though it cannot be characterised by words, is perfectly distinguishable by those who have felt them all, from hunger, heartburn, or cramp. It infests the enervated and the indolent. The muscles that move the limbs appear to be at the same time in the same powerless state. For you will find them flabby, if you grasp them with your hand. And when the stomach is relieved by eating, a weary aching sense shall be felt in the extensive muscles of the legs and thighs.

In these circumstances the mischief is seldom long before it spreads from the centre to the circumference. But even when it confines itself to the original spot, a comfortable moment is scarce known the whole day; and the night passes in watchfulness, disturbed dreams, or wearying sleep. This misery may arise from various causes. Ill-contrived cloathing and a sedentary life are conspicuous among the number. Animals at large seem to enjoy great exemption from stomach complaints, which in human life produce a much greater share of distress, than in proportion to the space they occupy in the nosology. Wild animals in quest of food, and domestic animals in that spirit of wantonness, with which they are filled by the wholesome delicacies of a rich pasture, each exert themselves sufficiently to preserve the moving and digestive organs in good plight. Those that are fed within doors are frequently worked in proportion. It would appear that they hold the privilege of a light and easy stomach upon this condition. For that pampered parlour-boarder, the lap-dog, is hardly less subject to queasiness, inappetence and vapours, than his mistress. He has indeed one security more than she has. Fashion never offers violence to his chief digestive organ. It is never cased and corded down to its smallest dimensions: but after a repast is at liberty to rise from its resting place upon the interior of the vertebræ, and to take that free swell, without which the process of nutrition cannot be properly begun.

CHEMICAL FUNCTIONS of the STOMACH.

The proper digestive power may be called chemical, inasmuch as it seems to depend on the admixture of fluids, furnished by the stomach, with the aliment, and on the entire alteration they produce in the interior constitution of the received aliment. The chemical and mechanical powers appear to go hand in hand. The more vigorously the walls of the stomach apply themselves to its contents, the more speedily and perfectly are the other operations carried on. The sense of heaviness above mentioned most frequently seems to arise from the fatigue of the muscular fibres, vainly exerting themselves against materials, not prepared for being propelled through the lower of the two orifices of the stomach into the duodenum, or that bowel, which immediately follows in the alimentary tract.—What this preparation is, has been in a tolerably satisfactory manner ascertained.

In the ordinary process, whatever has been received among the varieties of proper digestible food is found reduced in a certain time to a pultaceous mass, and passing into the duodenum. This time we have reason to believe in health seldom less than three, and seldom more than six hours. For different experimenters, in consequence of taking a vomit about so long after eating, have found the contents of the stomach of a fluid creamy consistence, though in part previously solid; and in other animals, supported by mixed food, the same appearance has occurred upon opening the stomach after the same space of time. However this calculation is subject to the greatest uncertainty; the stomach having been found empty in two hours after feeding,* and ordinary articles of food having remained undigested for days and even weeks. Besides, there is a great difference, depending on the difference of substances.

Among the chemical properties of the stomach, that of consolidating certain liquids is one of the most remarkable, and one of the most anciently known from the effect of animal rennet on milk. The human stomach possesses this property in a very eminent degree. Half a dozen grains of the human stomach, steeped in water, afford an infusion, which will make an hundred ounces of milk *come* into curd, as the people of the dairy express it. The change into curd of a particular part of the milk is instantaneous, supposing there be already a provision of that particular liquid, which, in health, is always thrown out by secreting vessels, and after

* Marshall BIRON, who used to eat with an excessively keen appetite, was killed by a ball an hour after dinner. The surgeons on opening the body found not the least trace of food in the stomach. This however is no demonstrative proof of rapidity of digestion. For the gastric liquor continues its solvent action as well after death as out of the body. The dead stomach is dissolved by what it has itself secreted. Earth-worms will live in gastric liquor at the temperature of 60°. At 80° they die, and are dissolved. The heat kills them first. Young rattle-snakes are found alive in the stomach of the old ones. It is here they take refuge.

death may be extracted by maceration. But if there be not a provision of this liquid, or if it be not of a proper quality, coagulation will not instantly take place, as will be rendered more than probable by some facts to be related by and bye. So that the mere membrane has not the power of producing curd.

This species of coagulation is believed to extend also to solids; that is to say, the solid is rendered of a firmer texture by the application of the coagulating juice; nor is there any difference between the operation on solids and fluids; it being in both cases, according to the late accurate Dr. Fordyce, the abstraction of water from a material, which, when acted upon by the stomachic liquid, or by any other coagulating power, is exactly the same, whether it come from the pure serum of the blood, the pure coagulable lymph, the white of an egg, the mucus, the skin, a tendon, a cartilage, a membrane or a muscle. Later chemists have indeed found that these substances differ more than in the proportion of water they contain. Three distinguishable principles *gelatin*, *albumin*, and *fibrin* having been found in some of them; and others consisting solely of one of these principles. But we may assume on the authority of Dr. FORDYCE that they are similarly affected by the liquid thrown out by the stomach. Though I know not if this has been ascertained of gelatin.

Much uneasy feeling, if nothing worse, arises, when the work of consolidation goes on improperly; as it is so apt to do in the weak stomach of infants and in the depraved stomach of those adults, who have dealt unmercifully with this powerful, but tender organ.

In infants, and occasionally in others, the coagulating power would appear sometimes excessive, as we may judge from the strongly curdled milk, which is sometimes rejected. Whether this depends upon the conjoined effect of an acid and of the fluid gastric rennet, or upon some fault in the secretion of the latter is not certain. In either case, the addition of an alkaline substance, as of lime-water, will, I believe, prove an adequate corrective. Or if there be any scruple against the use of lime-water, alkali saturated with carbonic acid, which is now commonly sold under the title of *super-carbonated kali,* and is a perfectly bland neutral compound, may be used with the utmost safety. From one to two drachms in a pint of milk can produce no constitutional bad effect, but will prevent this and other species of indigestion, incident to infants. The same admixture will tend to obviate the bad effects of water-gruel, arrow-root, and biscuit or bread diffused in water, which substances it is, I understand, now fashionable to use as the food of early life, though their use be contrary to every chemical and physiological principle. For these vegetable preparations are not more nearly allied than animal substances are to milk, the natural sustenance at this period, and they undergo prejudicial changes where the digestive organs are weak. Dr. CAMPER constantly gave his children biscuit pap with an addition of Venice soap. Soap does not appear the most eligible alkaline substance, but was fixed upon in deference to BOERHAAVE, who ascribes to it extraordinary virtues. We have it on the father's authority that they all grew up stout, and seldom were troubled with sourness.—As to adults, when they have injured the stomach by receiving into it matters of a pernicious quality, they will find no permanent comfort but in a change of diet. The subsequent considerations will explain what this change ought to be.

In the digestive process, the next step to coagulation is solution. Whether the same liquid perform both offices is uncertain, because it has been impracticable to obtain separately the fluids, poured into the cavity of the stomach by its rich and varied apparatus for secretion. It appears unquestionable that the aliment, though first rendered solid, afterwards undergoes a solution as perfect as that of sugar in water. For hollow perforated silver globes, filled with flesh and other articles used in diet, have been recovered empty in a short time; which could have happened no otherwise than by the solvent effect of a liquid, that had found its way through the perforations, and acted upon the contents of the globes. And if farther proof were wanting, it has been obtained by procuring the gastric liquor from the stomach, and applying it to the same substances at the temperature of the body, by which treatment they have been observed to undergo perfect solution.

The idea of placing a window in the breast, through which the thoughts might be manifest, has been sometimes started. Disease has really afforded opportunity of directly ascertaining in a great degree what passes with regard to food in the interior of the stomach. In the old medical observers we find accounts of persons, living with a preternatural perforation in this organ; but little credit was, I believe, attached to the reports, and no information drawn from the accident. However, two persons, within these few years, have exhibited themselves in two of the capital cities of Europe, with holes or wounds through the skin and coats of the stomach, by which that viscus could be filled from without or emptied at pleasure. Nor did the opening seem to impair the power of digestion or the general health. One was at Dublin. This man had received a wound penetrating into his stomach. The wound left a permanent opening about the third of an inch in diameter, which the patient kept plugged up. He lived to the age of 65, that is, twenty seven years after receiving the wound. He was excessively intemperate, and having suffered much hardship, died in the house of industry at Dublin of a gradual decline. He had been received into the Dublin charity at a former period in good health, but his sudden departure prevented any observations on his digestive process. All that was collected from him was: "that "on removing the plug after taking milk, "part of it, quite pure, escaped through the "opening, and when his stomach was empty "of food, a whitish sweet-tasted fluid adhered "to the plug on its being withdrawn. He "never felt pain in the opening nor incon-"venience from any particular food." The other patient with a perforated stomach was received into the general hospital at Vienna in 1798, and after a vain attempt to close the opening, examined by many of the medical faculty there, and presented to the students of medicine at the public lectures. Dr. HELM relates the following among other observations. When any particular things, as sour crout, gave the person (who was a female, and if not very lately dead, is still living) uneasiness, she had recourse to an expedient, which many a glutton might envy her, and which consisted in taking the oppressive food with her finger out at the wound. She then washed out the viscus with water, and this always relieved her, and revived her appetite.—Milk instantaneously curdled, except when the patient had carefully swilled out her stomach with water. In this case, says Dr. HELM, the coagulation did not take place, till after some time, for want of gastric liquor; as I incline to suppose." But I could always, he adds, expedite this phænomenon by irritating the inner surface of the stomach with my finger. Asses' milk coagulated with most difficulty, requiring several minutes on occasions, when cow's milk curdled on the spot.

Among the different articles of diet, flesh-meat was constantly digested with most ease and expedition. Eggs and cheese—it is not said what kind of cheese—were very quickly digested. Vegetables in general were much longer in undergoing this process. But potatoes, cucumbers! and carrots soonest passed into the intestines.

But coagulation and solution are by no means the whole of digestion. A great deal more preparation is necessary to provide for repairing the waste of the body. The succession of steps may be compared to a very complicated manufactory, where the material, before it can attain the necessary perfection, passes through the hands of a multitude of artisans, each of whom, in his department, must do justice to the fabric; or else the whole labour will be as good as lost. What happens further to the solution in the gastric liquor is very obscurely known. It is undoubtedly propelled through the lower orifice of the stomach into the intestines in proportion as the solution is completed. In the first intestine, the bile and the pancreatic juice and the liquids, secreted by the intestine, are gradually mixed with it. But the effect of this mixture no one has ascertained. The addition of the bile is not necessary to the main purpose of digestion. For the formation of chyle (or of that fluid which replaces the blood, as it is consumed in feeding the various solid parts of the body) would appear clearly independant of the bile; since the access of the bile having been artificially prevented, chyle was nevertheless found in its proper receptacles. The same thing has been ascertained with regard to the pancreatic juice, which, in ordinary circumstances, flows into the duodenum to be mixed with the substances that have undergone solution in the stomach, the full discovery of which step by SPALLANZANI, is by far the most considerable, hitherto made on the subject of digestion.

The only way we have of attaining any thing like a satisfactory notion concerning what happens to the alimentary mass, after it has been dissolved, is to consider the nature of animal and vegetable substances. These substances appear to be in general distinguished from inorganic bodies by their susceptibility of change. Their elements may be considered as united in a manner peculiarly loose, and as ready, on the smallest provocation, to start into new combinations. An apple or a piece of flesh placed in a temperature, too low to have any ascertainable effect on a pebble shall be altered in its taste, consistence and whole constitution. Some of the elements, after undergoing new combinations, shall be given out in the form of air; some in a liquid state; and the remainder shall be unlike any thing the body contained at first. The alterations shall be different in a temperature some degrees below the ordinary temperature of the atmosphere from those which take place some degrees below it, or in a degree of cold sufficient to freeze the juices. In short, there will be a sensible variation of product, correspondent to variations of circumstances, apparently very trivial. Organised bodies, the matter of aliment, have a very complicated as well as a very changeable composition. Hence they yield great variety of products. Held in solution in the gastric juice—having other juices of various quality gradually added to them from the interior surfaces of the intestines—acted upon by the heat of the body—and perhaps influenced by the living membranes in some unknown way, remotely analogous to the effect of the metals in the galvanic pile, their disunited floating particles assume an arrangement, which art has not yet acquired the power to imitate, and form chyle;—a milky matter, consisting, when it is taken out of the vessels, of a fluid, which coagulates spontaneously, of another fluid coagulable by heat, and of white globules, that give the whole its whiteness and opacity.

The principal means by which DIGESTION *is impaired, and the* DIGESTIVE ORGANS *injured.*

All the materials therefore which are capable of serving for food must have one common property. Each must contain the principles of chyle. Though it may be more easy to bring some into the state, in which their elements will coalesce into that particular species of fluid.—Hence we learn the task imposed upon the stomach and its appendages. By them it is necessary that the agents, on which the several steps in the transmutation of food into chyle depend, should be supplied from day to day in sufficient quantity, and possess the requisite qualities. Otherwise the whole system will languish for want of a renovation of that part of its substance which is continually wasted, and the region of the stomach will be the seat of a perpetual succession of nameless, harrassing sensations, rendering existence cheerless, and afterwards by means of certain associations, as we shall presently see, becoming scarcely supportable.

If the chyle be the fountain of the blood, the blood is the fountain of the juices, instrumental in the preparation of the chyle. The blood must not only furnish to the secretory organs the elements of these juices, but it must circulate through them with a certain degree of vigour. No fact belonging to animal nature is proved by a greater number of examples than both the permanent and occasional connection between exertions attended by a vigorous circulation of the blood, and a keen appetite with a correspondent digestion. Not only do those occupations produce a contrary effect, which are carried on in a posture, unfavourable to a free expansion of the stomach, as weaving, but those also that scarce allow more range of motion than the oyster enjoys, though there be nothing injurious in the posture.

Who would not desire to have an accurate acquaintance with the substances, especially if there be any in common use, which impair the coagulating or solvent powers? We cannot make direct observations upon the interior of the human stomach; but we have animals, very closely resembling man in having a single and equally thin stomach, and in feeding on flesh and vegetables promiscuously. And what we can observe of the effects in our own species agrees perfectly with the phænomena in animals, so nearly a-kin to him in this respect.

The habitual ingesta or substances received into the stomach, that prove most hurtful to its operations, are intoxicating or fermented liquors. In order to observe the first effect of these liquors, when of considerable strength, I caused an equal quantity of the same food to be given to two young dogs of the same litter. Immediately after feeding, three drachms of spirit of wine of commerce, mixed with a single drachm of water, were poured down the throat of one of the animals. In five hours both were opened within a very few minutes of each other. The animal to which the spirit was given, had its stomach nearly twice as full as its fellow. The bits of flesh were as angular as immediately after they were cut by the knife at the time of feeding. They were also as firm in their substance. In the other dog, these angles were rounded off, and the pieces throughout much softer. Opium, the intoxicant of the countries where fermented liquors are forbidden, has been observed to have precisely the same effect. "In dogs who "have swallowed a large dose of opium (as "three grains) the peristaltic motion of the "stomach and intestines is much lessened or "totally suspended; the food last received "into the stomach remains indigested, and "the bowels are more than usually empty.*" Strong liquors are equally productive of indigestion in man. Many hours, and even a whole night after a debauch in wine, it is common enough to reject a part or the whole of a dinner undigested.

On the Rhine and in Bavaria, during the years 1797, 1798, and 1799, the scarcity of fodder and the prevalence of disease rendered it necessary to destroy a great number of horses, employed both for the purposes of agriculture and of war. Mr. Pilger, an active experimenter, availed himself of so favourable an opportunity for collecting a number of interesting observations on the effect of poisons or strong stimulants. His experiments were made upon near two hundred horses, with arsenic, corrosive sublimate, barytes, opium, belladonna, cicuta virosa, laurel-water, distilled spirits and with wine.†

---

* Whytts Works, 4to. p. 181, &c. Kauu impetum faciens.

† Versuche durch den Galvanismus die Wirkung verschiedener gifte und arzneimittel--zu pruifen—Giessen. 1801.

Mr. PILGER found that forty grains of arsenic, given at once, produced no other apparent effect than a slight transitory pain in the stomach of a horse. He then administered four grains a day, in doses of a grain each, for five weeks, and upon destroying the animal at the end of this time, found the stomach shrivelled; the villous coat dry and rough; that important organ the liver, as also the spleen, altered in their organization.

Belladonna and undiluted æther occasioned a shrivelling of the innermost coat of the stomach as considerable as if it had been long held close to a brisk fire.

"I have," continues the author, "given "a number of horses wine and brandy so as "to raise them to a sensible degree of exhi-"leration. Upon these occasions I constantly "found that brandy gave the interior of the "stomach a shrivelled appearance; that it "had encreased the orgasm of the blood-"vessels of the intestines,* and produced "congestions in the brain—which I did not "observe from wine in this dose."

---

* The blood-shot appearance of the intestines indicated that such an orgasm, or violent action of the blood-vessels had occurred.

"But when I gave either wine or brandy "enough to cause intoxication or staggering, "to induce sleep or to make the animal "drop, and in this state had it killed, the "brain was always much charged with blood, "the orgasm in the fine vessels of the intes-"tines strong, and the stomach shrivelled."

Except in case of a fatal stroke of apoplexy, immediately brought on by intoxication, we cannot, as I have said, procure sensible evidence of the primary operation of wine and spirits on the structure of the healthy human stomach. But that they impede digestion, as in the example of the dog, and that they alter the texture of the fine inner membrane of the stomach, as in that of the horse, we have every reason to believe from the history of those, who indulge in their use. For how often do we see our drunkards and sots so far deprived of the power of digestion, that almost every portion of food shall be thrown up nearly as swallowed! Now in what way shall we apply the knowledge we have acquired concerning the gastric processes, to the explanation of this phænomenon? Is it not rational to conclude that what, on the very first occasion, changes the appearance, while it impedes the action, of a very delicate organ, must, by many repetitions, establish an alteration of structure, and destroy the action altogether, or at least injure it permanently? The apparatus for secretion being no longer the same, the former coagulating and solvent gastric juices can no longer be poured out upon the food. The comparison of the facts respecting the two perforated stomachs seems to shew that milk curdled less readily in the Dublin patient, who made so extremely free with intoxicating liquors. I propose to ascertain by precise experiments whether a mixture of spirits and of wine with the food of animals will not impair the coagulating power of the stomach.

SPIRITUOUS LIQUORS in what state MOST INJURIOUS.

The stronger the worse. The injury from any fermented liquor is to be measured by the quantity of alcohol or ardent spirit, which is to be obtained from it on distillation. In the relation of Mr. PILGER's experiments, it may be observed that brandy or distilled spirit, when it was given only to the first stage of intoxication, produced phænomena, which wine, given according to the same measure, did not produce. In their immediate action upon the habitual drinkers of both, gin, rum, brandy and strong distilled liquors of whatever denomination differ essentially from wine and ale of the ordinary average strength. The former, if very largely swallowed, will generally produce instant death; if less largely, a state of frenzy, with which, in that degree, mere wine-bibers are for the most part unacquainted. In common life, it is true, we do not meet with accurate comparative experiments; but from what we may observe, we may lay it down certain enough, that the same quantity of intoxicating fluid will occasion remote as well as immediate effects of a very different kind, according as it is taken more or less concentrated.

In the Southern states of Europe, public opinion is exceedingly severe on the use of liquors, that affect the head, undiluted, even when sparingly taken; and on the subject of intoxication it is nearly inexorable. With us, the strongest wines may be every day swallowed without reproach in a quantity scarce consistent with the regular functions of the brain. But the difference between drinking port or madeira and brandy is the difference between respectability and disrepute. What can be the foundation of the distinction? What other than the more palpably baneful operation of spirits? Every one condemning what no one can deny to be detrimental. Were the consequences the same, emptying a goblet of brandy would have been in all countries an act as free from censure as taking a bason of soup; and dram-drinking would be no secret practice.

The lovers of the bottle flatter themselves that care in selecting wines will exempt them from evils, which they cannot dissemble to themselves that wine produces. The convivial possessor of ample cellars, whose pipes regularly ascend to the honours of his table in the order of their standing, praises his good fortune, when he compares himself with young men at the university, condemned as they are to regale upon home-brewed and unmellowed wines. But the distinction, I am afraid, rests upon a difference, little in point as to health. It is the result of common observation, confirmed by unvarying medical experience, that among country gentlemen, citizens, the frequenters of the common and combination room, the associates of the mess, the farmer, the merchants' clerks and the artisans that croud the smoke-shops in places like Birmingham, a large proportion injure their digestive organs, whatever be the coarseness or tenuity, the purity or adulteration of the liquors, provided they be about the same strength and taken with nearly equal freedom. Dissimilarity of constitution, of occupation, of exposure and the like will produce a dissimilarity in the result; and but seldom the oil, the mucilage, extractive matter and other vegetable principles that accompany the spirit in the stimulants, habitually used. — The difference between the effects of the most clumsily manufactured and the purest wines, when of the same strength, is perhaps scarcely equal to the difference between ninety nine and an hundred. And what difference exists, is not the smallest consequence to our present enquiry. — Peculiar mischief is imputed to *new* rum in the West-Indies, but ought we to be persuaded that this is owing to its age, until old rum is given to soldiers and sailors with equal profusion ?—One thing only, on a comparative view, would appear to make somewhat in favour of spirits and of wines, long kept—They lose of their bulk; that is, they become weaker, and the alcohol is the part that evaporates. *Long* keeping however seems necessary, in order that this effect should take place in any sensible degree. For new musty liquors remain at least equally strong for a time by the slow conversion of their fermentable ingredients into spirit.

The more quickly also any considerable quantity of intoxicating liquor is drank, the more injurious to the organ to which it is immediately applied, and to those upon which it remotely acts. Healthy people, I presume, would be sensibly less injured by a course of vinous stimulants, if the quantity taken in the afternoon was regularly distributed through the day. In grave diseases of debility, we find that cordials and strengthening medicines are given at stated intervals; and in affections of less consequence, the same rule ought to obtain. Immediately after a meal seems in general the most improper of all periods, where there is any purpose of health in view. Proper food in proper quantity is exciting enough for the strong; and without caution, is apt to be over-exciting for the weak. Extraordinary stimulants should therefore be reserved for the seasons when the powers of the system begin to flag.

In whom Digestion is most liable to Injury from this Cause.

There are an infinite number of facts which shew that the organization of children is, in general, most apt to suffer from many classes of violent agents. The writings of anatomists are full of melancholy examples of the most monstrous deformity, evidently produced by pressure in the early and more plastic state of the several parts, of which the body is composed. Examples, somewhat less striking, occur every day in spite of the more rational methods of management, recently adopted. Fatal local ailments are constantly produced by blows and falls. The same law obtains with regard to the powers that, by their specific action upon the stomach as a living organ, throw the system into strong commotion. Medical practitioners, much conversant among the poor find them perpetually stinting the growth, and destroying the constitution, of their children, by their ill-judged kindness in sharing with them those distilled liquors, which they swallow with so much avidity themselves. Among the causes, fatal to the health of the higher classes, the allowance of wine, that is so often served out to the children, short as it may appear, deserves to be considered as not the least considerable. Mr. SANDFORD, surgeon at Worcester, in his useful and entertaining tract on wine and spirits, relates the following observation, which may be confirmed by thousands equally certain, though made with less precision. "A late

" ingenious surgeon, occupied for a great part
" of his life in experiments equally well con-
" ceived and accurately executed, gave to one
" of his children a full glass of sherry every
" day after dinner for a week. The child was
" then about five years old and had never been
" accustomed to wine. To another child,
" nearly of the same age, and under similar
" circumstances he gave a large china orange
" for the same space of time. At the end of
" the week, he found a very material difference
" in the pulse, the heat of the body, the urine
" and the stools of the two children. In the
" first, the pulse was quickened, the heat in-
" creased, the urine high coloured, and the
" stools *destitute of their usual quantity of*
" *bile*, whilst the second had every appearance
" that indicated high health. He then reversed
" the experiment: to the first mentioned child
" he gave the orange, and to the other the wine.
" The effects followed as before—a striking
" and demonstrative proof of the pernicious
" effects of vinous liquors on the constitution
" of children in full health." The deficiency of bile is full evidence of the injurious effect of the wine upon the digestive organs in this double experiment. And the result itself must be regarded as the more satisfactory, since the author, to judge from the terms in which he is

described, could have been no less a person than the late Mr. HUNTER.

If direct proof be wanting of the power of diluted spirit to destroy both appetite and digestion, it may be found in a very marked degree, wherever this kind of liquid is given to children a little freely. During the late distressful seasons, some poor families were driven to that dire expedient by motives of œconomy. A lady—Miss ELIZABETH SEWARD—met with a family of poor children, whose pale faces and emaciated bodies forcibly attracted her attention. Upon enquiring of the mother how they were fed, she was informed that "*they did not* "*eat much, and that what they did eat was not* "*sufficient to nourish them without gin and* "*water. It was scanty vegetable fare.*" The lady, after stating to the woman the pernicious effects likely to follow from such a regimen, advised her to purchase a little animal food with the money she expended in gin, and to give the children water to drink with their meals—"*Lord, madam,*" replied the poor woman, "*if I was to do that, I should never* "*be able to satisfy them in these hard times.* "*I was used to give them water. But then* "*they were always hungry: and I could not* "*beg or buy victuals enough for them.*"

The glass of wine, we ought to observe,

had a similar effect in the first of these cases to the gin and water in the second. Indeed its operation in a single week must make every person, open to palpable facts, shudder for the consequences of a long continuance of the same habit. Why then not profit by the example? The false shame of standing single amid a host of bad examples, is out of the question. Children know nothing of that factitious flatness of imagination, which in debauched adults calls for relief from the enlivening potion, with a voice as imperious as that, in which the natural appetites of hunger and thirst demand the means by which they are respectively aswaged. What happens so seldom in cases where the constitution is at stake—Propriety here stands against Impropriety, single and unsanctioned. As the greatest authorities are against wine; as there are none worth regard on the opposite side, and above all, as there is so little danger of being thought odd, why risk the early destruction of that organ, which may be regarded as the great regulator of the inward man? Since as that stands affected, so do we feel to ourselves. So also do those, with whom we live, find us; and how often does the vitiated state of the stomach and its dependencies express itself in the countenance as clearly as the disastrous tidings in the features of him, who had to tell PRIAM that his Troy was lost?

Doubtless there are cases where the violent stimulation of wine is not more than sufficient to bring back the digestive organs, and the general system of children, to their healthy condition. But even as a medicine, wine ought perhaps hardly ever to be administered to young people in their chronic ailments. Drugs may commonly be found, which will equally answer the purpose; and these would without difficulty be relinquished, when their continuance would be as hurtful as wine to a healthy child. Or it may be good policy to give a disagreeable drug in wine, that the vehicle may become hateful by association. If the case should be thought such as to render a substitute inadmissible, dilution may still be proper, and the utmost care should be taken to administer the diluted liquor so as to prevent faintness and chill, not to produce excitement and heat.

Children will not crave this dangerous beverage, unless they are instigated by example or have been taught to relish it by frequent use. But what shall we say to our young men, and to men of genius, who derive such exquisite delight from the glow of imagination, and who are therefore apt to embrace with so much avidity every means of kindling this glow?

According to the general feelings of mankind, (and these afford the only true criterion) that effervescence of the animal spirits which takes place just on this side of intoxication, ranks among the highest of human pleasures. It has the advantage of being enhanced by participation; and if the human frame were otherwise constituted, the free use of wine would not only be an allowable indulgence but might properly be inculcated as a duty. For there are none of the ingenuous arts, of which the cultivation has a more healthy effect on the mind, and a more humanising, than the frequent excitement of the chearful sympathies.

At some future period, a sufficient fund of hilarity will be discovered in social exercises, and in the communication of ideas, belonging to literature and science. The accumulation and diffusion of knowledge directly tend to render the human species independent of wine. Not an elementary treatise or course of lectures perhaps but is contributing towards the eventual restoration to CERES of that extensive domain which BACCHUS has been so long suffered to usurp. All that is wanting is a stock of materials, which may serve in common for the production of agreeable feeling without damage to the organization. That this is to be had, we may learn from the lives of many individuals of our own country, from WALLER the poet, to the poet and physician DARWIN. Some of these furnish examples of perpetual sprightliness with perpetual abstinence from wine; and others, what is stronger still, of sprightliness unimpaired by the disuse of wine after free indulgence. We need look but a little way abroad to discover nations airy and alert in proportion to their abstemiousness from strong fermented liquors. There remains to be conferred upon our own nation a benefit of which the author would deserve infinitely more than any legislator, warrior, or inventor, who has obtained celebrity among us. It is a simple, and would probably be an easy undertaking to persons of influence. I mean a plan of social intercourse, independent of the pleasures of the bottle. Such a plan would seem to be a natural consequence of the interest, which we have lately seen the higher orders of society taking in an Institution for the diffusion of knowledge in the metropolis. The interest by which people are to be drawn and held together could be drawn only from the objects of nature, the processes of art, and from the range of polite letters. The meetings should be made more promiscuous than our parties by the introduction of boys and girls at the age when they begin to use their reason. A few of the best informed inhabitants of different places must at first be the active persons, and provide entertainment for the circle. But the most passive part of the company would not find themselves more listless than at the places they now frequent. In many there would be kindled a zeal for information which would make them find a new pleasure in existence. And in all, the exclusion of the petty malignant topics of common conversation would tend to create a better spirit, and more happy disposition. To describe the probable effects of such a system of communication among families, *gradually* substituted for the present, upon temper, health, and morals, would be worthy the pen of our ablest author of fictitious biography. The state of society as it existed at Geneva before the troubles in Europe, would afford many traits towards the picture.

The period of majority by law has continued the same. But the period of majority by custom has been greatly curtailed. The good effects, which the reception of young people into the society of their seniors nearly on a footing of equality, have been perhaps hitherto more than counterbalanced by giving an earlier date to the habit of drinking a few glasses of wine daily. If it be but one or two, the observation will still hold in a multitude of instances. I am well assured that shortly before the war an ancient female at Calais, who had had opportunity of seeing most of the English travellers, remarked the latest flights to have had in general, a more puny and worn out appearance than their predecessors forty years before. It is what the habits of many of our boys would lead us to expect. The injury to the digestive and other organs from the error in diet, which I am here considering, is certainly not the sole cause; but it must very materially contribute to the difference, if it exist.

That *all*, who begin early and continue without intermission, the use of strong vinous liquors are not injured, is a certain truth; and we are perpetually reminded of the exception as an excuse for a practice so universally marked by medical observers as destructive. But neither do *all* who are exposed to its contagion catch the plague? And yet is not the hazard sufficient to induce every man in his sober senses to keep out of the way of infection? That every man shall become a valetudinarian, more or less miserable, if he daily drink a quarter of a pint or half a pint of port wine from his fourteenth, fifteenth, sixteenth, seventeenth, or eighteenth year, is to the full as probable as that he shall have a dangerous disease if he come within reach of its effluvia.

Among those of a weakly constitution there

are some who take wine as a luxury, and unexpectedly find it their cure. This, however, is a fact only in favour of wine as a medicine, and by no means as an article of diet. It is certain that were the most powerful productions of nature and preparations of the elaboratory, whether they rank in our estimation as drugs or as poisons, to be habitually used, they would have a deleterious effect upon the majority, and prove salutary in single instances.—For the former part of this assertion we have the experience of those extensive regions, where opium is in daily use. Our most authentic travellers leave us no doubt about the wretched state of the opium-takers in general. It may be more difficult to find direct authorities for asserting that this powerful juice is ever beneficial to the Orientals. But the analogy of cases at home will supply this defect; and if it were as liberally used in Christendom, it would also sometimes perform the same work of supererogation that wine does; it would not only soothe and exhilarate, but occasionally heal. In the countries where asa-fœtida is habitually employed in mixture with the food on account of its cordial operation and particularly to promote digestion, it must sometimes remove and keep off indisposition. The five-hundreth, and perhaps even the hundreth or the fiftieth part of a grain of arsenic might probably be substituted among us in the place of the usual allowance of fermented liquors without more detriment on the whole, and with as great accidental advantage in particular cases.

All considerations, we see, combine to shew that fermented liquors by their activity class with the most powerful and therefore with the most hazardous drugs. Of course, their use ought always to be a point for medical decision. From the healthy stripling, not yet habituated to them, they should, in every shape, if possible, be withheld. Healthy men, under the age of fifty, should gradually emancipate themselves from so pernicious a habit. Their determination should be more prompt as their constitution has suffered less from time, disease, accident or intemperance. As to persons more advanced in years, it would be happy if they could be prevailed upon to call in the aid of wine merely to prevent themselves from sinking into languor, and never to raise their feelings above the comfortable pitch.

Where the spirit is willing, but the flesh weak, I should propose a compromise between indulgence and health. One fourth part of the usual fermented liquor, if it be wine, or cyder, or mead (which old British beverage is still in use among certain classes in the counties adjacent to Bristol) should be at first withdrawn, and the measure made up by water; afterwards one half or two thirds. The mixture may easily be prevented from tasting raw or vapid by adding water impregnated with fixed air or carbonic acid instead of plain water. The art of preparing artificial mineral waters is now spreading in every quarter, and the alkaline or Seltzer waters would perfectly answer the purpose, and, as we shall presently find, correct a property in wine, by which it is perhaps even more injurious to the weak stomach than by its direct operation. Where the mineral waters are not easily procured or where they contain too much gas for the stomach, the common glass machine will impregnate water sufficiently for this purpose; and its expence would soon be saved by the diminished consumption of wine, an article now become so costly.—As to malt liquors, their strength is easily lowered by withholding a portion of the material that supplies the spirit.

Undefined INDISPOSITION, *and Specific* DISORDERS *produced, or whose production is favoured, by the operation of Vinous and Spirituous Liquors on the* DIGESTIVE ORGANS.

Some disorders are curable by a variety of means, which have nothing in common in their composition, and probably nothing in common in their mode of operation. Thus the ague will yield indifferently to Peruvian bark, to arsenic, and to an impression upon the mind. Conversely, the same disorder may be produced by a variety of causes, physical and moral. There is probably no one complaint produced exclusively by fermented liquors or by analogous stimulants. But they are the most frequent cause of many complaints, and, where they do not operate alone, they reduce the system to a state in which it is overpowered by enemies that otherwise would make no impression. Nor is it the *Sot's or Drunkard's progress* only that I delineate here. Great multitudes come to the same misery with the drunkard and the sot without ever numbering wine among the pleasures of existence, and who have always filled their glass scantily, in involuntary imitation of, or reluctant compliance with their associates. But there are actions, in which it is fatal to bear the smallest share or even to be present during their perpetration. Murder, and that kind of slow self-murder, which is committed by the abuse of diluted poisons are of the number.

To pine in a uncertain comfortless state before falling into a well-marked disorder, is not less frequent, than to linger on a sick-bed before being permitted to enter the asylum of the grave. The drinker of too much wine commonly finds himself what in medical phraseology is called *cachectic*, or as some familiar writer terms it, *I don't-know-howish* for a long time before he is compelled to resort to the practitioner of medicines. However trifling his daily allowance, let him not flatter himself with the idea that it is impossible what he feels can be owing to so small a quantity, but make a half year's trial by first diluting his vinous beverage, and then quitting it, if possible, altogether.

One of the first indications of mischief from wine taken constantly in moderate quantity, when it may be supposed to act as an alterative, is a sense of dissatisfaction and being ill at ease some hours afterwards. The young and sparing votary of Bacchus cannot be expected to tremble all over on first rising, and to exhibit to every spectator, in a lack-lustre eye and cheerless morning visage, the effects of his afternoon libations. But what the veteran dram-drinker is unable to conceal, the other will be sensible of, in a proportionate degree, on self-examination. He will probably awake hot, restless and heavy. The early sun will seem an intruder. He will shake off his drowsiness reluctantly, dress with languor, and be indifferent about food. The mouth will feel clammy, the stomach uneasy, till revived a little by the stimulant operation of warm tea or coffee. After stretching and yawning till the limbs are properly awake, he will eagerly close with any scheme, which promises to raise emotions or to relieve that listlessness, which dinner and the circulation of the glass are required completely to dissipate.

Habits like those of MEADOWS in *Cecilia*, will be created by the enfeebling power of small portions of wine, regularly drank. This indifference and apathy, so unnatural in the prime of life, must have been originally felt, not causelessly assumed. It may be convenient policy, doubtless, to render them fashionable. But juvenile ardour and alacrity must have been first extinguished in those, who set up that sort of demeanour, which implies the absence of both, as a pattern for imitation. Deformities of figure and of character must exist in people of influence, before their inferiors will agree to punish themselves by ostentatiously cramping the free play, and distorting the erect growth, of their nature. No mortal would ever have thought of making apathy the mode, but a worn out beau.

In the course sometimes of a very few years there takes place general want of comfort, accompanied by particular uneasiness about the region of the stomach, except shortly after that organ has been roused into temporary energy by more or less repletion. There is yet no severe or alarming symptom. Though scarce a day passes but the alimentary canal is sensibly out of order. These irksome sensations, however, come and go; and the person who experiences them, can continue uninterruptedly his business, his studies or his pleasures.

It is now that vinous liquor begins to act as a two-edged sword. By its first operation, it increases that indigestion, of which it has already so largely contributed to lay the foundation. Its second is little less pernicious to the enfeebled viscera. This depends upon the change into vinegar, which wine, however genuine, as well as ale, and other fermented liquors, undergo, as also the acescent extractive vegetable matters with which they are incorporated.

Vinegar, taken frequently and freely, we know to be destructive to the stomach. When slenderness of waist was particularly in request, many women totally ruined the digestive faculty by vinegar. I have known instances where a fatal atrophy, by the acknowledgment of the patient, was thus induced; and to physicians of longer standing such cases must, at one time, have been familiar. Since the common use of nitrous acid, medical practitioners must have had the most ample opportunities of observing that the action of this and the other mineral acids upon the first passages is totally different from the action of acid fruits. Perhaps it is different from that of the pure vegetable acids, for those are alterable by heat, and in some situations, analogous to the human stomach, where the others are not. Of vinegar in particular, compared with mineral acids, the action on the stomach of horses appeared to be highly deleterious in the experiments of Mr. PILGER.* This he himself considers as the

* "Concentrated vinegar, (says this observer,) acts in a far more stimulating manner than the mineral acids----

most remarkable result he obtained: his best informed readers will agree with him. And it is no inconsiderable corroboration of the other proofs of its power to injure the feeble human stomach. Indeed were a person, who is hourly reminded that he carries such a debilitated organ under his belt, to be offered ready-made vinegar alone, the idea would go near to bring on a fit of stomach-cramp. Yet he may be assured that the wine he drinks not only attacks the coats of the stomach as wine, but that it afterwards returns to the charge in an acidified form. And not only so, but it assists in turning sour the vegetable portion of the meal, upon which it is poured. Let it not be supposed that wine does not become sour upon the stomach, because no heart-burn or sourness in the throat is felt. The processes by which indigestion is promoted, must have been carried

---

given without addition of water, it excites the most dreadful spasms, and produces fatal symptoms. When a horse is killed by it, there appears [little or] no irritability immediately after death."

"To ascertain this, I drenched several horses, one after another with a pound of concentrated vinegar. If I had them killed, some convulsions took place upon arming the nerves. But they were most exceedingly weak, when the horse was left to die of the effects of the vinegar."

to a greater length than I here suppose, before these troublesome symptoms attend it, except on particular provocations. For a long time, there may be no risings; and the acid contents of the stomach may pursue their natural downward course, though with a degree of irregularity and inconvenience. There wants a series of experiments on the acidity of the recent alvine discharges after various kinds of food. They would throw great light on the different articles of diet, and be useful to dyspeptics, whether infants or adults.

Distilled spirits and water are held in great dread by many, who take wine and ale of the same strength, without scruple. The apprehension seems ill-founded. Mere diluted spirits cannot possibly do more harm by their primary operation: and the secondary does not take place. I do not notice this difference in absolute recommendation of *spirits*. The purpose, stated at the head of this essay, can be very seldom realised by taking either wine, or lowered spirits as the substitute of wine. But the public is deeply interested in being made acquainted with every fact, which may tend to destroy the superstitious belief in the innocence of genuine wine, as compared with other liquids, equally intoxieating. Opium, because it does not turn into vinegar, will often be useful in breaking the fall from wine to water, when just alarm shall lead any one to determine upon the reformation of his diet. But here medical counsel should interpose. There are very few dyspeptic persons, however, in the incipient stage, who will not receive benefit from vitriolic, nitrous or marine acids, mixed with water and sugar so as to suit the palate, and not strong enough to set the teeth on edge. The loss of fermented beverage will be little, if at all, felt on account of the grateful effect of these acids; and they are very powerful in checking that acescency, which the former so strongly promote. I know no reason why almost any sufferer from indigestion may not make trial of the *mineral lemonade.* To shrink from this proposal under the pretext of being unwilling to *tamper with strong things*, betrays too gross ignorance, or else very ill-timed prudery in those, who have not been afraid to shrivel up and scorch the most sensible and delicately formed of their internal parts, by the daily and immediate application of liquid fire.

Indigestion establishes itself by degrees in full form. It is attended by loathing, rejection of food, by constantly distressing flatulency, tremors, comfortless nights, emaciation, and decay of the intellectual faculties. A very great proportion even of the moderate drinkers of our stronger fermented liquors experience some of these evils. Those whom the poet describes as—

> Once fellow-rakes perhaps, now rural friends—

find that in place of being partners in pleasure, as in their better days, they have now only to condole with one another upon the torment of a bad stomach. You shall perpetually hear from one, when he is offered some of the delicacies that bring up the rear of our dinners, that he *never deals in pastry*—from a second that he *does not dare to venture upon sweets*—from a third that *he likes such and such a kind of fruit very well, but that it does not like him.* These are the common avowals of a depravation of the digestive organs, owing in great measure to the cause in question, if not altogether induced by it. Nothing is more fallacious than the common saying that every man of sense at 40 knows what is good for his constitution. The party himself therefore, though a man of sense, may not be thoroughly aware of the mischievous power of what is reckoned a moderate daily dose of wine--after all his experience; though it is most likely that he would have checked himself, if he had only been apprized, before his habits became fixed, of the conditions on which alone he could certainly expect the organs of digestion to continue to perform their functions pleasantly during life. SALLUST, in comparing man and brute animals, mentions as a mark of inferiority in the latter, that nature has formed them *obedient to the belly—ventri obedientia.* Now to understand and practise what is due to that central part, is really a great perfection, and such an one as few are happy enough to be endowed with. Indeed the interest of the palate, well understood, coincides with that of the alimentary canal, and of those remoter organs, which perform their offices well or ill, according to the state of the first passages. For besides the variety of uneasy feelings, with which the dyspeptic valetudinarian is harrassed both in his periods of wakefulness and sleep, he must either forego those things that have the most agreeable relish, and which perpetually furnish innoxious enjoyment to others, or else he must do penance by the hour for every minute of indulgence. In the constitution of the human body, there is surely no reason why the grown man should not be able to feast, with as much impunity as the growing boy, upon the fruits of our own seasons, upon those of countries more favoured by the sun, and upon the most savoury preparations of the culinary and confectionary arts. The impediment arises, in the majority of cases, from the injury, done the stomach by the products of the vinous fermentation, with frequent assistance no doubt from unhealthy occupations and uneasiness of mind.

Too light cloathing seems to have no inconsiderable share in weakening the stomach. In sedentary people the belly is apt to feel cold from insufficient covering; and in this respect the antient materials and fashions of dress have an advantage over the modern, particularly in men. It is impossible that the region of the stomach should be habitually chilled without injury to digestion.

One troublesome circumstance belonging to indigestion in all its gradations deserves particular elucidation. This is the production of air in the intestines. It must be considered that the quality and quantity of the matter, exposed to the powers of digestion, are no less entitled to regard than the state of those powers themselves. In the heat and moisture of the stomach, both kinds of food run into fermentation; vegetable substances very rapidly, and with prodigious generation of air and acid. When cattle eat too greedily of dewy clover, notwithstanding the undepraved condition of their organs, they are sometimes burst by the abundance of elastic fluids, thrown out in the fermentation of this plant. But it is among the properties of the juices of the stomach to check, and in ordinary circumstances, to prevent the attractions, upon which this fermentative process depends, and to induce others of a quite opposite nature. From the concurrent testimony of many observers it appears, that in the healthy state, when the organ is not overloaded, and nothing indigestible has been employed as food, neither air nor acid are to be discovered in the alimentary tract.

Flatulency, therefore, may be assumed as the measure of indigestion. When arising from vegetables, it never perhaps fails to be accompanied by acidity, even though that acidity does not shew itself by any sensation. The gastric fluid restores the sweetness of putrid flesh; and that of the human stomach among the rest, seems to have great power in preventing animal substances from undergoing the putrid fermentation. So that the most dyspeptic are hardly ever troubled with flatulency, when they adhere strictly to an animal diet; that is, when they avoid sugar, wine, bread, and every thing of vegetable origin. Such a regimen, however, requires excessive self-denial; and is, by no means, to be adopted without the most urgent considerations. For though it may prevent inflation for the time being, it may produce other bad consequences, and certainly will have little tendency to remove indigestion, except as it implies the absence of one grand exciting cause of this malady. No one should be content with his stomach till it has recovered that power of digesting vegetables, which it possessed in the light and joyous spring of life, and which it retains to old age, when uninjured by accident or imprudence. It is this class of substances, if either of the two, which seems to have the fairest pretensions to be regarded as the natural food of man.* However this may

* Of all animals the ape comes nearest to man in the structure of his mouth, teeth, tongue, gullet, stomach, and abdominal viscera. This implies a corresponding similarity in the function of digestion (*Daubenton*). And wild apes live exclusively on vegetables. So likewise do men in some of the most populous regions of the globe. To balance these arguments, the Africans on Orange river are asserted by Dr. Fordyce to live solely on animal substances. And there are other instances from the Ethiopians of Herodotus downwards.

be, as few of us will, even supposing we could, do long without vegetable food, the interest we have in possessing digestive organs, capable of subduing its disposition to ferment, does not require further illustration.

No process in human life is more common than sinning against the stomach and repenting shortly afterwards. Many people, with digestive organs beginning to fail, sit down with a resolution to avoid such and such things, from having lately found them always to disagree. And if the articles they have reason to avoid were never to come into sight, they would hardly excite a wish. But example unites its influence with opportunity, the constant subduer of human virtue, and the firmest purpose proves glass in their way.—*I had made up my mind not to meddle with any of those green gages to-day—I must try one however, they look so very fine.*—One does but instigate desire. Another and another succeeds, till a sensation of cold in the *epigastric* region, approaching to an ache or a cramp, gives warning of what is to follow, and the alarmed glutton desists from gratifying his palate at the expence of his stomach. After a short time he cries out in his anguish—"*I wish I had not suffered myself to be tempted by those vile plumbs.*" At least this penitential sentiment is felt almost as often as the *desert* comes before them, by a number of sensible and considerate people, if it be not always expressed.

There is perhaps but one effectual way of correcting this inconvenient weakness of character. The greater part of mankind seem not only to lose the benefit of the experience of others, but of their own. They suffer their health to be undermined and their ease to be perpetually broken in upon, for want of fixing their attention stedfastly upon their feelings, and connecting the circumstances, upon which they depend, strongly together in the mind. Ideas, stamped upon the memory with great distinctness, have undoubted power in deciding the will, and very frequently they prove capable of resisting the seductive tendency of impressions, made by present objects. To pass our table-transactions in frequent review, as the golden verses ascribed to PYTHAGORAS recommend with regard to our whole conduct; to dwell upon their consequences, particularly their disagreeable ones; to call up in lively colours before the imagination that delightfully free and unencumbered state of all the faculties, which accompanies an easy digestion; to compare what is lost and gained by throwing into the stomach materials that puff it up like a balloon, is our best moral preservative against the danger of becoming dyspeptic; and without this, our physical resources will hardly get fair play.

Among the physical, none is more essential than the proper regulation of food. The usual English breakfast is nearly as bad an one as it is possible to contrive. Let those who doubt, make a mixture of sugared tea with saliva and bread and butter, and keep it in a temperature of 96 or 98 degrees. If the stomach be so far enfeebled that its juices cannot restrain the fermentative process, the result will be exactly the same, whether the materials lie in a living pouch or in an earthen pot. Every morning regularly a degree of that misery will come on which I find delineated in somewhat broad strokes, by a foreign medical writer, under a fictitious signature.*

* "I am a person who could spend his days in decent "tranquillity along with his family, were we not almost "constantly plagued with one inconvenience. When I "have taken my tea in the morning with my wife and three "daughters, each of us repairs to his employment. I go "to my window and watch the people pass. My wife "stations herself with her work-bag in an opposite window; "and my three daughters occupy the middle of the room

The general still life of one sex, and of many of the other, causing the circulation to languish, the extremities to chill, and the secretions to be imperfectly performed, the alimentary canal in a few hours after breakfast begins to suffer from the sharpness of the fermented mass; and volumes of air, moving about in all directions except in the proper one—a sense of hollowness within—gnawings at the pit of the stomach—torpor of the bowels, and other miseries, pervading the whole system, but too various to admit of description, fill up the principal part of the time between breakfast and dinner.

"with their tambour-frames. Scarce have we seated our-"selves, when there arises a murmur, as if the room were "a Jew school. Sometimes I ask my wife:—*what is it* "*you say, my dear?* and the reply always is—*nothing; it* "*was only the wind*—. It often rumbles as loud in my "daughter Charlotte's body, as when an engagement is "fought in a raree-shew box; the youngest will raise a "croaking as if she carried a number of frogs in her pockets; "and in the middlemost it chirps like a hatch of young birds "----We are ashamed of going into company. For "when we are seated in the circle with the utmost deco-"rum, one stomach or another sets up such an uproar, that "the persons present are obliged to bite their lips for fear "of bursting into a loud laugh. We dread nothing so "much as the usual pause in conversation when a topic is "done with."——URBAN FLATUS.

In order to understand clearly the effect of an ordinary dinner, it will be necessary to consider the gradations of debility in the stomach. If any quantity of acescent food, (as for example, that usually taken for breakfast) be imagined to be divided into a number of portions, in the first degree of indigestion one of these portions may be supposed to follow its propensity to ferment, while the rest are subdued into a mass, fit to recruit the body. With the increase of inability in the gastric juices, these proportions change, and always to the unfavourable side. Along with the impaired faculty of the stomach to furnish proper digestive juices, the other faculties of the intestinal cavities become impaired. The muscles, in particular, grow too feeble to perform the office of propulsion.

In the early stage of indigestion, therefore, the effect of dinner will be to give relief for a time. The muscles will be invigorated and the elastic fluids, the most moveable part of the contents of the bowels, will be driven downwards *. In the sequel, however, the

* The same thing happens also at breakfast; soon after which more than elastic fluids usually passes, when persons are tolerably healthy and regular. For the bowels have been

former grievances will recur sooner or later, and with more or less aggravation, according to circumstances. The souring of fermented liquors is one of the principal determining circumstances. Another, is the addition of a meal from the garden, to a meal (probably too full an one) from the kitchen. Common observation has long since ascertained the peculiar unseasonableness of fruit after dinner. But as long as the digestion continues weak, fruit in general, (except perhaps the best ripened oranges we get in this country, and those in sparing quantity) may be set down as improper, for there cannot but be injury from converting the stomach into a vinegar vat every day.

If the dinner and after dinner medley do not work sufficiently, there is a provision in our customs for promoting the fermentative process. The evening repast of tea and coffee will act upon them like yeast added to sweet wort.

---

all night labouring to forward the refuse of the food of the day before, which waits but for a little stimulation and a voluntary effort. From the effect of protracted sleep too, the alimentary canal is more susceptible of being moved in the morning than at any other time of the day.

In that stage of indigestion, in which a large portion of the breakfast escapes the action of the gastric liquor, dinner does but add to the distress of the patient. The first mouthfuls make an ungrateful impression. There is some approach towards an effort for their rejection. The body suddenly feels bloated, and it often becomes necessary to let the dress loose. It would seem that in this case the stimulus of the food produces a reflux of the flatulence from the intestines into the stomach. It is common for muscular fibres, when much weakened, to act in the direction contrary to the healthy; and we in fact find, in the worst dyspeptics, the greatest tendency in the wind to discharge itself in this way, that is, to be propelled upwards. One or more glasses of wine will considerably alleviate this uneasiness for the moment, and brandy still more effectually. For these violent stimulants rouse the muscular fibres into more vigorous action: the stomach contracts and expels the elastic fluids, in consequence of which the fulness is relieved. This temporary advantage, which seems to have introduced the pernicious custom of drinking wine repeatedly during dinner, is much more than compensated by the consequences, even within a few hours—

not to reckon the progressive injury to the structure of the digestive organs. A diseased follows different laws from a healthy part. It would therefore be possible that strong wine should cause more gastric liquor to be given out by the coats of a stomach, injured by wine, though the contrary is apparent in a sound state. There is however no experiment to countenance this possibility; and the increase of the peristaltic motion, as explained above, together with the grateful glow and temporary exhilaration produced by wine, when the dyspeptic take it along with their meals, is sufficient to account for all the phænomena.

The correction of this troublesome and dangerous infirmity is rendered difficult only by the influence of fashion, and the obstinacy of habit. For this purpose such a regimen should be adopted as suits the debilitated viscera; and such measures should be pursued as will restore to them their due tone.

For breakfast, there is the choice of milk diluted with one third or one fourth of its bulk of lime-water, or milk with mephitic alkaline or Seltzer water, supposing the gas not to disagree—or of milk and water with supercarbonated kali dissolved in it—or milk thickened with rennet before the whey and curd have separated, in which tender state it is eaten in the Western counties under the name of *junket*. An infusion of some of the spices, as ginger-tea, or nitrous acid diluted with water and rendered palatable by sugar, will answer the purpose of a diluent. Sometimes one and sometimes another of the above liquids may be used; and that persisted in which is found to sit best, without the production of wind, of sourness or oppression. The lower their temperature the better, provided no chill follows. In the summer, they may soon be reduced to the temperature of the atmosphere without inconvenience.

Bread in its best form is a very objectionable article of diet for this class of valetudinarians. When new and impregnated with melted butter, it is nearly the most abominable that can be contrived. Bread, toasted thin, and eaten with unmelted butter, has rather less acescency than in other states, and in slighter cases this acescency will be corrected by several of the preceding liquors. But when the symptoms are severe, it must be altogether abandoned. Nor will biscuit answer better. As a substitute, flour baked grey,* and then boiled with milk to the con-

* This preparation of partially charred flour is described

sistence of a thin custard, will frequently afford sufficient sustenance, without exciting any commotion in the first passages. Various persons of weak stomach have found another very eatable substitute for bread in a pudding of suet, milk and flour, cut into thin slices and toasted. The most unexceptionable substitute for bread, however, is a sort of cake, prepared from the whites of eggs without any flour.

There is no invalid of the description in question, that cannot make out from his own experience a list of fruits and vegetables, which incommode him soon after he has swallowed them. I would most strongly advise every invalid to make out such a list: and indeed I am of opinion that to mark all the familiar occasions of pain and pleasure by a distinct memento would answer as useful a purpose in the art of living as buoys in that of navigation.

No two fellow-sufferers would find their accounts exactly to tally. In general, stone-

in Dr. Underwood's Diseases of Children. The flour should be put into a jar, partially covered, and be taken out of the oven several times, and stirred up from the bottom and sides of the jar, that it may not run into hard lumps, but the whole be equally baked.

fruit, nuts and other oily kernels, pease and beans would be noted as the vegetable substances most sedulously to be avoided. The character of turnips, greens, and cabbage would not stand much fairer. Raw vegetables would be altogether proscribed; and even potatoes be allowed in very sparing quantity. The grains in general rank with bread. So that till the stomach has received some accession of strength, a very small quantity of vegetable matter is admissible at dinner; and this should be thoroughly dressed, and would probably be better if stewed in gravy. The mineral lemonade will prevent the souring of this small quantity; and so will alkaline drinks. But the acid I think preferable, because it reconciles the palate to a longer continuance of animal diet, because it has considerable stomachic virtue, and also, because it will perhaps enable the constitution better to bear the subduction of vegetable matter.

There is one mistake under which I find dyspeptics commonly labouring, and into which I believe they are partly seduced by the treachery of sound. To a scheme of diet, framed according to our most accurate notions of the functions of the stomach, they will object that they can bear none but *light* things. By these they frequently mean things specifically light, as mealy potatoes, and very porous bread, not things easily digestible, especially by them. Mealy potatoes and porous bread may indeed have some advantage over bread not well fermented and mere paste in solubility in gastric liquor, within the body or without, principally perhaps because it can get at a greater extent of surface. But they turn sour full as readily. The same idea of lightness frequently misleads the dyspeptic to give an imprudent preference to liquids over solids, and sometimes to a liquor so very acescent as water-gruel. But a little attention to their feelings will soon convince them that liquid food, even that least disposed to acescency, is oppressive in comparison with solid, and that it should therefore be sparingly taken. The kinds of liquor, however, specified above, being either coagulable with ease or preventive of acidity, or invigorating, will bear very well to be used in moderation.

It is far from being a good rule to think what we fancy always the most suitable. The palate is apt enough in general to be influenced by opinion concerning the wholesomeness or unwholesomeness of different things. In many complaints it becomes miserably depraved, as we see in the dirt-eaters of the

West-Indies, and in our own charcoal and chalk-eaters. I have never known a thirsty person wish for red herring; but I have known red herrings cure habitual thirst. The dyspeptic ought not then to rely upon the palate as an unerring guide. They may find the things most craved disagree in the most violent manner. They will often find the things they relish very little at first, agree best.

I have already stated the mischief sustained in early life, from unfrequent and insufficient meals. Examples of young people perishing from atrophy in consequence, have occurred to me. And within the last five weeks I have been consulted by a lady, who believes herself to have been the favourite of the heads of one of the most numerous seminaries in England, because she was the least eater, and who declared to me, upon being curious to observe the exact consumption of meat in the school, that she saw forty girls fed two successive days, upon a single leg of mutton. It was not that the allowance of food was professedly limited, but the calls of appetite were suppressed from shame. Thus the censure of superiors has often more effect than positive law, and it may be sometimes as cruel to employ ridicule as the rack.

Dyspeptics, full-grown and become their

own masters, commonly need restraint as to the quantity of food to be taken at one time. They will find in their feelings an infallible rule. As they should never fast till they feel faint, so they should never go on eating after the smallest sense of fullness or distension comes on. The CYRUS of XENOPHON professes to rise from table as alert as he sate down. Similar temperance may be recommended to the dyspeptic: and according to the old rhime, they should refrain from exertion an hour after dinner. But, in truth, the wisest plan for this class of invalids is not to bind themselves to the times of eating of the people in health. They should take a little food often; as near going to bed as they can without inconvenience, and even in the night. Sleep should by no means be indulged just after eating, unless it be evidently refreshing and unattended by feverishness. It will be much better to walk slowly in the open air in fine weather. I say *slowly*, since Dr. HARWOOD's experiment, in which a dog, that was hunted immediately after feeding, had made scarce any progress in digesting his food, while his fellow, that remained at rest, had nearly compleated the operation, is rendered entirely applicable to our purpose by frequent experience in human invalids. From such experience we learn, that moving a couple of hundred yards with a brisk pace, immediately after a full meal, will occasion an indigestion. Any effort of the mind will have the same bad consequence. At other times, that quantity of exercise, which renders the whole system most vigorous, is best calculated to prevent or remove the evil under consideration.

The objections formerly made to hot rooms, might be repeated now. They are just as pernicious to the stomach as the lungs. By weakening the skin in particular, and rendering it susceptible of chill from the contact of the external atmosphere during a great part of our year, they impair the digestive organs in proportion; and besides bringing on habitual debility, expose them to sudden fits of torpor from the influence of that degree of cold, which the natives of a climate like ours ought to be inured to bear without shrinking. There are persons now living, who when students at our Universities sate over their books up to the knees in baskets of straw. I have endeavoured to shew in the preceding Essay, that the difference between these students and their more comfortably lodged successors, turns out altogether in favour of the viscera of the chest.

We may believe, upon the faith of every analogy, that the same must hold with regard to the viscera of the belly.

Dr. C. L. Hoffman, the most eminent physician now in Germany, as far as professional eminence is determined by professional profits, gained great credit in the case of the first ecclesiastical elector of the empire, by advising him to suppress the tendency to eructation, when his electoral highness found himself distended with wind. Medical writers have published the same advice. It rests upon the idea of resisting the inverted motion of the fibres of the stomach, and upon the effect of pressure in stopping the extrication of gas. Thus if diluted acid be poured on calcareous spar, and the phial closed, the effervescence will cease before the acid is saturated; and it will go on again, if the condensed gas be set at liberty. Fermentation may be impeded in the same manner. I was unwilling to withhold from my reader so philosophical a method of easing the qualms of *stomachic pregnancy*, and preventing its recurrence, because its beneficial effect is strongly attested. But it is upon a more radical plan that I wish him to rest his hopes. Daily and repeated friction about the stomach, as well as on the sides and back, has been frequently and justly recommended. It ought to be practised by every one subject to this species of complaint.

When the transactions at the Pneumatic Institution are given to the public, the effects of the red sulfate of iron in simple indigestion will be stated; and from the result of hundreds of cases, it will be seen how far this deserves the preference above all other preparations of iron, and above *tonics* in general. This medicine almost infallibly cures in the dose of a few drops. Smallness of dose is always a happy circumstance. In dyspeptic cases, when I formerly had recourse to those slight bitter infusions, for which I never have occasion now, the fear of being overwhelmed with the quantity was a frequent obstacle to success; and in other cases, where it is necessary to administer drugs with a somewhat bountiful hand, the same apprehension often stands in the way of recovery. The greater the knowledge of the world, and the stronger the habit of reflection, the more intractable is erroneous opinion upon this point. The desire of the apothecary to swell his bill, and the complaisance of the physician towards the apothecary are justly believed to create infinite abuse in our profession. Hence, on all occasions, where medicines are not put into a small compass, alarm is indiscriminately taken lest this abuse should be carrying on; or perhaps it is thought that sufficient attention is not paid to the delicate state of the invalid's stomach. People seldom give a physician credit for common sense, whatever they may think of his skill. But on their own crude and chance diaetetical notions they rely as if they had the certainty of instinct. *It is impossible*, cries the mother, *he should eat any thing, if he must take all that quantity of trash.* Her commiserating eloquence improves into nausea that inappetence, which a proper medicine would remove. Half a pint of bark decoction in the day is rejected as intolerable, when a pint of wine in two hours is nothing thought of; and this merely because wine, though many times more violent in its operation and more palling to the appetite, happens not to be sold at the same shop with bark.

INDIGESTION with BAD SPIRITS.

Infants seldom utter more melancholy cries than when they are pinched with hunger. At all ages, a peevish, fretful or splenetic disposition is apt to come on, when the stomach is out of order from whatever cause. The predominant feelings of the dyspeptic being therefore painful, their impressions and ideas, as they occur in succession, become associated, each in their turn, with these feelings. Things present become the objects of aversion or indifference; things absent of fear. The smallest undertaking is shrunk from as an Herculean labour; every mole-hill assumes the dimensions of a mountain.

Each individual, subject to indigestion, will also be subject enough to low spirits. But till both these affections recur with a certain frequency and violence, they are not held to constitute the disorder, denominated hypochondriasis, though there is no precise criterion by which to distinguish it. There can be little doubt but the sensitive part of the system is greatly altered. A patient of this class has confessed to me that the movement of a pen over rough furrowed writing paper has brought on a undescribable jarring, like a general tooth-edge, which seemed to pass to the bowels and proceed there with redoubled and excruciating violence. Another writes of himself—"I could not look at a "large cypher, as a 6 or a 9 for example, "without an internal shock; a shelf, on "which only three or four books were stand-"ing, threw me into a qualm, and I could "not feel easy, till it was filled up."—What confirms the supposition of a gradual alteration throughout the whole nervous system, is the observation that in rare cases, the disorder begins in the mind and afterwards involves the stomach, though originally strong and at no time exposed to any serious cause of injury. Doubtless also by their constitution some persons are more susceptible than others of the operation of these causes.

The hypochondriac sufferer always finds language fail him, when he gives vent to his complaints. After vain and unsatisfactory efforts, his conclusion generally is—"*In short* "*you see before you, the most miserable wretch* "*upon the face of the earth.*" And in his secret thoughts and his confidential communications, he incessantly paces round the circle of his miseries, making only a brief pause at this heart-breaking summary.—He tells you he has heart-burn, dreadful flatulence, rising of a clear watery liquor in a morning. Then there are stretchings, tearings, and a heavy load in the region of the stomach under the short ribs, on both sides or only on one. They alternate with a dragging down of the navel, with prickings, startings, and most distressing throbbings in the belly. These sometimes fly round to the back; and the feeling, lately a cholic, now from its changed seat, creates an apprehension of stone in the kidneys. Irregularity in the body attends, but the costive state predominates. When otherwise, the evacuated matter often looks yeasty, like wort in full fermentation. After eating, the abdomen becomes distressingly tense. But the distress, referred to the chest, is a great deal more severe. Overpowering fits of anxiety and breathlessness supervene, palpitations arise, and in the midst of all, there is a painful cramped sensation which makes the patient feel as if the heart had its sides violently squeezed together. Pulsations in the neck and at the temples, difficulty of swallowing, ringing at the ears, giddiness on suddenly looking up, on turning round or stooping, are in the train of the hypochondriac's torments. A circumscribed spot in the head shall be as cold as if ice was melting on it;—another will burn like a live coal; and there is often a rending within, as if the head was about to burst. The signs of derangement in the sensitive power are in fact innumerable.

The intellect itself is subject to fits of intermission. There is a frequent incapability of mustering the ideas, or of finding words for the thoughts. The memory, at such moments, is extinguished; the fountains of the imagination are dried up, and on a hundred little occasions in life, which demand an expedient or present a choice of difficulties, invention and judgment are both totally at fault. "The head is a desert," as one who speaks from experience, expresses it.

A great hypochondriac once confessed to me that, one summer's evening, being desirous to breathe the fresh air, he waited till dusk that he might sally forth unseen; but scarce had he walked five paces from his door, when he saw his own servant moving towards him. He instantly turned about with all the anguish of guilt, and stole back into his apartment. He could not bear to meet any human countenance! Another, a very religious patient, relates of himself that at a time when he was apprehensive lest he should be unable to refrain from laying violent hands on himself, an old woman accidentally mentioned one, who had committed suicide on Good Friday. "Scarce had I heard the "story, when I was seized with violent "terror, lest I should come to the same end "on the same day.—I could not behold a "knife without shuddering. At dinner, "I grasped it with all my might, lest I should "suddenly turn it against myself; and when I "had cut my meat, I fed myself with my "fingers. From an impulse that arose more "suddenly than the passage of an arrow, "I set to my throat, in idea, the pen I was "holding, the knife with which I mended it, "the tobacco-pipe in my mouth, the snuf-"fers, my sword, the church-steeple opposite "me. Between sleep and awake, methought "the room was full of knives, and I fancied "that I heard the noise of their whetting." The same person says "he had locks put to "his sashes, lest he should jump out at the "window by night."

Deplorable as his day may be, it is the night the hypocondriac dreads. The night is his season of peculiar agony, and its horrors are often deeply imprinted on his morning countenance. His sleep is so short and so broken that he seems to himself not to have had a wink, though, on taking pains to ascertain the fact by a repeating watch, the veteran naturalist DAUBENTON found that under similar circumstances he had slept a good deal. But refreshment, the ordinary and proper evidence of sleep, he feels none.

After having committed the smallest sin against his stomach, he will struggle for a time with frightful dreams, and awake suddenly with a stifled scream. Then the greatest possible quantity of horror is accumulated upon every object that passes before his mind in review. The merchant fancies his ships all wrecks, and his debtors all bankrupts. The most pious communicant accuses himself of having partaken of the Lord's supper unworthily. The physician cannot believe but he has poisoned his most interesting patient by a slip of the pen, and, in every noise of the night, he hears a summons to the consequences of his oversight. The imagination mixes up ruin and guilt in the largest dose, which can be administered to existing circumstances. Every brother in affliction is ready to cry out with the Ayrshire poet—"*Miserable perdu!* "*that I am—here I lie, a monument of the* "*vengeance laid up in store for the wicked,* "*slowly counting every stroke of the clock, as* "*it slowly—slowly numbers over these lazy* "*scoundrels of hours, who, damn them, are* "*ranked up before me, every one at his neigh-* "*bour's backside, and every one with a portion* "*of anguish on his back to pour on my devoted* "*head—and there is none to pity me.*"*

* To the horrors of this disease none can do justice but

Among hypochondriacs, I look upon the *reformed rake* as one of the most remarkable

---

a great poet—himself a sufferer. The faint outlines and cold colouring of medical artists may serve another purpose, but scarce that of morality or health—I mean of the preservation of health. Abroad it seems much more the custom than with us, to let the public into the secret of mental sufferings. This was the case long before Rousseau, and seems not to have arisen in imitation of St. Austin. But of various confessions and histories of single valetudinarians, the greater number belong to *melancholy*; and of the rest, from which I have selected a few traits, however the authors may have laboured under the pangs of genuine *hypochondriasis*, they seldom possess the poet's power over the signs that express and awaken feeling.

A few loose touches of the master may perhaps work more upon kindred spirits than an elaborate design. If so, the letters, written by Burns in the latter period of his life, are still more worthy the attention of men of genius than his best poetical pieces. "His feelings at times could only be "envied by a reprobate spirit, listening to the sentence "that dooms it to perdition." After exhausting in reflection every topic of consolation, he compares himself "to "Judas Iscariot, preaching the Gospel.—He might melt "and mould the hearts of all around him, but his own "kept its native incorrigibility." "His spirits are fled! "fled!" and he "has tried every thing that used to amuse "him but in vain." One cannot help wishing that Burns's father could have anticipated this extremity of hypochondriacal wretchedness; and when the heart of the future poet had been warm with the exercises of devotion, so feelingly described in the *Cotter's Saturday Night*, that he could have impressed upon it sentiments, that might have operated as varieties. There is reason to think that ladies, and sometimes the best of the sex, are seized with the ambition to reform rakes, and that rakes, professing to be reformed, find especial favor in their sight. It was in part to expose the hazard of so tempting an enterprize, that RICHARDSON wrote his immortal *Clarissa.* It would be hard if many women had not profited by Richardson's admonitions; and for others it would be happy if they could be persuaded, that such reformation by no means implies the kind of merit, which is entitled to the high reward, so often sought under that plea.—I except one case only; and that is, where the pursuit of gross pleasure is abandoned, while the bodily

---

a preservative against intemperance. In this case, the son would not have added another name to the list of men of the highest talents, whom our age has seen sinking, destitute and despairing, into an early grave. To raise the feelings is an art, without which young people cannot be brought up to good. But it is farther necessary to connect these feelings with *determinate* ideas. This implies appropriate knowledge on the part of the institutor, and a well-timed application of that knowledge. On no other terms, can the warm wish of BURNS in behalf of the peasantry of Scotia, ever be fulfilled with regard to any class or country—

Long may thy hardy sons of rustic toil
Be blest with HEALTH and peace and SWEET CONTENT!

organs retain their freshness, the appetites their ardour, the imagination its radiant hues, and the nerves their first exquisite faculty of thrilling with delight through all their filaments. But if the horse is content to stretch himself along a spot of the pasture, over which he used to bound when a colt and to find it too narrow for his spirit and powers, let us examine whether he is tame, only because his limbs are foundered, before we commend the change in his disposition.

It can hardly be conceived that debauchery, when it is followed by disrelish, by dejection, and impotence of stomach, should not produce peevishness also. This is not only the case, but the expression of peevishness must on several accounts be most bitter towards the persons, with whom the sufferer is most familiar. Upon them, every eruption of ill humour is discharged, and their most kindly intended offices will often provoke the eruption. Should they entertain the senseless vulgar notion, *that hypochondriasis is not a real, but an imaginary disease,* they will infallibly aggravate it, and draw down upon themselves all the consequences of the aggravation.

Few that have the misfortune to be domesticated with an hypochondriac, but must live in constant fearful expectation. The mine is laid under their feet, and every step it may be sprung. The sudden sallies of impatience appear, on reflection, unaccountable even to him, from whom they oftenest burst. Thus, I find it recorded in the journal of an hypochondriac—"This day, a friend offered me "a liquor to allay that uneasiness of the "stomach, which I so often find the occa-"sion of my dreadful humours. Unluckily, "that very instant, the idea crossed me—"*that it might be poison!* - - - The sudden "origin of my fretfulness I am unable to "explain - - - I often shew a forced coldness "towards the most obliging overtures of "those about me, and make them no other "return than rudeness."

There is another thing which her guardian sylph should whisper into the ear of every woman—that she may pause before she unites her fate to that of a man, who may have abandoned a rakish course, because he has reduced himself to a state, in which his old gratifications are stale or impossible. Whatever signs of gloom or fretfulness may break out, they are infinitely less than the reality; and whatever the reality may previously be, the fits will grow stronger and more frequent after marriage. He, who has been but whimsical to remoter relations, will become bitter towards a wife. "After every conjugal "endearment," says the penitent just quoted, "I become horridly peevish. I fancy every "body has a wish to murder me. I dread "being turned out of my office, and dying "for want. My fellow men, towards whom, "at other times, I feel so much kindness, "then become intolerable to me. I go out "of my way to avoid my most intimate "friends. I look upon my poor wife in a "much worse light than she deserves - - - "Oh! had I never ascended the marriage-"bed, and above all, had I repressed the "early sallies of a voluptuous imagination, "I had probably been among the healthiest "persons upon earth, whereas now I feel as "if every day would be my last."

The physician to a hypochondriac would be nearly as ill off as the wife, if the connection were equally indissoluble. Patients of this description, as long as they can endure society, wind themselves up by voluntary efforts and by the incentives to merriment, for a few convivial hours, to sink the deeper into despondency when alone or in the bosom of their family. In like manner, on hearing the name of a new physician, they muster up what hope they can, and repair to him in expectation of sudden deliverance from their calamities. But in this they must be disappointed. The idea of the physician and his advice present themselves again and again in some unlucky moments of painful sensation. They are not long before they share the common lot of every object that occurs to thought and sense. Though lately sought with ardour, they are soon regarded with indifference, and at last, with aversion. Nor can any thing but some fresh name light up a short gleam of hope in the mind. Hence we see hypochondriacs running from one medical counsellor to another, till perpetual disappointment and progressive loss of spirits end in universal distrust of others, and compleat despair as to themselves.

SEVERE STUDY.

In accounting for the injury which we so frequently see done to the sensitive and digestive organs, much is imputed to intensity of voluntary application. There is, undoubtedly, a degree of mental contention, incompatible with florid health; and in every generation of men, perhaps a dozen individuals may justly impute the ruin of their constitution to this cause. But it is of very limited operation; and by tracing back the history of the pale, tremulous, unnerved and dyspeptic student, we shall commonly arrive at a period of his life, when his devotions were openly or secretly addressed to divinities of a very different stamp from Minerva and the Muses. Accordingly it is observed by the modern editor of Ramazzini's work on the diseases proper to different vocations, that "the man of letters, who, in the season "of manhood, is plagued by hypochondriasis, "must either have been brought up in an "effeminate manner, or have laid the foun-"dation of his malady during his residence "at seminaries of education. We therefore "find this complaint far more frequently "among such literary people as have been de-"bauched in their youth, than among those "who have been placed in circumstances, "where this source of mischief could not "reach them. The same remark is applica-"ble to artisans whose occupations are "sedentary. Their mode of life may dispose "them to hypochondriasis; but the disease "seldom forms, unless they have enfeebled "themselves during their youthful years by "some kind of excess. The labour of the "mind is very seldom sufficient of itself to "create hypochondriasis, but in a debili-"tated system, it easily becomes the occa-

"sional cause of that disorder; the weak-"ened nerves being thrown into a state of "irritation and exhaustion by any unusual "exertion of the intellect."

There is another common species of excess besides too free indulgence in spirituous liquors, which has, unquestionably, a great share in producing the miseries of hypochondriasis. But as far as it is possible, consistently with an upright and impartial investigation of the practices, by which the weakly periods of childhood and old age are brought almost to touch each other without any intermediate season of vigour and enjoyment, I would avoid scandalizing the over-delicate* among my readers. I content my-

* I call those *over-delicate*, who would sooner commit a great than a small crime—sooner break a rule of morality than of decorum—I call Lady Macbeth *over-delicate* when she urges her husband to commit a murder merely to save an oath——

——I have given suck, and know
How tender 'tis to love the babe that milks me:
I would, while it was smiling in my face,
Have plucked my nipple from its boneless gums,
And dash'd the brains out, had I so sworn, as you
Have done, to this——

I call the archbishop of Novogorod *over-delicate*, who thought assassination less odious to heaven than three Lutheran churches. I call those *over-delicate*, who rather than that mention of certain things should arise, would risk self therefore with referring back to the fourth Essay, and with adding that those who would not be haunted themselves, and who wish to save others from being haunted, by the foulest fiends from the cave of Spleen, must be attentive to the doctrines, inculcated in the longest section of that essay.

MERCHANTS' CLERKS.

By information from London, Liverpool, Birmingham, and other mercantile and manufacturing places, I learn that a custom has been lately introduced, and is now becoming general, which must add greatly to the ravages of all chronic diseases among the stronger sex. It seems the next thing to reviving the practice of that ancient people, whose heart commerce hardened, till they came to sacrifice their offspring to *Moloch*. Boys are now frequently placed out to business at a very tender age. This is done with a view to advance their fortune; and in that view, no doubt, the measure is judicious. But in respect to health it is so pernicious, that

---

the danger of their inexperienced children debauching themselves to death. Those I repute truly delicate, who are studious of decorum, except when some duty calls upon them to violate it.

were it possible to gain the unprejudiced ear of any man of common sense and humanity, he must immediately reject, with horror, the proposed advantage, as being, in many instances, the price of blood, and at best, as the purchase of that well-being, which is above all price.

I am indebted, for the following observations, to a merchant, whose eyes were opened to this custom by its consequences—not by the hypothesis of any medical speculatist. It is not surely too much to hope that his testimony may open the eyes of others.

"In compliance with your request, I do "not scruple to repeat what I dropped in "conversation. Two fatal cases have oc-"curred in my own counting-house in the "course of a very short period. I do not "doubt but many other counting houses "could afford you others equally melan-"choly, if the principals, by swelling the "list of the unfortunate victims of avarice, "were not unwilling to record, at the same "time, their own neglect and severity. I "am persuaded that most of the diseases "incidental to the sons of commerce may "be attributed in a great degree to the fol-"lowing causes, *viz.*

"1st. To the want of proper conveniences

" for writing—the tall and short being gene-
" rally employed at the same desk.

" 2nd. To a sufficient time not being
" allowed after the meals, so that digestion
" is retarded or impeded by leaning over the
" desk with a full stomach.

" 3rd. To a necessary change of occupa-
" tion not being attended to, particularly in
" favour of young and weakly persons, and

" 4th. To the frequent long sittings in the
" same attitude.

" I have also observed that it is now-a-days
" much the custom with parents to endea-
" vour to exonerate themselves as soon as
" possible from the cares, which the health
" and education of their children demand,
" by placing them too early in commercial
" monasteries; alledging as an excuse that
" young men cannot too soon contract habits
" of industry, and a taste for business. Too
" frequently, however, these prudent parents
" are only actuated by a wish either to put
" their children in the way of gaining their
" own livelihood soon, or to remove them
" from home as troublesome impediments to
" their own gay and fashionable mode of
" life. Such an unnatural practice is attend-
" ed with the most injurious consequences,
" not only with regard to their health by

" confinement in dungeons of counting
" houses, at the period of the developement
" of their bodily and mental faculties, but
" also with regard to their minds, which
" become so contracted by the early habit of
" viewing every thing through one medium,
" that there is no room left for more
" generous sentiments and liberal ideas.
" Another fatal consequence attending the
" premature seclusion of young men in
" counting houses, is, that they are rendered
" too soon independent of their parents, to
" whom they cease to pay due respect—filial
" affection being then superseded by the love
" of getting money.

" It cannot be too strongly recommended
" to those who are entering on the vocation
" of trade, to endeavour as much as possible
" to write in an erect posture, and on desks
" sufficiently high to reach above the pit of
" the stomach. Foot-stools may be easily
" procured if the desk should be too high,
" and if too low, a light moveable desk may
" be placed upon it. They should also obtain
" at least a full hour's respite after every
" meal, and never remain too long in the
" same attitude without taking some positive
" exercise. I knew a stout, athletic man
" who by long confinement to writing on a

"low desk became extremely thin, was "troubled with pains in the side, and found "his health gradually declining, until he "procured a higher desk and wrote standing, "when he very soon felt the benefit of the "change in the return of his former health "and vigour. My own stomach was once "so debilitated by the toils of the desk, that "it frequently refused to perform its func- "tions, and soon after dinner regularly re- "jected the food nearly in the state it "was when swallowed. I had recourse to "travelling, and recovered in a great degree "after a long absence; but even now the "organs of digestion have not completely "regained their proper tone; for I am ex- "tremely subject to indigestions, and to "that disagreeable sensation, vulgarly called "the heart-burn. I. L. M."

HYSTERIA.

Many discerning medical writers have confessed their inability to lay down any marks, by which hysterics in women can be distinguished from hypochondriasis in men. To enter into this question does not belong to the plan of the present work. Hysterical women, it is certain, are liable to be discomposed by causes equally trivial with hypochondriacal men. The slightest moral or physical impulse gives a shock, which is felt with equal violence throughout the system. There is therefore the same inability to cope with the common incidents of life; and there is the same soreness of feeling. Nor will an hysterical person be easily found, who is not subject to flatulence, acidity, oppression, and all the other evils consequent upon feebleness of stomach. Many hypochondriacs too, it may be observed, find nothing so troublesome as a difficulty of swallowing, a choaking and rising in the throat, like the hysterical ball.

For the evils of indigestion and the common propensity to hysteria, women are indebted to their wretched education, and the life they afterwards lead; to the neglect of that exercise, which the human animal was formed for taking; to close confinement in unwholesome apartments; to books that have too great power to melt; and to other habits and privations, already described at large.

The quantity of wine, small as it may appear, which many women allow themselves, deserves separate mention; as it is probably not without its influence in rendering them dyspeptic and low-spirited. If it be, in any case, true, that the effect of intoxicating fluids is not to be estimated by measure only, but also by the state of the person who takes it, it must be true in regard to inactive, delicate and nervous females. In them, the digestive organs may be as much injured by a glass (suppose two ounces) of wine, as in a robust man by a pint. I have repeatedly known the head, in such females, to be most disagreeably affected by a small glass of port-wine and water. The operation of the mixture has been compleatly *narcotic*, as much so as that of opium or digitalis ever is; not the slightest exhilaration having preceded the sense of heaviness and stupefaction.—The parties thus affected, have always found themselves better under a course of total abstinence from vinous liquors.

Women in genteel life must be particularly subject to the secondary bad effect of wine. In scarce any instance, perhaps, are the gastric fluids sufficient to prevent the acetous fermentation, to which wine is subject; there is a concomitant change of the whole vegetable portion of the food into a similar sour mass with the extrication of an immense quantity of air. Though taking one, two, three or four glasses of wine at dinner, cannot be considered as intemperance, because that term, as applied to liquids, generally relates to their inebriating operation, yet the smallest portion of fermented drink may be productive of other inconveniences of the most serious nature. And no female of the class in question, who has to complain of want of power in the stomach, should be content till she has utterly discarded such articles of diet. They may be gradually discarded; and spices may be taken in place of the portion omitted. There is reason to suppose that spices, used as freely as by the inhabitants of hot countries, would be far less injurious than our usual wines, if indeed injurious at all.

PREVENTIVE REGIMEN, with some Observations on injurious means of temporary relief.

The most compleat general security consists in such a system of living as is best calculated, according to the principles already laid down, to maintain the digestive powers in full vigour. A strong stomach is an infallible amulet against blue devils. They do not easily enter into a body so fortified, or if by chance they should, they are without difficulty expelled. For there are accidental and temporary cases of hypochondriasis, of which the practitioner of medicine should never lose sight, and with which it may be useful to acquaint the general reader, because they demonstrate with peculiar force, how essential a healthy stomach is to good and even spirits. A physician on the continent relates that at one and the same time, a very considerable number of students from Lower Saxony and Westphalia came under his care in an university of Upper Saxony. They were dreadfully plagued with flatulence, and laboured indeed under every symptom of confirmed hypochondriasis. The most active medicines were prescribed, and a scrupulous diet not only enjoyned, but faithfully observed. Yet the situation of the patients grew more deplorable every day. Relief did not take in a single instance; but, on the contrary, heavier complaints were continually made. In this state of things, accident led to a plan of treatment, which, says the reporter, " was followed by a happier result " than all my medicines. One of my patients " received from home a supply of suet and " pumpernickel. These customary articles " of food tempted him prodigiously. But " he durst not touch them from an idea that " they would double his sufferings, since

" the lightest diet agreed so ill. I was obliged " to use much reasoning before he could be " persuaded to a cautious trial of the very dish " he longed for. However no sooner had he " indulged his inclination than all his distress " vanished at once, and he recovered the " firmest health. The rest followed the ex- " ample of their countryman with equal " success. They had every one been habi- " tuated to this coarse fare from their youth " up."—Perhaps the woman with a perforated stomach digested cucumbers quickly from habit.

When uneasy feelings have associated themselves with indigestion, one may frequently perceive a principle of action, similar to that, which so forcibly actuates the sailor on being discharged from his floating prison. The hypochondriac, with a well-covered table before him, is apt to make himself amends for his uncomfortable morning. So he fixes, with desperate eagerness, upon the delicacies of each course, and afterwards crams himself with fruit like a hungry school-boy, who has just carried off the spoils of an orchard. This is a very natural sort of imprudence; and it is impossible not to pity those, who have such an excuse for suffering themselves to be seduced into it. I know not how they will find means to curb the palate, but by considering into what an irritating, foul, oppressive mass, these delicacies will presently be converted, and forming to themselves a lively image of the consequences, as suggested in the preceding part of this essay. To oppose reflection to sensation is the only effectual resource in this, and in many other cases of temptation.--And can any other argument be thought necessary to evince the universal propriety of physiological information? What is reasoning, but to trace the order and relation of events? and how is it possible for any one to be rational with regard to himself, if he be not made sensible of the precise result of a given line of conduct? Without this, how shall any one avoid throwing himself into situations of extreme misery with as little consciousness of his danger, as is shewn by the calf, when the butcher is carting him to the slaughter-house?

When the inconveniences of intemperance begin to be felt, it is common enough to attempt to relieve the stomach by emetics, and still more so by quack preparations of different denomination—Emetics cannot be often used without a degree of mechanical injury, though so violent a case as that of the Dutch admiral Van Wassenaer may be extremely rare.* But without laceration or excessive distension, substances which excite an enfeebled stomach into preternatural action, cannot fail to be destructive to its fine organization. I have never seen debauchery make so quick work with the constitution, as when the debauchee was in the habit of taking analeptic pills and similar compositions. I have noticed this in the most robust habits; and I will venture to say that there is no day in the season, when various wretched examples of the truth of the remark, could not be met with in the pump room at Bath.

There is no error more common or more mischievous among dyspeptic, hypochondriacal and hysterical invalids than to suppose themselves *bilious*. *The bile! the bile!* is the general watch-word among them; and

---

* Whenever Admiral Van Wassenaer gorged himself too full, he was accustomed to drink tea of the *carduus benedictus*, and then irritate his throat with a finger. Once upon having recourse to this method of relief, he felt as if something in his body gave way.—The most dreadful pain and oppression immediately followed. The physicians were equally at a loss to comprehend the cause of his distress, or to alleviate it in any degree. The patient expired in the greatest torture. On dissection, the superior orifice of the stomach was found to be torn off.

they think they can never sufficiently work it off with aloes, magnesia, or salts. I once heard the maid of a lady, whom a brisk course of this kind, determined by the dread of bile, had brought very near to death's door, observe that "*it was no wonder—for her mistress was never easy, but when she was on the* chaise perçée!"

Many women are subject to returns of sickness. They first eject sour or glairy matter. After much retching, a little bile is squeezed up. This only shews that, like the rest of the world, they have a liver, and that this liver secretes bile, as it ought to do; though, not improbably, it secretes too little. But sometimes, even without this *occular proof*, the idea of an excess of bile arises merely from those uneasy feelings in the abdomen that accompany indigestion; from a foul taste and furred tongue on awaking; and from that sallowness of the skin, which is usual in various kinds of chronic weakness. But none of these circumstances afford evidence of an overflow of bile. The uneasy feelings and the foul taste are owing to the general bad condition of all the organs of digestion from the mouth to the farther extremity of the alimentary canal. More stress seems to be laid upon the yellow hue of the complexion. But an unhealthy state of the cutaneous glands is alone sufficient to produce this appearance. The bile may have no share in it. When this fluid is diffused over the surface of the body and through the blood, that is, when the jaundice is actually present, the white of the eye will also be tinged; the urine will dye rags yellow, and the stools will either be light coloured or deeply stained with bile. Without such symptoms there never can be reason for suspecting an habitual excess of secretion in the liver. It is true that the liver, in common with the other viscera, will be subject to irregularities of action. In general, it will furnish a scanty portion of imperfect bile; and at times, when irritated by imprudencies in respect to diet, by exposure to cold, or from sympathy with the nervous head-ache, it will run into the opposite extreme. But this disposition will only be aggravated by the frequent employment of purgatives with a view to remedy the evils of an inactive life and an improper regimen.

The dietetic rules for the class of invalids under consideration are precisely the same as for the dyspeptic. But in all stages of the complaint, moderation in quantity and abstinence from acescent and intoxicating substances are the chief means for obtaining temporary respite and eventual deliverance. "Regimen," says an old physician, once sufferer enough from hypochondriasis to write his own epitaph, under the persuasion it would soon be wanted, "is the most im-"portant part of the treatment. By diet "alone is it possible to alleviate the extremity "of distress, and to hinder the disease from "getting compleatly the upper hand. With-"out attention to diet, it is quite in vain to "hope that any patient can be rescued from "it. To employ the best remedies, while "regimen is neglected, is to build up on one 'side and pull down on the other. When "the patient is brought into the most pro-"mising state of convalescence, the whole "advantage will be destroyed by a SINGLE "CONSIDERABLE ERROR IN DIET; and the "treatment will be to be begun anew."—(*Dr. J. C. Tode Unterricht fur hypochondristen. Copenhagen,* 1797.) Accurate observation will shew that ease and health are strictly attached to this seemingly hard condition. A bowl of water-gruel, a plate of fruit, will put the invalid ten degrees backwards; half a pint of strong wine twenty.

"The hypochondriac," says Dr. Tode, "should make no meal without flesh, be it

" roast or boiled, salted or smoaked. Animal " food, as the great antagonist of acidity, is " and must remain the soul of his diet. In " this complaint, the great secret consists in " guarding against acidity, and in recruiting " weakness. Both purposes are answered by " animal diet." The pernicious change which bread and vegetables are sure to undergo, should be prevented, as formerly recommended, by such a quantity of carbonated alkali, as on experience shall be found adequate to the effect. The time of eating is not indifferent. I do not merely mean that hypochondriacs should feed at short intervals, but that they should never dine late. Healthy animals may digest best in their sleep; but unhealthy men only become feverish, and are exhausted by sleep after meals. Our late dinners therefore must be very mischievous to hypochondriacs. It is probable too that this and many other classes of invalids indulge too much in sleep. It may be adviseable to try to reduce the allowance to five or six hours. Several hours may be spent, at the same time, in reclining on a sofa awake.

It is obvious then that if the physician does not enjoin a proper regimen, or if the patient secretly deviates from the physician's injunctions that nothing will be effected by medicine. Most hypochondriacs are men who have led irregular lives. Their unhappy habits are too apt to overpower the strongest resolutions, of which an enfeebled mind is capable. This is precisely the greatest difficulty in the way of a cure. Otherwise the greatest benefit is to be expected from drugs. They will strengthen the stomach, and in consequence improve the spirits. It may not always be in the power of the practitioner to restrain himself to one article of the materia medica. But the same red sulfate of iron, which I have recommended for indigestion, will generally prove sufficient, where hypochondriasis is not complicated with enlargement of the heart, or with some other irreparable organic injury. When it is, it will procure many hours of tranquillity. A few drops of this preparation put an end to that terrible inward struggle between the disposition to suicide and the counteracting motives, which some hypochondriacs avow, and all perhaps experience—

> Hic motus animorum atque hæc certamina tanta
> Pulveris exigui jactu conposta quiescunt.

Few phænomena belonging to human nature are more wonderful. The very same are produced by the same power in persons afflicted with hysteria.

It is unnecessary to say that the hypochondriac must be particularly careful to keep the region of the stomach and the extremities warm with cloathing and exercise.

In treating of a disorder, which so deeply affects the mind, much may be expected concerning moral remedies. The expectation is perfectly reasonable. But what I have to say on this head, I shall refer to the following essay, in which various nervous complaints, or deviations from a sound state of feeling and intellect, will be considered. I shall only here caution the friends of hypochondriacs against improper management of the mind. Indeed, the doctrines laid down by the most judicious writers on education are in great measure applicable here.

Every thing sour and severe, all general maxims and sage apophthegms, should be avoided. As young people cannot be lectured into good conduct, so neither can hypochondriacs into cheerfulness. Ridicule, reproach, and advice, will only tend to ulcerate a heart, already inflamed. Exhortation ought to be very sparingly employed. The most cogent arguments will avail nothing. This is what provokes the friends of the invalid;

this it is that drives *them* to provoke *him*. They cannot bear that motives, in *their* apprehension so irresistible, should have no influence upon *his* resolutions. But the very suitableness of a set of considerations to their inclination and judgment, is the reason why they should not expect it to suit his. His mind has undergone an alteration in form and quality; and therefore every thing must act upon it differently from what it does upon minds, more soundly constituted. No one takes it amiss that a person in a burning fever does not feel disposed to make a hearty meal on roast beef, yet this would be altogether as rational and humane. Indeed in barbarous ages, and among the vulgar of all times, diversity of palate is a serious ground of ill-will. There is however now a large portion of society, cultivated enough to be tolerant in meats and drinks. When the range of human nature in regard to ideas and sensations is more generally understood, the same toleration will be extended. The sick in mind will give no more offence, and incur no more censure, from the inevitable symptoms of their malady, than the Greenlanders for ranking train-oil as a delicacy, or certain Asiatics from admiring asa fœtida as a perfume. Let the friends of hypochondriacs therefore not suffer themselves to be scandalized at their caprices and imbecility. Let them not dream that any remonstrances of theirs can excite an effort of the will strong enough to throw off the disease. Let them not presume too much upon their own skill in the great art of regulating the ideas and affections of others, unless they can bear themselves witness that they have studied it with as much assiduity as any other art or science requires. The spells commonly used for exorcising those dreadful tormentors, the blue devils, put them but more securely possession of their man. It is given to few to pour oil into the wounds of the mind; to allure the harrassed thoughts into the shady groves and pleasant fields of imagination; to instil inclinations, which shall scorn the spontaneous suggestion of him, by whom they are adopted; to set about the thing they would have another do, in such a manner, as by help of the principle of involuntary imitation, to ensure what aukward managers strive in vain to accomplish by tedious entreaty and harrassing solicitation. These, and such as these, however, are the only arts, by which those about an hypochondriac can contribute to his recovery. To repel his communications when he desires to unburthen his soul of its griefs; to treat every new complaint as ideal, and to tell him that he is well enough if he would but think himself so, is to take the most likely way of rendering recovery hopeless. Nor does it shew want of understanding less than of feeling, because, from the state of his organs of thought, he cannot think of himself but as he does.

LIVER COMPLAINTS.

If the size of an organ be a criterion of its importance, the liver will be entitled to rank high, it being the largest gland in the human frame. The intelligence concerning the incidents of life, that passes daily from mouth to mouth, sufficiently bespeaks its importance in a medical view. Nothing is more common than to hear deaths announced in consequence of an unsound liver. After the years of childhood, I know not if greater mortality arises from any other cause among the higher classes, except only from an unsound state of the lungs. Nothing more compleatly destroys the faculty of pleasureable enjoyment, or provides a larger space for all the modes and degrees of misery, that can stand in the interval between untainted health, and the last expiring agony.

The inhabitants of warm climates are known to be extremely subject to complaints of the liver. Violent external heat has the power of stimulating that organ to excessive secretion, as is proved to demonstration by the abundance of bile in the matters evacuated from the stomach and bowels both upwards and downwards; and also by the truly jaundiced appearance of the skin and eyes, and the power of the urine to give a yellow dye. These undoubted facts have given rise to much groundless apprehension, and have contributed to bring the bile into greater disrepute than any fluid, furnished by the secretory apparatus of the body. For general languor, nausea, foulness of the tongue, want of appetite and indigestion, *accompany* or *follow* the excessive action of the liver, and therefore even by medical writers these complaints are said to be *produced* by too large a quantity of bile in the first passages. This opinion having been spread by the publications and conversation of physicians, those who feel languor, squeamishness and all the other symptoms, only not those of real jaundice, ascribe them without scruple to bile upon the stomach, and, as I have already remarked, continue without mercy to scour the intestinal canal, till they reduce it to an almost total incapability of performing its functions. It is probable that the smallest portion of these effects is owing to the bile. The violent stimulation of so considerable an organ as the liver is alone sufficient to produce the full measure of debility. The same cause, which excites the liver, is very capable of occasioning nausea, and of disposing the stomach to reject its contents. It is not to be supposed that any of the parts concerned in digestion can be strongly affected while the rest continue in a healthy state: and, without the lodgement of bile in the stomach, sickness will arise from a disease of any neighbouring organ, as from inflammation of a tract of the intestines, and from the irritation of the kidney by calculous matter. Indeed, the symptoms mentioned above will all, except yellowness of the eye, be felt, where the bile is deficient or imperfect: moreover, inspissated bile is a favourite remedy for a bad stomach in many countries, and may be taken largely without sickness. After the heats of our own summer, slighter causes will bring on a violent and often dangerous increase of the function, proper to the liver, with vomiting of bile. Under these circumstances, the instant before a portion of bile is about to be discharged from the stomach, as far as can be judged from the feelings, a little ice or cold water will stop the retching, and by repetition at proper intervals, cure the whole complaint, without the discharge of another atom of bile. Whence it would seem that the presence of bile is not the essential, or not the only condition: and that the stomach, on such occasions, must be at least as much disposed to be offended, as the rank and unhealthy bile is capable of offending it.

I was informed by an officer from the spot, that nearly all the men belonging to a regiment of horse in India, died within a short space of time of a liver complaint, because the commanding officer would persevere in exercising them in the face of the sun. Were evidence to the point wanting, nothing could more clearly shew the effect of excessive heat on the European constitution. Whether the natives of hot climates often suffer in the same manner without the assistance of intemperance is doubtful. We can hardly suppose so large a body of men as a whole regiment to have been intemperate; and we know that our countrymen in the East and West Indies, without transgressing the bounds of moderation, as those bounds are usually traced, sometimes carry home a liver compleatly disorganized. It is most likely, however, that by observing the abstemiousness, necessary to some constitutions in the torrid Zone, they would escape.*

For it is still the same abuse of inebriating liquors that falls as heavy on this great appendage to the stomach, as we have seen it falling upon the stomach itself. The classes that take inebriating liquors strongest, and in largest quantity, other circumstances being equal, are most subject to complaints of the liver: the accidental antidotes in some taint of the constitution, or in the mode of life, are precisely the same in both cases; and the exceptions are equally seductive, in the way of example—from the propensity of mankind to draw rules, agreeable to their inclination, from single instances, how-

---

* It may be imagined what havoc heat and vinous stimulants, united, must produce. The facts completely answer expectation. I lately saw the last days of a West Indian, who died with jaundice, and at the same time, with excessive evacuation of bile both upwards and downwards. I was informed that he had belonged to a large club of *bon vivants* in the islands of slavery and pestilence; of whom the greater part had perished from complaints of the liver, at an early period of life.

ever out of the common course of things those single instances may be.

It would be desirable if any particular feelings or appearances could be pointed out as certainly indicating when the liver begins to be in danger. But in this country, the change as depending upon a particular course of life, is usually very slow. In the most sensible organs, slow changes will take place without pain or alarming disturbance of the functions; and the liver has very little sensibility.

By persons, in any degree debauched, and indeed by the drinkers of fermented liquors in general, the faintest standing signs of indigestion may serve to indicate that there is something amiss with the liver. Organs, so intimately connected, must be expected to sustain damage in common. The liver has sometimes been found indurated after death, without any symptom but that of indigestion.

Luckily, however, in most cases, peculiar signs do shew themselves before it is too late to avert the most terrible part of the consequences, though seldom early enough to enable the physician fully to effect that renovation upon the aged in constitution, which Medea is fabled to have accomplished upon the advanced in years. A sense of heaviness in the upper part of the belly, an obtuse pain below the ribs on the right side, with flatulence or acidity in the first passages, are just grounds of apprehension. When a person, upon whom the ordinary causes of hepatic complaint have operated, is sensible of any of these symptoms, and cannot lie with ease on the left side, no time ought to be lost in reforming the diet. On closely questioning invalids with palpable disorganization of the liver, you sometimes find that they can recollect the period, since which they always found themselves upon the right side on awaking, though that was not the case at a period still more remote. Faint inward sensations, raised probably into greater distinctness by sleep, occasion this position to be regularly taken. Just as similar sensations of a nature, not yet ascertained, seem to occasion the general preference of the right hand, though contrary sensations, in some individuals, are so much less faint as decide in favor of the left. Let not, however, any one pronounce himself sound because this propensity is absent. Equal facility of lying on either side and in any posture, is no proof of the good condition of the liver. Sallowness of the skin, and particularly a light yellow colour of the forehead; accompany disorganization of the liver. A morning cough followed by ejection of a little froth from the mouth, or what is vulgarly called *spitting six-pences,* is another strong indication. The same is to be observed of pain in the right shoulder-blade. Lastly, the liver can be occasionally felt hard or enlarged. Then there can be no doubt. But I hope few of my readers will be so philosophical as to supend their belief till the touch furnishes this degree of evidence.

Nations, advanced to a certain point in policy, offer brute animals to their divinities. At a more barbarous period, human victims are offered; and in the conduct of the most civilized communities, about which history furnishes information, we have a singular proof of the observation that extremes meet. For among them also, human victims are sacrificed. It does not signify though there be a little difference in point of ceremony. The result is the same; only that in refined states the number of victims is incomparably greater, and at least in proportion to their population.

Every apartment devoted to the circulation of the glass, may be regarded as a temple, set apart for the performance of such sacrifices. And they ought to be fitted up, like the ancient temples of Egypt, in a manner to shew the real atrocity of the superstition that is carried on within their walls. Then, and not till then, will every person be put fairly in a condition to judge whether he will join in the celebration of these abhorred rites. Among the subjects, which may with propriety be selected either for representation, or for exhibition in nature, the disorganized liver, and the complaints consequent upon its disorganization, would form some of the most appropriate. It has hitherto been very much in vain that anatomists have taken pains to preserve in their museums, and to describe in their writings, the changes, which can be induced upon this viscus:—how it is sometimes indurated in its substance, starred on its surface, and has its edges crisped till they bend forwards; how it is sometimes rendered preternaturally tender; and at others, studded with white, yellow, brown, soft or hard tubercles. It is, I say, very much in vain, because these common incidents in the human body, and effects of the common mode of living, are kept out of the sight of those who would often avoid the mischief, had they that ocular demonstration of its existence, and knowledge of its origin, which the physician can seldom turn to any considerable account for the benefit of the diseased.*

---

* I have observed, on a former occasion, that in the annals of humanity, there is no trace of any rich man having had the heart to promote or diffuse the knowledge of his own nature with the spirit, with which the late Duke of Bedford promoted the important, but very subordinate science of agriculture. It may be added that among our shoals of artists, there is no vestige of any one bending his talents towards the same object. Yet in this walk, what good may be done with a fiftieth portion of the genius of Hogarth. A set of sketches, contrasting our customs with those of savages, in their tendency to disfigure, enfeeble, and demolish the human frame would, I suppose, require scarce any thing beyond decent skill in managing the pencil. These sketches, with accompanying explanations, should be offered to the public in all shapes, and at all prices—from that of poor Richard's Almanack, to the style of magnificence, in which the engravers of different countries have degraded their skill to deck the senseless effusions of the fanatic, Lavater.

An infinity of subjects present themselves on the first glance. What is there, for instance, in the refined Briton with his features all obliterated and his face all carbuncles, that renders him unfit to be a companion to any tattooed antarctic savage? If the East can produce its Faquirs with their legs and arms distorted and rendered useless by constant confinement to the same position, cannot the West match them with figures, equally disabled by their own endeavours? For my part, I know no right our crippled bacchanalians have to set the pleasures they derive from the process, by which they have been reduced to their present

In this country it is less common to be immediately destroyed by a disease of the liver. Other fatal complaints arise at the same time, or are the consequences of this. One kind of jaundice, and that the worst, is the result of the disorganized structure of the liver, or of that of the neighbouring parts concerned in

---

state, above the satisfaction the Faquir finds in being followed and feasted by his countrymen and countrywomen. The motives of the Faquir of BRAMA seem of a more refined and respectable character than those of a Faquir of BACCHUS; implying greater elevation and force of sentiment.—Among this or that uncultivated tribe, it may be the fashion to squeeze the head into an unnatural form —But we—do not we also take pains to derange the interior full as much as they can disfigure the exterior?—The hordes of the wilderness, it is true, sometimes practice horrid mutilations. But then they suffer the residue of the frame to enjoy its full measure of vitality. But we drain off from every fibre its spirit and strength, leaving a vain image, stuffed out to the human dimensions with the dregs of nerve and muscle.—In fact, whenever the Genius of civilivation shall take it into his head to compare notes with the Genius of barbarism, he may adopt the language, in which some modern versifier makes the muse of painting address the muse of poetry—

"——— A son of mine
"Has more than followed every son of thine."

I doubt much if the Genius of barbarism can justly pay himself the same compliment.

digestion, which are liable to suffer from exactly the same causes.

But dropsy is the usual severe fate of those drunkards, sots, or moderate wine-bibers, who have persevered till an alteration has taken place in the substance of this master gland. Many writers on pathology have taught that when it has become dense and hard, so as, if cut into slices, to exhibit a more solid texture than natural, it presents an impediment to the free transmission of the blood. By the accumulation of blood in one set of vessels, another or those destined to throw out the lymph into all the cells and cavities of the body, are thought to be stimulated into more vehement action. Hence an accumulation of lymph or water, as it is commonly called, in the belly; or in the chest, of an enlarged liver press upon the diaphragm. Others place the origin of dropsy, accompanying a diseased condition of the liver, in torpor of the lymphatic vessels, induced by reiterated excitement. On account of this torpor, the lymphatics return less liquid into the circulation than the exhalants throw out; and what is thus continually left, accumulates till it forms a swelling in some outward part, or in others, inaccessible to the senses, acts an impediment to the functions of vital organs, and puts a lingering and cruel termination to life. According to this view dropsy would arise from the same over-stimulation of the lymphatics, which weakens all the other parts of the body at the same time.

Nothing could be better adapted to apartments in which the orgies of Bacchus are celebrated; nothing more like to preserve those, who unwittingly join in the celebration, than bloated dropsical figures, some overwhelmed by death-like languor, some starting out of their sleep under those horrors, which water in the chest brings on, and others in one of those gasping-fits, which come on with greater and greater violence, till the lungs are entirely overwhelmed by the increasing inundation.

It is curious to observe the difference of degree in which the two sexes are subject to disorders of the digestive organs, and to their consequences. Inactive women and effeminate men are almost alone subject to the more curable species of jaundice. The passages, through which the bile flows, have little power of propelling this fluid. Its discharge depends on the contraction of the nearest muscles and of the intestines. Now whatever hinders the propulsion of the bile favours the production of jaundice. For the same reason, the slighter jaundice arising from the sluggishness of the parts, yields so readily to emetics, to purgatives, or electrical shocks; and its return is so easily prevented by bodily exercise. But incurable jaundice, depending on induration of some of the contents of the abdomen, would be confined to the intemperate, if the stomach and its dependencies were less associated with mental agitation. In like manner the sex, more habituated to drinking, is more subject to those liver-complaints that precede dropsy. And if women are not totally exempt from them, it seems to be, in part, because the small quantity of fermented liquor, which women take, in unthinking obedience to custom, has sometimes as pernicious an operation upon their viscera as more copious draughts upon those of men. The seasons for tubercles in the liver and in the lungs are very different. Tubercles in the liver are rare in the young, and progressively more frequent as life advances. For this, the assigned reason perfectly accounts. Violent stimulation, often repeated, ensures the effect. However, as the mortality from intemperance now usually happens at a more early date than in former times, we must suppose the preliminary steps also to occur earlier. If intoxicating beverages continue to make a part of our diet, and if our habits of sedentary indulgence go on increasing, the schirrous liver will become common in the prime of life; and perhaps, in men, diseases of the chest will be altogether supplanted by diseases of the abdomen. What pity that so little can be gained by the change!

Other causes disorganise the liver. For example, repeated ague-fits. But as the fear of having an ague too soon cured, is banished from almost every breast, it would be useless to insist upon this point. In some of the American states, I suspect that neglect of agues cuts short many lives by leaving an opportunity for the formation of abdominal diseases.

Mercury is well enough known speedily to destroy subjects, predisposed to scrophula. But its effect in laying the foundation of remote fatal disorders, has not been sufficiently remarked. I am disposed to think that it sometimes counteracts intoxicating liquors, but much more frequently affects the structure of the abdominal viscera. In repeated experiments on animals, whose food and manner of digestion resemble those of man, I have witnessed its prodigious power to stimulate the liver; and I believe there is no organ—not even the salivary glands—upon which mercury has more action. This is known indeed from its medicinal operation. But that operation is by no means always salutary. In persons, who had avoided fermented liquors of every species from their youth up, I have known the liver become as much indurated after the inordinate use of mercurials, as in any dram-drinker. The deleterious effects of this mineral, whether immediate or remote, ought to interest the friends of young people more in the anti-venereal power of acids. For myself, I continually see, and continually receive confirmations of this power; and immediately on the conclusion of these Essays, I intend to proceed to a public trial, which I hope will contribute somewhat towards checking the endless mischief, resulting from mercurials, in cases which can just as successfully be otherwise treated.*

WHAT expectations can be formed under the preceding Disorders.

A great cloud is left upon the powers of most substances, employed in medicine, be-

---

* See *Testimonies respecting the use of Nitrous Acid*, collected by the Author, and the other Tracts, relative to this controversy.

cause physicians have been at little pains to record the history of patients, subsequent to their recovery! Three things, therefore, of the first importance remain very inaccurately known; the first, what collateral injury is sustained from different remedies; what probability there is of the return of different disorders; and thirdly, what degree of restoration the medical art is capable of bringing about in given cases.

Diet and drugs seem to be efficacious, nearly in the order in which the several diseases have been enumerated in this Essay. By sobriety, exercise, and other means, alacrity and strength may be almost compleatly regained, while there is yet no very sensible alteration in the structure of the digestive viscera. But this becomes less and less practicable in proportion to the change. The liver can certainly, in many instances, be reclaimed after every sign of disorganization has shewn itself. The symptoms, at least, can be removed, and the usual consequences kept off for a considerable time. For a doubt will remain upon this point, till the diversities of which the human organs are susceptible, and the causes of these diversities shall be taken up as a general concern, and not left to the inadequate resources and opportunities of one profession. By multiplied and continued experiments on animals, it would be easy to attain a high degree of probability concerning what takes place in the interior of the human body. Thus if instead of a pack of hounds for the miserable purpose of running down a hare or a fox, a man of fortune were to keep an equal number of omnivorous animals with a membranous stomach, and to diet part of them as many men of fortune diet themselves, a liver-complaint would be produced in the majority. We are pretty certain of this from the history of hogs and horses, fed from the distilleries, and also from the arts employed to provide large livers for pies. Different lots of the animals, so subjected to experiment, ought to be put at different periods upon a healthy course, and some might be medicated. The comparison of the viscera of those that had been always healthy, and of those that had suffered in their health, would inform us much better what we ought to believe respecting the variations of the human liver, than we know at present.

It is often found impracticable even to relieve the dropsy of intemperance. The dropsical can have no reasonable expectation of being again enabled to enjoy the pleasures of existence in full measure. In that dreadful complaint, dropsy of the chest or lungs, the fox-glove in particular, and sometimes other medicines, will often procure a respite; and the patient will seem to himself quite renovated. But the gleam is generally short. The tide flows back. The distress recommences. The same means indeed commonly procure another interval; but it is less perfect and shorter. At last, it comes to be as on board a ship on springing a leak that cannot be stopped. No sooner do the pumps cease to work, than the water rises in the hold. If medicine discharges the water one day, it is collected in as great quantity the next. The absorbents however now soon begin to be insensible to the spur. Then the horrors of slow suffocation commence, and a succession of spectacles are presented, at sight of which the reflecting bystanders may well regret being endowed with animation, and may envy the very stones under their feet for their insensibility.

Some hydropics are happy enough to be saved a part of these miseries by sudden death. Instead of a slow accumulation, the water seems now and then to collect all at once; and dropsy of the chest begins and ends in a few hours or minutes. At other times, after an attack has been relieved, a second is thus anticipated. There is either an instantaneous effusion, or the organs in the thorax, having been injured by the same causes that led to the production of dropsy, cease from their functions in a moment. This accident has been falsely ascribed to medicines, formerly taken. But it is proper to the complaint, and was observed before the medicines, to which it has been ascribed, were brought into use.

DESTRUCTION OF NERVOUS POWER.

Loss of muscular power and of feeling is occasioned by a vast variety of causes. Instances are related of palsy, immediately subsequent to exposure to severe cold. Medical writers remark that fishermen, and persons in in general who are much exposed to wet, are subject to the same disease. But perhaps they have not sufficiently distinguished the loss of motion, subsequent to rheumatism, which is a complaint very different from that, which I mean to consider in the present section.

Metals in vapour and in other states, as lead, arsenic, and mercury, are often the occasion of that real, original palsy which seems to begin in the nerves themselves. But the public is sufficiently warned against these noxious powers. The effect of violent blows on the head need not be insisted upon here. Respectable authors mention a paralytic affection of the whole side from violent coughing, excited by a morsel of food in the wind-pipe *. Weakening losses of blood are attended by the same consequences. The passions of the mind have no less power to deprive the nerves of their properties. Joy is adequate to this as well as fear. A foreign writer records the case of a man, whose system was shaken by unexpected good news, and whose articulation was impaired on every similar occasion, till at last he fell into compleat hemiplegia or loss of one side (*Scheidemantel v. d. Leidenschaften.*) A young sturdy peasant is mentioned by Tissot as having had a stroke of one arm and leg on falling into a passion, while he was in liquor. Other causes are of less general occurrence. But on examining the life, which paralytics have led, we find, in an immense majority of instances, that they have been given to the abuse of fermented liquors. In the set of experiments, formerly quoted, on horses, it appeared on dissection that those which were killed under the influence of spirits or of wine,

---

* Hofman, Med. System, IV. p. 4. obs. 13. This person however was well prepared for the stroke. He drank spirits, had suffered much anxiety, and had even brought on the jaundice by fits of anger.

had an unusual degree of redness on the surface, and in the interior, of the brain. These traces of excessive action in the blood-vessels correspond to phænomena in common life, too well known to admit of description. Considering both together, we cannot wonder that the temporary apoplexy of intoxication should, by many repetitions, be at last converted into immediately fatal apoplexy, or into such an attack as leaves hemiplegia behind it.

Between compleat health and this catastrophe, the interval is frequently long. The changes too from the time when the nervous energy begins to be impaired, to its compleat abolition, proceed insensibly. Strictly speaking, whoever has less feeling or voluntary motion than he would have had at any given period, if no noxious power had operated upon his nervous system, may be considered as an incipient paralytic. But it may be of use in the way of caution to note the symptoms that more certainly prognosticate, and more immediately forerun, a violent seizure. Among these, the principal are feelings, chiefly referred to the head, such as deep-seated, heavy pains, particularly behind, that give an idea of pressure, and are not symptoms of any other known complaint. These are more portentous, when accompanied by ringing in the ears, by difficulty in moving, by slow, clumsy, difficult, or unusually stammering articulation. Inclination to sleep with hearing or vision impaired for a time, and then returning to their ordinary acuteness, are among the number of the harbingers of apoplexy. Accessions of giddiness, with loss of memory, imply that condition of the organs of sensation in which, on a slight occasion, an apoplectic fit may supervene. For of different individuals, equally pre-disposed, it is to be remembered that the fate shall be different according as they are exposed to certain accidents or otherwise. An exertion of the muscles, an emotion of the mind, an acute disorder as a malignant fever or the small-pox, shall overset one person, while another shall escape, because no such trying occurrences befel him.

A passing numbness of one side shall precede hemiplegia many years. The greater number of apoplectics conceal such feelings, though it is probable that they generally raise apprehensions. A person, who not long since died suddenly, mentioned to me the frequent return of those partial numbnesses for years before. He was not insensible of their signification, and once feeling so while the taylor was busy about his person, said "that he probably should never want the "suit of clothes, as he distinctly felt Death "taking measure of him for his shroud." Scheidemantel tells of a patient, who felt numbness of his hands and feet, with formication, or a sense of ants creeping upon his skin, full ten years before the fatal stroke. I think that great liability in old people to lose their speech, or the use of their hands in severe weather, must be regarded as a disposition towards palsy.

Frequent nausea also or vomiting in old people *(vomitus grandævorum)* is a suspicious sign. So, at an advanced age, is frequent mastication without food, in the judgment of some; frequent yawning; deep laborious respiration; depending upper eyelid; slabbering of the saliva; bleeding at the nose; involuntary tears; flashes of light or dark spots before the eyes. But several of these indications must be deemed much more uncertain than the former, which are in fact slight paralytic seizures. It is of the utmost consequence to attend to these warnings, because by a due and cautious change of regimen, assisted by medicine,* the dreaded event can almost always be prevented.

---

* The chief of these remedies I believe will prove to be

The first fit of apoplexy, it is well known, sometimes proves fatal on the spot, sometimes in the course of a few hours. But when the event is not at once fatal, it may be useful to know the subsequent changes, in order to be able to judge of the value of the common means of cure. Few, who have suffered an apoplectic fit, emerge from it entire. In the great majority, we observe a more or less perfect deprivation of loss and motion in some part, or of perfection in some of the senses. The lessened mobility or sensibility of one side of the body is, a common consequence. These faculties, however, very frequently return in some degree; and the

---

nitrous oxyd, or *gazeous oxyd of azote*, of which Mr. DAVY has given the public so masterly an account from the Pneumatic Institution *(Davy's Researches, Johnson)*. This gas is always safe, generally highly agreeable and salutary to paralytics. Its action has nothing equivocal. In the first trials, it was occasionally withheld from the patient, in which case his disorder was stationary or retrograde, though he had been improving under the use of the gas, and improved again upon returning to its use.—But the most decisive and astonishing proof is a phænomenon, which various persons have witnessed with me.—Paralytic patients have had the perfect command over their diseased limbs, while breathing this gas, and the moment after have returned to their former state—till by continued use of the gas, the limbs have been more or less restored.

invalid is able to make considerable use of the affected parts, till he is overtaken by a new seizure; and this will be either distinctly marked by an entire and instantaneous abolition of sense and motion; or it will happen unperceived in the night, and be known only by its effects.

Hence, when, under any course, some paralytics insensibly recover their bodily powers to a certain point, while others continue stationary, and the remainder are rendered worse or destroyed by fresh accessions, the probability will be that the treatment itself produces no effect, good or bad, but that all the changes belong to the mere progress of the complaint, as it varies in different individuals. This probability will arise near to a certainty, if, in collateral phænomena, there be no evidence of a specific virtue in the means employed.

This character seems to apply to a number of remedies once in vogue but now discarded. It suits none perhaps more exactly than one still in request—I mean the water of the hot springs at Bath. Some have condemned this water as bringing strokes of apoplexy on the paralytic. They have known such invalids relapse under the use of the water. But so they would have done at home.

Others have founded a favourable judgment on the improvement they have known to take place in certain instances. But occasional improvement takes place independently of all external or internal applications. So that we may acquiesce in the judgment, which Dr. HEBERDEN, an observer, above all praise, faithful and diligent, founds upon his long experience:—*that these waters are neither in any way detrimental nor of the least use in palsy.**

The other terminations of paralytic or apoplectic seizure are less applicable to any practical purpose. They are very various, as epilepsy, impaired faculties, or madness.—In extremely rare cases the senses have become more acute; epilepsy has ceased; and I knew one person, whom a first stroke deranged, and a second restored to his intellects. Paralytic complaints are more frequent in advanced life, and among the male, a may be expected from the nature of that very common exciting cause, intemperance. That they are not confined to any period or sex is only what must be expected, since the system at all

* Posthumous Commentaries—p. 303, of the latin edition.

periods is liable to sudden shocks; and any of these may bring on palsy.

GOUT.

In attempting to speak of this disorder, one feels as if on *tabooed* ground. The public has unlearned an infinity of false opinions with regard to other disorders, but the prejudices of the darkest ages cling as fast as ever to the idea of gout.

It is the result of large and sure experience that the gout is very much confined to those who lead a life of indulgence and intemperance in regard to spirituous liquors. This intemperance must, as in all other cases, be relatively considered. The condition is not absolute. It is taking too narrow a view of human nature to suppose gout producible only by hard living, or by some use of fermented drinks in the descendants of gouty ancestors. One of the perfectly abstemious patients above-mentioned as affected with a liver-complaint, afterwards had regular fits of the gout without any change of diet or any medicines, which could be considered as equivalent to alcohol. As to labour, it doubtless exempts multitudes from the gout, and has been known to remove it.

Among four thousand poor *chronic* invalids, scarce two instances occur of gout in this part of England. One of the greatest martyrs to the gout I know, told me that in the quarter of a century and upwards during which he has been gouty, his freest year was that of a warmly contested election, at which he was candidate for a county. He both drank and exerted himself infinitely more than at any other period of his life.

What clear experience, without any hypothesis, teaches respecting the nature of the gout, is in brief as follows.

It scarce ever comes on before the æra of puberty. It seizes the ball of the great toe, which soon becomes of a deep red colour, and afterwards swells. The first paroxysm seldom endures above a day and night with unabated violence, but, when it is considerably mitigated, it again seizes the same or some other part of the foot, and lasts for an equal length of time. These paroxysms go on with more or less intermission for a time seldom short of nine days, and seldom exceeding a month.

Not unfrequently, when the swelling has nearly subsided, and the patient looks towards recovery as an event near at hand, another fit shall come on, and run through its course, as before.

A long interval of ease is frequently enjoyed after the first fit. Sometimes this interval extends to four or five years; it is seldom less than two. This shall go on for a second and a third accession, and, on all these occasions, the gout shall steadily continue in the foot in which it begins.

After several repetitions the fits occur more frequently, and the affection shall fly from foot to foot; from a foot to one of the knees; from the knee to the wrist. It shall at times transfer itself to the chest, the neck, or to one of the senses. Thus gouty pain in a joint shall cease, and instant deafness come on, which again new pain shall sometimes remove, though it sometimes becomes permanent. Lastly, the stomach shall be seized with dangerous violence.

In an old case, the lower extremities shall be so debilitated that the patient shall not have his feet eight months in the twelve out of gouty shoes. The legs are left crippled, and crutches become necessary. As the lameness begins to go off, the patient is cut down again by a fresh fit. In this debilitated state, the nature of the attack shall change; so that once in six weeks there shall be a swelling without pain or discolouration. This is sometimes called the *dumb gout*—I suppose, because there is not the usual agony to extort groans.

The *sinews*, as it is said, *become contracted* at this latter period; that is, the muscles are disorganized, and no longer capable of that contraction and dilatation, upon which voluntary action depends.

In the mean time, the gouty, besides having to struggle with every thing that pain has of severe, are annoyed with all the inconveniences that attach themselves to a bad stomach. No hypochondriac is subject to more terrible depression of spirits. Meanwhile concretions of a chalky consistence are generated, and sometimes in such abundance, and of such size, that it seems as if the skin had been stuffed with plumbstones.

Thus it is that athritics exist, when the gout thickens upon them. Is their death more enviable than their life? It happens sometimes, after many alarms of the same kind, in consequence of a stomach seizure. This is not however their most common, though I think clearly their most desireable end. The greater part are shared between dropsy of the chest and palsy. One or other of these disorders is commonly the fate of those, who leave the disorder to its own course, without resorting to any means of relief, except perhaps the use of the Bath waters.

If such be the usual origin, progress, and termination of the gout, can it really deserve to be regarded in the favourable light it usually is? Is it reasonable to desire, and wise to cherish such a complaint? Or is the eagerness, with which we hear it sometimes invoked, to be placed among the most egregious examples of the *vanity of human wishes?*

Either the gout itself, or the gout conjointly with the regimen that first brought it on, begins at the circumference, and proceeds to the centre of the constitution, injuring every organ it affects, and at last rendering them incapable of carrying on life. This, we see, is the clear result of the history of those whom it affects.

Many arthritics are healthy during the intervals of fits, particularly if those fits be regular. This I apprehend is one principal reason why the gout has been able hitherto to maintain so advantageous a character. But the fact really proves nothing to its advantage. The character was established among the rude and hardy ancestors of the present inhabitants of Europe—men of an iron frame, which it required the long-continued operation of luxury to soften, and of debauchery to wear down. Does any one think it so obvious a truth that these men could not continue robust to a good old age, or even to the middle term of life, without paying so severe a fine for the renewal of their lease of health?

But perhaps it is imagined that *the dreadful disorders which follow in the train of gout, are kept off by its occasional recurrence.* There is nothing on the face of the facts to persuade a person, who reasons with the smallest degree of caution, that this is more just than the opinion directly opposite—namely—that *such and such person become dropsical or paralytic, because they have suffered so much from the gout.* Certainly, the gout seems to possess every qualification for rendering either the lymphatics or the nerves incapable of their functions.

Analogy appears strong against the preventive power of the gout. I have already said that the ague will be followed by dropsy in some habits, if it be suffered to hold on its course unchecked. It may be added that, in other habits, it will be followed by apoplexy. Now when the Peruvian bark was first introduced, mankind were weak enough to suffer themselves to be persuaded, in opposition to all experience, that this dropsy and this palsy were the products—not of the diease—but of the remedy—of that very remedy which, if used properly and in season, is sure to prevent the disease from producing them. Some other disorders of frequent recurrence are followed, like the gout, by dropsy and palsy; and I do not believe any one would assume, without the smallest evidence, that they exert the preservative power attributed to the gout. —Mrs. ——, aged 72, had returns of erysipelas for above twenty years. No considerable difference was made in the treatment of the different attacks. For a few years last past, she appeared to be drooping; and a year and half ago she had a dropsical swelling of the legs. This was removed by diuretic medicines. But not long afterwards, she was seized with an apoplectic fit, out of which she was recovered with the loss of her right side. Such was the tone of her health, when she was first seized with erysipelas that any competent judge would, at once, have been satisfied that there was then no call for the interposition of erysipelas to save her from dropsy. So also would competent judges decide as to a large proportion of our country squires and citizens, when the gout *first* begins to precipitate them, before their time, into the autumn of life. There is no room for supposing that such subjects would have any dropsical or paralytic disorder, if they had not the gout. There is no room for supposing they would have had the gout, if they had not done all they could to bring it upon themselves.

But perhaps if the gout does not operate as a preventive, it does as a cure. It may remove worse, or at least some disorders, actually present.

It does not prevent or remove cancer. I have at present a female patient more than enough harrassed with gout, for any medicinal purpose it could answer, and yet she has lately been obliged to submit to the extirpation of a cancerous tumour of the breast. It does not remove induration of the liver. That may very commonly be ascertained. It does not remove asthma, as I have had occasion to see. It does not generally remove or relieve indigestion, as I think most arthritics would bear me witness. When it is thought to do so, the following fact, recorded by Dr. Tode of himself, will shew on what terms it renders this service. He was one of those who wished for the gout. "My case is singular," says he. "As long as "I was hypochondriac, and had to contend "with the concealed gout, I could not bear "the smallest quantity of rye-bread. It "caused excessive sourness. - - - Afterwards, "when I was visited by the regular gout, "this disagreeable effect ceased at once. "From time to time I have suffered at a later "period from an intermission of gout, and "also therefore from acidity. But within "about a year, that is, since *I have had a* "*palsy of the tongue and of one hand*, and "have almost got over it again,* I feel not "the least inconvenience from rye-bread." So that according to this account gout and palsy were pretty nearly upon a par in restoring the digestive power. Perhaps the difference, at any time, is little greater than that between laurel-water and *acqua toffana.*— Another disease which the gout does not remove is stone and gravel. We see people labouring under both; and both frequently derived from the same intemperance. I have known some patients worse in respect to pre-

* No wonder the palsy is better. This we have seen is common, till the next stroke. Besides, poor paralytics are apt to flatter themselves not less than the consumptive.

vious diseases, after strong fits of regular gout. But for the present I shall suspend the relation of my own experience for the sake of introducing an extract from Dr. Heberden, wishing that the passage may induce candidates for the gout to consider the whole of what he says under that head and in various other articles, concerning the insufficiency of gout as a preservative.

" Although the gout recurs but seldom at " first, in time it grows more familiar, and " the fits are not only more frequent, but " last longer, so that no interval remains for " the enjoyment or business of life. The " patient, while in pain, is as helpless as a " child, though there are many discourage- " ments from administering the help he " needs, with as much alacrity as to chil- " dren. It scarce deserves to be numbered " among the evils of gout, that after poison- " ing all the pleasures of life, and wasting " the bodily faculties slowly, or abolishing " them at once by a paralytic or apoplectic " stroke, it occasions premature death.

" Few however are either ashamed or sorry, " though in such dreadful plight. On the " contrary, many court the disease; and, if " they feel themselves disordered, are pleased " at the idea that gout is at the bottom,

" and console themselves with reflecting that " they shall some time become gouty. They " even take much pains to secure the event, " though luckily for themselves, they are " often disappointed in their wishes.

" It is not easy to say on what account, " in certain countries and ages, certain dis- " orders have been deemed reputable, while " others have been looked upon as disgrace- " ful, though in the disorders themselves, " and the manner in which they are con- " tracted, there has been no ground for " either sort of estimation. — — — Some " affections have become honourable, be- " cause they happened to some particular " great man, or because they chiefly befal " the fortunate mortals, that abound in lei- " sure and all the good things of life. When " LOUIS XIV. had a fistula, many of his " subjects would have it that they had a " fistula too: and the surgeons of those " days had no small trouble with these " *malades imaginaires.* Had there been in " France, a mineral water which could have " given the fistula, people would have " flocked thither, as they now flock to Bath " to get the gout. For this is the complaint " in esteem among the English. It is this " which is desired by those, who are unac-

" quainted with it; which is boasted of by " those who believe they are on the point " of an attack, but which those seriously " deplore, who really have to contend with " so cruel a calamity. Hence, by a singular " sort of contingency, more pains are be- " stowed to contract than to drive off the " gout. For so much honour is paid to this " malady, that arthritics ascribe all their " other ailments—not to their having had " too much, but to their not having yet " had enough of gout. So that the gout " not only escapes blame as being the cause " of these ailments, but is wished for, as if " the only chance of health consisted in " repeated fits.

" From attachment to life or dread of " death, most men are loath to feel or own " the signs of a shattered or ruined consti- " tution. These signs are therefore assigned " to any cause rather than to approaching " and inevitable death. Hence, when the " powers begin to fail, they easily suffer " themselves to be persuaded that their ill- " ness is owing to *latent gout*, and that " nothing is wanting to their entire resto- " ration, but a regular fit. But, not to " notice the slight probabilities on which " the hope of latent gout is founded, I

" should be glad to know, whether that " which is so confidently talked of is al- " together certain—I mean, *whether regular* " *gout expels* * *all other complaints? and* " *whether one can produce or excite the gout,* " *without heaping new evils upon the old?*— " In countries where the itch is the most " common disorder, it is also held to be " salutary. Formerly the ague was believed " by many physicians to be the handmaid " of health. Some think it so still. There- " fore they congratulate themselves and " their friends on its arrival, and take par- " ticular pains that it should not quit them " too soon. Time has nearly destroyed " these opinions among us. And should a " remedy for gout be discovered, as effica- " cious as those we employ against ague or " itch, we shall learn at least how safely and " advantageously the gout also can be for " ever removed."

"To learn that gout is an effort of " nature to throw out I know not what viti- " ated humours, we must appeal to expe- " rience, as the only adequate arbitress of " this question. Some, I confess, are de-

* " If the gout prevents every thing else, would not " one have something that does?" *Hor. Walpole's letters.*

" lighted at the first onset of gout, as if it brought " all sorts of happiness in its train. With " nothing but entire restitution in their heads, " these people easily fancy themselves wonder- " fully benefited—a common sort of persuasion " on resorting to a new physician, or a new " remedy. But, not to dwell on such con- " siderations, let us consult those who have " been sufferers enough to know what gout is. " Of such I have seen a great number, and " noted down their cases; and there have been, " to say the least, TWICE AS MANY who have " either found no alleviation of former com- " plaints, or in whom the gout has occasioned " new ones, as of those to whom it has ap- " peared to be of use. In my judgment, too, the " mischief imputed to it was a great deal clearer " than the benefit. Doubtless, we sometimes " see other disorders subside on accession of " gout. But the same thing, as I have often " observed, happens with regard to fever, palsy, " asthma, small-pox, and insanity. Yet nobody " would call these salutary disorders.

" People, seldom and slightly affected, are " scarce hurt. So, when epilepsy is of rare " occurrence, the patient dies of old age, not of " the complaint. But some constitutions are " damaged from the first onset of gout, and " suffer more and more till they are completely

" demolished. Should we doubt whether it is " the gout that does the harm here, we must, at " least, be very certain that it has not been " serviceable.

" In gout of long standing there is less pain. " Hence may have arisen the idea that the " decline of the gout may have occasioned that " ill-health, which is simply owing to increase " of years, to other complaints, or to repeated " attacks of the gout itself.

" I remember a person, who for 15 suc-" cessive years had both gout and asthma. At " length, his constitution being very much " broken, he was five years without either " complaint. For so long did he linger be-" fore he died. Nobody will ascribe the ruin " of his health to the absence of asthma. " Why should it then any more be ascribed to " the absence of gout?

" After a few fits, the genuine gout returns " every year or oftener. But after several yearly " visits for 10, 12, 14, 20, and even 27 years, " it has intermitted for 3, 10, 12, and even for " 30 years; and this without the smallest bad " effect that could with any fairness be ascribed " to absent gout. That the regular return of " fits is not necessary to health is demonstrated " by such instances. For here we see arthritics " missing them, even after the complaint has

" become so fixed, that, according to the vulgar " notion, neither health nor life could go on " without its aid.

" But however useful it may be to bring out " the gout when latent, or to ripen it when " crude; every physician must own that the " symptoms of gout in this state are very ob- " scure, and that there are none indeed by which " we can recognize or justly suspect the disease, " except there be pain, redness and swelling of " the foot, or unless the patient himself or his " parents have been subject to the disease. " But, without any of these circumstances, it is " common to refer the beginning of many " chronic and of some acute disorders to gout. " But a mistake in this matter, where it is so " easy to mistake, will occasion a great deal of " mischief. Many have been led to drink far " too much wine and strong liquors. I have " seen some patients be nearly intoxicated for " two or three days together, because they " fancied themselves in want of a fit of the " gout; and this they thought the most likely " way to procure it. I grant, indeed, that a life " of debauchery will readily produce this wretch- " ed disease. But cramming and hard drinking " for two or three days will certainly reduce " the strength, whether the gout be at hand " or not; and few will deny that such ex-

" cess is hurtful, provided another disease " be at hand, which has been erroneously ap- " prehended to be gout. I have much too " often met not only with carious bones, schirrus " and cancer of the brain, and with other chronic " diseases, but also with some acute ones, as " inflammation of the lungs, of the stomach " and other viscera, under the denomination of " gouty affection. The consequence of this " denomination was to take spices, wine, and " other stimulants, till it was too late either for " a remedy or a palliative. This kind of treat- " ment, in my opinion, does not render so much " service in true gout, as it does mischief by " increasing pain and fever when this disease is " wrongly suspected.

" I do not propose to enquire into the power " of the Bath waters, and heating remedies com- " posed of wine and aromatics, to excite gout, " if, in any case, this should be desirable. " After a little trial, they will, if I mistake not, " be found to fall short of their character. Those " slight, wandering pains which are felt on " taking anti-arthritic remedies, and are styled " *rudiments of the gout*, would perhaps never fail " to be felt, if attention were paid, in the decline " of life.

" It has been thought that a large quantity of " sheer wine is good for arthritics. I should

" suppose this opinion to have grown current, not
" on account of its justness, but because it was
" agreeable to the lovers of wine. These should
" however recollect first, that free use of wine
" of itself injures the stomach and impairs di-
" gestion; and, secondly, that persons addicted
" to wine are those who chiefly bring on and
" cherish gout: whereas the poor, who drink
" scarce any thing but water, have a keener ap-
" petite with a better digestion, and scarce know
" any thing of gout. I have seen very few
" perfect cures of this disorder. One or two
" wretchedly debilitated patients, on total abs-
" tinence from animal food and wine, have re-
" covered so firm a state of health, that their life
" was no longer a burthen to themselves or use-
" less to others *. But these examples are too
" rare to give me much hope that others will be
" so fortunate.

" It is a question whether the patient ought
" immediately to give up to the gout and lay
" himself on his back to receive it in flannel,

---

* I have known three instances in medical men, and four in others, where the gout ceased and the health improved after leaving off fermented liquors only. I should suppose a temperate regimen, with a remedy announced below, would very generally remove the disease, or rescue invalids from their misery.

" fleecy hosiery, or eider-down—or to strive his " utmost to move the affected limbs, and shake " off the foe, if possible. The more it is in- " dulged, the longer does the gout keep its " station: a thing the patient wishes for, and " what indeed he tries to secure by medicine. " A gouty matter is imagined to be produced in " certain parts of the body, and to receive daily " accumulation. Now unless this matter is " allowed free egress, it is apprehended the con- " stitution will be liable to all sorts of maladies. " Others, on the contrary, from impatience of " pain, rather than from the influence of any " opinion, have left nothing unattempted to " stifle the fit in its very infancy. The great " Harvey, as I have heard from his relations, " was accustomed to remove the pain by im- " mersing his foot in cold water, as soon as he " felt the disorder approaching. I have known " others make incessantly efforts with the af- " fected limb till the pain went off, and its " power returned, in which they have at length " succeeded. Nor was this repulse of the dis- " ease by violence followed by any serious bad " consequences. How much may be done to " throw off the gout, if no endeavour be left " untried, appears from those undoubted his- " tories of invalids, who, when they appeared " unable to stir a single limb, have forgot their "gout in an instant, and betaken themselves to "their heels with abundant alacrity, on an alarm "of fire or the enemy.

"I would not propose Dr. Harvey's example "for imitation to others, though he attained to "eighty, and died of old age rather than dis- "ease. I have not however found that people "have done themselves any injury by preventing "the gout from fixing, by exertions of the limb. "If indeed, during a fit of the gout, the body "were disburthened of any matter of disease, "how comes a purge soon after it is over to "occasion a relapse of equal violence? This has "occurred not only to me, but to other physi- "cians of the first authority. The same thing "sometimes happens from a blow, a strain, or "the puncture of a gnat. By these facts, the "doctrine of an accumulation of gouty matter, "as occasioning the fits, is rendered extremely "difficult of conception."

Such are the sentiments bequeathed to posterity by a physician far remote from the love of paradox. They will some time enlarge the public stock of health and virtue. They may at present make less impression, on account of the inveterate prejudice that *gout performs the good office of carrying something hurtful out of the body.* It would be in vain to ask for a definition of that *something*. But perhaps we may be referred to the concretions for sensible evidence of its existence, or to the foul urine, where there are no concretions. In like manner, it was long believed that in all exanthematous disorders, in all disorders terminating in sweat, or in any other evacuation, that the humours were defæcated, or cleared of a certain scum, by these discharges. Numbers are now living, who, to promote despumation in the small-pox, were quartered in hot apartments; a measure which no doubt contributed to their health, as much as a place in the *black hole* at Calcutta did to the health of those who survived confinement there. No reader, at this day, need be told how little it is to be desired that the eruption in small-pox should be profuse. The Peruvian bark and the mineral solution expel no morbific matter in intermittent fevers. There is, on the contrary, all the reason in the world for being satisfied that the matter in small-pox, the concretions and urinary sediment in gout, are formed by diseased action at the expense, by no means to the relief, of the constitution.

"What," says an invalid, "is the cause of the "sudden shifting of the gout from one foot to "the other in the twinkling of an eye? What "is the subtle æthereal matter that passes through "the whole muscular system like lightning?" I scrupulously copy my correspondent's words.

But what occasion is there for any matter to pass? Why may not actions be communicated in the human frame without transfer of substance? Will not certain applications produce an immediate sensation in a distant part? Do not some persons sneeze, as soon as cold strikes upon the hand or foot? Does not warm flannel on the skin ease the colic? An impression upon the eye from a disgusting object excites sickness. I have known a sudden loud sound occasion pain in the back. There is no necessity to suppose a transition of matter in these instances, and certainly not in any of them a transition of *morbid* matter.

One circumstance, which seems to have confirmed the prevailing prejudices, is the manner in which some desperate arthritics have gone about to relieve themselves. I have obtained evidence, which I cannot doubt, of two cases where the abovementioned expedient of the great HARVEY proved fatal. The terror, inspired by these events, is kept up by the natural transition of the gout from the extremities to vital organs. It is, upon the whole, concluded that this is a disorder, with which it is impossible to interfere without the most imminent peril—the *noli me tangere* of constitutional affections. When the habit is vigorous, and the gout has never shifted during the same fit, it is probable that cold *may* be freely applied to the inflamed joint. In opposite circumstances, it seems as if this could never be done without danger. But as there is no moment, at which the stomach or head may not be susceptible of the gouty action, it must be rash *in any case* to reduce pain in a limb by means, that increase the susceptibility of a vital organ. Such means are cold water or substances that abstract heat rapidly. When the bowels are torpid, putting the feet upon marble, or wetting them, has expedited the operation of purgatives, or, of itself, brought about the same effect. This will explain what I mean by cold applied to the foot increasing the susceptibility of gouty action in remote vital parts. Physiology offers such analogies in abundance.

But the ill success of one method does not prove the inadequacy of another, quite different. It would be hard, if culprits, now in confinement, were to be pronounced guilty, without further enquiry, upon the evidence which convicted their predecessors fifty years ago. The mischief of cold water, for example, can be no objection to the application of leeches to lower the local inflammation, or of calomel to excite the stomach, liver, and other intestinal glands.

A superstitious habit of thinking on the subject of the gout seems, therefore, to act fatally in opposite ways. Those, to whom the pain is intolerable, blindly grasp at the nearest chance of relief, while others unnecessarily submit to have the constitution broken down, because their dread of innocent or effectual remedies predominates over the violence of their present tortures.

In the activity of research, and the great accessibility of men's minds to reason, which distinguish the present times, it is not to be expected that an affection, bearing a great analogy to complaints which we have already learned to treat with success, should be passively submitted to, as of old. I am fortunate enough to have it in my power to offer a reasonable expectation of relief. Something less than a twelvemonth ago, an entire stranger offered me an anti-arthritic remedy of a character entirely new. He said it was prepared from a vegetable, which had not been introduced into any *materia medica.* It was described as having the property of dispelling the fits of gout with as much gentleness of operation as the bark exhibits against intermittents, and as having been in use several years with great benefit to the constitution, and almost total reduction of the disease upon the whole. I received a number of references to persons who had tried it with success. The discoverer also acquainted me with the name of the plant. He added from Dr. Bradley, the well-known editor of the Medical Journal, an attestation, purporting that the preparation would " speedily and *safely* " remove the pain of gout."

On laying this statement before gouty invalids, I found some willing to hazard the trial.

In no instance has the smallest untoward appearance shown itself, though some of my patients have been mere human ruins, threatening to fall at the moment of approach, and certain to be overturned by even a slight shock. The gouty symptoms have, in general, been removed with the greatest ease and rapidity. Many of those who have taken the remedy speak with a warmth of enthusiasm, which is unusual except after some great deliverance. The disease has been sometimes, at once, dislodged from the interior; and it seems as if, by stopping at a certain dose, worn-out invalids might have what is called a healthy fit. I have been struck with surprise at the signs of improved health at the end of six months after the *occasional* use of the medicine; for it has been seldom given, and seems only necessary, at the time of the fit. In the mean time the experience of Dr. BRADLEY has been equally favourable. There lies before me a note, in which he declares himself to have used the anti-arthritic medicine in several cases of gout as well as obstinate rheumatism. "It gives," he adds, "great relief to the patients, and generally seems to produce a perfect cure;—he has never seen the slightest ill effect arise from its exhibition; and he has reason to believe, from a case or two, that it will prevent a fit of gout as well as cure one."—If the medicine shall continue to justify its pretensions, the discoverer expresses himself ready to disclose his secret, on obtaining a compensation adequate to the benefit he shall thus confer on society. In a very few months, he informs me, it is his intention to bring the whole subject before the public.

In irregular and ill-formed gout, no better counsel as to regimen can perhaps be given than that under the articles *dyspepsia* and *hypochondriasis*. Whether it is that the disorder has declined from the truly inflammatory form—whether it never appeared but in tran ient redness, pain and swelling of the foot, and soon removed to other seats—or (what is less characteristic) persons of gouty descent have felt gout-like pains without any affection of the foot, or well-marked inflammation in any of the extremities :—in these cases, as well as where the attack is regular, the digestive organs are much out of order. Wine and fermented liquors of equal strength are sometimes so soon and so distinctly injurious, that, without laying any stress upon other considerations, it seems as if it might be laid down as a rule for arthritics, *to wean themselves as soon as possible from their use* *. To this rule, puny arthritics owe implicit obedience. Cautious use of vegetable sub-

* " I have known several arthritics who could *never in-* " *dulge in a few glasses of wine* without bringing on (gouty) " symptoms, but who, by a rigid and persevering abstinence

stances is equally to be recommended. Dr. READ has lately given the interesting observation of a gouty convalescent, who was repeatedly threatened with a relapse, "from having "barely *tasted* a vegetable acid." (*Monthly Magazine for May* 1802.) By persons unaccustomed to cyder and perry, I have been assured that a small quantity of either has brought on a paroxysm. On this principle we may perhaps understand why the weaker wines of France are reputed *more gouty* than those in common use among the English. The former more certainly turn sour on a depraved stomach, or on one that is weak from inheritance. Otherwise, as port (according to information from those who assure me they have repeatedly ascertained the fact by distillation) may be estimated as containing one-fourth part of its bulk of brandy; and as claret contains but an eighth, the result ought to be very different, if we suppose the stomach healthy at starting, I should expect four, if not eight times as many winebibers to be thrown into gout, dropsy, or apoplexy, by equal large quantities of port as by claret.

I know one person with a diseased liver, in

"from wine, kept off any fit. The late Dr. BRODBELT, of "Spanish Town (in Jamaica), was an eminent instance of "this." Dr. DANCER's *Medical Assistant*, 4to. 1800.

"I only drank a little wine two days after being very "much fatigued in the house; and the worthy pioneer began to "cry SWEAR from my foot the next day." HOR. WALPOLE.

whom a glass of port wine and a cup of strong tea equally excite pain in the right hypochondriac region. And I suppose tea in some habits, if it be of any strength and fineness, a promoter of all the disorders considered in this essay. Men, as in every disease of similar origin, are oftener gouty than women. But I think the proportion of imperfect gout in women greater than in men; the disease in the weaker sex oftener beginning with being irregular, and sooner becoming so.

In some places on the Continent an imperfect gout, which however sometimes assumes a regular form, appears to be quite epidemic among females. And as they are females to whom our genteel ranks will feel no pride in being assimilated, a knowledge of the fact may have its influence in helping to render the gout unfashionable. " The washerwomen among us," says Dr. J. C. G. Ackermann of Altdorf, " are " far more subject to gouty complaints than in " other places. Many houses are unprovided " with a wash-house, secured against the imme-" diate influence of the atmosphere and capa-" ble of being heated. Hence, in the coldest " days of winter, these poor people are obliged " to wash in the open court in the most cutting " winds. They are exposed alternately to cold " and hot water. The cold moisture inevitably " penetrates through the clothes. While the

" upper part of the body perspires from steam,
" the lower extremities are frozen. They en-
" deavour to counteract these inconveniences
" by coffee, brandy, and the like. The result
" is, that it is scarce possible to find a washer-
" woman above forty-five who is not a sufferer
" from gout, though the irregular kind is by
" far the more common of the two."

The power of temperature to give such certainty and determination to the gout-creating action of stimulants is remarkable; and the relation excites a wish for further particulars. One can the less suppose that the author misnames the complaint, as he appears throughout his paper * well acquainted with the varieties of gout. The same identical cause, indeed, may often be observed to produce different complaints, according to the condition in which the mode of life places parts of the body. The diseases of intemperate hackney-coachmen and butchers differ from the diseases of intemperate taylors and shoemakers.

By attending to the influence of exposure, our opulent arthritics may escape many of their worst visitations. The gout observes exactly the same law as to its season with other inflammatory disorders. It is most common when pleurisy and rheumatism are most common too. It

* HUFELAND's N. Journal, iv. 3. 31.

is rare for a person not yet reduced, and subject only to strong fits, to be seized in the fine part of the summer. This is the lot only of those whose system has become so liable to gouty action, or, in vulgar language, *is so penetrated with gouty matter*, that a stream of air, a little variety in food, the twist of a limb—agents, in fact, too slight to be sensible in any other way,—shall stir up the slumbering gout. The consequence of a catarrh is so serious, that arthritics may look upon the directions for escaping and mitigating this inflammatory affection as drawn up particularly for their use.

There are still many other maladies which persons, blindly eager for sensual pleasure, bring upon themselves; and some of them of a very harassing nature. It is thus that the stone and gravel are often occasioned. The piles befall scarce any but the debauched or the indolent. Indolence and intemperance commonly go together, if, as I have repeatedly inculcated, those things are to be considered as violent stimulants to the indolent, which are scarce felt as gentle cordials by the regularly active.

It may be asked, *Since all men must die at last, what signifies it whether the catastrophe come from intemperance or otherwise?* Doubtless I think it wise to dread the termination of existence less than its continuance in misery. But is it nothing that luxury has almost banished *euthanasia* from among men, and deprived us of all the consolation we might draw from the good old comparison between SLEEP and DEATH?

It does not require to be professionally conversant with the sick to be sensible, that the multiplication of painful passages to the grave must equally multiply the objects of terror.

I do not, indeed, think disease much more terrible than that blank state which so often intervenes between the offices of the medical tribe and just health. This is the state which more particularly requires to be delineated to the public eye. But, on looking back, I perceive that I want pencil and colours alike. There is, I think, however, one way in which the reader may attain a just idea. Let him well consider a knot of children at play, of whom the boys are yet unbroken by the tyranny of PRISCIAN, the girls by that of the GRACES. Let him then transfer his eye to a fashionable boarding-shool, to a rout, a mess-room, or singly inspect our youth newly risen from repose. Let him employ a month in analysing these objects into their minutest differences; and then he will have some faint comprehension how it is with thousands, who, under a gay exterior, have the springs of the body all broken by violent action, while those of the mind have little left of their elasticity—unless, what often happens, they have become rusty from disuse.

END OF VOL. II.

APPENDIX

TO

VOLUME II.

[*For the following observations, the author is indebted to* M. FITZGERALD, Esq. *a gentleman well known to have practised long and very extensively at Madras.* Mr. F. *not only had endless opportunities of seeing the disorder of which he speaks, but himself was a severe sufferer: and it was during a residence at Clifton, in consequence of the injury his health had sustained, that the author had an opportunity of learning what an ample fund of accurate information he had treasured up during his residence in India. He sincerely hopes that the public will one day be the better for papers, which bespeak a degree of industry rare even in a temperate climate.*

The reader will perceive that the following remarks must be placed as an appendix at the end of the last volume. Should these essays wander into our East or West Indian possessions, or into the hands of persons going thither, the masterly description of the approach of the liver disease of those regions will be found particularly worthy of the notice of every European. I would not disjoin what is said of the cure, because that goes to demonstrate the necessity of early attention to the complaint.]

T. B.

NOTES ON THE LIVER ESSAY.

1. I HAVE perused with attention Doctor Beddoes's valuable Essay on Liver Complaints, and can answer from much experience of the facts, that all he has therein advanced, corresponds with the appearances observed to take place in persons, prone to the accession of this disease in the climate of India, particularly on the coast of Coromandel, where derangements of that viscus are most frequent.

Heat, and the abuse of spirituous and vinous stimulants, are certainly the great causes of the mortality occasioned by hepatitis among the officers and soldiers of his Majesty's regiments, when recently arrived in India; and as many men still continue to suffer from a total disregard of these material points, it is of the first importance that it should be generally known to the commanding officers of corps, proceeding to India, that abstemiousness, as Doctor Beddoes strongly inculcates, is certainly the best preventative of this disease; and what ranks next to it in importance as an established rule, is carefully to avoid exposure to the sun, especially unnecessary exposure.

A detachment of the royal artillery which joined Lord Cornwallis's army in the year 1790 in the Mysore country, soon after its arrival in India lost most of its men by liver complaints, in the course of the ensuing campaign; and his Majesty's 76th regiment of foot, from similar causes, continued during the whole of that war to be the most sickly corps in the field, having suffered prodigiously from hepatitis and bowel complaints dependent on disorders of that organ.

---

2. The natives of India, Doctor Beddoes may rely, sometimes suffer in very hot seasons from acute inflammation of the liver, *without the assistance of intemperance*, to which they are very rarely addicted: but I have remarked that Sepoys of the Mahometan cast, native Portugueze, and Pariahs who are prone to abuse

of spirituous potation, are frequent sufferers from liver complaints, in seasons of the ordinary temperature on the coast of Coromandel.

---

3. A very frequent mode of the approach of a liver attack in India, is by irregularity in the state of the bowels continued for several days, nay weeks. The patient complains at first of a slight purging, attended with gripes: this continues without material aggravation for several days, after which the stools become mixed with a little blood, and there is an increase of the griping immediately before going to stool. All this time the patient seems otherwise to enjoy every appearance of health, and is not alarmed at the purging, from its not being attended with fever, loss of appetite or strength, or any particular irritation about the rectum. There soon however come on slight heats in the palms of the hands, a disrelish for particular kinds of food not observed before, and a gradual diminution of strength, accompanied by profuse sweating about the breast and forehead, after every attempt at the usual exercise. In this state the patient becomes alarmed, and fancies he is about to be threatened with dysentery,

which he endeavours to check by doses of ipecacuanha and opium.

Under this treatment the griping frequency of stool, and appearance of blood, are somewhat relieved; but the purging soon increases, and a fever comes on, attended with a distressing short cough, nausea, distension of the stomach, flatulence in the intestinal canal, and total disrelish for animal food. Upon questioning the patient now, his fear of dysentery is extreme, and his desire to be relieved from the purging very great. He denies the existence of any pain in the side and shoulder, or any incompetency to lie on either side; but on placing him upon his back, with his head low, and then examining the region of the liver, by slightly though pretty firmly pressing about the left lobe towards the stomach, and desiring him to draw a deep inspiration at the same time, you find he flinches from the touch, and complains that his breath is interrupted by a pain stretching up to the right clavicle, the top of the shoulder, or shoulder blade.

He now conceives instantaneously the nature of his own complaint, and calls to recollection many instances of having felt slight pains about the clavicle and shoulder before, but considering them as rheumatic, he gave them no further attention.

It is to be observed that during the purging above described, there is, except at first, but a slight appearance of bile in the stools: even this soon disappears, and the stools become greyish, and afterwards of a lighter colour still. Upon a further continuance of the disease, however, the evacuations become slimy, offensive, and sometimes tinged with bile, though in most instances the fæculent part retains the light colour throughout.

There seems some reason for distinguishing affections of the convex from the concave side of the liver: the latter is the part affected in most instances of the disease above described, which makes it so necessary to have recourse to examination by the touch; a measure, which at least in the hot countries, should be continually recommended in all suspicious and protracted complaints of the bowels.

It is from affections of the convex side, that the pain is most acute in the shoulder and side; but the difficulty in this case, is generally in lying upon the *left* side, whereas in the other instance, it is more painful to lie upon the *right* side: there is also not unfrequently a pain in the *left* shoulder, sometimes at its top, at others under the clavicle, and occasionally in the shoulder blade. When this is the case, there is rarely any affection of the corresponding shoulder; though this remark does not always hold good, for I have known both shoulders equally affected at the same time.

There is no disease that yields more decidedly to mercury than this, when recourse is had to it, in any reasonable time after the accession; but if its nature is to be disregarded until the pain in the side and shoulder come on distinctly, which it seldom fails to do, perhaps a day or two before the formation of a liver abscess, it becomes then extremely difficult to manage, and for the most part terminates fatally. The actual formation of matter is sometimes very sudden, but is usually preceded by rigors of different degrees of strength and duration: Hiccup is also an attending symptom in this case, particularly after the stomach is distended by any sort of food or drink.

There is, however, an abuse of mercury to be feared. It manifestly stimulates the liver to a high degree, and in several instances I have known hepatitis actually *commence at the moment a ptyalism* was produced for the cure of syphilis. In other cases it is common for persons in

India, on exposure to the sun, *after courses of mercury*, to be laid up with dangerous liver complaints, who never had the smallest affection of that viscus before. It is a fact that mercury can rarely be made to affect the salivary glands, when matter has actually formed itself in the liver.

As auxiliaries in affections of the liver, it should be remarked that the pain in the side and shoulder, and difficulty in breathing, is almost always powerfully relieved by the application of a blister to the region of the liver. In plethoric habits, blood-letting is first proper, and in all cases, a lax bowel is required, previous to the use of the blister.

M. F.

HYGËIA

Essays Moral and Medical

Introduced by
Robert Mitchell
Duke University

THOEMMES CONTINUUM

Printed in England by Antony Rowe Ltd, Chippenham

HYGËIA

Essays Moral and Medical

Volume 3

Thomas Beddoes

THOEMMES CONTINUUM

This edition published by Thoemmes Continuum, 2004

Thoemmes Continuum is an imprint of
The Continuum International Publishing Group Ltd

Thoemmes Continuum
11 Great George Street
Bristol BS1 5RR, England

http://www.thoemmes.com

Hygëia
3 Volumes : ISBN 1 84371 098 6

Reprinted from the 1802–1803 edition owned by the
Rare Book, Manuscript, and Special Collections Library,
Duke University

British Library Cataloguing-in-Publication Data
A CIP record of this title is available from the British Library

Publisher's Note

The Publisher has gone to great lengths to ensure the quality of this reprint but points out that some imperfections in the original book may be apparent.

This book is printed on acid-free paper, sewn, and cased in a durable buckram.

HYGËIA:

OR

ESSAYS

MORAL AND MEDICAL,

ON

THE CAUSES

AFFECTING THE PERSONAL STATE

OF

OUR MIDDLING AND AFFLUENT

CLASSES.

By THOMAS BEDDOES, M. D.

VOLUME THE THIRD.

BRISTOL:

PRINTED BY J. MILLS, ST. AUGUSTINE'S BACK,

For R. PHILLIPS, No. 71, St. Paul's Church-yard, London.

1803.

CONTENTS.

* The clergy are very apt to be affected with hypochondriacal complaints, perhaps from an idea "that they do not occupy their proper rank in life." *Dr. George Fordyce.* Had the author seen the manner in which young men, destined for the church, live at Oxford, he might, perhaps, have accounted for the injury to digestion from intemperance; their low spirits are often lowered by anxiety and disappointment about preferment; next goes the understanding. But I protest against garbling. Let no one, who combines his own principles with mine, hold me responsible for the result.

ERRATA.

Essay ix. *p.* 5, *l.* 6, *for* have *read* has. *p.* 30, *l.* 17, *for* he *read* the patient: *p.* 36, *l.* 25, *dele* if: *p.* 42, *l.* 28, *read* as I had lately done: *p.* 60, *l.* 26, *read* interest of money: *p.* 64, *l.* 1, *read* author of the case of epylepsy: *p.* 79, *l.* 7, *read* at the age of seven: *p.* 95, *l.* 14, *for* natural *read* nocturnal: *p.* 96, *l.* 21, *for* any *read* an: *p.* 98, *l.* 3, *read* it would be nothing singular though: *p.* 104, *l.* 9, *for* no *read* not too: *p.* 106, *l.* 4, *for* days *read* evenings: *p.* 108, *l.* 2, *read* awake: *p.* 110, *l.* 15, *for* scarcely *read* no such: *p.* 113, *l.* 14, *for* or *read* and: *p.* 121, *l.* 9, *for* and *read* but: *p.* 136, *l.* 16, *after* do *read* not: *p.* 148, *l.* 10, *for* shews *read* shew: *p.* 166, *l.* 9, *read* when the mind and body are most easily injured: *p.* 175, *l.* 3, *after* in *read* not: *p.* 185, *l.* 8, *dele* all: *p.* 190, *l.* 11, *for* readily *read* suddenly: *p.* 191, *l.* 28, *for* women *read* mistresses.

Essay x. *p.* 5, *l.* 7, *after* of *insert* that: *p.* 36, *l.* 23, *after* for *insert* justly: *p.* 40, *l.* 15, *for* untutored *read* less tutored: *p.* 56, *l.* 18, *to* unusual *add* ly: *p.* 70, *l.* 27, *for* hypochondriases *read* hypochondriacs: *p.* 82, *l.* 2, *after* on *read* the.

Essay xi. *p.* 55. Change the references to the two French notes: *p.* 65, *l.* 3, *before* than *read* other.

In the essay on stomach complaints, there is a blunder respecting cakes made without flour.—The *Echaudées*, which consist of much white of egg with flour do not turn sour in some stomachs, on which bread does.

ESSAYS

ON

THE MEANS

OF

AVOIDING

HABITUAL SICKLINESS,

AND

PREMATURE MORTALITY.

ESSAY NINTH.

Printed by MILLS, St. Augustine's, Bristol.

Quodsi qui mentes hominum verà sapientià ac virtute imbuere student, præter cætera cogitare etiam debent de impedimentis aut adjumentis quæ varia corporis constitutio adferre potest: haud dispar ratio exigit ut illi quoque, quorum officium est corpus humanum ad sanitatem dirigere, quid animi in illud potestas addere aut opponere suis conatibus valeat, diligenter considerent.

If those who labour to imbue men's minds with true wisdom and virtue, are, among other things, obliged to pay attention to the helps or hindrances, arising from this or that condition of the body, there is equally good reason why those, who have to conduct the body towards health, should carefully consider the influence of the mind on their endeavours.

GAUBIUS, *on the Management of the Mind.*

ESSAY

ON

THE NATURE AND PREVENTION

OF SOME OF

THE DISORDERS,

COMMONLY CALLED

NERVOUS.

GENERAL IDEA.

IN every part of the study of nature, it is common for ancient distinctions to vanish, as a greater number of objects become known, and as every object is more compleatly investigated. The intervals are gradually filled up, and productions, once held the most dissimilar, are seen to gain a degree of affinity by the successive modification of the particular features of the extremes in the members that fall in to compleat the series.

Proofs of this observation may be found in the attempts which have been made to distribute diseases into classes: and I hope to afford a striking, as well as an useful confirmation of its truth, in the course of the present essay.—In these later ages, an infinite number of experiments have raised the brain, the spinal marrow, and the nerves connected with these, into a degree of importance which they did not possess in the estimation of ancient physiologists. It has been found that where the nerve going to a limb is cut, tied, or injured, the empire of the will over that limb is destroyed, and applications to it produce no sensation. What more natural therefore than to impute loss of motion and sense to an altered state of that portion of the animal structure, of which intentional alterations are followed by the same disability to feel and move?—Again, pinch, prick, or otherwise irritate a particular nerve, and the muscles, into which that nerve passes, shall be thrown into irregular action. When the same irregular actions manifest themselves without the interposition of any experimenter, it is simple and reasonable to infer that the nerve suffers as when irritated on purpose, and that the cause of the diseased or irregular action lies in the nerve. Here therefore we have a foundation for a distinct class of diseases. This class gains extension from other phænomena. That compression of the brain, which accidents perpetually produce, and which sometimes the want of the scull or bony cover in the human subject has enabled observers to effect at pleasure, have demonstrated that a certain condition of this principal member of the nervous system is an indispensable requisite to perception, loco-motion, and to the other exertions, characteristic of the healthy, waking state. The anatomy of the organs of sense in different states has proved how essential their nerves are to the proper performance of their offices. If the optic nerve be entire, other parts of the eye may be removed, and vision remain. The apparatus provided for conducting the sonorous vibrations of the air to the auditory nerve, may be totally destroyed, and yet sounds may be conveyed through the teeth. But hearing ceases, or is impaired, upon any disorganization of the nerve of hearing. It is in vain that the other parts continue perfect. To these direct proofs, it may be added that a vast quantity of nervous substance is appropriated to each sense, and that it is distributed in a manner, which always raises admiration when exhibited. The nicety of the organ also corresponds to the apparent care bestowed in the distribution, as is particularly manifest on comparing the interior of the nostrils in animals which differ in acuteness of smell *.

These observations give us two additional kinds of complaint, properly to be denominated *nervous.* When the hearing is impaired and the auditory nerve altered in its structure, we have *nervous deafness*; and *nervous dimness of sight* or *blindness*, under similar circumstances of vision and of the optic nerve.

From slight numbness up to utter insensibility, and from premature loss of power to total immobility of a part, there are many gradations; but the whole appear to be steps in the same progress. Hence the first come to be referred to the same head as the last. And in fact corresponding causes, capable of acting on the nerves, are often discovered in the slighter affections. Mineral poisons make the limbs tremulous first, and afterwards motionless. Fear and anger are, every day, observed to make persons *quake all over.* They are occasionally observed to bring on perfect palsy. Indeed, common language marks the analogy of these affections. One

* Harwood's Comparative Anatomy.

has often been called the *shaking*, the other the *dead palsy*.

So far the deduction appears to proceed satisfactorily. We may follow it yet another step without danger of being puzzled. Supposing a complaint, in which a considerable part of the symptoms shall be, from the first, exactly such as some of those above enumerated, and another part of a different nature and not by any certain experiment or observation, referable to that particular part of the organization, yet we may venture to term the whole disorder nervous. We do this with more confidence, if the symptoms of each description occur together or alternate.—Hysteria may be taken as an example. The same hysterical person shall tremble at one moment, and become blind or deaf, or lose the use of a limb, at another. Then a sneezing or retching of the most excessive violence shall supervene. Not only so; but the paralytic limb shall be restored, in the twinkling of an eye, to its full powers, upon the commencement of sneezing or retching; upon a fit of laughing or crying, or the rising of a ball in the throat, or upon the occurrence of any other of the symptoms, which shew themselves in this infinitely diversified complaint. Now, having referred the universal trembling and the partial deadness to a certain change in the state of the nerves, I refer to the same constituent parts of the body, the suffocating ball in the throat, the retching, the sneezing, and the other symptoms, though I may not be able to excite them by any operation upon a dissected nerve. But they appear to be intimately connected and interwoven with the others; and indeed seem to make up one disorder, as different strands are twisted into one rope.

But from this point our track becomes more uncertain. By extending the term *nervous* upon the strength of one or two circumstances of resemblance, we are in danger of losing all meaning, and reducing the most heterogeneous affections to one head. It will therefore, if we would keep clear of embarrassment in language, be necessary to have a word entirely new, or some addition to that which we have hitherto retained, in some cases where nervous symptoms occur. For there occur also symptoms of a different character; and as these predominate, they require to be marked. In the beginning of the small-pox for example, and also of some feverish complaints without eruption, there shall be irregular action of the muscles, or *convulsions*. Along with the convulsions too, there shall be other appearances exactly the same as in diseases strictly called nervous. Yet the small-pox passes not either with nosologists, or with those that follow popular language, for a nervous complaint. The reason is obvious. The nervous symptoms do not generally appear; and in hysteria, if there had been only the retching and the ball in the throat, probably it would never have been considered as a proper nervous disease. The strong pulse, heat, and violent affection of the breathing in bad cases of small-pox, are more common, permanent, and dangerous. They would therefore probably give the disease its denomination, and occasion it to be referred to the *inflammatory* class, did not the pustules offer so striking an appearance, and furnish the matter, by which it may be propagated.

There are certainly no diseases of whatever name, in which the nerves are not affected; and there are probably no nervous diseases, in which the other parts (if they can be considered as destitute of nerves) are not affected. But to call only certain complaints *nervous*, may enable the writer to be more concise, without being less perspicuous, which is the proper use of general terms. I shall therefore, according to the foregoing explanation, continue to employ this word for disorders, of which some of the most remarkable symptoms can be imitated by applications to the nerves, and for others appearing akin to these from unconstrained analogy. They are disorders, in which the limbs move irregularly or without the direction of the will, in which the organs of perception suffer, and the intellectual functions are disturbed; and this, in most instances, without any preceding or concomitant symptoms of a different nature, either general or partial.

It is not easy, and perhaps not possible, to invent a term, which shall point out any one set of disorders to the exclusion of others. But, in common life, there would be much less confusion and somewhat fewer mistakes, if the term *nervous* were not indiscriminately employed, both where there exist or have existed symptoms of a different and peculiar nature, and where there do not. A young person in consumption feels herself low, and is hysterical in various ways; she is therefore persuaded that the cough is *nervous*: that is, she and her friends persuade themselves there is no danger, or that a treatment very different from that likely to answer is requisite. A gouty veteran becomes tremulous throughout his whole frame. Hence he concludes that he labours under opposite or antagonist maladies; and that if the gout should gain the ascendancy, it would deliver him from his other enemy. But, in the generality of cases, though it be true that the patient is nervous, it would be more exact to consider him as nervous, because he has been gouty. He is only secondarily nervous.

According to a vulgar phrase, when such an one is said, and truly said, to have a *complaint of the bile*, the physician can frequently ascertain, to which of two opposite states the given case belongs. For in some the liver secretes too much bile, in others too little. Perhaps, some such opposition may be hereafter discovered in the variety of nervous complaints. The convulsions, to which children are so prone, and the loss of power in old age, would appear to arise from contrary states;—the one from excess, the other from defect of nervous energy? But nothing of this kind can be generally established. Our present acquaintance with the subject does not entitle us to divide the family of nervous ailments into two branches;—one *positive*, and another *negative*. In the same day an hysterical person shall be troubled with painful acuteness of hearing, and with deafness, and be in the intermediate or natural state. The limbs shall be palsied one hour, and the next, a strong propensity to motion shall impel the invalid to place all the chairs and tables in a circle, and to skip from one to the other, till quite out of breath. Poetical effusions shall follow a fit of stupidity in a person who is neither dull nor brilliant in her ordinary mood.

In all departments of knowledge, the nomenclature must be as the information. Where one is deficient, the other will be vague. In medicine it will, I fear, for a long time, be equally easy to object to a general term, and difficult to replace it by another, free from just objections.

Peculiar Perplexity respecting NERVOUS AFFECTIONS.

All diseases, nay, every sort of injury, will, on strict consideration, be found to affect the mind. Some difference in the inward state will correspond to the slightest external difference. I shall have no difficulty in shewing by and by, that disorders, which, like the lumbar or psoas abscess, proceed to a considerable head without being in the least suspected, must affect the individual as a rational and sensitive being. But in many instances the bodily change is so much more permanent and considerable than the mental, that this is overlooked. In a pleurisy, the history of the disease and of its treatment, is thought to be fully comprehended in an account of the pain, of the pulse, of the breathing, of the number of bleedings and so forth. In this and other febrile affections, it is deemed sufficient just to note whether delirium occur, or at most, to mark whether it be a delirium of the low or the frantic kind.—Again, the parts concerned in many diseases have been so often examined by dissection, that the physician can read the state of the interior as plainly as if the body were a book, the alteration in organs accessible to sense, and the patient's account of his feelings, serving him for letters.

In nervous complaints, every thing is different. Changes of the organization, it is true, have been sometimes detected in nervous complaints: but then the same symptoms equally appear, when no such changes exist; so that here morbid anatomy is at present of little or no use in pointing out the correspondence between the obvious and hidden workings of the machine. The intellectual faculties bear a considerable part in these mysterious and often terrific scenes. The members, long since reduced under perfect obedience to the will, throw off all controul. In ruder ages, it was a general persuasion that the furious agitations which sometimes take place, were still voluntary, but proceeding from the will of a malignant spirit, who had possession of the patient, and managed his members as the master of a puppet-shew does those of his diminutive representatives of man. Among the lower ranks of society there may still be found some, who are haunted by the same pernicious belief.

For these irregularities of thought, feeling, and action, which constitute the essential character, if it be possible to assign the essential character, of this class of disorders, we cannot hope to account, without some insight into the laws of the human mind. What terrors there are inherent in this subject, different people will probably disagree in attempting to determine. But one thing is clear. It is a subject of enquiry equally shunned by the faculty, and by those who are not of the faculty. Yet it is obvious that the investigation of these laws bears exactly the same relation to the diseases in question, as the comparative anatomy of sound and morbid bodies does to many other diseases.

This neglect accounts, in some measure, for our unequal advances in the art of medicine. We continually see improvements made in the treatment of fever, of scrophula, syphilis, consumption, diabetes, and sore legs. Public rewards are given for methods of destroying contagion and exterminating the small-pox. But scarcely does even a rumour of any aid for man, when he is suffering in his higher powers, stir abroad. The attempts made for his relief have nothing in them *rational.* At least they are not guided by any researches into the nature of his sufferings. So that if we are destined to find only in proportion as we seek, we shall have a remedy for cancer a hundred years before we arrive at any tolerable arrangement of the appearances attending mental alienation, or are provided with a set of safe rules for the medicinal employment of the passions.

Considerable discoveries have however been made in pneumatology. If grown people do not take as much pleasure in becoming acquainted with the operations of their own understanding, as children in informing themselves concerning the instincts of animals, is it not in part because the former have not yet had the advantage of being delineated by a BUFFON? But however it may be to be wished that some luminous and elegant writer should take up the subject of the human mind, we need not remain in a state of inactivity till that happens. Some principles appear to be equally easy of exposition, and capable of application to the design of this tract, namely, to a practical elucidation of the nature of a harrassing set of diseases.

The following sections will serve to shew how far this opinion is well founded. For the sake of perspicuity I shall select one of the most grievous of nervous maladies, as an example of the whole class. Of this I shall first briefly describe the obvious appearances. I shall then enter more at large into its effects upon those operations, with which none but the individual himself can be acquainted. Last of all, I shall endeavour to point out the difference between the healthy and regular manner in which these operations are carried on, and the disturbances to which they become liable. The result, I hope, will turn out very fertile in applications to complaints of familiar occurrence, and which, though separated in our technical books, are yet nearly allied according to the process of nature. For this we are provided with the most ample and authentic materials, persons affected, in the midst of their distress, having had the fortitude to try to decypher, and the patience to transcribe, their whole interior. This illustration will be found to extend to analogous complaints. For it is remarkable that the same invalid will run through almost the whole range of nervous symptoms. Thus we have seen that the hysterical will become occasionally paralytic. The paralytic will no less become occasionally hysterical.

EPILEPSY.—Name, Definition, and obvious Appearances.

No other disease is so strongly marked as a subject of terror by the language of mankind. From the idea that it was a punishment inflicted upon offending mortals by angry divinities or dæmons, it acquired in ancient times a title much like that of Attila the Hun. Long before the consternation excited by the ravages of that barbarous leader, the epilepsy had been considered as the *scourge of God (lues deifica: morbus divinus)*. It was likewise called the HERCULEAN complaint, an appellation which medical etymologists are puzzled to explain, but which was perhaps derived from the difficulty experienced in subduing it, and the violence of its attack, somewhat as particular medicines have been named *heroic*. In token of abhorrence, and perhaps * too as a charm to keep the agitating dæmon away, the by-standers used to spit upon the epileptic. Hence it is mentioned by Plautus and other ancient writers, as the *morbus insputatus*; *morbus qui sputatur*. Among the Romans it produced so much consternation, that the popular assemblies or *comitia* were dissolved, if any person present had a seizure. Hence we find it designated as the *comitial disease*. The afflicted were relinquished to their misery by their friends, as if they were objects of guilt. Connected with this impression, we find the terms *scelestus* and *sonticus* applied to the disorder. Its present accepted name has nothing superstitious at least. For it is derived from the suddenness of its onset. But were any attention paid to its etymology by those who use it, this name would appear on many accounts objectionable. Its onset is often not immediate: and under such a denomi-

* Despuimus comitiales morbos, hoc est, contagia regerimus. *Pliny*.

nation it stands apart from affections, to which, as we shall afterwards see, it owes its origin, with which it is apt to alternate, and from which, in fact, it differs but in degree.

It is no wonder, however, that before its source and affinities were known, this circumstance should have been among those which principally fixed attention. Indeed in the list of its hundred names, it would have been very natural to find the *tyger-disease*, from the propensity of that formidable animal to spring without notice upon its prey. And what in reality can be more terrifying, and more calculated to humble the spectator by inducing a sense of the helpless and precarious condition of human nature, than to behold a person, to all appearance stout and healthy, in the midst of his pleasures or amusements, deprived instantaneously of the command of his limbs, of his reason, and the use of the senses; with all his muscles rigid at one moment, and the moment after actuated by the most furious convulsions, while the saliva, worked into froth by his breath, oozes from his mouth during the continuance of his agitations? There are many people for whom the horror of such a sight is too much. They are obliged to turn away from it, and that under pain of being seized in a similar manner—an occurrence which has sometimes taken place.

Those who can bear to view the phænomena of an epileptic paroxysm attentively, will see the patient either fall upon the ground without any previous effort; or if he happens to be on his legs, he will suddenly rush two or three steps forwards, or turn round in a circle, and then fall. All to whom nosologists allow the term *epileptic* to be applied, become insensible at the moment of their fall. Frequently they utter a shrill cry, of which none but those, who by great efforts have gained an unusual power of self-observation, retain any remembrance upon the resolution of the paroxysm.

No one account will apply to different epileptics; nor indeed are the observable circumstances of different accessions in the same case absolutely identical. The most remarkable are the *spasms* or fixed contractions of the muscles; and the *convulsions*, or the strong contractions alternating with relaxations. From spasm of the muscles under the covered skin of the head, the hair stands an end as in fright. The eye-brows are often drawn down and brought nearer as in vehement anger. The eyes appear prominent and fixed, all from extreme rigidity of the muscular fibres that move these parts, insomuch that the medical spectator has sometimes been led by the expression of the countenance to mistake the effect of a beginning paroxysm for passion. The neck swells so much that there would be danger of suffocation, if the collar were left unbuttoned. The head is sometimes drawn backwards and the body bent like a bow. At other times the head comes forward, and the chin is fixed down to the top of the breast bone. Sometimes the neck has been so twisted that the chin has come almost upon the point of the shoulder. The same stiffness extends to the extremities. The thumbs in particular are brought forwards into the palms of the hands.

Convulsions are more common than spasms. Dr. Cullen, whose definitions of diseases have perhaps more authority than value, after some older writers, makes epilepsy to consist in convulsions and insensibility (*convulsio cum sopore*). The fixed contractions either cease for the most part after the fit has continued a short time, or the fit consists principally in agitations without rigidity. The scalp and the forehead are then apt to work incessantly. The eye-lids, though nearly closed, are in a state of perpetual tremor. The eye itself will roll with the utmost wildness. The other muscles of the face are no less in motion. The lips are pointed so as to form a beak, and the next instant drawn so forcibly back as to bring the corners of the mouth near the ears. Boerhaave obeserved this species of convulsion so rapid in a Jewess, as on being attentively observed to occasion giddiness in spectators of great sensibility.

The jaw shall be either locked (and that so forcibly as to break the teeth) or it shall be so convulsed as to be thrown out of joint. It is then that the tongue is in danger. For being protruded in the general orgasm, it gets between the teeth and is severely wounded, and in some cases almost chopped in two. The foam, as it works out of the mouth, being reddened by the blood, adds greatly to the terror of the scene. Some observers have found the smell of this discharge insupportably cadaverous *(Tissot sur l'epilepsie)*; and others notice an acrid odour, which renders it impossible to remain in the chamber without opening the windows, though this does not appear to take place in consequence of the fit, and therefore has probably nothing to do with the disease—*(Doussin-Debreuil de l'epilepsie, Paris, an. 5.)*

The nerves having as much influence on

the secretory organs as on the muscles themselves, no wonder that the secretions should, on some occasions, become excessively offensive. I know two persons, in whom the breath turns instantly disagreeable on receiving any chagrin. This depends on a sudden change in the action of the glands of the mucous membrane of the chest.

The head is tossed to and fro with the utmost velocity by the numerous and powerful muscles of the neck. The trunk and limbs are thrown into every species of movement, of which their structure renders them capable. In children the feet have been seen bent with so much force that the great toe has come nearly in contact with the heel.

The internal parts are no less convulsed. Sobbing and hiccuping, which are convulsions of the diaphragm, accompany epileptic fits as also violent retching. The bowels work with equal violence, as is proved by a tremendous rumbling noise. The bladder contracts with almost inconceivable power. Its contents, have been remarked in children, to form a jet of ten feet. Other evacuations take place in consequence of the involuntary action of the muscles, upon which they depend.

It has been asserted that epileptics, during the accession, will emit sounds resembling the voice of many different animals. It might seem as if this was a circumstance super-added from the imagination of the terrified spectators. Nothing, however, is more true. Witnesses by no means superstitious, and perfectly calm, attest the fact. And if we consider that no opera-dancer can come up to the contorsions of the larger muscles, we shall not wonder that those of the larynx should be affected by the most extraordinary agitations, and these will produce tones, equally extraordinary.

Parts of the skin frequently look black from a sluggishness of the circulation of the blood. Sometimes the discoloration becomes permanent from an effusion of that fluid under the skin; and this, if I do not mistake, by no means merely from the bruises which the patient may receive from his own hands during the epileptic commotion; but from bursting of the vessels, or dilatation of their mouths. I believe there is a species of pulmonary hæmorrhage, purely epileptic. I have at this moment under my care a patient, who has long discharged large quantities of blood from the chest, without apparent injury to the lungs. This hæmorrhage alternates with most violent and long continued retching, not excited by coughing or by any thing offensive in the stomach; and there are, at other times, fits of insensibility, attended by a locked jaw with convulsions, and going off, as I shall immediately describe those of epilepsy to do. Loss of blood has been frequently noticed from the stomach and bowels, of which the muscular fibres, from the internal noises often heard, seem to be convulsed like the external; and there are phænomena proving distinctly enough that particular cavities, destined only for the occasion.. reception of blood, are sometimes gorged with it at the time of the fit.*

After a certain number of seconds or minutes or hours (for nothing can be more indeterminate than the duration of the paroxysm, though from fifteen to twenty minutes has been reckoned an ordinary average period) the agitations begin to subside, and the fixed contractions to relax. The respiration becomes less hurried, and the patient gradually recovers his perception and consciousness, very commonly amid deep sighing. The accession in some instances leaves so little vestige of disorder that the patient cannot imagine that any thing has

* Constat insultibus epilepticis priapismum sæpenumero se addere comitem.

happened to him. But in general he finds himself excessively heavy and languid for hours and even occasionally for days. Every fibre of his body will tremble for a time afterwards; and a simple fainting-fit occasionally comes on. Sometimes he seems to sleep profoundly, and awakes with a dread ful head-ache. There is not unfrequently a difficulty of swallowing, a squinting or blindness, or stammering, left behind by the seizure. The intellect itself scarce ever holds out against frequent or severe returns of epilepsy. The memory is weakened; the apprehension grows dull. The organs of sense are impaired. The complaint seldom lasts long without destroying all cheerfulness, inducing a disposition to solitude, and at last is succeeded or accompanied by idiotism, apoplexy, or madness.

Infants are more liable to simple convulsions, but still sufficiently subject to proper epilepsy. This however is a complaint, of which the origin belongs rather to youth than infancy. The epilepsy of infants often ceases in the first four or five years. The later the complaint begins, the more difficult of cure is it generally reckoned. When it occurs in early youth, medical men, on finding their prescriptions unavailing, sooth the patient and his friends with the idea that a natural cure will take place at the period of puberty. To relieve present distress by pointing out a brighter prospect may be excuseable on the ground of humanity; and a concern for the dignity of his own character is apt enough to prompt the physician to exercise his ingenuity in providing his patient with a consolation, when he is obliged to leave him without a cure. But it would be difficult to understand how the doctrine of the spontaneous cessation of epilepsy at the time of puberty, could have found its way into the writings of so many medical authors, did we not know that the most outrageous falsehood comes to be believed by dint of frequent repetition. Women past the prime are flattered in their turn with the idea that this and many other disorders will cease upon the next change of constitution. And it is surely natural that the afflicted should credit and promulgate such welcome predictions! They fail however, and bring their author into discredit. But a new medical prophet arises, and he is listened to with equal credulity.

There are persons of mature years, who experience convulsions with unconsciousness, once in their lives, and never afterwards are attacked. When some of those exciting causes for example, which will presently be noticed, operate upon a particular occasion with extraordinary vehemence, a fit will be produced without the formation of a diseased habit. In the usual cases, that is, where they do return from time to time, nothing can be more irregular. Sometimes there are several in one day; sometimes one in a week, or in a month, or in a quarter of a year. Instances have been remarked where they supervene at stated times, corresponding to the phases of the moon, as at every change, at every full or both.

It seems to be agreed among observers that boys are equally liable to seizure with girls. But as to its comparative frequency in the two sexes after the æra of maturity, there prevails a difference of opinion. Some of our own most experienced practitioners of medicine represent the disorder as most common among men.* The contrary representation is found among physicians of the largest practice, and among the most learned

* Quanquam pueri et puellæ juxta corripiuntur, fæminæ tamen rarius quam viri in eam incidunt. (*Heberden Commentarii*, 142.)

compilers, on the continent*. The customs of different countries may perhaps go some way towards accounting for the difference. Females have probably, from the peculiarities of their habit, a stronger predisposition. Where vinous liquors are too freely used, and where disappointments in life and mental depression are very common, men may well be more apt to forfeit the advantage of their firmer constitution. The former circumstance is not foreign to our situation; and in no country perhaps has the pursuit of gold occasioned so much anxiety; to which must be added the contention of mind, excited by other objects, and the adventurous spirit, that leads to many dangerous accidents.—We shall be reduced to conjectures on this and many other still more interesting questions, till medical practitioners shall either voluntarily resolve, or or be obliged by law, to preserve, in some public record, what experience teaches them

* Le temperament des femmes est, en général, plus foible, plus mobile que celui des hommes; et je me suis assuré par ma propre pratique que le nombre des femmes epileptiques est plus considerable que celui des hommes. Tissot, p. 38.—La difference se trouve dès l'age de sept ans. p. 38.

Children, and women especially when pregnant, are more disposed to epilepsy than adults and males.—*Sprengel's Pathology* iii. 331.

individually concerning the frequency and production of different complaints. Upon such documents might be constructed tables similar to those of births, marriages, and deaths, from which the state of the public health could be deduced; and it is surely much more desirable to know this, than the mere chance any one has of continuing in existence, whether it be a happy existence or otherwise.

Expectation from the drugs hitherto brought into use, diminishes in proportion to the opportunity of trial. A late imperfect catalogue of anti-epileptic remedies, with the titles of the works, in which their effect is described, occupies an hundred and fifty pages*, and scarce one of the number is used by any considerable portion of the faculty with the smallest confidence. I have known the nitrate of silver—a medicine lately introduced, or rather revived among us, by a physician of great respectability—administered in twenty cases without succeeding once. Indeed instances of success are so rare in proportion to failures, that it must remain uncertain whether accident or nature does

---

* *Friderici Henning analecta literaria epilepsiam spectantia. Lipsiæ*, 1798. 4to. p. 107—156.

not produce the benefit, though it sometimes happen under the employment of medicines. The accessions, at least, we know will sometimes terminate of themselves. The lengthening and shortening of the intervals without any assignable cause, increases the difficulty of judging. Dr. Heberden mentions a patient, who became more severely and frequently epileptic again after an interval of thirteen years. Stahl speaks of a young man, who had a fit in his earliest infancy, and remained afterwards quite free till his eighteenth year, when he was seized upon being awaked by his master at three o'clock in the morning, and continued to be regularly attacked at the same hour once a month.

The just dread of so terrible a disorder, and the difficulty of its cure, are motives equally obvious and strong for leaving nothing unattempted for its prevention. But it will, I think, further appear evident, that great numbers are placed in the mid-way between firm health and epilepsy, though they may never be exposed to causes capable of exciting the exact symptoms, referred to this head in books of physic. Each approximating state, however, is attended with a multitude of dismal sensations, and indeed occasionally renders almost every instant of life comfortless. A long course of imprudence is often necessary to effect this unhappy change in the organization; and it is certain that many persons proceed without suspicion of the point to which they are tending.

Interior State of certain Epileptics before the Disorder is fully formed.

When the intervals are very long, and the onset happens without warning, the intermediate state may be either little different from that of persons in health; or, if otherwise, we may have no sufficient means of ascertaining the difference. We must therefore have recourse to cases, in which the complaint itself originally comes on by degrees, and the single fits are announced by definite feelings. From these only can we hope to determine that irregularity in the springs and wheels of the animated machinery, which answers to the violent disturbance in the movements of the external parts. What occurs in other cases may well be supposed similar, though it pass with too great rapidity to admit of observation. Or if this reasonable supposition should be rejected, the explanation will not only hold to a certain extent in epilepsy, but apply to a great many valetudinarians besides, which circumstance more than doubles the interest of an attempt to analyse this particular complaint.

It has sometimes been observed that particular impressions of terror remain long in early life, with those who are destined to become epileptic, and that these impressions are revived in sleep and in solitary situations. The same is exactly true of those who suffer from nervous head-ache and other affections, allied to epilepsy. One shall believe himself followed close by a lion or a tyger, from having heard some frightful adventure, in which these animals have borne a share. It will be sufficient if the idea of these animals cross the mind, when it is depressed by fear from any other cause. Indeed the feeling and the idea do not usually occur originally together, but become connected in a manner to be hereafter explained. A patient, whose journal of his illness seems to me to give a clearer representation of the ebbs and flows and eddies of the mind in epilepsy, than all the writings of medical men * put together, relates that, about his seventh year, he was

* Physiche & physiologische Geschichte einer siebenjahriger epilepsie. 1798.

told that he should see an apparition the succeeding night. He was at the same time conjured, under pain of death, to keep the communication secret. He was quit for an afternoon of horrible expectation; for his parents, perceiving his agitation, prudently forbore to force him to a disclosure, and carried him to sleep in their own apartment. From that time a confused image established itself so firmly in his mind, that in his eleventh year, whenever he fell into a general dry heat, it would occasionally awaken him out of his sleep. Through the whole period of his adolescence, upon becoming hot from the small-pox, measles, (which he had at 16) or from any other trifling indisposition, the same image occurred to him in dreams. Of the composition of this image, he remembered at 33 that a waving flame formed part. Secret imprudences *, though they were not

* "My good parents had provided me with a common-"place tutor. But alas that I could have had a Theophron "for my companion and monitor! All the influences that "under proper guidance would have mildly fertilized the "warm soil of my constitution, broke out into wild "inundations, which tore down every dam, and only "retired within bounds at that period of later than legal "majority, which we do not attain but by reflection on "the mischief we feel we have done ourselves."

continued for a long time, added to the injury occasioned by this impression, which was further aggravated by exertions in a public office, made with a view to promotion in the state. In this employment, an unusual difficulty of composition was sometimes experienced. During particular efforts of attention or imagination, there was, within the head, a sense of something moving upwards on the right side. After long sittings, or in company, a weariness of mind frequently came on. At this period, which was eight or nine years before any epileptic attack, he was subject to frequent sneezing, but this by no means took place to any unpleasant degree. Marriage succeeded, and in less than four years the birth of three children. The labours of the cabinet were still persevered in, and within about half a year before the first accession of the disease, more than one subject of continued chagrin occurred.

By degrees, without any considerable pain in the head, a weight and indistinctness began to be sensible. It a good deal impeded thinking, and still more speaking and hearing. At many moments the memory refused the commonest words, and the organs of speech withheld their service altogether. On making efforts to force the expression, he stammered many times, or one syllable was drawled out after another, and this not above a sixth as fast as usual. Becoming sensible of this incapability beforehand, the expedient of remaining silent and hemming for a couple of minutes, while the cloud was passing over the head, was adopted; and this was found sufficient for rallying the powers.—At the time of such qualms, it was often impossible to collect the sense of the plainest thing uttered by another person, even with the utmost attention. The situation resembled that of a person who hears a language, of which he is but little master, fluently spoken. The words had an obtuse, muffled sound: and yet the ear had acquired so much internal sensibility, that the beat of a drum near at hand, and even the falling down of childrens' little houses, built of pieces of wood, or a rough human voice, were productive of pain. The mornings were more uncomfortable than the evenings; but this inconvenience was obviated in some measure by a more substantial breakfast.

If sleepiness, which sleep does not remove, a frequently returning sense of intoxication, independent of any intoxicating beverage or drug; unsteadiness upon the legs, and actual slight staggering have often preceded epilepsy for many months, or even for years. In some of these cases of gradual approach, patients have remarked to me that they had perpetually been accustomed to misplace the strokes of the letters in writing, and almost daily to use one word for another, bearing some resemblance in sense or sound or frequently occurring together, but immediately corrected themselves as soon as the sound struck their own ear. For example, *every body feels very languid this* WET *weather — I mean, this* HOT *weather :*—or, *come who will sit down to supper ;—here is only cold meat and pudding ?—I mean,* PYE. These blunders, and many similar, I have known come from a person previous to epilepsy. The propensity to commit them is much greater fasting. A nutritious meal will add equally to the firmness of step and certainty of tongue.

Some experience an incessant restlessness, like that of the wandering Jew. In whatever place they may happen to be, they are seized with an irresistible desire of removing to another, where they hope to find comfort. But no

—— resting finds the sole
Of unblest feet——

This restlessness is usually the symptom of a

serious disorder already formed, as hypochondrasis, or the fore-runner of one still more serious. At the slightest, it shows a thorough derangement of the sensitive organs.

The mind partakes of the same sort of inquietude. Going through a train of close reasoning is felt as an undertaking absolutely impracticable. Indeed, to dwell upon any one thought steadily is a task; and a task too that can only be gone through at long intervals. Some acute observer has remarked of the great king of Prussia that his conceptions were quick, but that on contemplating a subject, he grew confused. Whether it be true in the particular instance or not, the observation holds of many individuals in the way to become epileptic. They are generally those, who somehow or other, have tampered with their sensibility. They seize a question dexterously. But their strength is exhausted in the first assault. If you try to make them grapple with a difficulty, they immediately flinch. To any proposition requiring them to contemplate a number of ideas stedfastly, they will yield a flat unintelligent assent; or to mask their want of bottom, as I think the jockeys term it, they will endeavour to fly off to another topic, and so on perpetually. To conceive the condition of the head in such cases more distinctly, we may recollect how it fares with the eye, when weakened in such a manner that the instant it is cast upon an inscription, the characters are perfectly plain, but that in a little time they seem to run into one another, then become undistinguishable, and at last vanish altogether. From misconduct of the understanding, all frivolous people must be troubled with some flightiness of attention. We need no other reason to enable us to understand why it becomes requisite, in polite circles, to change the topic of conversation every second minute.

Flashes of light before the eyes, headache of various degrees, and occupying different parts, violent rushings as of blood towards the brain, dizziness, and a great variety of unpleasant sensations are felt by persons, who afterwards become subject to epilepsy. The actual occurrence of these symptoms, and particularly of excessive sensibility, with transient diminution or suspension of the intellectual powers, giddiness, and faintness I should esteem more certain indications than any particular colour of the hair, eyes, or skin. Human beings have not such fundamental differences, but that a single strong unfavourable impression, or a succession of slighter ones will deprive any temperament of its original advantages.

Accession of Paroxysms, and State in the intervals.

The artists of antiquity compounded their portraits by selection of features from many individuals. On the same principle the describers of diseases make up an ideal form; and this, though it may not correspond to any succession of symptoms, past, present, or to come, in the case of an individual patient, is designed to represent them all. It may be questioned, perhaps, whether even in writings for professional men, a well-chosen series of cases would not more than make amends for its prolixity by greater instructiveness and fidelity. On the present occasion at least, a *generic* description would not, I think, be entitled in any respect to the reader's confidence. To serve the purpose of analysis, the account must necessarily be minute; and the more particular the delineation, the greater danger of introducing false strokes without perpetual reference to a prototype in nature. Moreover, language has not yet been adjusted with any degree of exactness, to our inward feelings. Hence medical reports, where these feelings come in question, stand a double chance of inaccuracy. The invalid, with whom the representation must originate, may express himself ill; and the physician may misconceive him if he takes him simply at his word, or by trying to help him out, may substitute his own ideas. How little then can we depend upon generalisation of such insecure data! I think it therefore better to depend on the labour of the before-mentioned intelligent individual, whom seven years practice in studying and describing himself with all the assistance he could derive from physicians, must have carried to a greater perfection of talent than we can expect to find elsewhere. I have, however, checked his narrative by examination of others similarly affected.

The attacks were not of any unusual kind. Between August 1788 and 6th January 1795, there were sixty-five fits in all. At this period, the absence of concomitant sensations, and the suspension of the fits for above three years persuaded the patient that he was cured. The first fit is thus briefly described. "August 14, 1788, after a disturbed, debi-"litating* night, and after working harder

---

* Hoc nomine designatur pollutio nocturna.

" than usual in my office till one o'clock, I " played two games at billiards, a thing I " had not for some time done. This strained " my eyes a little. In going home, I still " remember to have looked steadily in at the " window of an acquaintance to ask him " how he did, and soon afterwards to have " lost myself in thought and in a dreaming " weakness. Twenty steps further on, I was " found senseless—It was said that I uttered " a frightful cry."

The second attack occurred on the 22nd of October—" At one o'clock, I took only " half a meal at home, intending to make " myself amends somewhat later at court, " where foreigners were expected. Instead " of going thither, however, I attended a ' conference from two o'clock till eight. " The vivacity of the conversation engaged " me too deeply, especially as I forgot to " take my quassia or any thing else. In this " state I hastened to court in foul weather, " where unfortunately it was ten before the " trumpets sounded for supper. I mistook my " heavy languor for mere sleepiness, and " remember to have entered with the train " into an apartment, where I lost my senses " as suddenly as lately in the street. Accord- " ing to the credible report of those who

" hastened to my assistance, I roared twice
" before I sunk. Afterwards convulsive mo-
" tions were remarked. Awaking as out of
" a profound dream, I returned apt answers
" to various questions, though they betrayed
" entire ignorance of what had occurred to
" myself. Nor did I quite come to myself
" again, till I had vomited undigested the
" spinach I had eaten at one o'clock. It
" was now eleven. - - - - The 3rd and 4th
" days after I was still crippled in the loins. The
" belly and stomach were distended, and I
" had distinct pain on being touched under
" the ribs."

July 19, 1789, half an hour after a second breakfast, there came on an attack so suddenly that "I could not speak, and had but
" time to seize my servant by the arm, and
" seat myself quickly on my bed. So far I
" proceeded voluntarily. As soon as I had
" sate down I perceived many images, that
" had occurred to me during the last days
" whisk backwards and forwards in my head.
" Amid these I lost myself. My servant
" says, that I lay about 2 minutes on the
" bed, looked at him steadily as if I had my
" senses but could not speak, then suddenly
" cried out aloud (but did not bleat, as I
" had done in a fit on the 16th of December

"preceding) and repeated the cry several "times in a sinking voice—that at the first "cry I stretched both hands stiff along my "sides, drew my thumbs inwards, fixed my "jaws together, with my lips open, beat about "with my feet, threw out large bubbles of "foam at my mouth, and had my eyes "distorted—that I looked black about the "eyes and lips, that I continued to bend "my head more and more back, and that "my neck and chest were quite stiff;—that I "then raised myself with force, the convul-"sions continuing in the legs, but the body "being less stiff, and that I made efforts to "get off the bed, on which he could with "difficulty bring me to lie down; that he held "*eau de Cologne* to my nostrils, and without "being conscious of what I did I pushed "the bottle away, closed my eyes, which "had all along continued open and distorted, "and that while some of this spirituous "water was rubbed on my temples, the "colour of life returned to my lips and "countenance. But a deep sleep continued "for a quarter of an hour longer. At last "I opened my eyes, could see but dimly, "or speak but as in a dream, with little col-"lectedness. My memory was quite lost. It "no longer, for instance, occurred to me

" that I had made an engagement to break-
" fast at the Crown inn, and I repeatedly
" asked *where the Crown was?* of which my
" servant talked, and *where I was myself?*

These particulars will I suppose satisfy the most scrupulous medical writer that the case was truly epileptic. Respecting the other seizures, therefore, I shall only notice some of the more remarkable observations, the account of the state at the intermediate times being, in my opinion, more interesting.—Dec. 31. " a debilitating night—much
" sensibility to impressions next day.—In
" the evening, a shivering came on from
" washing in too cold water; and in a
" minute afterwards, a feeling of contrac-
" tion in the head; but not the usual viva-
" city of hurrying ideas. I rubbed some aro-
" matic water on my head, but in vain. I
" placed myself on the sofa close at hand,
" gave my servant the signal agreed upon,
" and immediately lost all consciousness.
" The scream, distortion or turning up of
" the eyes, and convulsions followed."—March 10.—On thinking too deeply on some moral propositions when the stomach was empty, had a seizure, with but little of the moving images, though there was notice enough to close the book he was

reading, and to ring for the servant.—August 25.—After superintending a solemnity for five hours, and speaking much, though without apparent exhaustion, could not "get the Turkish music out of his vibrating head" and in five hours after had a seizure.—October 18. After speaking too earnestly, "the revolution I felt in my "head, and of which I knew the consequences, instantly waked me. I was not "able to speak at the time, but had presence of mind left to give notice with my "voice, to swallow cold water, and take "snuff. Amid these efforts I gave up to "the inability, and laid myself on my left "side as if to sleep. Then consciousness "immediately forsook me."—December 30. "Observed the confusion rising in the head; "sprang back to the cup-board, in which "my cordial liquor was kept at night. By "standing still and reaching out for it, I "perhaps committed the oversight of not "struggling against the fit with my hands "and feet. Not finding the liquor, I sprang "back to the bed, and still made the reflection that I might hurt myself in falling "against the snuff-box, which was in my "night-gown. So I placed it safely on a "small ledge. I had but three steps to the "bed, and reached it head foremost. At "that moment followed the loss of con-"sciousness and the bleating sound."— — Jan. 14 and Feb. 1792, was waked out of sleep by the confusion of the head; and the last time was not sensible of the flitting images.—June 18, exertion in conversing—at night, unpleasant feelings in the head—the same on awaking—which continued ten hours—a very offensive evacuation—and in spite of putting the feet into warm water, rubbing, elixir of vitriol, a fit took place "while losing myself in thought and amid "lively pictures of the fancy, which at such "hours I have less in my power. — — — "The cry though protracted was not heard "in the house: but it seems remarkable that "I heard it quite plain myself, in conse-"quence reflected that a relapse was "at hand, and lost all inward sense."—Sept. 19. "Soon after supper I remarked a "great sensibility to all noises. Between "eight and nine a relapse occurred rapidly, "and with the usual hurrying images. — — "After the convulsions, being raised up and "assisted, I walked tremblingly, and with-"out consciousness, about the room. This "I learned afterwards, for I had no recollec-"tion of the circumstance, and I then sunk "into the sleep, which usually had followed "the fits without intermission."—Aug. 29, 1794, a seizure immediately preceded by ideas of a kind that had not occupied him before. They hurried, as it were with violence, across the mind.

Shortly before the invasion of compleat epilepsy, and during the intervals of the paroxysms, there were other species of nervous seizures. These occurred very frequently—not less than seven thousand times between August 1788, and the 6th of January 1795. The following extracts will give an idea of their nature, and of their relation to the full-formed epileptic paroxysms.——"There are moments, wherein my ear finds "no reasonable sense in the words that are "spoken. At the same time, my organs of "speech are incapable of expressing a thing, "which I nevertheless perfectly conceive. "But frequently the ideas also are clouded "in the same manner. I then endeavour to "stop all intellectual action, and to recruit "myself by quietly refraining from think-"ing. Still according to my feeling some "single nerve continues to work on in my "head, and carries me, as in a dream, or in "a state of intoxication, through obscure "visions.—June 4th, 1789, I was waked

" before I had had my rest out. Indeed, for " a short time previous, I had only slum- " bered. I was in a dream at the moment, " and my fancy continued too busy after " waking. All at once I felt, while lying " in bed, that suspicious crusade of a num- " ber of ideas against one another, which " has heretofore preceded the most violent " attacks. My sight was weak and dizzy, " and a little commotion in the empty " stomach took place at the same time. I " therefore turned from the back on which " I was lying to the left side, closed my eyes, " and in a few minutes this little attack was " over — — July 20, 1790. On an airing I " disputed a moment with warmth. On " returning home, a number of people " pressed to speak to me, and I heard of " intelligence that had been just brought " by the post.—I went to bed not suffi- " ciently composed, and at six o'clock was " debilitated; soon after this I awaked, and " had a stupor which lasted four hours.— " When this was over, I had a sensation " almost of pain over the right temple and " was disagreeably affected, when any little " thing in the room fell — — — In dictat- " ing a letter at four o'clock, I had the

" dangerous stupor which precedes loss of " consciousness. I dispelled it however by " rubbing, tickling, and vehement scraping " with my feet. My wife observed that at " the moment of stupor I entered the cham- " ber with a very flushed face.—8th June " 1793 I awaked, after a good night, with " a clear head. But as I was putting on my " cloaths in bed, which on account of the " time taken up in wiping away the perspi- " ration was not very quickly done, a phrase, " which the day before had struck my at- " tention, recurred; and during frequent " *involuntary* repetitions of this phrase, I " lost the rein with which I am, as it were, " obliged to hold in my thoughts, and im- " mediately a dangerous stupor came on. " The approach was rapid, and the apparent " congestion up towards the head as con- " tinued as when a compleat epileptic pa- " roxysm follows, though this has gone " off without further effect.—The spasm in " the head seems, at this period of the dis- " order, to be so far diminished, that during " the time, I can not only find the proper " name of many objects (and this is the test " to which my understanding, still con- " tinuing unimpaired, immediately resorts), " but I could also enuntiate them distinctly "enough. Therefore the organs of speech "were not so much affected.—At the same "time I had such morbid sensibility of the "extremities, that a shivering came on in "that leg, which was still under the bed-"cloaths, but without a stocking. By dint "of rubbing, shaking, kneeling, and with "the help of the smelling-bottle I got "through, but in the succeeding minutes "was very languid and breathless."—17th March preceding. "To-day the bowels were "not sufficiently open; this however was "fully remedied by a pill—after the opera-"tion of which I sate down to rest myself "with a comfortable feeling in my head, as "I believed, when a stupor came on with a "hurry of ideas, that lasted a couple of "minutes. Yet it was rather the protracted "vibration of the string once put in motion, "than any variety of images, my under-"standing remaining clear, and suggesting "instructions by signs to the persons pre-"sent how to assist me."—March 22. "I "went to bed feeling well: but at three in "the morning being sensible, while slum-"bering, of something amiss in my head, I "was waked by the continuing stupor. I "immediately took preventive essence, walked "about, and applied cold water to the head.

"In this manner the stupor went on for half " an hour, but was less severe than the one " preceding. For I had now no nervous " shivering, and at short intervals could " utter words in succession, though between " whiles involuntary ideas came forward, as if " broke loose, yet this time they were tolerably " connected; and twice during the stupor, " spasms in the head began to be alarmingly " strong."—June 3. "After a reasoning " dream, a stupor came on, such as I never " afterwards experienced in the night. I " felt, the whole day through, slight stupors, " and not only had in the evening, shortly " before falling asleep, a stupor somewhat " more severe, but also at the very begin- " ning of my slumber a momentary convul- " sion, or twitching of all my limbs."

December 1791. "Two dangerous stupors, " from which however I rescued myself by " long and incessant struggling with hands " and feet. It is singular that amid these " confusions of the head and the hurry of " involuntary ideas, the intellect should " remain clear, hovering as it were in a " region superior to that, in which the other " ideas and images are tossed to and fro, as " in a storm. In dangerous stupors I can " still regulate my external actions by my "judgment—as for example, I can alter the "position of my body, reach for snuff or "anodyne liquor, and at times save myself "from a fit."

13th and 17th July 1790. "In the night "debilitated—on the 14th tolerable, as I "was very cautious. But towards the even-"ing too much speaking and writing. At "night — — — after getting into bed with "the idea—*thou wilt have no alarm this* "*time*—I actually had one of the most dan-"gerous kinds of stupor. Reasonings, which "I could neither distinctly comprehend nor "retain, chased each other, and I was cer-"tainly very near a compleat relapse. But "very hard rubbing, applied in time, brought "me a little to myself. The spasms had "advanced so far that my pulse was "hardly sensible, and the extremities quite "cold — — — — —

"On the 17th my feet, on awaking, were "not warm as usual. I could do little to "restrain the hurry of my ideas, and had "the stupor (incapability) of speech and "ear. It did not quite go off from five to "ten o'clock. Then I had a somewhat "freer hour, but the head remained heavy "and ached a little.—From 11 to 12, I "slumbered, and even then the ideas at

" first shot rapidly through each other. But
" I awaked more composed, and in the
" evening found myself well."

" As soon as the ideas get into commotion, I try to ascertain the degree of danger by my ability to utter words of more or less easy pronunciation. If I fall on such as I cannot articulate, a sense of anxiety appears to come on. This renders my situation worse, and stirs the ideas still more quickly about. On the contrary, if I fall on an object, whose name I can speak, the nervous irritations seem to be allayed. — — — I have provided myself with a dozen short, easy words, the object corresponding to which may instantly strike the eye, and thus contribute to the fixing of my mind — — —

" I have several times since been confirmed in the efficacy of this method of fixing the ideas. But on the 4th of August 1790, after dispelling a stupor in this manner, another very alarming supervened in the evening, before I fell asleep. On looking at my hand, it was not possible for me to think of the words HAND and SKIN, much less to pronounce them. The mind grew more and more clouded, as is the case a moment before an entire loss of conscious-

"ness.—I was merely capable of stroking my "head downwards, while two persons rubbed "the spine hard. The critical state lasted "about a minute, when I perceived it was a "little better with me. I constantly kept "turning one hand backwards and forwards "before my eyes to elicit the idea (of the "words) HAND and SKIN. In a minute "more I succeeded in gaining them very "distinct, and immediately afterwards in "pronouncing them. My night was good.*"

---

* It is curious enough to find that different classes of ideas have a different force of connection with the organs of speech. A paralytic patient could not, with all his efforts, read aloud from his book. But he could fluently repeat the words of the same book, when first uttered by another. Is the association between hearing and speaking the less easy of the two to be destroyed, because first formed and oftenest repeated? If an instructed deaf and dumb person were to recover his hearing by any miracle, and then to fall into the situation of the paralytic patient, he ought to be able to articulate from a book, and not by ear, the impressions of his eye having been earliest connected with the movements of his vocal organs.

In epileptic threatenings, where the struggle between the diseased and the healthy actions is difficult, the pronunciation of easy words by the assistants may contribute more to excite the organs of speech than bringing the object to be named before the eye. It will appear below that the recovery of speech will probably stop some fits.

Among the most remarkable seizures I reckon that which stands reported under date of Feb. 2, 1790. "During a meditation, by "no means deep, several involuntary thoughts "got loose in my head. I not only had a "stupor, but immediately apprehended some-"thing worse. So I went hastily from the "inner into the outer chamber; not being "able to speak, I hemmed aloud by way of "signal to my people, and laid myself on "the sofa. Notwithstanding a pulling of "nerves connected with the hurrying ideas, "as strong as had ever yet preceded a con-"vulsive fainting, and stronger than any that "had passed off without it, I had never-"theless so much obscure consciousness as "to determine, this time, to resist, and "to stroke myself from my face down-"wards. My wife came to my help with "aromatic water and rubbed my back. While "this was going on, I lost all consciousness; "and on coming to myself, enquired *whether* "*I had uttered a scream and had convulsions*, "and received the consoling answer that "*I had not*. I was told besides that I had "been but a few minutes in all upon the "sofa, and had stroked myself without inter-"mission. I could therefore have been only "for a very short time overpowered by lan-

"guor. The pulse was in good order: and "neither hands nor feet, and nothing but the "very tip of the nose and fingers, cold. The "face heated a very little indeed, and it had "not been particularly pale during the "seizure. I afterwards found myself not "extremely weak: I moved about somewhat "less than half an hour, and felt only a small "weight on my head, which went off after "a tolerably composed night." This degree of affection shews how nature ranges wide of the definitions of nomenclators.* On the

* I shall explain myself fully hereafter. But not to be misunderstood, I must here observe that I do not agree with the Brunonians in proscribing nosology. Yet it would be very desireable to bring the medical nomenclature nearer to the actual uninterrupted series of aberrations from the healthy state. Nosology has been founded in inattention to these aberrations, and when prized above its desert, it fosters this very inattention, than which nothing can more obstruct advances in the cure and prevention of disorders. I have repeatedly endeavoured to explore the space between a just performance of the functions and pulmonary consumption. I would fain do something of the same here with regard to another set of complaints. A good deal is to be found to this purpose in medical authors, respecting dropsy. There the deviations are marked and gradual. Who will bring to light the first tendency and gradual progress towards diabetes mellitus, towards calculus, madness, and twenty other complaints?

30th March 1789, a seizure nearly the reverse took place. In that, last described, there was a slight loss of consciousness without convulsions. The following exhibits slight convulsions with scarcely an entire loss of consciousness. "I am subject to single "convulsions or startings in sleep, but not "more considerable than is usual in plethoric "people. But yesterday on placing myself, "when I was somewhat heavy, on the sopha, "I fell nearly into a slumber, but was alarm- "ed out of it, every minute or two, by a "convulsion sometimes in one part and "sometimes in another. In the evening "before supper I slumbered again and had a "stronger shock than before, which passed "as it were through the brain. But in this "there was nothing like what occurs in my "stupors. I rather found myself alert at "at the moment, and slept quietly through "the night, except that I had a single fit of "starting."

These extracts will be illustrated, and the account of the epileptic phœnomena rendered more compleat by the following catalogue of the sensations described at large in this valuable seven years' journal.

Stupor. Stumpfheit. By this term is understood—*defect of hearing and of speech, with heaviness.* The author describes himself as having experienced three gradations. These he calls 1. the slight daily stupor—2. the tedious, and 3. the alarming stupor. His feelings bespeak them to be all alike "epileptic ebullitions;" for the state immediately preceding an actual fit was precisely the same as during the alarming stupor; and he is of opinion that this, without opposition from art or some accidental favourable circumstance, would always have passed into a paroxysm. The first compleat attacks in 1788 and 1789 were not ushered in by such rapid approach of the heaviness as occurred several times afterwards, when the danger was repelled. They amounted to epileptic faintings under the quiescence of the inexperienced patient. And after the approaches of spasm had become slower, the attacks by day, better admitting the use of counteracting means, did not go beyond dangerous threatenings, and most of the compleat paroxysms grew out of stupors, that came on shortly before or shortly after sleep, or in a dream. The subsequent feelings, when the danger had risen very high, consisted in a slight internal thrilling or tremor which diffused itself through all the members in a couple of minutes. Sometimes a little distension was felt in the head. Sleep seldom followed, and only 7 or 8 minutes repose was necessary to recruit the frame—the unpleasant consequences of the slight stupor were merely more inclination to heat in the head or to a repetition of the same, when he could not keep still for a short time afterwards. They came on oftener when he lost himself in thought or wrote or read or spoke, long or with warmth. Their number in a day was from 2 to 4, and once 8. The tedious stupor was the slight one continued.

Dr. Spalding, the celebrated divine of Berlin, who was never epileptic, gives an account of an attack, very similar to these stupors with hurry of ideas. It may be worth while to compare the reports of two such intelligent self-observers. That of Dr. Spalding was written on the very day it happened, viz. Jan. 31, 1772. He had to speak to many people in quick succession, and to write many trifling memorandums concerning very dissimilar things, so that the attention was incessantly impelled in quite contrary directions. He had, last of all, to draw up a receipt for interest. He accordingly sate down and wrote the two first words requisite, but in a moment became incapable of finding the rest of the words in his memory, or the strokes of the letters belonging to them. He strained his attention to the utmost in endeavours leisurely to delineate letter after letter with constant reference to the preceding, in order to be sure it suited. He observed, and said to himself, that they were not the right strokes, without being able in the least to conceive wherein they were deficient. He therefore gave up the attempt, and partly by monosyllables, and partly by signs ordered away the man who was waiting for the receipt, and quietly resigned himself to his state. For a good half hour there was a tumult in part of his ideas. He could only recognize them for such as forced themselves on him without his participation, He endeavoured to dispel them to make room for better, which he was conscious of "in the bottom of his thinking faculty." He threw his attention, as far as the swarm of confused intruding images would permit, on his religious principles, and said to himself distinctly that *if by a kind of death, he was extricated from the tumult in his brain, which he felt as foreign and exterior to himself, he should exist and think on in the happiest quiet and order.*

With all this there was not the least illusion of the senses. He saw and heard every thing about him in its real shape, but could not get rid of the strange confusion in his head. He tried to speak, for the sake of finding whether he could bring out any thing connected; but however vehemently he strove to force together attention and thoughts, and though he proceeded with the utmost deliberation, he soon perceived that unmeaning syllables only followed, quite different from the words he wished. He was as little master now of the organs of speech, as he had before found himself of those of writing. "I therefore" says he, "contented myself "with the not very satisfactory expectation, "that if this state should continue, I should "never all my life be able to speak or write "again; but that my sentiments and prin-"ciples, continuing the same, would be a "permanent spring of satisfaction and hope, "till my compleat separation from this im-"portunate ferment of the brain. I was "only sorry for my relations and friends, "who in this case must have lost me for "duties and business and all proper inter-"course with them, and looked upon me "as a burden to the earth. But, thank "God! this melancholy prospect did not "last long. After the completion of the "half hour, my head began to grow clearer "and more quiet. The uproar and vivid-

"ness of the strange, troublesome ideas "diminished. I could now carry through "my process of thought. I chose subjects "for reflection from my own fund with less "interruption from them, and somewhat "more regularity. I wished now to ring "for the servant, that he might request "my wife to come up. But I required yet "some time to practise the right pronuncia-"tion of the requisite words. In the first "conversation with my family, I proceeded "for another half hour slowly, and in some "measure anxiously, till I at length found "myself as free and clear as at the begin-"ning of the day, only I had a very trifling "head-ache.—Here I thought of the receipt, "which I had begun and knew to be wrong. "Behold instead of *fifty dollars for half an* "*year's interest*, as it should have been, I "found, in as clear and straight strokes as "I ever made in my life—*fifty dollars through* "*the sanctification of the bri-* with a hyphen, "as I had come to the end of the line.—I "could not possibly fall upon any thing in "my previous ideas or occupations, which, "by any obscure mechanical influence, could "have given occasion to these unintelligible "words."

This will throw light upon the *hurry of ideas*,

so repeatedly occurring, which the author describes, as image after image, chacing each other, like the trees, houses, and animals by the road, when one drives fast ; or like the figures in a magic lanthern. The understanding, he says, was a passive spectator, retained its consciousness and occasionally felt surprize, and even anxiety, at the situation. "In saying "*occasionally* I allude to the total absorption "in thought, which sometimes intervened for "a few seconds, and even for a whole minute. "Do such different mental operations occur at "the same time, *as it seemed to me*, or were "they successive with the velocity of light-"ning?—I dare not absolutely presume to de-"cide this point, from my own experience.

When he speaks of *being lost, bewildered in thought*, he is not to be understood of this hurry or tumult of ideas, coursing each other, but of such attachment to one object, and often to a single phrase as to forget every thing else about him.

Reasoning dreams are regular though strangely wrought up scenes, not different from the lively dreams of a person in health, but they proved injurious to this particular patient and received an appropriate name in his journal, because they did not awaken him, as the involuntary quick succession of ideas did, out of his illusion. When the dreaming invalid makes entire harangues in his sleep, overcomes difficulties with honour, or performs an adventure agreeable to the current turn of his imagination, the operation strains the weak nerves of the head too much. In this particular case, the sensibility of the head and the number and vivacity of these dreams declined much upon obtaining better sleep by artificial means and acquiring more vigour.

In the *waking dreams* the bodily and mental powers sink and sometimes vanish. This seems to precede the common paroxysms when the attack happens to go that length ; otherwise it is termed *weakness* : "on these occasions, I often asked "myself, *art thou only dreaming ?* or is that "which *thou so faintly seest or hearest, actually* "*present ?* I was only sensible to a degree of "uncertainty of what I said myself. Years "before my epilepsy, I was liable to fall into "a similar state. I read many passages in "reports of my own composing, without "apprehending any clear meaning ; and while "the disorder was forming, I had weak moments when my voice sunk, during long read- "ings, to an unintelligible stammering. This "was the prelude to the subsequent stupors, "and it was the accession of one of these very "waking dreams that, in 1788. for about a "minute, preceded the first attacks of the then "unknown malady."

The *headache* was never considerable in this case, as it so often is, but there were two species of sensibility or soreness of head—the one diffused itself over the crown. It arose from cold in the night, and came after the fit—the second seemed to lie deeper in the brain, and was more on the right side, on which the author usually fell in his paroxysms.

Throbbing was common on reclining the head to obtain a respite from troublesome feelings and at lying down in bed. General external heat of the head did not always accompany the throbbings nor *vice versâ*.

Loss of speech, as we have seen, belongs to the state of stupor: but a wrong word was also frequently uttered; and this had no connection in sound or sense with the word intended, nor was the blunder immediately perceived.

The ear. In 1787, common tones often had a muffled sound, and a moderate jarring or screaming voice gave pain. A thickness of hearing sometimes did and sometimes did not accompany this sensation. A confused noise, from a multitude of men even at a considerable distance, produced a sort of intoxication. These morbid affections of the ear, the author considers as one of the first steps towards stupor, and it is also a just description of incipient *nervous deafness.*

The external ear was very subject to burnings —sometimes it was cold to the touch ; scarce ever of a due temperature. The right ear was most frequently heated without the other, and sometimes the left. " The sensibility of these " parts is so extreme that I not only internally " feel the heat, that is about to penetrate into " the ear,* while that is yet cold to the touch, " but the smallest breath of warm air, nay, even " single hairs coming in contact with the ear, " will make it burn;" an hour generally passed before the application of cold air or cold water could reduce this partial flushing ; and lively conversations at table, or exposure to the wind, bring it on still.

The *power of attention* was so weakened, that the author was obliged to make short pauses of rest, in conversation and business. But these were generally sufficient to enable

* This, if I understand rightly, is a phænomenon, common to all chilled parts, exposed to a much higher temperature. When the hand, chilled by snow, is held to the fire, the hotache begins, while the outward skin, especially the tips of the fingers, feel cold to a person in the natural state.

him to go on. During the decline of the complaint in 1793 and 1794, the memory did not proportionnally improve but rather appeared weaker than from 1789 to 1793. It was not, however as in the first years of childhood. Intelligence and expressions were not wanting pending any occurence, but impressions had little duration. The author was obliged to fly on account of the irruption of the French into Germany. In his flight he met with many scenes of a very striking nature, which shortly after appeared to him only as so many faint dreams. He could not recall the names of couriers whom he had very frequently seen, and the ideas of his own transactions vanished very soon. In 1795, his first year of convalescence,* the traces remained longer, except of what he heard, read, or thought during a dreaming weakness. In the second year of convalescence he, sometimes but more rarely, lost the thread of his discourse at one moment, and found it offer itself spontaneously the next; whereas at a former period, it must have been sought with dangerous efforts of attention or altogether in vain.

On the approach of the disorder and at the

---

* The last fit, it may be remembered, was the 6th of January, of that year.

period of the first fits, the perusal of poetry and poetical attempts, which were resorted to by way of salutary dissipation, had the reverse effect, for they excited a dangerous agitation of the nerves.

Attempt to explain the principal of these phænomena, and to apply them to the purposes of prevention.

I flatter myself that the reader begins to have some insight into the nature of epilepsy, and that it now occurs to him under a very different view than heretofore :—if he has merely chanced to see an invalid lying helpless upon his back with rigidity of one part of his muscles, and twitchings of the other, foaming at the mouth, black in the face and with the pupil of the eye drawn up into the head, and if he has turned away in horror and tried to drive the scene out of his head as soon as possible. I wish that the following observations may contribute to encrease the interest he may have taken in the subject, and put him upon his guard against the causes of those semi-epileptic qualms, with which more, I am afraid, than three out of ten in every genteel circle are not unfrequently overtaken.

There is no one that has not heard of

found out from his own experience, that association of ideas is a property of the mind. In fact, nothing is so easy as to perceive that the actions and feelings of the animated machine begin to settle into groupes from the moment it is possible to make observations upon it, and, no doubt also, from the first moment of its origin. If an infant be gently touched in the palm of the hand, its fingers will grasp the body, by which it is touched; that is, the impression on the skin is attended with a certain feeling (probably a pleasurable one); and this feeling occasions a contraction in the muscles, by which the hand is closed. The sense of hunger gets connected with the muscular fibres of the lips and of the œsophagus, all of which contract in succession so as to carry the food from the mouth into the stomach. Long preparation, we know, is necessary to establish the series of efforts, that are concerned in so common an action as walking, and to judge by the tedious apprenticeship, and the industry exercised from first to last, talking also must be an art of difficult attainment. Indeed, many of us never acquire this art in any considerable degree of perfection.

Impressions, ideas, pleasure and pain mix

and alternate in an endless diversity of ways, as every one must be sensible in himself and in others, though the order that presides over these phænomena, has not been, by any means, completely investigated. Indeed a compleat investigation would not only comprehend the whole history of human nature in all its states, but disclose the means of maintaining the system in its most desirable known condition, and perhaps of exalting it to a higher.

The union between the several sorts of animal action is confirmed by degrees. The child goes on to move with a firmer step, and articulate in more distinct tones. In like manner, the ideas, derived from the senses, attach themselves to each other more stedfastly. Each individual in his progress from infancy, finds himself in a capacity to call up a greater number of thoughts, in consequence of the occurrence of any one thought, just as he is able to move his limbs or his organs of speech for a longer time together at one certain period, than he was at a period preceding.

In all individuals, (unless such as happen to plant themselves within range of a cannon ball, or to be cut off by some acute disease) but more distinctly in some than in others, the reverse process takes place. There is commonly observed a second childhood as well as a first: and the observation is very much founded upon the circumstance in question. The old man totters like the infant. He is not able to call up the same number of thoughts he once was, in consequence of the occurrence of any one thought; and at some certain period he will find the groupes of ideas he is in a capacity to call up, and the trains of actions he can perform, sensibly diminish. This takes place universally in the natural course of existence, but in some cases an unusual degree of penetration or self-observation is required to discover it. In other cases, it strikes the most careless. A child to talk his best, must be still, and to walk his best he must be silent. So it is with decrepit old men, and with younger ones inclining to be paralytic. The readiest arithmetician existing cannot, perhaps, multiply so many figures together, when galloping full speed, as when reclining on a sofa. Yet he might probably come, by practice, to exert the muscles necessary to sit a horse at Newmarket and to work figures by head at the same time. Till the period of incipient decline, it is the effect of the natural employment of the powers, that each exerts itself more strongly, and that a greater number can be exerted at once. I do not now enter into any of the various disquisitions which belong to the knowledge of the improvement and combination of the intellectual and animal faculties. Such details would be little suitable to the purpose of these Essays, because as information and curiosity go at present, they would not be agreeable to one of their readers perhaps in a hundred.

Now whether we advert to the whole progress of epilepsy, or to a single paroxysm, we shall find the preceding observations, applicable to the phænomena. The same may be asserted as to the formation and exciting causes of the complaint.

The invalid suddenly falls or becomes previously uncertain of his movements. Now any one who will procure a set of anatomical plates, and make himself a little master of the figures exhibiting the muscles, may easily satisfy himself as to the concert of actions, which it is requisite to maintain, in order merely to keep the body erect; and as to the nice reciprocation in the contractions and dilatations in the several sets, concerned in every kind of motion.

Among the circumstances contained in the above extracts from the account of the interior feelings during the approach of the disorder, as also on many occasions in the intervals between the stupors and the paroxysms, and at the time of these two varieties of accession, some will appear merely so many interruptions of successive actions, or the absence of a part of such as used to be synchronous. To this head belong the gradual falling off of the pronunciation during the public reading of a state paper to inarticulate stammering;—the invalid's inability to comprehend those of his own composition, while he was reading them, (for here though the association between the visible impression, made by the writing, and the organs of speech remained, the ideas which at other times the sound or sight or the words, would have called up, now did not occur);—the still more distressing want of power to proceed from the visible impression of the *hand* to the pronunciation of the few letters, which constitute the symbol of this member, and even sometimes to recal its visible or audible idea, when it was held before the eyes, and made its proper impression on the organ.

To endeavour, in the manner already described, to insure the associations at the time when a fit threatened, was a rightly directed effort towards preventing the processes of the system from running into irregularity. A more effectual plan for this purpose will, no doubt, form part of the treatment, when we practitioners of physic come to direct our measures according to a right conception of the nature of the complaint *. It appears from accidental occurrences that whatever binds the animal actions to their habitual succession, puts off the paroxysm. Dubreuil relates of a porter at Paris, who was his patient, that having one day a very heavy burden on his back, he felt himself all at once in the crisis of a paroxysm, to the vio-

* The historian of his own case, so frequently quoted, proposes a separate institution for the cure of epileptics. He quotes an estimate, where the number of such sufferers is made to amount nearly to ten thousand in Germany alone. He adds that there is good reason to fear " if the enfeebling " of mens' bodies and the severe exertions, physical and " mental, of bodies so debilitated, should stretch any con- " siderable way into the nineteenth century, lest the disorder " should also grow more frequent."—The disorder prevails exceedingly in England, In a country where there are 10,000 compleat epileptics, we must multiply by a high term to find the number of persons similarly affected, and, in consequence, more or less incapacitated for being happy in themselves, or of service to their neighbours.—I know not if such institutions would be as much incentives, as they would afford opportunities, to explore that internal world, in which we medical men are such strangers. If so, there could not be imagined a better reason for their establishment.

lence of which, however, he opposed so much resistance as not to fall, which he did at other times. "I went" said the porter, "from "the point of St. Eustachius to the com-"mon sewer of Montmartre, without being "able to communicate my situation to my "wife, who was all the way beside me. I "could neither see or hear; and yet some-"thing which I cannot explain, but of "which I have a sort of remembrance, "seemed to direct me in my march, and "made me feel the importance of the ar-"ticles committed to my charge: My wife, "who could not possibly have any suspicion "of my state, because I still went on, does "not recollect that I struck against any "person on the way, although we met a "great number." (p. 22.)

Here we see a partial suspension of the animal processes, while a certain part went on uninterruptedly. The muscles, employed in bearing a burden acted perfectly, while those of the voice refused their usual co-operation with the ideas. This occurrence too in a person where the suspension usually extended to both kinds of action, contributes with many other examples of the same kind, towards shewing how affections, that excite much less horror, graduate into epilepsy.

There are very few animal motions that do not stop or run into disorder for a time, and then return to their usual rate, while the rest go on in the way to which they have been trained by long habit. Hence the varieties of imperfect epilepsy become innumerable. I have now under my care a patient aged 52, who for four years has had, and continues to have, numerous paroxysms about the full and change of the moon, in which her hands become violently convulsed, with insensibility, fixed eyes, foaming mouth, locked jaws, which in closing have severely wounded the tongue. Yet sometimes on her accessions she walks into the house, if abroad, or seats herself if within, and is said never to have fallen. Her fits are preceded for a day, by excessive distension of all the veins, and they are followed by the involuntary whirl of images, formerly described, with consciousness. The images, she says, frequently consist of ships and scenes at sea, on which she has never been. But a son caused her infinite anxiety by entering on board the navy against her will.

I lately saw a boy, who after being greatly terrified, lost himself almost every half hour in the day. He continued on the spot where he was seized. He had his eyes turned up, his features fixed, talked with so great rapidity, that it was not possible to catch above a syllable now and then of what he said; he sometimes whistled. All the time he preserved his equilibrium perfectly.—The case of another was stated to me, the course of whose ideas, in consequence of the same cause, came frequently to be suspended for a couple of minutes without convulsions or stiffness. Exactly the same attacks alternate with perfect epilepsy. Tissot saw a girl very healthy till the age of 7, when she was terribly frighted by a storm on the water, and alone of her whole party was not sick. Some days after she had convulsive movements of the eye-lids, and in four months severe and frequent epileptic fits. In the interval came on seizures, marked by slight movements in the eyes and loss of consciousness, which cut short what she was saying. Sometimes on recovery she finished her phrase, and at others had forgotten what she was going to say (l. c. p. 12.)—Tissot also saw (Sept. 1, 1769) a lad of 15 who, a few days after a fright, was seized with sudden speechlessness without the loss of any of his senses. He had a wild delirium, a countenance expressing fear, haggard eyes, and a large livid tumour between his eye-brows. The fit had come on the day before, when he emerged weak, pale, and oppressed with terror; and "it is "beyond all doubt," says the reporter, "if "he is not quickly cured that he will become "epileptic" (l. c. p. 175).

Mr. Ritter relates of his nephew, that at 7, on seeing a person in a nervous paroxysm (of what kind he has not specified, though probably epileptic), he fell into the same himself. In two years he recovered from these fits: and the year after became subject to a sort of lethargy, so that whether standing or sitting, he would fall asleep. In this state he would converse and name the things held before his eyes, though they looked shut. On being roused, he knew nothing of what had passed in the lethargic paroxysm; but he soon relapsed and then you might continue the conversation, which was going on with him in that state before. When roused again, he knew of what was said in his previous wakefulness, and so on. "He "seemed to have two souls, one for waking, "and another for sleep." (*Moritz Erfahrungseelenkunde* II. 2. 69.) The relation between the impressions on the senses, and the motions performed in articulation were not destroyed; but it was otherwise with the muscles, whose contraction keeps the body upright, for Mr. Ritter informs us that he always sunk, unless supported.——The following is the only way to get at an exacter knowledge of such seizures, that is, to learn their relation to one another, and to the average healthy state, for as to what they are in themselves, there is not at present the smallest hope of attaining so far. We must content ourselves with drawing up tables of the sets and courses of action, which are arrested or altered. We must learn what we can of the previous and subsequent feelings; and to make our information serviceable, we must not only explore the immediate exciting causes but get at the whole history of the patient. And in all this no regard must be had to traditionary names.

In a set of tables, shewing the various ways in which our actions may be disjoined from their dependance upon our ideas or upon one another, we should meet with cases, exhibiting a part of the symptoms described above, without the least tendency to spasm, convulsion or sleepiness *(sopor)*, which are made the characteristics of epilepsy.

The case of MOSES MENDELSOHN, the great Jew metaphysician,* deserves as much

* He is accounted nearly the finest writer of his language. His tract on the immortality of the soul is translated into English.

attention as any which I have met with.—It has been described both by himself and Dr. Bloch, the well-known ichthyologist; the account therefore must be received as both authentic and correct. Mendelsohn was a person so engaged in business as to be regularly confined almost all day in a compting-house. He was actuated by an intense desire of knowledge. He gratified this curiosity; and his extraordinary acquisitions were the sole fruit of his nocturnal studies. He was constitutionally weak; but his temperance enabled him for a long time to hold out under his very severe application.—At last he fell into a nervous illness, the attacks of which always happened on his awaking from a disturbed sleep. At these times, he says, "he was perfectly sensible, could "pursue any train of thought he chose with "perfect order and clearness, but was totally "incapable of voluntary motion. He could "neither stir a limb, nor utter a tone, nor "even open his eyes. Every effort he used "for these purposes was not only fruitless, "but increased the very disagreeable sensa-"tions that accompanied the inability. He "felt at the time as if something red hot "flowed from the brain along the spine, or

" as if his back had been scourged with red " hot rods. He was therefore obliged to " keep quiet till some impression arrived " from without, and, at that very moment, " he became perfectly master of all his " voluntary motions." (*Moritz*, l. c. 11. 3. 68.)—" Any thing" observes Dr. Bloch; (*Bemerkungen*, p. 61.) " which acted upon " either of the five senses restored him. A " sound, the bite of an insect, a light*, took " away the immobility of the limbs." These paroxysms were attended with anxiety and palpitation, which disappeared along with the immobility upon the slightest irritation. The whole of the following day was passed however in that state of headache and dejection, which so commonly succeeds a compleat epileptic paroxysm. And when, thirteen years afterwards, an increase of his mercan-

* I have had an opportunity of enquiring very minutely into the case of three ladies, who upon the accidental going out of the light that burned in their chamber at night, always awoke terrified and with a sense of suffocation and palpitation. This attack would go off on kindling the light again. I could not doubt the fact from their own and their husbands' concurring evidence. The effect of the light in restoring Mendelsohn seems an analogous phænomenon.—All three ladies had diseased sensibility, but no nervous complaint bearing a particular name.

tile occupations, and an engagement to carry on three literary works at the same time, brought back the same sort of attack, several new symptoms, common in epilepsy, made their appearance, as giddiness, throbbing, and flushings. The disposition to these at the latter period was such, that it was scarcely possible for him to read or hear a page, or enter into any literary conversation without provoking them. In proper epilepsy, the fits sometimes come at last to be so easily excitable, that riding over an extensive plain, or looking from a height, or reading a single line will bring them on, as I have myself seen. This too is the case with a large proportion among nervous invalids of all degrees, from epilepsy down to that tendency to mere fluttering or discomposure from a sudden noise, which has obtained no specific name.—There is a composition, well known to modern chemists under the name of *fulminating silver*. It is among the properties of this composition to go off on being touched even with the beard of a feather. If the system of the healthy and the vigorous may be compared to gunpowder with respect to its readiness to start into motion, that of the whole class of invalids at present under consideration, will be represented by the over-hasty and ineffectual explosiveness of fulminating silver.

As those things which tend to strengthen the associations, formed in the progress of life, appear to have power in preventing epileptic paroxysms, so we find the contrary not less true. These and similar affections are most readily brought on at the age when the associations are least confirmed. In infants, acidity, worms, or pain in the teeth, will excite convulsions with or without senselessness, much more easily than in grown people. The effect of sleep in this respect, is remarkable in persons of all ages. The greater number of epileptic fits * make their onset while the patient is asleep. There are few who are subject to them in the day, who have them not also in the night. In many cases they are absolutely confined to this period of our existence, and where a gradua decline is observed in epilepsy, and the attacks have been common both to the day and night, they appear, towards the close of the complaint, for the most part if not solely in the night. This appears to have happened to the person, from whose description of his

* Cum epilepsia vires acquirit per somnum, illi nimis indulgere impune non licebit.——HEBERDEN.

own sufferings I have made such copious extracts. The same law obtains with regard to accessions, that answer perhaps strictly to the nosological idea of convulsion and unconsciousness, but are not commonly considered as epileptic.—Miss R——, aged 26, complained that her predominant state of feeling was uncomfortable. She started at sudden sounds, though slight, and had partial headaches, tremors on any moderate exertion, with frequent flushings, throbbings and palpitations. She always found herself better towards evening. The principal cause of her distress, and that for which she sought medical advice, was her being accustomed to grind her jaws in sleep, and sometimes so violently, as to break her teeth. She scarcely ever awaked refreshed; and when her friends had heard the grinding most by night, she was sure to be more particularly fatigued, heavy, spiritless, and oppressed with headache the succeeding day. The nightly muscular action of the jaws in this case cannot, I apprehend, be distinguished by any so proper epithet as *convulsive.* — — The heart, of all the muscles, seems the most liable to convulsions, for such are in fact its palpitations, as throbbing is the convulsion of arteries. How often are nervous persons disturbed out of sleep by the violence of the sensations, occasioned by palpitations and throbbings? Yet perhaps for once that they produce this effect, they happen an hundred times without being noticed. Startings in sleep are to all intents and purposes convulsions. They particularly infest the nervous, though occasionally and in less force they may befal others, from some imprudence of diet, or some great mental agitation.

These phænomena seem to depend upon some cause besides the mere disunion of those groupes and trains which form during our waking hours. Or more properly speaking, the disunion is but the effect of that cause. Some change happens to the nerves during sleep. This the healthy may recognize in their lighter feelings and renovated powers. It is equally manifest in debilitated persons with too acute feelings, not merely from their awaking more harrassed than when they went to sleep.—For it might be uncertain whether the fatigue does not arise from the incessant activity of the imagination in dreams (whether they be remembered or forgotten), and from the exhausting orgasm of the heart and blood vessels that accompanies these nocturnal excitations of the mind.—But the influence of sleep is sometimes perceptible the moment it comes on. In the above quotations from the journal, we find our invalid one evening starting as soon as he began to dose. Mrs. W. a lady of a delicate and exquisitely sensible habit, miscarried in the seventh month of pregnancy, and lost much blood. For many days she was awaked, the instant she dropped asleep, by convulsive motions of her different limbs, as considerable as if she had received an electric shock.—Master W. S. had suffered considerably from convulsions while he was cutting his teeth; and when he grew older, he frequently started out of his sleep, so terrified that it took a considerable time to pacify him. His terrors arose undoubtedly, as in similar cases, from frightful dreams. The family resided not far from a manufactory, in which there were loud noises and large flames, issuing from the midst of a black and smoaky scene; and it was ascertained that whenever the child was taken within view of this scenery by day, he awaked in terror soon after being put to sleep.

The writer, who ventures to lay down any propositions respecting dreams, will be given up by every prudent reader, as one that launches upon an uncertain sea without compass or rudder. Yet upon the examination of many dreamers, nervous and others, I have found, and I think any one, who will enquire, must find, that the following circumstances are general. The images, that figure in dreams, are of more recent date, the younger the person is to whom they occur. Probable evidence of this might be derived from the manner in which the ideas of the human mind multiply, mix and decay. But I build the opinion upon the direct experience of dreamers of different ages. When the elder SCALIGER dreamed that Brugnolus reproached him for the omission of his name in a poem on the celebrated natives of Verona, and when the younger SCALIGER learned that this Brugnolus, of whom his father had not the least recollection, had really figured as a Veronese critic, it must be supposed (unless we admit of inspiration on an occasion so trifling) that the name was brought back by some association in sleep, though it had vanished from the waking memory. The name of Brugnolus or his portrait might have occurred to Scaliger in his early years along with a particular state of the stomach, a certain position of the muscles or groupe of ideas: and these modifications of the system recurring at the time of his dream, would easily call up the name or the portrait, formerly connected with them. For whatever we think of in sleep is apt to recur with its original or with more than its original, vivacity,* whereas the ideas in healthy wakefulness are fainter than the impressions from which they are derived. Dr. PRIESTLY has told us that "many persons "(and he is one of the number) seldom "dream of any thing recent."† The ready forgetfulness of fresh incidents among old people, and those whose memory is not more retentive than in old people, makes it probable that their mind will commonly reach back in sleep to the events of their earlier years, whether remembered while awake or forgotten, as in the case of Scaliger.

If it be true in children that the individual ideas of the day before recur more frequently in dreams, the same holds also, other circumstances being alike, with regard to nervous adults; and in both it is still more

* A boy learns his task imperfectly over night; the next day he can repeat it without fault. Here it is evident that sleep has introduced a stronger tie between the ideas constituting the task. And should the repetition make part of a dream, it will doubtless be more fluent than in the morning, as we all find ourselves much more masters of a language we are studying, when we think in it asleep.

† New York Med. Repository, V. 2, 128.

strictly true of the state of feeling than of the individual i eas. For if any thing has happened to discompose a nervous person, the discomposure will commonly be felt in sleep to a still greater extent; and though the very ideas last connected with the discomposed feelings, should not be reproduced, others that have formerly been so connected will. But generally there will be a mixture of both. For feelings that have accompanied ideas at different times, have prodigious power in bringing these ideas together;* and this is the chief secret for unriddling the inconsistencies of dreams, and the key to the boldest flights of lyric and dithyrambic poetry.

But Morpheus is not content with keeping the minds of his nervous slaves on the alert by the exhibition of his magic lanthern. His images rouse into action the muscles that ought to remain at rest during the night. The sleep of savages and of labourers, well fed and not overstrained, may admit, perhaps, of a pretty uniform description. But from the profound injury which the

* I only touch here upon the associating power of the feelings. It is the most neglected, and perhaps at the same time the most pregnant topic in the doctrine of the mind.

sensitive part of the human frame sustains in numberless members of civilized society, the modifications of existence come to be almost as numerous in what is usually termed sleep, as they are in the waking hours. Indeed sleep has become, from this cause, almost a word without meaning. In its vulgar acceptation it appears to stand for the want of those actions that are usually determined by impressions upon the eye. The best way of arriving at a steady use of the word would perhaps be to examine well the repose of persons, who do full justice to their animal faculties, and to apply it strictly to this state.

Talking in sleep must be considered as the sign of a diseased condition of the nerves. When it is considerable and frequent, it takes away from sleep all title to be looked upon as "chief nourisher in life's feast," and indicates such a disposition to the worst forms of nervous affection as should not be suffered to go on without correction. We have the well-attested case of a basket-maker in the principality of Waldeck, who being much affected by a sermon, repeated it the next night in his sleep; and ever afterwards preached extempore, from some internal impulse like a fit, and also from the effect of spirits. During the time he sate in a kind of stupor, with his eyes fixed, but without seeing, perspired and had oppression of his breath though he did not speak loud or long. He seemed much exhausted, sighed deeply and recovered slowly.—(*Moritz*, l. c. iii. 1. 44).

The grotesque, yet affecting scenes which sleep-walkers go through, and concerning which story-tellers have commonly so much to relate, exhibit a much more alarming union between muscular action and the imagery of dreams. The number of accurate observations we have on persons afflicted with this species of nervous malady is not very considerable. The partial retardation or interruption of some courses of action, while others run on (and often irregularly) in the channels to which they have been accustomed, when the first are in motion, seems to be common to all; and indeed, if no variation took place, there could, it is evident, be no distinction between sleep-walkers and waking-walkers. SAUVAGES relates of a girl subject to catalepsy, that one morning he found her with all her muscles rigid, and that in 5 or 6 minutes she set to yawn, and began the following scene, which she had often acted before. She spoke with a quickness and vivacity which she never shewed but in the paroxysm, for at other times she was timid and low. What she now said was in connection with what she had said in the last accession, Sauvages convinced himself she was not asleep by aiming to strike her in the face, and though his hand came quite close to her eye, she neither winked nor drew back, nor made the least pause in what she was saying. A burning taper was held almost close enough to singe her eye-lashes, without effect. A person approached her from behind and cried with all his might in her ear but equally in vain. Sauvages applied hartshorn and brandy to her mouth and eyes; he put strong snuff into her nostrils, pricked her with pins, turned her fingers backwards and forwards, which yielded like those of a puppet, and lastly he touched the ball of her eye with his finger without producing the least appearance of sensation. She rose out of bed, walked hastily through the narrow passage between the bed frame and wall without striking either against the one or the other, or against the chairs. After going clear round, she threw herself on the bed again, lay down, covered herself at first. In a quarter of an hour she came to herself just as if she had awaked out of a deep sleep. *Mem. de l'Ac. Roy. des Sciences*, 1742. 409.)

In the *Encyclopedie* under the article *somnambule* there is an observation, which shews a remarkable suspension of one series of affections, while others, the most intimately connected with them, were going on. The archbishop of Bourdeaux was at college with a student, subject to walking in his sleep. On planting himself from curiosity in the student's chamber, so as to ascertain his motions, he observed the young man sit down to compose sermons, which he read page by page as he committed them to paper, if it can be called reading, when no use was made of the eyes. On being dissatisfied with any passage during the recitation, he crossed it out and wrote the correction with much accuracy over it. The writer of the article saw the beginning of a sermon, in which was the following amendment. It stood at first *ce divin enfant*. On revisal, it struck the student to substitute *adorable* for *divin*. So he struck out the first word, and set the second exactly above it. But remarking that the article *ce* could not stand before *adorable*, he very nicely set a *t* after *ce*, and it then stood *cet adorable enfant*.

To satisfy himself that the somnambulist, in all these operations, made no use of his eyes, the archbishop held something under his chin, sufficient to intercept the view of the paper, on which he wrote. But he wrote on without being interrupted by this obstacle in the way of his sight.—To discover how the night-walker knew the presence of objects, the archbishop took away the paper on which he wrote, and pushed other papers under his hand. Whenever they were of unequal size, the student was aware of the change, but when they were equal, he wrote on and made corrections on the spots, corresponding with his own paper. And it was in this way that possession was gained of some of his natural lucubrations.—Among other manuscripts, which the archbishop put into the hand of the encyclopedist are musical pieces written with tolerable exactness.

One night, having dreamed that he was beside a river into which a child had fallen, he went through all the actions tending to its rescue, and with teeth chattering as from cold asked for brandy. None being at hand, a glass of water was given him instead. But he immediately remarked the difference, and with greater impatience demanded brandy, saying he should die if none were given him. Brandy was therefore now brought. He took it with pleasure and said, as he smelled to it, that he found himself already better.

All this time he did not awake, but, as soon as the paroxysm was over, lay down on his bed and slept very composedly.

It was remarkable, and it coincides with the effect of the slightest irritation in recovering Mendelssohn from his inability, that when the imagination of this student in his somnambulism ran upon melancholy ideas, he could be turned to a different subject by stroking his lips with a feather.* The feather, I suppose, by tickling the very sensible skin of the lips, induced a pleasurable sensation, which brought along with it a train of ideas, formerly associated with pleasure of the same degree, and dispossessed the intruders of a melancholy cast.

The power of distinguishing different sized pieces of paper may at first appear almost incredible without the assistance of the eyes. But one can conceive that this might arise from the touch joined to any

* Immediately on writing this sentence I touched the back of the hand in a teething child, while asleep. The whole arm, the head and the body, though quiet before, moved several times. The child had only its skin rather heated from the teeth, otherwise did not suffer, though it must be supposed to have the sensibility of its nerves exalted during dentition: but not more so than the nerves connected with the motions in a somnambulist.

exact habitual sense of the space, which the pen had to traverse every line. In the same manner, we may understand how the girl, observed by Sauvages, was able to walk briskly round her bed without running against any thing in the room. The same accuracy of step has been often but not always observed in night-walkers, when the most perfect use of their eyes could not have availed them; and why should not the arm of a student accustomed to write have as much? The accuracy with which actions are frequently performed in sleep, I have often known exemplified.—A consumptive patient of mine slept in a bed with a flat head, not above three feet high. Upon this he used every night to place a glass, two thirds full of water, for the reception of his expectoration. His attendants were frequently changed, and they all assured me, that he got up in his sleep when he had any thing to expectorate, took down the glass, and regularly replaced it on the bed's head. Nor did he ever spill any of the water except one night, when a bottle had been placed in the way of his hand to the glass, after he had fallen asleep. As there was plenty of light in the room, his not being sensible of the bottle shews that he made no use of his eyes in taking down the glass.

It is nothing singular that the eye should be incapable of being affected as usual by light, while other parts are in the contrary extreme. Such phænomena are among the most common of those that are observed in nervous accessions. The stomach, for instance, shall be often so insensible that a glass of distilled spirit shall produce no more effect than a glass of cold water at another time; and this, in the case of a person, who has been all his life an utter stranger to the use of liquor of that strength. At the moment when the stomach is deprived of its natural feeling to such a degree, the mind shall be all alive to the groundless fear of immediate dissolution, and some of the senses shall be so ready to be acted upon, that a slight excitement shall produce a starting of the whole body. Other contrasts, equally strong, are observed in the indispositions to which the frame is liable. Who, for example, has not known the feet benumbed with excessive cold, while the cheeks were all in a flame? It is not, however, certain that the retina or expanded optic nerve, is ever torpid in such cases, the experiment of throwing a sudden light upon the eye, and observing whether the pupil contracts or not, having seldom been made.

In some cases, the susceptibility of visual impressions has undoubtedly remained, because the somnambulist has been sensible to some present objects of sight, though not to others. This partial attention seems clearly to have existed in the case described in Zoonomia (I. 19.) since the patient drank tea, when it was set before her, and smelt to a tuberose, which last action might, however, have arisen from accurate recollection of the place where it stood. In the remarkable instance, frequently observed and described at large by PORATI, the same circumstance occurred. A young man in his employ, of the name of Castelli, who had often before suffered from nervous affections, took a candle and lighted it. Instead of going to bed, as was supposed to be his intention, he went down stairs, reached a book from a shelf in his master's study, and set to read. Porati several times desired him to go to bed, but received no answer, upon which he took away the light. Castelli upon this got up, and opened the window, observing—*it is cloudy, we shall have rain to day.* The master now replaced the light, and Castelli immediately walked up to it, and began to read aloud. The people of the house collected round him, but he observed no one. They called to him, but he did not hear. The light being again removed, he got up, rubbed his eyes, and said "*I am not well; I must go out into the air.* He accordingly went into the apothecary's shop belonging to his master. Here, other methods having been fruitlessly tried to rouse him, some spirit of hartshorn was held to his nostrils, when he immediately sank to the floor and fell into convulsions. After some minutes he had a tetanus or universal spasm of his muscles, which by degrees relaxed, and he slept. Presently, he awaked as out of a sound sleep, but knew nothing of what had passed.

Another time he fell asleep in the shop, but had continued only a short time in that state, when he opened his eyes, muttered a few words, lighted a candle, went into the study, and set to read. He was at this time studying French, and had to translate a passage out of the Italian. He ceased reading, took up the dictionary and made his exercise as well as if he had been awake.

They snuffed out his candle, when, without noticing other candles that were burning in the room, he went into the kitchen and struck a light, by which he rekindled his own candle, and set to his task again. It was snuffed out again. He shut the window, as supposing the wind had blown out his candle, which he lighted anew; when Porati blew it out. He now grew impatient. He stirred the embers, took some matches in his hand and said—"*Is that fire, or is it not?*"—He lighted the candle and went into the study, where he took a dispensatory, examined by it some plants that stood ready for distillation, and found one right after the other. Porati now spoke to him about the plants. He immediately heard and answered as rationally as if he had been awake. He then fell asleep, soon awaked and was sent to bed.

On another occasion, a remark he made on a book which he was reading, excited in in one of his comrades a suspicion that he was feigning. To ascertain this, the comrade held the flame of a candle to his hand, but he did not withdraw it. Next morning, he complained of a pain in the place, but did not know how it came.

He was repeatedly observed not to know any person or thing but what coincided with his passing train of ideas. When he went into the shop, if faulty recipes were brought to him, he would soon detect the error, but did not know the persons who brought them, though they were his own comrades, but took them for servants who came for medicines.—He often tried to light a candle at a lamp standing under a glass bell, of which he was not aware; and it was observed that he could not distinguish two powders of which one had a strong smell and the other was without any. It was accidentally discovered that the waving of a fan threw him into a quiet sleep, and this expedient was frequently employed to stop him when he attempted to do what his master did not wish. The effect of the moving air was to arrest him suddenly, upon which he would sink down and sleep a short time, but his paroxysm went on afresh afterwards.*

* The case of NEGRETTI, as described by PIGATTI, in the *Journal Encyclopedique*, *Juin*. 1762, is among the most celebrated of those that have occurred in modern times. This man had walked in his sleep from his eleventh year, though only in March or part of April. He went through his whole business as a servant with the utmost exactness. He lighted a candle, tripped down stairs, went to the door as if to attend a visitor, who was going away, briskly remounted the stairs, put out his light and set it in its place. When in the pantry, he searched his pocket for the key of the cupboard where the glasses were kept, and

MONTAIGNE in his infancy used to be awakened by soft music; and we have common sayings, which bespeak the influence of the feelings, that predominate at the moment of awaking, upon the temper during the day. But it would appear that the frame of mind in which we close our eyes is of still greater consequence than that in which we open them; and the chapter upon this subject, which is commonly to be found in books of practical piety, ought to be transferred to those of medicine.

---

not finding it, called up the footman, opened the door, took down a salver and glasses, filled them with water, and handed them about. Upon their being returned to him some time afterwards, he carried them back and locked up the cup-board.—If the bore in the key was purposely stopped, he knocked it against the wall, and if that did not make it free, he looked for a splinter, with which he took out the matter with which it was stuffed.—At other times, he laid the proper table, dressed the sallad, fetching every thing that was necessary from the usual places—but if any thing was set before him in place of the herb for the sallad, he did not observe the difference.—He sometimes went very nimbly a couple of hundred yards to an inn, asked for wine, in place of which they would give him water, which he contentedly drank for wine, promising to pay next morning.—In this state he once told how he would pretend to be walking in his sleep; and laughed very heartily at the idea of thus imposing upon some person.

If we would procure composed, refreshing sleep, and avoid those seizures that make night so dangerous to the nervous, one thing to which we ought to attend, is obvious. That part of the organization which is concerned in mental operations continues, we find, to be affected for some time after impressions are received, even when there is no great intensity of feeling. For if this were not the case, the school-boy could not repeat his task better for a night's rest. Sensations also are revived with greater force, as well as ideas strengthened, during sleep. For agitations, which the objects themselves could not produce, have arisen from their nocturnal images. I had an epileptic patient, whose first fit succeeded a dream, in which there had occurred the idea of precipices in the Alps, which he had seen the day before. Thus nurses when they play with infants of a lively temperament, late in the evening, so as to raise strong emotions, frequently find that they soon awake disturbed; and people inclining to too great sensibility with weakness, either cannot compose themselves or else they dream much after the glare and noise of a ball-room or any affecting representations on the stage. Nor indeed is this confined to the weak, who should therefore be particularly studious in the evening to avoid all objects which powerfully stimulate the senses as well as all ideas that may raise a tumult in the bosom. How faithfully do we find the consequences of this rule exemplified in the people around us, and especially in females who join in the train of fashion! What a satisfactory explanation does it afford of the effect of circumstances, that otherwise might appear too trivial to be followed by any distinct consequence! I have frequently heard it remarked by men of penetration, accustomed to the great world, and it is, I suppose, a perfectly notorious fact, that at the close of the season in London, women appear worn down, haggard and spent. During their stay in the country, their shrunk countenances regain a degree of plumpness, their muscles recover their tone, and they really feel somewhat of that flow of spirits, which they often afterwards so miserably affect. Persons, it is commonly seen, who are governed by different habits, conceive towards each other a species of antipathy, so quick and inveterate, that it may almost pass for instinctive; and many a fine lady, on hearing the œconomy of her time, arraigned by the sedate, has persuaded herself that the censure proceeds from this feeling, and not from any foundation in reason. But it is only necessary for her to revert to the influence of her days upon her nights, to be convinced that the sober part of mankind do not barely arrogate to themselves a superiority, such as any clan or cast may arbitrarily found upon some insignificant distinction, but which no other need acknowledge. On the contrary, the advantage they enjoy is not less solid and permanent than the laws, according to which Nature has ordained that the human frame shall be affected. The throngs, by which the followers of high life are perpetually pressed; the dazzling scenes which they frequent in quick succession; their unceasing hurry of body and mind; the anxiety (to say nothing of the mortification) which every candidate for admiration must undergo; all contribute to stir up a correspondent tumult of imagination as soon as they are sunk into slumber. When they have tossed themselves awake, how can they help feeling more weary by half than on going to bed? Without spirit to raise their head from the pillow, even if they suspected this to be the best measure they could adopt, and in spite of constant disappointments, they still hope to find refreshment in an additional nap. But another and another leave them but in more languid plight. We cannot therefore be at a loss to comprehend the tendency of that mode of existence, to which we see our countrywomen so emulously devoting themselves. The night is the season in which the vulture of fashion flies abroad for prey. Many of the primrose cheeks and aspen constitutions, which are to be met with so abundantly in the great world, exhibit the consequences of his secret depredations.

Before delivering themselves up to repose therefore, those who find or fear injury in this delicate part of the organization, have, not merely, to shun occasions of strong impressions upon their senses or their mind. It should also be their study to still the vibrations into which any inevitable impulses of the day may have thrown their nerves. A simple and natural life affords many resources for the purpose. From spring to autumn, the occupations attached to the cultivation, or at least, to the superintendance of a garden, will preserve the mind tranquil, or help to quiet its commotions. After a light early supper, a walk will diffuse that complacency over the system, which is the best preparation for sleep in all cases, and which will by degrees bring the nervous valetudinarian to wake with the lark and with somewhat of the lark's alacrity. The dejected often feel as if the fresh air ventilated the mind. Early rising will reconcile us to early retirement to rest; and for at least five months out of twelve, it should be a rule with the invalids, whose case we are here considering, to go to bed before candle-light.

Whatever engages without anxiety should be appropriated to the evenings of the whole year. A tribute of winter-amusement might be levied upon the experimental sciences and mechanical arts. In the state to which they have now been carried, it requires but small ingenuity to render many of them perfectly domestic. A variety of tools and some philosophical apparatus, might be as conveniently placed in a house as many articles of our furniture; and they would be as useful as part of that is pernicious, or rather they would act as antidotes against the indulgence to which that invites, and make even our cushioned chairs and bolstered sofas salutary in their turn. I should not be surprised if, in a few years, it should become as common for persons to go about to instruct private families in chemistry, mechanics, in tangible geometry and various sorts of manufactures, as it now is in music and drawing. The lessons of the former sort of masters would help to save many from falling into nervous indispositions; and they would prevent others from suffering so severely by them—things, which can scarcely be said with equal justice of the professors of the fine arts. For their pupils not only suffer, as formerly mentioned, by the application necessary to excel, but in the case of music, they are often tormented into indisposition, by their very excellence. What young lady, till she is hardened by frequent exhibitions, can come out, though ever so conscious of her skill, to entertain a company without some flutter of spirits? and when she has acquired confidence, must she not find herself chagrined at the smallest deduction from her accustomed measure of applause, though that applause shall have ceased to afford the pleasure it gave when fresh—just as the dram-drinker is wretched on being deprived of his potion, after it has lost its ancient zest? Whereas witnessing processes in philosophy or the arts, or in rural œconomy, brings with it that composure which the nervous so much need. The spirit therefore that has just gone abroad in favour of useful pursuits, may lead mankind round to health, and exempt the rich from the necessity of having recourse to the irritations of vanity to make themselves feel that they are alive. Hence it is to be wished that vanity may interfere as little as possible with this new system of amusement. For wherever vanity takes the lead, there will be danger that the constitution, as well as the character, should suffer; and indeed, in some habits, the inquietudes of vanity are alone sufficient to excite disease. This is a truth, which it might be well if those who devise and superintend plans for forming young people would keep more constantly in mind. Otherwise the objects of their concern will scarcely be gainers by the substitution of solid science for superficial accomplishments.

The scenes, presented by a new and beautiful country, have long been known to possess a soothing power. They, in fact, keep the mind in the medium state between anxiety and listlessness. Some knowledge of antiquities, agriculture, manufactures, mineralogy and botany, adds greatly to the beneficial effect of travelling. By diversified cultivation therefore of the understanding in youth, we lay up agreeable remedies in store against the evils of future years. I am persuaded every one, who has made the trial, will join me in bearing testimony, how hopeless it is to set about to amuse a person who has had but a single pursuit, when bodily or mental infirmities once disqualify him for following it.

Whatever brings back the sensations of a healthier period, will more or less compleatly recall that action of the organs with which the sensations were once attended; and a degree of rejuvenescence may be the recompence of a well conducted journey. How can we understand the cure of a number of students by pumpernickel and suet, mentioned in a former Essay, unless we suppose that by carrying back the imagination to past pleasures, they occasioned the coats of the stomach to throw out the same healthy juices, which had flowed from them during the enjoyment of those pleasures? It is less likely that the direct application of this coarse and heavy fare to the stomach should have produced a change in its secretions. Upon the same principle, the cantatas of Bernier are said to have cured a French musician of a fever, though Rousseau declares them capable of giving one to a musician of any other nation. These cantatas were to the Frenchman's organ of hearing what the pumpernickel was to the palate of the Germans—a memorial of the days of other years. We have abundance of examples, in which the mere sight of the spot, where a disgusting object had produced sickness years before, has re-excited it; and the smell of tar will occasion nausea in delicate persons who have suffered much at sea.

In joining my predecessors to recommend journeying for the purpose of health, I ought not to omit that the design is often frustrated by injudicious arrangemeut at our places of resort for invalids. The amusements are such that the ladies might almost as well have continued to frequent the routs and assemblies of their own residence; and the gentlemen their counting-houses. I would therefore advise those who are in earnest to get rid of indisposition, to shun places of glitter, parade and jostling. If there could be a paradise for loungers, BEAU NASH might perhaps have laid it out. But neither he nor his imitators could be expected to have talents to plan for the sick.

Persons who *must* be occupied, and yet have heads that can bear but little, differ so widely in habits, circumstances, and degree of indisposition, that concise directions for their use will appear trivial and commonplace. I know indeed that the art of impelling the body of thought and feeling into a brisk current whenever it is inclined to stagnate, and of restraining it when over-rapid, would be well worth explaining ever so much at large, that it may be applicable to the diversity of exigencies. But I doubt whether there be any person, at once sufficiently acquainted with human nature, and with the sources from which lenient amusement might be drawn, to execute a task of such various knowledge and nice combination. But by the contributions of the ingenious, we may acquire something in addition to common games and light reading, which often do not answer, or of which the latter is always too soon exhausted. I know not if stories, such as are said still to be told, extempore, in the coffee-houses of the East, would not prove excellent *nervines* and *anodynes*. If well chosen, they must have all the advantages and none of the disadvantages of theatrical exhibitions. Our books, whether perused or listened to, are not, in some respects, near so good as the things they have banished.

The intelligent historian of his own epilepsy, who persevered for nearly ten years in observations and experiments upon himself, may be thought to have amassed less instruction for the benefit of his fellow-sufferers than in proportion to his opportunities and attention. And as there is a class of men who despise what they do not understand, so there is another, much more numerous, who scout, as fanciful, evils which they have never felt. To these, many articles in the valetudinarian's code of regimen will appear to belong to the same predicament as the late Emperor Paul's regulations respecting his subjects' hats, and the members of his soldiers on parade. But those, who know how often whole nights and indeed whole weeks, months and years, of patients with disordered nerves depend upon the mere turn of a straw, will not think the minuteness of some of the following measures so absurd.

Our epileptic found it of great importance to remove any obvious cause of wakefulness before he lay down to sleep, and to let its effect wear out, even if that kept him up an hour later. Washing the head for some time with tepid water before lying down, when there was any danger of flushing, throbbing, or noises in the ears, brought on sleep sooner and made it more composed. He at first resisted drowsiness, when it befel him immediately after supper, but found the resistance injurious to his feeble nerves, and unfavourable to rest on going afterwards to bed. In this case, he found it best to yield to the inclination, but instead of lying down, to sit and recline only the head. At the end of a twenty minutes slumber, or sooner, in case of feverishness, he paced the room, chatted, and listened to a book, but by no means suffered himself to be raised to any pitch of vivacity.—After he had laid himself down for the night, friction from the head down the back and along the uppermost arm, for one or two minutes, by an attendant, seemed to have a composing effect. It was necessary to dispel the slightest chill of any part either by the warm bottle, rubbing, or additional bed-clothes. The nice adjustment of the temperature of the head was of the utmost consequence. On waking in the night, or lying long before falling asleep, he usually took a few mouthfuls of food, as the slightest sense of hunger occasioned great disturbance in the nervous system.

" A few mouthfuls of some kind of rather " strong beverage had the effect of improv- " ing the heaviness of head, which more " frequently came on in the evening, into " compleat sleep —— I am extremely careful " not to make this a daily custom, and " prefer the more natural expedients above

" mentioned; however I keep warm water " ready in bed, to mix with cherry brandy, " as also mountain wine."—Rubbing the temples with spirituous waters seemed to have a soporific effect, when there was a considerable weakness of the head.

The distress and mischief arising in epilepsy, nervous head-ache, and all kindred ailments, from that violent excitement of the blood-vessels, which produces the flushed face, throbbings in the neck or at the temples, and ringing or burning at the ears, is such as to justify all the care this invalid bestowed on their prevention. These symptoms foster the disease from which they spring. The thickening of the scull and induration of the brain, which have been found in many epileptics, are I suppose the consequence, and seldom or never the cause, of the complaint. They must take place on the same principle as the increase of bulk in the muscles of a blacksmith's arm. Similar alterations in the structure of the parts would, I doubt not, be discovered indifferently in all nervous persons, greatly troubled with symptoms about the head, whether they were exactly epileptic or not. Epilepsy might, in one sense, be said to be the consequence of such kinds of disorganization.

Frequent throbbings, flushings, and heats about the head, the tokens of too great action in its arteries, must, from all analogy, be regarded as capable of thickening the bones, and producing an alteration in the structure of all the parts, which the arteries feed. After a certain time, the parts altered, may produce epilepsy, by irritating those in their neighbourhood. Accidents have the same effect, and there is no difference in the two cases, except that injuries from violence do that in a second, which the other cause brings about in years.

It might assist considerably in the cure of disorders of this species, if instead of taking a short nap with the head reclined, while the body remains erect, the whole time of sleep were passed in that manner, as the force with which the blood flows to the head would then be lessened. Though the muscles in general would be less relaxed in this posture, yet the whole system might be more recruited; and the back could be supported, which would ease the principal muscles.

I have thought it a better plan to cool the head by wet sponges attached to an oil-case cap, or by putting over it a moist cotton-cap, than by applying water to the head itself. By this contrivance the cooling process can be stopped the moment the point is gained, and without trouble or interfering with the invalids' comforts, it can be kept up for any length of time, which is of the highest importance, as every body knows that, after slight refrigeration, the tormenting heats return with tenfold violence —In very distressing determinations of blood, as they are termed, I have sometimes advised a bucket with a cavity at the bottom, to receive the head. The bucket is filled with cold water, ice, or a freezing mixture, and suspended by a weight over a pulley. In hectic and nervous night-heats, the blood-warm bath, some hours before bed-time, has great preventive power, if it be not used when it is likely in these habits to excite feverish heats, as often happens, if it is taken immediately after a full meal.

The means which our self-observer opposed to his formidable *hurry of ideas*, may serve the purpose of evening composure with persons similarly constituted and circumstanced, if they will condescend to petty attentions in their own behalf. I shall therefore give the substance of what he says on this head. The facts are at least curious in a view to pneumatology. Nor can any thing more strongly paint the situation to which we may be reduced, by letting the sensitive part of our frame get out of order! For what must be that man's misery, whose means of relief consist in speaking aloud and somewhat slowly, at those times of imminent danger, when he can speak at all. This, he says, seemed to moderate the random velocity of the mental power by distributing its action into the time, necessary to thinking and speaking, and into more space, by reason of the synchronous movement of the tongue, lips, and other organs.*

* This seems a fanciful explanation. Nor shall we understand how such efforts prevent a paroxysm, till we know the exact difference in the condition of the organs when the patient feels well and ill. And of this we have not the most distant notion. All we can say is that the ideas and connected play of the muscles, on a successful attempt to articulate deliberately, set the head right by the power of association. In an epileptic accession, there seems some derangement of the brain and nerves, with which such efforts are incompatible. So that if the one state can be maintained, the other cannot take place. It is worth while to add that compression of arteries that carry blood to the head, has been found to put a temporary stop to nervous fits in a case where palpitation of the heart, headache, coldness of the feet, occasional shivering followed by extreme heats, particularly about the face and head, locked jaw, convulsions of the muscles of the neck and body, difficulty of breathing, stupor and delirium succeeded each other. In an attack of delirium, when the brow was contracted into

When he was not capable of uttering fresh sentences, consisting of his own thoughts, the recitation of couplets, which he had long had by heart, would succeed. So did going on to count the fingers, or some other objects in sight, aloud. He took care at first to enuntiate slowly and only a syllable at a time; but afterwards he went on quickly, and added objects and syllables.

In moments still more critical, when he was unable to pronounce above a single word, it was, as above stated, a few times of use to fix the attention upon some object, de-

---

an immoveable frown, pressure on the right carotid artery, a little below the larynx, occasioned this expression of the countenance to vanish, and restored the patient perfectly to herself. When the pressure was taken off, the frown returned. The headache, difficulty of breathing, and convulsions were equally removed by the pressure. An epileptic fit, taken on its approach, was put off entirely for some hours. In a few cases of nervous and bilious headaches, it equally answered. It relieved the symptoms of inflammation of the brain, and of chronic mania. Curious effects were observed in disorders of another class. Compression of the temporal arteries immediately, for instance, relieved a gouty pain in the head which had come on in consequence of an inflammation of the same kind in the foot nearly ceasing; but the pain in the foot returned the same instant with violence. But this was removed on pressing the artery going to that foot.—(*Dr. Parry in Medical Memoirs, Vol.* iii.)

noted by one or at most by two syllables, (as *watch*, *chair*, *table*,) and to have it held up to the eye at the same time.—But these exercises were never pushed so far as to exhaust.

In the stupors, which often lasted for hours with lucid intervals of about a minute, when he was not only unable to articulate himself, and when also the sounds, uttered by others, failed to call up their accustomed ideas in his mind, he was most relieved by the contemplation of sensible objects. These did not require so much exertion as pronouncing a word, and yet they roused the mind so much that a wildly rushing stream of ideas was often arrested, dispersed, and thus far weakened in its dangerous course. This sort of occupation answered better—1. When the objects were strikingly new—thus wild beasts, troops on march, a full market-place, several times produced sensible alleviation.—2. On practising something quite easy, but not perfectly mechanical. To this head belong the simplest games with tables or at cards, an exercise in writing a fine hand, the copying of a precedent.—3. Oftner still, though not always, the morbid course of ideas was obviated by an impression, for which he had already conceived a predilection; for instance, by receiving sensible tokens of affection from his friends, and by returning them; by pantomimic play with his children; by the mechanical preparations belonging to a favourite affair, as putting a cover on interesting letters.

He was master enough of the harpsichord to play a dozen tunes, that had remained at his fingers' ends since his younger days. But the repetition of these was too mechanical to drive different ideas out of his head, and learning new pieces required too much attention.—Listening to a vocal performer, to the nightingale or the harmonica, enraptured him, only on the first impression, enough to serve as a remedy against the press of ideas, and this end it sometimes answered; but after a few minutes, the diseased affection got the upper hand. He judges however that any one with musical propensity and ear, sufficient to follow the order of the tones, would continue longer to be diverted from other thoughts.

When the hurry of ideas came on in the night, but was not so violent as to make him rise to walk about, or play at draughts, he had no resource (unless he took medicine) but in going over, slowly in thought, couplets or forms of words which he had learned by heart. He often found himself put out and overpowered by the irregular course of ideas, but by beginning the same form again, or going to another, he commonly composed himself so much as to fall asleep. He was told by another person that silently rehearsing the Lord's Prayer for a number of times, serves to dispel thoughts that tend to keep off his sleep.

Difficult as it may prove to give people of too sensible nerves effectual instructions for preparing themselves properly for rest, there remains a task which seems placed still farther beyond the sphere of human regulation. This is to provide that their rest shall be free from danger and real. In slighter cases it would be requiring too much of the healthy that they often should give up the comforts of their own nights, for the advantage of their nervous friends: and yet it is very manifest the effect of any services that can be rendered them must depend upon their being exactly timed.—Bed clothes could be made of cotton for that part which covers the trunk, and of woollen for the feet, while an additional linen sleeve might be drawn over the arm. When inordinate heat came on after sleep, as is so common with the feeble, it would be easy to contrive an apparatus, which should draw off the coverlid at a given hour. Only we cannot be certain that the changes shall observe certain periods.

That sleep is not absolutely beneficial and in proportion to its duration, is a truth which invalids will not readily admit. There are however two cases, in which it proves the more weakening, the more it is prolonged. The one I have frequently mentioned. It is when the pulse is greatly quickened and the skin heated. The heat, no doubt, often depends in part on external causes, as a close bed chamber, or an excessive load of covering. But it will take place under opposite circumstances; and then an alarum might be set to the particular hour, when the excitement is likely to have come on, that the body may be cooled by lightening the bed or by a few turns about the chamber. The second case which often occurs in conjunction with the first, is, where a disproportionately large quantity of urine is passed by night. Both are found to take place in the dyspeptic, especially after indigestible food, and in people inclining to be dropsical, as well as in the more strictly nervous. Every medical practitioner must meet with the latter, and a hundred authorities may be quoted for its existence in an enormous degree. ISENFLAMM mentions a person who became hysterical during confinement with her eighth child. The attacks grew gradually more frequent and distinct. They were worst by night, when the navel was drawn in to the vertebræ of the loins, and there was the sensation of a round body, rolling about. This patient had another very troublesome symptom, which she called a screwing of the eyes. It was as if they were screwed together and then out of their sockets, whence a few drops of liquid flowed at the conclusion of the seizure*—By night she was obliged to make water five or six times. At one hour, she would feel an irresistible impulse to discharge it as quickly as possible. At another, though in the same night, she could part with it quietly and even retain it without inconvenience. The first, she named the spasmodic; the second the natural urine. The first was limpid; the second deeper than that of a person in health. "In "the evening, I regularly found five or six "glasses, standing full; each of about eight "ounces, and could easily tell by the colour, "that the third and fifth, for example, had come "away with a violent desire, and so was spas- "modic, while the rest was natural: nor did "*I ever fail* to distinguish them properly, as

---

* In this case, there was no doubt a spasmodic contraction, and perhaps an unequal one, of the muscles that move the eye-ball. The liquid consisted of tears and mucus, forced out from the glands.

" the patient herself assured me. She slept very " composedly between whiles, till she was " awakened by the vehemence of call from the " spasmodic." *(Isenflamm prakt. anmerk. ueb. d. nerven.* 2 4 3, 1774.*)*

In cases somewhat similar, patients have found themselve most heated, when most pressed to evacuate the bladder. And it probably was so here. For it seems not to be in these nervous affections as it is in inflamatory complaints, where the secretion by the kidneys is scanty when the fever rages. But the nightly flow of urine goes on with the heat of the skin ; though in the day, large quantities of watery urine shall accompany the chilly paroxysms, to which nervous people are so very subject. I have known an evident reduction follow a better regulation of the temperature of the bed-room. " I have " had patients, says Dr. Whytt, who after a " long fever or some other tedious and weaken-" ing distemper, made a great quantity of pale " water in the night, but in the day time no " more urine, sometimes less, than usual, and of " a natural colour. This increased secretion did " not generally begin at any certain hour in the " evening, but soon after going to bed ; and " in the morning after getting up, it gradually " abated. Nay a gentleman, who had been in " use, for ten or twelve days, to make from five

" to six english pints of pale water in the night, " finding himself greatly weakened thereby, re- " solved to try what getting out of bed would " do ; and accordingly, at two in the morning, " after having that night passed about a pint " and a half of urine at twice, he rose and sat " up for two hours, and then was able to make " but about half a gill. After this, he went to " bed again, and in two hours more made near " three quarters of a pint (12 ounces) of pale " water. This experiment was repeated some " nights after with the same event. Those," continues the author, " who are troubled with " this flux of limpid urine in the night, find " themselves stronger, cooler and in the best " spirits in the evening, at which time their " pulse is slowest. But soon after going to " bed, their pulse becomes quicker, they grow " warmer and begin to make water in greater " quantity. They are not refreshed with sleep, " and in the morning, they feel thirsty and " languid, and have a quicker pulse than at " other times."

Early going to bed will be of the more advantage to nervous invalids, as it must afford their friends an opportunity, in the first part of the night, of examining their state without inconvenience, and altering the bed-cloths if necessary. But a bed-fellow, or a person lying in the same chamber, may be of the most essential service by rousing them, when parched with heat or when agitated with frightful dreams, and sometimes on the approach of different kinds of paroxysms. On this, as well as on so many other accounts, it is so desirable to make an acquaintance with the human body and mind, a branch of household knowledge. The wish to be useful would confer the faculty of observing; and the attentive wife of an epileptic husband would be able to give his physician instruction, which he could derive from no other source, and to which the patient himself might add. Twenty journals drawn up by as many attentive and ingenious couples, would I believe, do more towards the art of procuring composed sleep, and consequently towards breaking the force of nervous complaints than an hundred volumes by physicians, who have no better opportunity of information than the ordinary course of practise affords.

It is well known that when any one has urgent reasons for unusually early rising, if these recur frequently and with force the evening preceding, he very seldom oversleeps his time. The purposes of the day may, I believe, in almost every case be brought effectually to bear upon the mind in sleep. By repeated endeavours, I have very little doubt but perceptions may be gained of the condition of the system at any moment, and that muscular action may become connected with these perceptions. Thus, the wish to preserve his nightly expectoration for my inspection gave the phthisical patient before mentioned perceptions, which he would not otherwise have had; and to these perceptions, the movements of the arms and legs, necessary to his rising and taking hold of the glass, associated themselves. The history of the seven years' epilepsy affords another instance directly in point. "The following were the few expedients," says the writer, "which were occasionally of use to "prevent wanderings of thought in sleep. I "endeavoured, before going to bed, to rub, as "it were, one or two favourite ideas into my "mind. I sometimes read over tranquillizing "passages, which I had previously marked in "different books. I secured myself, as far as "I could by the posture I took at going to "sleep, against getting to lie on my back. I "requested my wife, whenever I uttered a word "or tone, betraying perturbation, to admonish "me, and I readily heard her voice. *Several* "*times while dreaming, I succeeded in calling* "*to mind the determination I had formed,* "*when awake, to put down for dreams any ex-*"*traordinary scenes which should occur by night,* "*and in which it was impossible I could be en-*" "*gaged at the time.*" By the same rule, were a person to fix it strongly, by frequent meditation, in his head, that he is apt to burn excessively during sleep, and that he is determined to awake, when he is in this state, it would probably often happen as he desires. It is so common to exert the will or to reason in dreams, that it would be impossible to account for the contrary opinion in a late sagacious author, if his arbitrary and untenable division of the sensorial power had not led to it. And if we will, and reason at all in sleep, why should we suppose that this faculty may not be strengthened and employed to advantage?

Further Remarks on the Affinity of NERVOUS COMPLAINTS.

The proneness of noctambulists to loss of consciousness and convulsions, appears from more than one of the preceding examples. It might be confirmed by others. The lady, whose reverie or somnambulation is described at large in the Zoonomia, is one among the number. This did not escape the author, who remarks "that reverie is a disease of "the epileptic or cataleptic kind, since the "paroxysms of this young lady always

" began and frequently terminated in con-" vulsions." It may seem singular that at the moment our trains are so rivetted together that the most violent impressions on the senses, as shouting in the ear, cannot disarrange them, they should be on the point of being broken by a sudden change in the nervous system! But there are several analogous appearances in the human frame.

The near approach of that very frequent and very diversified affection, absurdly termed *hysteria*, to epilepsy, is still more evident. Nor can there be a stronger proof of their resemblance, than the necessity under which the definers of diseases have found themselves to blend their signs. Thus Dr. CULLEN, who, as we have seen, makes epilepsy to consist in *muscular convulsion with sleep*, applies these very marks to hysteria, and accumulates others upon them in order to establish the distinction: hysteria, according to him, being characterized by *noises in the belly—the sense of a ball rolling about there, ascending to the fauces, and then producing a sense of suffocation—by sleep (fainting, or loss of consciousness)—convulsions—profusion of limpid urine—by great, but not spontaneous, changeableness of mind.*

It continually happens that patients come

to medical practitioners with stories like the following. 'I was brought up in the manner 'of most other young ladies. I read, played, 'and painted a good deal, that is to say, my 'amusements and occupations were almost 'all carried on within doors. I was excessively apt to be chilled, particularly in my 'feet during winter. In sultry weather, I 'was as much in the contrary extreme; being 'overpowered by the heat. From my childhood I have felt feeble, and slight causes 'would always bring on a sensible degree of 'indisposition. Among these slight causes, 'I reckon a little too much dancing, or 'exertion of any kind, tea somewhat over 'strong, buttered toast, fruit (particularly 'after dinner), the sudden arrival of a person 'in whom I was interested, the entrance 'of a stranger without notice, or any thing 'which came upon me unexpectedly. And 'the hurry I felt on such occasions would 'keep me indisposed at least till the following day. After a ride on horseback, or 'even a drive in a carriage, I had such an 'increased weakness in my limbs, as inclined me to stagger, especially if I had 'fasted some little time longer than usual.

'A change of constitution took place 'early: I was at those seasons particularly 'languid and unwell, and troubled with 'severe pain in the loins.—The complaints 'which I at present consider as most grievous, 'took place some years afterwards. They 'consist partly in an aggravation of the 'preceding. Besides, I am alternately sub'ject to a fixed pain or uneasiness in some 'spot of my head,—it is often a distressing 'sense of cold on the very crown,—and to 'giddiness. My ears will frequently set to 'ring without any apparent cause: and my 'eye-sight will grow occasionally dim. I 'have at these times palpitations of the 'heart, oppression attended with deep sigh'ing, and extreme dejection of spirits. I 'am obliged to let loose my cloaths after 'eating, by way of relief to a sudden swell'ing of the stomach, and this too is always 'more troublesome, if I have gone without 'food half an hour longer than ordinary. 'An aching pain or sense of distention comes 'on in the hollow of the side under the ribs. 'I am in general under the necessity of using 'aperient remedies, but their effect is at 'times very violent. This state will indeed 'every now and then take place without the 'use of any medicines. There is often a 'most pressing call to void urine; it is 'colourless like water, and the quantity 'incredible. My appetite is unequal, some- 'times very bad. My nights are seldom 'refreshing. It is common with me to be 'awakened by startings of the limbs, and 'I have long observed small twitchings over 'my skin almost every day. This is part of 'what I more or less habitually feel. But 'when I am particularly unwell, I am struck 'by a deep and sudden anxiety. My breath 'is very much oppressed indeed, so as to 'induce a dread of being stifled; my heart 'is as it were violently squeezed together, 'and my feet are like absolute marble. My 'power of speech is at the same time sus- 'pended; with all my efforts I cannot collect 'my ideas or find words. I have a prodigious 'swimming in my head. I then totter to a 'chair, and feel quite distinctly a round 'body rolling upwards and downwards from 'below my stomach to my throat, while 'the part about the navel is dragged in 'towards the spine. At these moments, it 'is as if something was drawn tight round 'my neck, or a ball was fixed in the top of 'the throat. I sometimes do immediately 'get rid of these feelings, and sometimes do 'not, but fall into a state, in which my 'senses forsake me; and I lie, as I am told, 'in a swoon, either with a pulse hardly per-

'ceptible and scarcely seeming to breathe, 'or I am drawn awry by spasms, or agitated 'by convulsions, in which my hands and 'feet especially bear a large share. In this 'state my pulse bounds in a violent manner, 'and my heart palpitates so that its throbs 'may be seen through my cloaths. Meanwhile 'a loud rumbling noise commences in the 'stomach. After this has arisen to a con-'siderable height, the seizure abates, and I 'come to myself by slow degrees, but am 'apt to experience a distressing head-ache 'for many hours after; and sometimes I have 'perceived one of my arms to be particularly 'heavy and benumbed. These sort of at-'tacks I have observed to be brought on by 'any considerable unpleasant emotion of 'mind, by sudden noises, particularly when 'I have risen unrefreshed and tremulous, by 'the scent of certain flowers and perfumes. 'They often begin by a violent sneezing.

'The intervals become gradually more 'distressing. There is a constant sense of 'relaxation, or rather of all the flesh dissolv-'ing from the bones. On making an attempt 'to perform the smallest office, which requires 'a little more force than usual to be em-'ployed, the head seems full, the temples 'and neck set to throb, the breath instantly

'grows short, and an aching comes on, pre-
'ceded by an universal tremor of the limbs.

'All the time of the cold weather in winter I am miserable. The least breath of frosty air makes me shrink and sob, and, except in a very close room, I feel as if skin-bound for days together.'

What physician but would put down as hysterical a case, marked by such circumstances as these: and what reader but must see the very close resemblance between such a disorder and epilepsy? The following view of some of the principal phænomena in nervous affections, will render the resemblance more striking. There are few collectors of observations who do give instances (and in some they abound), of hysterical people becoming epileptic. The same is also true of hypochondriacs. The fact would not indeed apply as decisive evidence in favour of the opinion that both complaints possess a common nature, were it not for the similarity of symptoms. For a cutaneous affection, suddenly disappearing, has frequently been seen to be followed forthwith by an internal disorder of a character totally different, as the itch, for example, by a fit of the asthma. But when hysteria is followed by epilepsy, the general state of the invalid between the paroxysms is much the same, and the nervous symptoms are only increased in violence, or varied in some of those modes, of which all nervous ailments are so peculiarly susceptible. It has been said by some writers that violent fear, as that of instant death, constitutes the sole perceptible difference, this feeling being the characteristic of hysteria in ambiguous cases. But I have seen two cases; one very lately, in which convulsions and swooning or sleep, with undiminished action of the arterial system, followed upon the sight of a person in an epileptic paroxysm, and returned with the hysterical ball and limpid urine, but no terror; and another, in which the seizures continued the same in other respects, only they come after some time to be preceded by alarm for immediate safety. The latter, if any one chooses, may be considered as epilepsy lowered down into hysterics. I believe, whatever names be used, terror will always be present, where the stomach is much out of order at the time of the fit. And it is sometimes curious to see in hysterical epilepsy, when medicines have removed the coldness and violent noises of the stomach and bowels, that the other symptoms, as trembling, torpor of the feet, spasms, and convulsions, shall supervene without the terror of dissolution.

Epilepsy is not to be distinguished from other nervous affections by being defined, *a convulsive nervous malady beyond the reach of art.* Epilepsy is not unfrequently cured by medicines. And there are many forms of nervous affection which hold a middle course between hysterics, epilepsy, and what has been termed painful epilepsy, *(epilepsia dolorifica)*, because it comes on with pain as intense as any to which human nature is liable. And these have defied medical means, as much as any cases of epilepsy.—A lady is well known to some physicians in London, Edinburgh, and Bath, as also to the author of this Essay, who has some of the principal symptoms common to the three complaints, and whose variety of symptoms would make a pamphlet. I have seen her speechless, and with her mouth drawn to one side, by the most excruciating spasms, and imagined her insensible at the time, but she convinced me of the contrary, by afterwards repeating what I had said. She has never become insensible, except on bathing in the sea and on a few other occasions, but is left with her faculties entire to contemplate and feel her own wretchedness.

It is difficult to choose among the endless variety of examples. Yet in order to persuade people to avoid the common causes of all, it seems essential to impress, by a sufficient number of examples, the tendency of nervous complaints to run into each other. PEILL, LEIDENFROST, and KORTUM (*Beitraege z Arzney-wiss. Gottingen,* 1796), relate the case of a young lady, given to sedentary devotion and much subject to pulmonary symptoms and to spitting of blood among the rest. At twenty-three, depressing passions were followed by a strong hysterical fit, of which she had no return for half a year, except some slight, transient feelings. But at the end of that time, she was thrown by the same causes into a still more violent paroxysm. This left her in such a state three or four times every day, that a dreadful accession came on from provocations absolutely inevitable: such as the entrance even of a very familiar acquaintance into her apartment, the falling of a spoon, the unexpected approach of a cat, the noise of boys in the street.—Sometimes a sense of cold in the extremities, murmurs in the head, anxiety about the chest precede, but the patient commonly falls down without notice. Immediately upon this, violent convulsive movements begin, in which every part of the body is affected. Sometimes spasms bend the trunk forwards, sometimes backwards; sometimes the upper extremities perform strange contorsions, sometimes the lower; sometimes quick motions actuate the tongue and organs of voice, and then strange inarticulate sounds are uttered. Sometimes the whole body is agitated at the same time, sometimes the head only; and the convulsions are so prodigiously quick and powerful, that a bystander would expect every joint to be dislocated. Yet she is generally sensible all the time, and if the tongue be free from spasm, answers every question naturally. Sometimes, though seldom, the head is affected, and she is delirious in the midst of the paroxysm. On the remission of the symptoms the patient complains of cold of the extremities, of excessive anxiety and oppression of the chest. At the conclusion of the report, she had continued for twenty years subject to these paroxysms though their frequency diminished. She lost nothing of her flesh all this time.

In the first part of Zoonomia, similar instances are briefly related, with a view to prove that delirium consists in convulsions of the fibres, which the author ingeniously supposes to exist in the organs of sense. But though the convulsions often stop when delirium comes on, the above statement proves the succession not to be universal. The disorder of the nervous system is sometimes great enough to disturb it through its whole extent; and the apparent substitution of one symptom for another, goes hardly a step towards proving them to be of the same kind.

LEVRET relates the history of a female who during her first pregnancy had the hysterical ball in the throat, but during her next pregnancy, no hysterical symptom affected any of the internal organs, except that the voice was a little weakened. But she had convulsions of the external muscles, and these, when they came on, lasted several hours. About the fifth month of her pregnancy, she had two fits a day, which between them lasted eighteen hours. Upon the whole, she was convulsed half the time between conception and delivery,—that is to say—near 5 out of 9 months. The comparison of her two pregnancies may serve, among a thousand other examples, to shew how simple convulsions and hysteria play into one another. Uneasy sensations in the head, confusion, flying chills and heats, faintness, languor, palpitations, startfulness, disposition to be terrified, exist more or less strongly, and are compounded by the peculiarities of constitution, and by accident, into all imaginable shapes. So that perhaps no combination of these affections could be feigned, which cannot be matched in faithful medical reports, and the reality exceeds whatever the most fertile fancy could invent in readiness to be affected, in violence, in the whimsical manner in which the symptoms are grouped, and in the rapid changes from one state to another. In this last respect, what is fabled in romances and masks concerning the power of enchanters' rods to induce, in a moment, the stiffness of a statue, or to restore the spell-bound person to motion, is much exceeded in nature; and it is possible that fiction took the hint from this species of reality, since it always must borrow from one species of reality or another. That the dropping of a hair-pin on the floor should make a person start from her seat, and fix her in a preternatural posture, by occasioning preternatural, fixed contractions of the muscles, or agitate her by contractions and relaxations equally preternatural, till she sink into insensibility, from which she awakes into vehement delirium, is hardly credible to those who are conversant only with the healthy, and the sorts of sickness to which the robust are subject. On comparing an individual liable to these sad varieties of being, to the engineer who stands unmoved amid the thunder of a battery, to the seaman who maintains his footing upon the deck or ropes of his vessel, reeling under the shock of the elements, or to the Indian who exhibits the signs, and probably feels the throb, of intense delight, while the flames are preying upon his flesh, how astonishing do we find the range in human susceptibility to the effect of the powers by which we are surrounded! how important is it to consider the causes of the difference, if on the one hand we should have as much reason to suspect, that resistance to pain may be united in the highest degree to capability of pleasure, as we have, on the other, to be persuaded that those who have become in so high a degree sensitive, are nearly lost to all but painful emotions, and that if their organs are like wax in being impressed by external appulses, they too often resemble adamant in retaining what impressions they may receive!

Remarks on some principal NERVOUS SYMPTOMS, separately considered.

CONVULSIONS AND SPASMS.

These symptoms are among the most common of all. In the usual order in which indispositions of this class arise, they begin by occasional stiffnesses in some of the limbs. These are owing to increased tone of the muscles, not yet strong enough to be denominated spasm. At the same time, palpitations of the heart from slight sensible impressions or emotions are frequent. A small number of the fibres belonging to a single muscle, act quickly and forcibly, and give a feeling as if the skin was alive. On going to sleep, a sudden disturbance is produced by the shock of a whole limb. The days grow more and more comfortless; the nights less refreshing. The stomach gets out of order; and females are distressed by variety of irregularities and weaknesses in the functions peculiar to the sex.

There are two powerful reasons, which ought to induce every invalid in this degree to recur to measures, fitted to restore the constitution to its healthy state. In the first place, he may promise himself that his measures will be successful: and, in the second, there is reason to apprehend from the close affinity between nervous complaints, that the continued operation of the same causes will aggravate the present slighter symptoms into some grievous malady; and that any powerful occasional cause, though it might operate upon people of firmer temperament with little injury, will plunge those, who are thus prepared, into the horrors of convulsion, epilepsy, or some kindred, but anomalous complaint.

An account of restorative remedies does not belong to my plan. But though with moderate skill in their application, they are almost certainly efficacious, it will be of little service to be restored to a healthy state of feeling, unless the necessary endeavours be employed to secure by regimen the advantage that shall have been procured by medicine.

Many of the general directions in the foregoing numbers apply to this purpose. Of this kind are the contents of the latter part of the fourth essay, upon which too much stress cannot be laid. It is equally necessary to guard against the errors in living that lead to consumption; for it is perhaps more generally true, that those who become consumptive in our climate are previously nervous, than that they are previously scrophulous. The reason will easily be understood from the discussions, which this point has undergone. Such are the vicissitudes of the atmosphere of these islands, and such the tendency of our habits, that constitutional weakness will very frequently be visited upon the lungs; and, more frequently still, will constitutional weakness shew itself by tremors, startings, wakefulness, fatiguing sleep, and by other indications, enumerated in the course of the present Essay.

Our most attentive observers have noticed the connection between these two classes of complaints. Thus I find it remarked "that a "*phthisis pulmonalis* may be also the conse-"quence of nervous disorders, when the "morbid matter producing them falls chiefly "upon the lungs, or when the vitiated chyle "or blood forms obstructions in that organ." I only wish my readers to take the authority of Dr. WHYTT, perhaps the best writer we have, certainly the best of the older ones, on the present subject, in reliance upon his long and attentive experience, without paying any attention to his hypothetical explanation of the fact.

It is a striking coincidence, that cold, slowly or suddenly applied to the feeble, will in proportion produce either or both diseases. We have seen how this cause acts upon the lungs. Twitchings of small bundles of fibres and startings of whole sets of muscles, are perpetually produced by imprudent cold-bathing, together with depression, confusion of head, and a whole tribe of kindred symptoms. The lady alluded to above, as uniting painful and common epilepsy with hysteria in the same person, and who never spends a day, and seldom an hour, without severe distress in some shape, had all her symptoms aggravated by immersion in cold water. I have known slight passing stupors converted into epilepsy by the shower-bath. A remarkable instance of the bad consequence, resulting from the vulgar persuasion of the bracing effect of cold bathing in a nervous invalid, was lately mentioned to me in conversation. It was the instantaneous production of the worst of convulsive diseases.—A nervous lady, who, contrary to the advice of her medical attendants went into the sea, was seized with epilepsy on the spot, though she was not only free from apprehension, but full of expectation of benefit. The fits continued at intervals, as I was assured by Mr. Thwaite, junior, a very ingenious apothecary of Dublin, by whose permission I here produce an example, so capable of affording a salutary lesson to invalids.

CATALEPSY.

Before I quit the head of spasm and convulsion, I shall add a few words concerning a curious affection of the system—not that I consider it as of particular importance to my readers, but lest it be supposed that I have overlooked what I ought to notice. This is *catalepsy*. The popular names of herbs frequently betray ingenuity, and sometimes elegance of fancy. But in chemistry, anatomy, and medicine, many denominations imply unsuccessful, or at least misplaced, attempts at wit, which shews the taste of the learned to great disadvantage, in comparison with that of the unlearned. As epilepsy means the disorder that suddenly seizes, catalepsy, I suppose, means that which both seizes suddenly and holds fast. Catalepsy may be regarded as rudimental epilepsy. The muscles have a contractility which seems to tend towards spasm or convulsion, and, in the continuance of the disease, rises into those affections, or alternates with them.

Catalepsy often comes on with some slighter nervous affection, as a sense of weakness in the whole body, tremor, nausea, pain in the nape of the neck, cramp in the stomach. The patient retains the position in which the attack finds him; if he is standing he remains so; if he is gesticulating, he preserves his posture. The vocal organs are arrested in the midst of a word, and often finish it afterwards. The trunk and limbs are perfectly passive. If the arm be raised to an angle, it does not descend in the smallest degree. Consciousness is suspended. The pupil of the eye has been observed, in different cases, not to contract on the approach of a strong light, or to expand in the dark.—But it is in vain to expect an exact continuance of the same state of symptoms in any considerable number of nervous cases, or that the infinite number of moving and sensible parts in the human frame shall act and feel alike. I have above mentioned a boy, cataleptic in the large muscles that keep the body erect, but convulsed in those of the voice.—Sometimes during the quiescence of the external muscles, deglutition takes place when food is put into the mouth. I have lately heard, on good authority, of a person subject to short seizures, who answered, after recovery, almost any number of questions proposed while he was affected. Here it is evident that the ear was duly affected by its appropriate impressions, but that a suspension took place in some part of the series of animal actions between the affection of the external sense and the motion of the vocal organs. It is to be wished that the ingenious practitioner, who had the care of his patient, will have availed himself of so rare an opportunity of ascertaining different particulars concerning the operations of the human intellect. From the facts, as above nakedly stated, it would appear that the effects of sensible impressions are of some, not inconsiderable, duration; and perhaps the exact time of their remaining at a certain intensity in given states of this individual might be ascertained. This, as I shall have occasion presently to point out, is a question of some importance.

The paroxysms are often short, but some are described as lasting almost twenty hours. They go off with uncomfortable feelings, as languor, dejection, drowsiness, headache and vertigo. Catalepsy occurs in the intervals, and apparently as the substitute of hysteria and similar affections. Patients have carried the disorder about them for life without apparent detriment, as has sometimes happened too with regard to epilepsy. But catalepsy, when not cured, as it generally may be with great ease, will pass into a more alarming form of nervous affection, and has been seen to end in madness itself.

The very same state of the internal organs has been observed, previously to accessions in catalepsy as in hysterics and epilepsy. And on dissection after death, the very same alterations have been discovered in the scull and the parts which it contains, as those already described in epilepsy.

GIDDINESS.

Giddiness, or vertigo, has been the subject of many elaborate disquisitions by medical writers. Among the last who have distinguished themselves are Doctors HERZ, WELLS, and DARWIN. I shall have opportunity in a future work to state the facts they have observed, and to bring forward my own sentiments. The reader versed in physiology, will have no difficulty in collecting from what follows, that I have not been able to adopt the sentiments of the illustrious Darwin. It is impossible not to feel the powers, which he has exerted on this question, but there are few where his ingenuity appears to me to have been less successfully exerted—an observation which I introduce here, lest it should be supposed that I neglect a great authority merely for want of having considered it.*

* The fault of Dr. DARWIN's theory of vertigo, as happens in other instances of erroneous philosophizing, lies in bring-

As to the nature of giddiness, I refer to simple feeling. Let any one, who is unacquainted with it, turn round on the same spot

---

ing together as cause and effect phænomena, which do not belong to one another, and where they do, in putting the effect for the cause.—Thus it is observed, "when a child "can stand erect, if you gain his attention to a white "handkerchief, steadily extended like a sail, and afterwards "make it undulate, he instantly loses his perpendicularity, "and tumbles on the ground."—(*Zoonomia, article Vertigo*) Now I doubt not but this simple experiment has frequently sounded decisive to readers, not over-scrupulous in receiving opinions. Yet it remains to be proved that the fact belongs at all to the subject of vertigo. A child can hardly explain himself as to his feelings. Certainly there appears no sign of vertigo in the eyes of children. suddenly falling as they have just begun to walk, or in their locomotive powers immediately afterwards. So that the affection would seem very slight and transient, if it exist at all. Other phænomena, exactly analogous to this, render it probable that the falling child is not giddy. In the subject of the experiment, certain sets of muscles, by supposition, have just acquired the habit of moving together, and others in succession. But such new associations, we know, are dissolved with the greatest readiness. The intrusion of almost any new impression is sufficient. Thus, when a child has just use enough of his hands to hold any small object, if you attract his eye to something else, he instantly lets go his hold. In a month or two more, he will at once retain the first object and observe the second. But what has the one state or the other to do with giddiness? No one, I presume, suspects that it has any thing. Yet, where is the difference in the two experiments? In the unencumbered arms of children, the for a time, and he will understand the import of the term. I am concerned with it only as a concomitant of nervous affections. It attends them all, that is, it exists in a considerable number of epileptic*, cataleptic, hysterical and other nervous ailments, though in certain other cases under each denomination, it may be seldom felt. Swimming and confusion in the head seem to be sensations nearly allied to vertigo, only inferior in degree. They all diminish or dissolve the connection between the ideas and muscular actions, as well as those between the different sets of muscles themselves. This is matter

---

muscles gain the power of acting together so as to grasp, sooner than those of the legs, trunk, and neck, so as to keep the body erect, and to move it forwards. When an impression upon the eye destroys the co-operation of the one or the other, it subsists in nearly the same force in them respectively; and many other impressions upon the eye will succeed just as well as an undulating white handkerchief. So will many impressions upon the ear. And I apprehend there can be no just suspicion of vertigo upon these occasions. Some horses are never so liable to trip or fall as when slightly startled. Here is again the same effect—a stop to the usual progress in the combined action of the muscles. Have we any reason to suppose the horse feels giddy?

* Vertigo, et maculæ nigræ, ab oculos volitantes, a quibusdam epilepticis vix unquam absunt. (Heberden, 140.)

of familiar observation. Every one feels less firm upon his feet, less able to muster his thoughts, and less free in the use of his powers in general, when the head is confused; when it swims or feels dizzy. Aptness to these sensations in any degree, is nearly the same thing as aptness to become nervous. Whoever has been educated, or manages himself so, as to escape this propensity in all its gradations, will have great certainty of exemption from the larger number among nervous disorders; nor will any thing perhaps induce the calamity, unless it be one of those overwhelming accidents, which bear down the moral faculties, as the fall of a building crushes the body. Probably indeed the mind can be more entirely withdrawn from the dominion of accident.

The doctrine of temperance in respect to meats and drinks, will strike every one in the first place as essential to his security against habitual vertigo. It follows nothing more frequently than those states of the stomach, which arise from the opposite extremes of gluttonous debauchery and abstemiousness. When the digestive organs are reduced by either means to that state, which is accompanied by nausea, retching, acid eructations, heart-burn, foul tongue, bad taste in the mouth, flatulence and irregularity of the bowels, it rarely happens that the head continues steady. As the hypochondriacal and the hysterical are always subject to be plagued by these miseries, it is no wonder that giddiness should be one of their most frequent complaints. Dimness of vision, which is so nearly allied to giddiness, has often been seen to depend in a very curious manner upon the state of the stomach. Observers of undoubted fidelity have met with cases, in which persons subject to sourness in the stomach, have seen every thing as indistinctly as if a thick smoke or mist were before the eyes, whenever the sourness arose to a greater height than usual; nor did clear vision return till the state of the stomach had been changed by absorbents, vomits or bitters. The same is the case with regard to noises in the head. I know a person, who if he waits too long for breakfast, is subject to these noises, and is unable to hear his clock strike from his sitting apartment; and in whom eating removes the noises and restores the faculty of hearing. It follows that fasting should never be so long continued as to be succeeded by so troublesome, and in many cases, so dangerous a forerunner of nervous disorders as giddiness.

It is too well known to be insisted upon at large here, that the frequent use of strong liquors renders giddiness almost a standing symptom. Drunkards not unfrequently fall into the unhappy necessity, as it appears to them, of recurring to their bane; for they find themselves scarcely able to perform so common a motion as walking, or even standing, with steadiness, till they swallow a certain allowance of the liquors that have reduced them so far to the condition of second childhood. I know not if this opinion arise from want of information or from the mere weak desire of indulgence. But whatever be its origin, it is an error, and a dangerous one. The inability may, and ought to be taken off at the moment by means that have not only no subsequent tendency to increase it, but the power of restoring to the limbs their proper firmness.

Intemperance of another kind directly tends to induce that deplorable vertigo, which verges on the most incurable forms of nervous complaint. In speaking of the effects of mens' own conduct, I feel that I ought not to pass this source of disease unnoticed, but, to avoid giving offence, I shall content myself with quoting from a writer of vast experience and unquestionable veracity, the following account in a dead language. "Jacturæ nimia seminis gravissimos et sæpe insanabiles vertigines facit: tunc præcipue, ubi semen turpissimo vitio, quod vinculum illud, quo sexus sexui jungitur, divellit, et licitæ veneris tædium ingenerat, manustupratione scilicet, effunditur. Alia occasione de morbis ex manustupratione ortis agemus. Id notasse modo sufficiat, morbos ex hac impurâ scaturigine ortos aut nunquam aut vix curari. — — Illud quoque hic adhuc addam, curam nunquam a sola pharmacia inchoandam esse; etenim tanta est nervorum per iteratos actus irritabilitas, mobilitas, ευμελαβησια tanta, idearum allicientium vivacitas tanta, tam prompta actionum huc pertinentium successio, ut medicamentorum roborantium nervorumque usus multum impar sit, quia irritabilitatem ipsam, quæ corrigenda esset, vi suâ nervirâ irritent magis. (*Max. Stoll prælectiones in diversos morbos chronicos.*)

It will easily occur to the intelligent reader that this is to be extended beyond the single habit, here described as injurious. It is common to see all the nervous symptoms enumerated in the course of the present tract, and nervous atrophy, originating in the marriage state, and especially among the weaker sex, from a cause little suspected by either of the parties. The cause alluded to, like others which agitate the system with strong sensations, is the more mischievous, the more the system is already inclined, by excessive sensibility and weakness, to nervous disorders. So true it is that *on a abusé d'un petit nombre d'observations pour en conclure que le mariage est le remede à tous les maux des jeunes personnes. Il n'y a rien de plus faux.**

Whatever other causes may be concerned in the production of a confused or dizzy head, there is one universal among the more cultivated members of society. This cause is seldom left to operate singly, but if it were, it would be adequate to the worst effects that are observed in nervous cases.

We have already seen that a feeling of vertigo seldom fails to be experienced, when a number of impressions are received in rapid

---

* Zimmermann a vû une jeune femme, que s'etoit blesseé plusieurs fois après des coliques spasmodiques, très-fortes, et qui avoua enfin que ces coliques etoient la suite des devoirs conjugaux remplis trop souvent par son mari—ce qui lui occasionnoit une extrême foiblesse, et ensuite ces douleurs atroces et insupportables.

J'ai été consulté pour un mari et femme, chez qui les mêmes excès occasionnoient les mêmes accidens, peut-être meme plus forts chez la femme.—(*Tissot.*)

succession by any of the senses. It is most sensible in the weak, but never fails to affect the strong, if long enough continued; and by frequent repetition will lower them to an equality with the weakest. "I was ac-"quainted" says Dr. HERZ, "with a man "whom the whole country admired. His "conformation rendered him liable to accu-"mulations of blood in the head, and by "intense exertions of the mind, he had con-"tracted an excessive weakness of the nerves, "during which he could not withstand the "conversation of a certain friend, who pos-"sessed great loquacity, and always let loose "a very rapid stream of words, without ex-"periencing a vertiginous seizure on the "spot." Dr. HERZ adds, that he had once the care of a young man, so weakened by the habit, spoken of in the preceding latin quotation, that at last he could not listen to a person, speaking fluently and in a moderately energetic manner, without becoming so giddy as, for fear of falling, to be obliged to sit down or lay hold of some neighbouring object. We have seen that slighter occasions still will bring on fits in some epileptics.

The continued noise of a waterfall, or the uniform sound of a mill, excites a slight dizziness, says the same author, as any one may convince himself by his own experience, and this without particular predisposition.

A great number of objects, viewed in quick succession, breed indistinctness in strong-headed people, and dizziness in others. So do a number of ideas, quickly following one after another, whether excited by associations that take place in some diseases, or irritations of the senses.

A certain rate of presentation, therefore, of impressions and ideas, is necessary to clearness of comprehension; and if we do not dwell upon each for a certain time, it will not only fail to retain its place among its associates, but do some violence to the intellectual organs. Of this violence, the feeling of embarrassment of thought seems to be one degree, and giddiness a higher. The familiar experiment of whirling a fire-stick, shews the senses to continue to be affected for some time after they are struck; and facts, already laid before the reader, seem to indicate that the parts of the nervous system, connected with the senses in the performance of mental operations, are in the same predicament. It would appear, therefore, that the power of these organs to perform their appropriate functions, will be impaired by stimulating them into one kind of action, while that, excited by another, keeps up in force. But if so, a very principal part of the nervous system will be deranged in its movements; that is, some nervous malady will manifestly be induced, though we may not be able to trace the morbid progress any further, in the altered state of the organs.

The improvement of the mind, resulting from geometrical studies, is universally remarked. I know not if it have been tolerably explained. But it is evident that one of its effects is to render the impressions distinct, by the necessity of dwelling for some time upon each. For it is notoriously impossible to proceed along the steps of a demonstration, without a clear conception of the preceding. A diagram like that of the fortieth or the forty-seventh propositions of the first book of Euclid, appears a perfect chaos of lines to the inexperienced eye: but use in contemplating such objects, confers the faculty of singling out every square and triangle, as if it were not traversed by the lines of any other.

But not only is each impression more perfect from the compulsory pause of thought, but the links in the ideal chain are more firmly closed; or, in other words, the associations are stronger.—Beginners are always awkward in designating angles. They puzzle themselves by saying *this* angle and *that* angle: but are at last obliged to mark every one by its denominating letters. There can hardly be a more profitable exercise in association. Others, as the experimental sciences and the law, are good. They have nearly the same requisites, as their terms do not disturb the clear current of thought by stirring up the mud of feeling. Either, gently administered, may be efficacious in the complaints, of which I am now treating. A maddened brain has been composed by mathematics, and I think the whirl of some vertiginous ones may be controuled by the same charm.

In many mechanical arts, the more deliberately the apprentice makes his first movements, the nimbler does his hand finally become. It is with the head as with the hand. A person, by once receiving his impressions slowly, will come to call up a multitude of distinct ideas in a short time. He will be able also to bring premises along with remote consequences into a single point of view—much like the surveyor, who by pushing his triangles, one by one, over the face of an extensive country, gets all the objects it presents, in their relative situation, within the compass of a few sheets of paper, and finds his labour as efficacious as if he had actually enlarged the sphere of the human senses.

Thus it is to attain quickness without confusion. But it is the misfortune of almost every one in circumstances to study his own amusement, by proceeding too rapidly from the first, to miss the quickness and to get the confusion. This misfortune is the parent of a thousand others.—*Did you see the papers to-day? Have you read the new play—the new poem—the new pamphlet—the last novel?*—You cannot creditably frequent intelligent company, without being prepared to answer these questions, and the progeny that springs from them. So you must needs hang your heavy head, and roll your blood-shot eyes over thousands of pages weekly. Of their contents at the week's end, you will know about as much as of a district, through which you have been whirled night and day in the mail-coach. The STATIRAS of one stupid fiction join with the ALEXANDERS of another, to mob the memory; and between them all, they breed as great a hubbub in the brain as that which prevailed in AGRAMANTE'S camp. What wonder then that we should hear complaints against the age as wanting energy of feeling and compass of mind? What wonder that while idea reels against idea, we should so often experience an analogous unsteadiness of footing?

It had fallen in my way, several times before, to speak of novels, as baneful to health. I hope now that I have led the reader to a situation from which he can perceive the mischievous inward workings, occasioned by quick, desultory reading in general. Concerning the effects of fictions that kindle the tender passion, there remains a great deal more to unveil, if one durst exhibit them in their nakedness. A modern author of good credit assures us that he knew a female, so affected by the reading of a romance, having for its subject a disappointment in love, as to be deprived of her senses and fall into convulsions. *(J. P. Michell, M. D. in his answer to the question by the Utrecht society).* Who has not found how much the ideas of the mind influence the bodily members? How quickly, for example, does the thought of a favourite dish make the mouth water, that is, occasion the salivary glands to act with increased vigour? Other parts offer a handle to the mind equally with those glands. Pursue this principle, and you will understand how possible it is that a variety of prevalent indispositions, as *fluor albus*, tendency to miscarriage, and even a dropsy of the ovarium, may be caught from the furniture of a circulating library.—Thus, then, do disorders of a very different nature connect themselves with nervous disorders by community of origin. The connection may be traced a great deal further still.—In natural men and women, whose pulse heaves slowly into a vigorous stroke, the vital motions proceed in an even and placid tenor. The face is not flushed. The feet are not frozen. No excrescences arise. The most accurate anatomist cannot feel out any disorganizations. These evils are shared among the luxurious and the indolent, whose artificial modes are for ever destroying the balance of action in the system, and reducing one part to death-like torpor, while in some other, as if to make amends, they excite a mischievous activity, or kindle a spurious sort of inflammation.

The last century witnessed a great alteration in the culture of the youthful mind. Wise men joined in condemning the jargon of the books used in the first stage of tuition. At the call of these wise men, a hundred writers seized the pen in behalf of the tender mind. And a new species of literature was created by their generous zeal. I shall not dispute that the adoption of intelligible books is a great improvement in early education. But I think myself certain that there remains a greater still; namely, to lay the use of books pretty nearly aside, and, during the season of the mind and body, to trust principally to the senses, directed by the living voice—that living voice, which, as Quintilian says, affords substantial nourishment, while books (it is certainly the case with much reading) do little perhaps but stimulate. As years advance, the mental diet may be gradually changed. Book-knowledge may be mixed with oral and sensible instruction.—The juvenile library! With submission, I must consider it as little better than a repository of poisons. What acrimonies will it not engender in the habit! what obstructions in the body! I know the ardour with which the little students seize these intellectual delicacies. I have seen how the parents delight in their eager application, reflecting all the time how much better off their children are than they were themselves at the same age, and gratified by recollecting their own disadvantages.

But do they not over-rate these disadvantages? Worse books were put into their hands, it is true. But then they had a great deal less of books. It was an article, of which they took only just what they were forced. So the quantity perhaps will compensate for quality.—At least, if little people are to be encouraged to read as continually as those of larger growth, one thing is but fair. The parent, when he brings a daughter or a son, "morbidly alive in every nerve" to the physician, should be candid enough to tell, how many dozens of Lilliputian volumes he devoured within a few years after he learned to read, and of how many hours' use of his limbs and senses he was daily defrauded in consequence.

It is not here the place to explain how to render education, at its outset, less dependent upon books. One expedient is obvious. Parents have read themselves. Let them then serve out the information they have collected in proper portions to their family. But methinks I hear parents own that they have read enough, but have lost all their reading. Well then, why do they wish the children to read so much?—It is plain it does not improve; and it cannot be for mere amusement, as might have been their own case when grown up. Running about is as amusing to children, and wholesome into the bargain. Children will remember much better what they hear, if well told, than what they read. Let parents then henceforward read on, but let them read so that they can relate. They will, I know, go over less than one tenth of the pages, but they will be twenty times as knowing, and in trying to preserve others from the imbecillity, under which they labour themselves, they will recover a share of the vigour they have lost in literary dalliance.

I have heard it remarked by foreigners that the English gentry make, in general, less use of their limbs than those of any other country similarly circumstanced. It is certain that the customs of our female gentry are, in this respect, exceedingly unnatural, and, without subtilizing too much, one may conclude that this circumstance will render them much more subject to vertigo, and to all the evils, usually complicated with it. One consequence of suffering their natural powers to languish in inaction can scarcely be doubted. These powers will perform their office imperfectly. The muscles of locomotion, like troops insufficiently disciplined, will be easily thrown into confusion; and the accidents, that tend to disunite our trains, will have the worst effect on the correspondence in the action of whole sets of muscles, while they break the connection between the ideas and the single muscles, of which the sets are composed. The experience of instability will confirm the apprehension of falling: and this apprehension will be attended by an eddy of thought, which will in its turn act upon the sensitive and nervous systems with force sufficient to produce fainting, and perhaps a convulsive fit.

Chagrin or apprehension, in any other circumstances, will often occasion giddiness, or a standing disposition to giddiness. During these states, a torrent of ideas rushes upon the mind; and the result is the same as too quick a succession of impressions upon any of the senses. Hence there is a class of valetudinarians, particularly affected on looking down from a height—a common cause of giddiness, but into the manner of whose operation I do not here propose to enquire. This class includes all the feeble and the low-spirited, the hypochondriacal with the hysterical. Thus STOLL, who was a slave, and afterwards a victim, to the extensive practice of medicine at Vienna, found a very large proportion among the monks, from mere dread of a fall, incapable of preaching from an elevated station. No one can be surprised that hypochondriasis should be the endemic disease of monasteries.

Giddiness is the well-known consequence of raising many convalescents into an erect posture, as it is also that of various modes of conveyance with some individuals. But I pass over these facts as little subservient to my purpose of pointing out the manner, in which diseased habits are contracted, and that in which they may be avoided. I do not always particularize every variation even in the circumstances, that strictly belong to my plan. For that would too much dilate the execution; nor will readers of ordinary attention and sagacity find it difficult to supply the omissions by the help of analogy.

CHILLS—SHIVERINGS—HEATS.

Nervous people cannot be said to be subject to the "fierce pains" of temperature, but their uneasinesses are, in return, frequent, and indeed almost continual. It is with these invalids worse than with children, who in sultry weather are wan with languor, and in frost, unless well protected, become lumps of ice. When young persons, who live much within doors, begin to have too great susceptibility of cold, this irregularity of sensation, though only the first stage of disease, and commonly neglected as such, should be considered as a sign that the constitution is in danger. In general, other symptoms will not be long in making their appearance. As the same temperature of the atmosphere, which used to be felt as comfortable, has now become the reverse, so will it happen in a short time with all the other powers that act upon the system.

Unpleasant feelings will arise in most of those situations, in which an agreeable sense of existence used to be felt. A burning in particular parts, as in the palms of the hands and soles of the feet, flushings, heaviness, headache, twitchings, pain and startings from noises formerly borne without inconvenience, disrelish of food and of occupation, and lowness of spirits, will arise in regular succession to indicate that so many sets of nerves have lost their natural or healthy faculty.

The patient's comfort for life will be totally lost, unless a thorough alteration take place in his habits The most essential is to introduce an increase of muscular exertion. A little digging, early in the morning, or turning on the lathe, a few strokes of the saw, or a brisk walk in quest of some interesting object will excite the arterial system sufficiently to guard against that debilitating chilly accession, which makes weak people all day good for nothing,—unless it be to pore over some production of the Minerva press by the fire-side. I suppose the cloathing in this case to be what the season requires for warmth, and the directions formerly given, respecting the times of eating and quality of food, to be exactly complied with. The plan of gentle and encreasing exertion need scarcely ever be interrupted, since there are few days of our year, whose mornings are not sufficiently cool. It will confer the additional benefit of putting an end to that tendency to *take cold,* with which nervous people are so constantly plagued, and it will restore to the objects of sight that amenity, to those of taste that flavour, to those of thought that interest, to all nature that grace and life, which they were losing apace.

Dry, unkindly heats are so general a consequence of chills, that the nerves, from being too little irritable by warmth, at one minute or hour, seem to become too much so the next. Either the chill or the heat is sometimes so slight as to escape inattentive valetudinarians; but by more exact observation, the reality of both states will be ascertained.

A flush, however, may be easily brought on independently of previous coldness. In the account of the epileptic journalist, we have seen that a loose hair touching his ears, would occasion them to burn. This invalid further relates that at night he wore for prevention, a cap tied close behind his ears; besides keeping the hair very short cropped. He says that he was even obliged to be cautious of going into the open air immediately after shaving, as he was liable to a chill in consequence, and that going to bed with his beard at all grown sensibly heated him about the face and head. I have known the strokes of a blunt razor bring on a fit of violent heat about the same parts in nervous invalids with a tender skin; and this without any vexation on account of the badness of the instrument.

Persons, arrived at so diseased a pitch of delicacy, require to be most pitiably minute in their measures for preserving an equilibrium of temperature. Till the art of the physician has restored to the habit somewhat of its original power of resisting the action of the elements, their toilette may, without reproach, apparently equal that of Sardanapalus in effeminacy, and their conduct be as scrupulous as that of MOLIERE'S *Malade imaginaire*, without affording just scope to ridicule. They should cover themselves so as to obviate every feeling of discomfort; they should warm their cloaths in cool weather, before changing; and in every house have license to accommodate their feelings without drawing down upon themselves that observation or officiousness, which to the weak and the sensitive are more distressing than any corporal infliction—when such accommodation implies nothing that need in the least disturb or offend others. But such is, at present, the horrible inhumanity of fashion, that great nervous invalids are often cut off from social intercourse, when they could enjoy themselves with the utmost advantage, and amuse others. For who, for instance, would dare to appear in well-dressed company, without one of those monstrous cravats, which with persons that regard only the rational ends of dress, have a title to rank in deformity with the monstrous craws, and perpetually occasion heats, throbbings, noises in the ears, headaches, and fits? Who could think of laying his feet in a foot-warmer out of the bosom of his own family, or applying his hand round a water-bottle to stop an impending chill? How would people that pique themselves on setting every body at their ease, be shocked if they knew in what penance many of their feeble visitors are kept, and the injury they sustain, in allowing themselves certain trivial indulgences. because they cannot hope to escape a scene at the moment, or being unmercifully laughed at as soon as their back is turned. That Hannibal should not have suffered his soldiers to melt down their vigour in the delights of Capua is a maxim, whose propriety every body sees. And it is one which every body is ready to apply and extend. But the misfortune is that nobody will be at the pains to understand the opposite case. Hence defence against unpleasant sensations and forbearance from disproportionate exertion, though equally essential in their turn, are not tolerated to any thing like the same extent as exposure, labour, and abstinence are enjoined. Yet those who exact efforts from the debilitated may easily convince themselves, and the debilitated who submit to the exaction, will find *that pretension beyond ability always turns out ill in the long run; but that false pretenders to bodily ability will be exposed the most certainly of all, and the soonest and in the most miserable manner.** Weak people therefore should not

---

* " A slight change in the temperature of different parts " of the body was to me particularly dangerous. To main-

only confess to themselves, and, as far as necessary to others, their own weakness. They should also study the disposition both of their

---

"tain an equality, there was room provided in my pocket "for a larger pair of socks, for covers for the wrist and "gloves, for sleeves, for a second handkerchief or for "paper to lay within the waistcoat before or behind, and "for a leathern cap when I step unobserved out of too "warm a room." ——"For entering or leaving my own "house, a vestibule, the inner door of which is much "open towards the next apartment, was particularly advan- "tageous, as on my return I could have the necessary rest "without being hurried, and as I found here the gentle "transition to a warm temperature during the first minutes, "which otherwise were capable of weakening me for hours "together. In other people's houses I made pretexts for "tarrying in the hall or elsewhere. — — The coolness of the "morning air was far more trying to me than that of the "evening. The nervous invalid is prone to shivering. "He sensibly feels the difference between the bed he has "just left and the room; he cannot do worse than take a "walk before breakfast, even in the garden, with a view "to get braced.—It is the warm-blooded man in health, "that cooling may be useful, and it therefore may brace "such a person in effect. In washing, the wrist may be "dried sooner or oftener, or covered with a sock in the "mean time—whenever I wash more than my hands and "feet, the warm-water bottle stands close by, and is of "the greatest use to reanimate the hands." After mentioning a vast number of other attentions, equally minute, the author concludes:—"It occurred to myself and to "many friends, whether all this care to avoid the least "tendency to an extreme of heat and cold might not get whole system, and its particular parts, towards heat and cold, and conduct themselves accordingly. It is not always that they suffer from what they knowingly do or omit. Very frequently indeed, it is for want of being aware of the operation of the most obvious causes. Many invalids, particularly girls, have told me that their feet were not habitually cold, when upon examination the contrary appeared; so that I seldom take an invalid's word for this

" into a bad habit, and bring on an aggravation of my tenderness—thanks to the pressure of disease; it left no time for rejecting visible help at hand in favor of distant possibilities! And now I tranquillize myself with considering that nature has placed man among the animals, that in the temperate zone need much warmth, and that he only, who draws a great deal streaming from within, can dispense with addition from without. In avoiding every excess of heat as also the slightest chill, my measures aim only at the temperature, accompanied by the feeling nearest that of health, and must therefore be in the long run most conducive to it. Besides on a solid addition of strength, the extreme sensibility to a trifling excess of warm or cold goes off. Now I have not so much occasion to be scrupulous in avoiding them, and from this refult, the apprehension that much warming for above two years must ruin the constitution, seems to be refuted with regard to mine, which had suffered in some degree from cold, before I went on this cautious plan——"

Hist. of the 7 years' epilepsy.

point. To know what external things are hurtful to them or otherwise, implies some talent for discovery; and this however trifling, is far from being universal. Medical attendants therefore can hardly go too close in their enquiries or directions. The common observation that some convalescents will faint if they be set upright in bed, when they do not even feel excessively languid on continuing to recline, proves what a vast difference in the system, is produced by an apparently trifling change in external circumstances. And the facts, adduced in the course of this tract, will not only shew that the same thing holds of nervous invalids, but instruct them towards what points they ought to direct their precautions.

TREMOR AND STARTING.

These are well known to exist in the slighter nameless variations of nervous affections, and indeed are among the principal characteristics of those affections. They attend hysteria and catalepsy; and are no less observed in epilepsy. Trembling is often brought on between the paroxysms, and sometimes immediately precedes the insensibility and convulsions. Of this I have an instance in a patient at present under my care. The causes which bring it on are sometimes almost inconceivably trivial.—Whatever occasions the slightest uncomfortable emotion is sufficient. "My writing," says the frequently quoted epileptic patient, "shews by its steady or zigzag strokes, how "it is with the inward man."—"One day," he relates, "I had dozed a little after a con-"siderable morning walk. I then paced "about with a head rather confused, and "was composing myself at my leisure. My "back happened to be turned towards the "outer chamber, when somebody entered "quietly, purely with the intention to look "after me, and addressed me at once in a "loud voice. I was struck all on a heap. "I felt all the nerves in my body, parti-"cularly those in my head, shaken. A little "nausea came on, but was removed by "taking some sustenance. After this shock, "my pulse beat very slowly and feebly, and "I felt weak, for a number of hours."

These facts agree with the others in shewing that the same disposition runs through the variety of nervous complaints. By what particular means these symptoms are to be prevented or mitigated, it is not so easy to say as it is of some others. Tranquillity of mind and keeping out of the way of disagreeable impulses, are means equally obvious to recommend, and difficult to practise. The ear is the sense through which most shocks reach the nerves. Perhaps at those times, when they are most capable of withstanding noises, it might be of great service to resort to some mechanical employment. The hum of a lathe we turn ourselves, and the strokes of a hammer in one's own hand, are less offensive; and they will enable us better to bear the same sounds from another quarter *.

* "What am I to do when the ear, the most sensible of "all my organs, cannot avoid the impression from a rat"tling carriage, the piercing cry of an infant, and other "sharp discords? To wear cotton constantly in the ears "stupifies, and is injurious in other ways besides. I felt "the same stupefaction from stopping the ear with my "finger at the moment, and if I withdrew it a little too "soon, I was still more cruelly affected — — — It was "not till the second year of my recovery that I accidentally "noticed (a thing others may be well aware of) that the "raising my own voice was the speediest, surest, and easiest "security against the too violent penetration of other "sounds. A sort of song of my own, occasions, in most "cases, a sufficient reaction — — — The nervous invalid "most safely enjoys a concert in the next room.—He "should make no scruple to request his acquaintance to "remove canary-birds out of the apartment for visitors." *(Hist. of 7 years' epilepsy)*. Whether from custom, or

A dress which makes an equal and gentle pressure upon the limbs, as laced stockings, and laced drawers suspended from the shoulders, will be sometimes found of very considerable use in restraining tremors.—It is worth while to observe, that when the brain is hurt or inflamed, there arise tremors, twitchings, and convulsions, which seem only different degrees of the same affection. This fact shews the propriety of considering them as nervous affections.

---

from the protective power of agreeable associations, nervous bird-fancying ladies experience no distress at home, while their friends, who are not at all greater invalids, feel their whole frame harrowed by the din, proceeding from the suspended cages. I know no protection against this annoyance, except carrying some loud instrument and playing it off against the birds.--No in-doors noise is so bad as that of birds; but habitual loud talkers often make nervous valetudinarians ill.

" Perfect stillness for a quarter of an hour is the best " remedy for a violent attack on the ear. If the stillness lasts " too long, I feel rather stunned, perhaps because the " thoughts ramble with less restraint; or, if I am reading, " the nerves are too uniformly affected. One may be so " much out of use, that the sudden buzz of a fly, and still " more, a return to society, shall be painful to the ear. " Let the nervous invalid, therefore, take care how he " carries stillness too far.—By this physical cause, the " hypochondriasis of many a recluse may be aggravated." (*History*).

VOCAL SOUNDS.

During the insensibility that intervenes between the reveries of somnambulists, and that which is common in hysteria, indistinct efforts to speak are not unfrequently perceived. They are exactly of the same kind, as a great deal of what is called talking in one's sleep; and this is just what might be expected in complaints so similar in their nature. That this happens in complicated catalepsy is evident from the case of the boy, related above, whose rapid mutterings a number of people heard in the patient's room at the Pneumatic Institution, while his large muscles were rigid. But this happens no less in epilepsy. The following is a passage in the account of the seven years' epileptic patient, who was surrounded by people of intelligence, anxious to watch him narrowly in proportion to his own wishes for information respecting himself. "I sunk "backwards insensible after a scream, "stretched out my limbs, and threw my "head backwards with upturned eyes. The "mouth foamed, and I lay for some mi-"nutes with my thumbs turned inwards, "and had feeble convulsions. Afterwards, I "threw my arms back with force, and they "were agitated more than my legs, while "the trunk and head made efforts to raise "themselves. This second act lasted some "minutes, and during its continuance, there "was heard *a muttering, which came near to "the natural voice, calling for help*, though it "was not distinct enough to be ascertained. "After this second act, I closed my eyes; "the thumbs ceased to be drawn in, and the "quiet insensibility (fainting, *sopor*) during "which I hem a great deal, continued."

The scream, here mentioned, is a very frequent symptom at the beginning of a fit. It seems to me to be merely a convulsion of the organs of the voice and of those concerned in respiration, not depending upon pain. The writer, from whom the preceding quotation is taken, repeatedly carried his attention far enough into the paroxysm to be sensible of the scream uttered by himself, but had no sensation of pain, to which he could impute it.—Whether any pain be felt during the epileptic paroxysm, we have no direct means of knowing; nor is it connected with my present design to enquire, unless the probability of its existence may operate as an inducement to greater care in avoiding the predisposing and occasional causes of this disease. The negative can scarcely be asserted of all fits, even though we should not believe the symptoms to be efforts to relieve pain. But it is with the state of the soul in the fit as after death. No one has come back to make us a report concerning the one, nor has any one recalled the ideas he has had during the seizure, as sometimes happens with regard to a dream, which on awaking we know we have had, though we cannot bring to mind the particulars, till some associated idea calls them up all at once.

DISUNION or WEAKENING of TRAINS.

This is general in all nervous complaints. The expression that a *person is fluttered*, means neither more nor less than that his thoughts and actions do not proceed with their usual regularity. Improvement of the understanding depends upon the accession of new ideas or upon fresh associations. Nervous invalids not only find the bond of union weakened when they suffer from their peculiar symptoms, but also during the intervals. The tremulous and over-sensible, the epileptic, the hysterical, the hypochondriac, very generally agree in complaining of a decrease of memory. One of the most judicious medical practitioners I have met with, whom hypochondriasis has drawn for a time from the exercise of his profession in a distant country, has this day assured me that his faculty of recollection, originally powerful, has greatly declined upon the whole, and at times has been such that he could not recall all his own name.

Fatuity, imbecillity, or idiotism, is principally characterized by the want of the power of association, or by loss of memory. The long continuance of nervous disorders always impairs this faculty. When the accessions are frequent and severe, it is often nearly destroyed altogether. In children it is weaker, and therefore more destructible. Accordingly, we frequently find very intelligent infants rendered ideots for life by convulsive or epileptic fits. Since the first part of this essay was written, there has been brought to me from one of the neighbouring villages a female child, eight years old, who being thrown into convulsions at sixteen months, during dentition, probably for want of having her gums lanced, became and has continued idiotic, with nightly paroxysms of epilepsy.—This is common; but it is most important to my readers to have what is most common in the origin of different complaints, distinctly exemplified.—"How "easily" says the describer of his own epilepsy, "might I have been rendered an "ideot, had the complaint lasted longer,—as "I judge by my feelings, and by the more "distinct indications that occurred in the "long-continued stupors. By dint of de-"creasing memory, and increasing imbecil-"lity of nerves, the lucid intervals must "have become shorter and shorter; conse-"quently also, the operations of the intel-"lect, which cannot go on without the help "of organs, at least here below.—How little "is the spectator capable of pronouncing "upon the internal worthlessness of any "human mind!"

Other nervous complaints no less bring on mental incapacity of every degree—sometimes blank idiotism, sometimes idiotism with intervening accessions of frenzy. Of the last variety, there stands upon record no example, superior in celebrity to that of Dr. JONATHAN SWIFT. Deafness, giddiness, want of vigour in the stomach, lowness of spirits are represented as ripening in this mighty genius, by just gradation, into fatuity blended with frenzy. His numerous biographers have laboured to develope the origin of ailments, which deprived his riper years of comfort, and rendered his closing scenes so humiliating. But it is with SWIFT as with illustrious men in general. We know the dates and occasions of their prominent actions. The effects also of their writings, speeches, and victories are amply detailed. But the man himself is hid in a cloud behind all this bustle and splendour: and of his prevailing inward state, we remain as ignorant as of that of an inhabitant of the planet SATURN.

Medical calculation may perhaps supply, to a considerable extent, the deficiency of positive information concerning the Dean of St. Patrick's. There is one hypothesis, and I think but one, which will at once unfold the nature of his ailment, and account for the more striking peculiarities of his disposition. This hypothesis regards his personal condition—that clue to the whole tenour of a man's conduct in familiar life, and often, where well followed, to the spirit of his overt actions likewise.

Extraordinary talents imply an uncommon portion of sensibility. Nor were the early studies of SWIFT of a nature to dry up the fountains of feeling, but rather to make them gush out in more abundant streams.* What then would be among the principal products of a soil, constituted and manured, if I so speak, like that of his mind? Is it credible that that appetite, of which the autumn of his days affords indications, that would be excessively ridiculous, had they stopped short of guilt, should not have sprung up, in its natural season, vigorous and impatient for gratification?

It is needless to expatiate upon the unfavourable circumstances, in which young SWIFT was placed just when he was the most likely to be assailed by the solicitations of passion. I do not conceive that any one can read, without commiseration, of his wretched state of desertion, or of his college solitude, at that dangerous period, without alarm. Who, that is capable of entering into the situation of young men of deep sensibility, shut out in any manner from society, and bereaved of the amusements of their age, can be at a loss in imagining to what

* "——In such a state, he could not bear to give the "necessary application to some of the more dry parts of the "academic studies, for which he had indeed habitually no "great relish, but passed his time chiefly in reading books "of history and poetry." *(Life by Sheridan.)*

means of assuaging desire they may be reduced, and what kinds of solace they may seek under the gloom of discouragement, and in a state of rejection by the world? Every one, whose sympathy and reason are not bound down by prejudice or superstition, must acknowledge that the errors, provoked by such tantalizing misery, and committed with perfect unconsciousness of their effects, are the objects of pity, and by no means, of censure.

The accounts we have of SWIFT bear that it was some time between the age of twenty and thirty that he began to be troubled with that giddiness (attended by coldness or weakness of stomach) from which he was never afterwards long free. Now, these are the very symptoms which a host of writers agree in representing as the most common consequences of the cause supposed —— That intense application should confirm the mischief, when the organs of thought and digestion were radically vitiated, is just what ought to be expected: and that SWIFT imputed his complaints to a surfeit of fruit, is of small account. He could scarcely conceal them altogether. He must say something. It is not perhaps impossible that he should have misjudged his own case.

SWIFT's whole conduct in his private relations so entirely bespeaks a debilitated man, and one too, debilitated by excess, that fiction, to be consistent, must make a similar character act upon a similar plan. Nay, a suspicion of his particular situation has not escaped his biographers, though they do not trace it back to its occasion and date.

Every body recollects the prediction of his being destined to die first at top, made in company with Dr. YOUNG. A striking emblem in nature might easily force out the dark expression of a secret, which had long lain heavy at his heart, but which was of such a nature, that his haughty spirit could not submit to confide it to any mortal ear. In his *Gulliver*, where his genius bursts out with such transcendant splendour, it still shines as if surrounded by a halo of malignant dissatisfaction. There is something in different parts of that work, from which the heart recoils. You would not accept its fame along with the feelings of its author. Cut off from the participation of pleasure, he might serve, envy, and hate his species, till advancing years, perpetually irritating his secret sufferings, plunged him into the madness of misanthropy.* Strong evidence,

* Mrs. WHITEWAY would sometimes steal a look at him, but did not dare to venture into his sight. His rage became absolute madness. *Sheridan.*

(I think as strong as circumstantial evidence can be,) is to be found in the delight he takes in images, "physically impure"—in the sentiments he professed at different times regarding marriage, and above all, in his behaviour towards several women, after he had attained the fullness of literary glory.

In whatever manner it may be that debauchery sullies the imagination, the thing is certain. We know that none so frequently shew a perverted taste as worn out debauchees. It would seem as if, in spite for the lost power of enjoyment, they vilified the things in which they once most delighted. Our travellers in the East are always taken for physicians. They relate, that those owners of Seraglios, who apply to them most eagerly for restoratives, are most eager to have beautiful women in their possession. The history of many faded gallants in our own country is, I believe, full of similar occurrences. But never creature appears to have hovered with more lingering fondness about the honey it could not sip. A dignitary of the church could not brave public opinion, as many laymen of fashion do. But though a kept mistress was out of the question, Swift's intercourse with women figures full as conspicuously in his life as his intercourse with ministers. His love, no doubt, was platonic; and perhaps his last friendly biographer need not have laboured that point so anxiously. But what, I pray, according to this very biographer and the internal evidence of his story, could it be else in the nature of things? SWIFT certainly did not attach himself in the manner of a person, intent only upon rationality in those women, with whom he became intimate. His eye appears to have guided him in great measure, nor was he at all less intent upon personal attractions than a youth, just beginning to feel the power of the other sex. Of his three mistresses, two at least had to boast of distinguished charms, (I do not know that the other was in any way deficient) and it is observable enough that the ardour of his attachment was in proportion to their youth and beauty. His passion for Miss VANHOMRIGH stands confest by himself in the most explicit terms. And though of himself and Mrs. JOHNSON he say—

> His conduct might have made him styled
> A father, and the nymph his child.

the assertion stands contradicted by what he calls *the little language*, "the playful "sallies of his undisguised heart, his expres-

"sions of tenderness and cordial affec-"tion:" all of which vanish from the journal, the moment his affection for STELLA is superseded by fondness for the brighter and more youthful VANESSA. That the ladies themselves looked towards ordinary, solid, matrimonial love is too shockingly evinced by the shortened life of one, and the embittered death of the other. Miss WARYNG, the first object of his adoration, not only entertained the same expectation, but in her the expectation was warranted by SWIFT himself, as is manifest from that memorable letter, in which he adopts the unworthy artifice of frightening her out of the idea of becoming his wife. (*Sheridan*, p. 278.).* Wherever therefore actions shall outweigh professions, will it be believed that SWIFT was "*naturally of a temperate constitution with regard to women*," however he may acknowledge it himself? As he was afterwards so fond of keeping the embers alive, is it not infinitely more probable that the flame once burned fiercely, than "that by the constant habit

* "The other thing you would know, is whether this "change of style be owing to the thoughts of a new mis-"tress. I declare upon the word of a christian and a gen-"tleman, it is not; neither had I ever thoughts of being "married to any other person but yourself."

"of suppressing his desires, he at last lost "the power of gratifying them." (*Sheridan.*) That he should have engaged the affection of three women (and of two by direct addresses) and that he should have escaped from one by a subterfuge, shaken off another by murderous violence, and dropped the third at the last *Amen* of the marriage ceremony, forms a difficulty in his conduct, which scarcely admits of any but a physical solution. Two are offered; and it will be difficult, I think, not to prefer that which is countenanced by analogy, while it coincides with the indisputable warmth of his temperament at one period of his life, fully accounts for the nature of his distressing complaints, and more naturally explains, why he entered the lists of love with all the flourish of a dauntless champion, but when he had proceeded to a certain point, threw down his arms, and turned his back like a recreant. I deem it satisfactory that something similar to my opinion had occurred to my predecessors, and suppose nobody will find it an objection that they, not being professional men, failed to develope the mystery fully.

That the understanding of SWIFT did not sooner tumble into ruins, is nothing extraordinary. What I have formerly called *the faculty of pleasurable sensation* suffers soonest.

The intellect often declines slowly, as we know from ascertained and even from avowed cases, which I suppose similar. And if recipitivity be in proportion to the recipient, what wonder that one of the sturdiest of mortal minds should offer a long resistance to the agency of destructive powers?

I am aware that this investigation of a malady, which has excited much curiosity among mankind, stands liable to reproach as an attempt to traduce the character of the patient. But I do not see why its nature should now for the first time be deemed an unfit object of enquiry; and the enquiry, if entered upon at all, doubtless ought to proceed frankly and without flinching from any consideration that may elucidate the problem. Nor can any thing done in ignorance affix a moral stain. SWIFT therefore will emerge pure from the suspicions of the dispassionate pathological reasoner, though they be ever so justly founded. It is what his friends avow, and what I deem strongest in proof of my opinion, that stamps him with indelible infamy—namely, his readiness to kindle, in the female bosom, hopes, which he never intended to gratify. To those who find the evidence I have brought together as conclusive as circumstances will admit, the case of this great man will afford posterity more than one useful lesson. It will teach how essential it is to check even slight-seeming nervous disorders in their commencement; and how such unwariness in youth, as a little information will correct, may proceed through protracted misery up to criminality, idiocy and madness.

Temporary palsy is brought on indifferently by epilepsy, hysteria, and anomalous nervous complaints. Epilepsy is frequently followed by compleat palsy or apoplexy. And in affections of inferior violence, the propensity to partial and temporary palsy is likewise aggravated into apoplexy, succeeded by death or loss of one side, by causes that would produce no more perhaps than a passing tremor in firmer constitutions. The causes I mean are such as great alarm, unusual exertion, or mistaken medical treatment. Instances of the former kind abound in a variety of medical authors. Of the latter, as I have observed that the sense of distention or fulness in nervous women, often excites a dangerous desire of being let blood, I shall produce from Dr. WHYTT the following distinct example. A delicate or nervous girl having chilled herself at the return of the monthly period, was next morning, at four o'clock, seized with stupor and difficulty of speaking and moving. She was soon after blooded and blistered. At eight o'clock she could neither speak nor swallow, had a hiccup, and was pale and cold, though her pulse and breathing were natural. About half after ten she began to breathe hard, and with a snorting noise. Besides taking medicines, she was now blooded again, and a third time in the afternoon, and died at ten o'clock, 18 hours after her first seizure. —This unmerciful use of the lancet proceeded, no doubt, from the groundless idea of the constant necessity of blood-letting in apoplexy—a practice sometimes fatal, and always perhaps of disputable utility. In the present case, an inability, which seems to have been merely one of those tendencies to palsy, that constantly come and go in the nervous affections of young people, was thus dreadfully aggravated by a treatment, exactly the reverse of what would have been proper. Never had mismanagement a more visible effect. The more the lancet was used, the nearer did the symptoms approach to that fatal apoplexy, which the third bleeding so speedily brought on. We clearly see diminished voluntary power, snorting or stertorous respiration, and death, following each other in just succession.

OTHER COMMON SYMPTOMS.

Some cases of epilepsy are distinguished by a remarkable feeling called *aura epileptica.* A thrilling begins in some distant part, as in one of the extremities, and proceeds gradually along, till it reaches the head, when insensibility and convulsions begin. This thrilling, however, is far from taking place always; indeed, the greater number of epileptics are free from it. Something pretty similar occurs in other nervous complaints. A feeling shall commence in the shoulder, rise up to the head, and be immediately followed by nervous headache. Again, a thrilling shall take place in one of the feet, ascend to the crown of the head, and pass down to the eyes, and produce only dimness or *obnubilation* of sight. These wandering sensations, as of a breeze passing along parts of the body, may have procured to nervous affections the name of *vapours.* In epilepsy, where the causes of these *auræ* or ascending vapours have been most industriously sought, they have been found to begin from some sudden injury or slow disorganization, and are well known to be sometimes cured by placing an issue in their course, as I have lately seen in a case of epilepsy, arising from an accidental hurt.

Sudden shifting pains, as well as thrills, are symptoms very general through this class of complaints. The sensitive organs being injured, we must expect the patients to be subject to variety of comfortless and painful sensations.

The stomach very universally suffers from flatulence, mawkishness, nausea, retchings, vomitings of viscid phlegm or bile. It is common for epileptics to be seized with sickness before and after the paroxysms. In hysteria, the stomach is never right, and violent efforts to discharge its contents are familiarly observed. Nervous headache is apt to terminate in ejection of bile: but is more frequently accompanied only by loathing of food or retching.

Periodical returns are not uncommon in nervous complaints; but they do not belong to every case of every species: nor can they be affirmed to belong to one species more than to another. We sometimes see the greatest regularity in the coming on of epileptic and hysteric fits. Nervous headaches have likewise frequently a period as exact as agues.

Frequent calls for food from a sense of internal sinking, not from genuine hunger, begin with the first origin of nervous complaints, and accompany them through their whole progress to the highest pitch of exacerbation. When you hear a person say that he always grows faint unless he eats something in the morning before dinner, and that after rising he cannot bear to be a moment without his breakfast, you seldom need be at a loss to divine the nature of his complaint. His nerves will have been shaken by some power that has more directly acted upon them, or have suffered in an indirect manner. These calls of weakness should not only be obeyed, but anticipated. Instead, for instance, of risking faintness before breakfast, something should be taken in bed. But enough has been formerly said on the care, necessary to supply the puny with sustenance in time.

OCCASIONAL CAUSES OF FITS.

The grand moral, deducible from the preceding too long and uniform discourse, is *the necessity of avoiding all predisposing causes of nervous disorders*. Many of these have been above enumerated with such an ample commentary, that I must despair of writing any thing useful to those, whom I have left uninstructed upon this head. I could make another essay not less extensive out of the *exciting* or *occasional* causes; and in so doing I should but follow respectable examples. But I own that I consider the topic as comparatively very unimportant, and I have in great measure anticipated it. Indeed, if the predisposing causes have operated with their usual effect, I am afraid that in this jarring and boisterous world, the poor sensitive human plant will be utterly at a loss to find an asylum. Wherever he retires, the occasional causes of his paroxysms, be they epileptic, hysterical, *cephalalgic*, or anomalous, will pursue and hunt him out. Earthquakes, volcanic eruptions, hurricanes, revolutions, mobs, he may have the good fortune, all his life, to escape. But he will find the whole cultivated part of the earth infested with heats, frosts, storms, thunder-claps, door-claps, creaking wheels, jarring windows, squeaking pigs, cackling geese, crowing cocks, and innumerable disorders besides in the apparatus of polished society. In spite of injunctions, the footman will let fall a tea spoon, or the cook will keep back dinner too late. A sister will send about the tea too hot or too strong. A child, forgetful in its sport, will utter a note too loud. Husbands will be inadvertent or ill natured, and contradictions will come to throw the whole frame into tremors, if not into convulsions.*

* J'ai vu, comme M. Lorry, une femme si mobile, que *la plus lègere* contradiction la faisoit evanouir; cet habile medecin cite le cas effrayant d'une femme honnête, qui ayant

If an invalid could entertain the desperate idea of flying into a desert from the possibility of these town and country plagues, in the desert, no doubt, others as terrible would meet him. It is the first concern, therefore, of a valetudinarian with too exquisite sensibility to seek—not how he may shun, but—how he may become once more enabled to encounter, the annoyances of life. This is only, in other words, remanding him to the care of the sons of Esculapius.

Till their operations shall have produced the desired effect, it will indisputably be the interest of the patient to suppress every species of vehement emotion. I have already explained how foolhardy I hold the least unnecessary exposure to the physical exciting causes of nervous paroxysms. Of the moral causes in particular, I know not that any thing can be said beyond what must suggest itself to the plainest common sense ; except perhaps a few words on the subject of fear.

If you ask people what threw them first into fits, they will answer, in a great majority of instances, that it was something, by which they were terrified. In a number of patients from

---

paru en public dans un habillement peu convenable, fut si irritée, par les propos indecens, que cet habillement lui attira de la part des jeunes libertens, qu'elle tomba morte.—(*Tissot.*)

humbler life, who have applied to me in succession, I have observed that there has not occurred any exception to this rule. One girl fell down convulsed on taking up a toad along with a pail of water; another from seeing a child attack his fellow with a knife; a third from being where a person was brought home lifeless out of the water; a fourth from passing by, as an epileptic person was lying, black in the face and foaming, in the street, a fifth from a black man's attempting to kiss her, and so on to infinity. A spider, an earwig, a frog, or an ant, crawling in the sugar bason, will often do just as much for the female members of a polite circle.

How well therefore has natural history been recommended for one of our earliest studies! and this, as far as it is most useful, can always be carried on by means of the objects themselves, which will exclude books any farther than for reference. Children always adopt the associations of those about them. I have known a timid mother, by well-acted horror, fill several children with insurmountable dread of crossing even ROWNHAM ferry. Conversely, whatever parent can but examine our insects and reptiles with interest, may be sure that it will never be in the power of the most loathsome among them to throw his family into fits.

But disciplining the mind against fear can be easily carried a great deal further. In a variety of cases, children should be made to act, over and over again, the proper parts in dangerous emergencies of common life. *Shew me what you would do, if your clothes were on fire ; or if you saw a person in that situation.—In a case of apparent death from drowning, let me see how you would go about to restore animation.* These and other processes, with the necessary explanations, should be gone through with some substitute object ; and the mind, thus habituated, would be nearly terror-proof.

In fresh cases, analogy would point out the line of safest or most serviceable conduct ; and it is clear, from the preceding ample explanations, that the hurry of thought, vertigo, and confusion of brain, which suspend consciousness, while the muscles are irritated into preternatural actions (convulsive or spasmodic) could not by any possible means come on. Even the coolly thinking over any new motion, as holding the fingers apart two and two, though less efficacious than practice, will render it more easy : and the maxim of Horace

Rem ante provisam verba non invita sequuntur—

will bear being extended to the consideration of such measures as the urgency of distress may call for. A whole book might easily be written on the art of hardening the mind against terrific impressions.—But I am too far exceeding my bounds, and I hope thus much will be sufficient. The only concern of prudence will be, in such exercises not to excite to wanton mischievousness.

Having this month tried the patience of my portion of the public by an unprecedented stretch of dissertation, I feel anxious that we should part in good humour at the end. I know not any thing more likely to contribute to my purpose, than a clear idea of the most desirable state of nerves. I choose an example, as thinking my lady and lady-minded male readers, who hate metaphysics*, will prefer an example to description. I dare say, however, they would puzzle long before guessing what pattern I mean to propose to them. But it is not my way to keep people in suspense. So, at a word, it is the

* I hate metaphysics too; that is, the school-learning of old and modern Kantianism. But the knowledge of the human understanding, resulting from the most refined and interesting species of observation, and capable of useful application, I prefer to all other kinds of knowledge. Every day I lament not being a greater proficient in it; but am afraid this is just what the readers abovementioned hate as metaphysics.

Savage of Aveyron.—This savage of Aveyron, if his history be accurate, enjoyed, in the happy pitch, to which his nerves were tuned, such an advantage, that I know not if we, his civilized fellow-tenants of this terraqueous globe, can produce any six equal to it. The advantage was this. His nerves were alive to pleasurable, and dead to painful sensations. He fired at the cracking of a nut. I need not tell any body, who has ever tasted the sweetness of a kernel, what associations the sound would excite in this denizen of the forests, with his fresh, unpalled palate. But if you discharged a pistol close under his nose, he continued as grave as my lord-chancellor on the woolsack. The crash of the storm had too often passed over his head without injury, for him to mind loud and sudden sounds; and perhaps he would have stood the discharge of a whole battery without emotion. At least, had the house tumbled about his ears, before he had been subjected to Dr. Itard's humanizing superintendance, I will answer for it that it would have been to be said of him as justly as of the virtuous sage of the Roman poet—

Impavidum feriunt ruinæ.

But we need not recur to this singular instance. Examples to be met with in every street and village, are sufficient to convince all who will give them consideration, that the opulent people of this country commonly render themselves wretched by pursuing *well-being* upon a false plan. The English are more remarkable than any other nation on the globe for the accumulation of *comforts*, and, indeed, unhappily we pique ourselves upon the distinction. I am not going to advise any one to discard his or her comforts all at once. I would only open a course of reflection, by which they may satisfy themselves that the reliance they place on externals is vain, and that those who have most comforts about them, are commonly the most comfortless of all mortals.

We have seen that, in the cold winters of Holland, people live with very little fire, and in apartments, into which the wind has free admittance. The people of our country, in similar circumstances, would shrink from the idea of such habitations, and indeed with good reason, as we should most certainly perish in them.—We know, however, that the Dutch suffer less from temperature than ourselves in their health, and therefore also in their feelings, when its effects do not quite bring them down to the sick bed. Were a traveller to explore the whole earth for the sake of comparing the feelings of its inhabitants, I will venture to predict that he will find the Hottentot in his grease, the Abyssinian with the reeking intestines of an ox wound round his body, the Kamtschadale in his underground den amid his putrid fish, the wild Irishman beside his pigs in his cabin, each infinitely better off, as to their internal condition, than the gentry of Great Britain with all their care about cleanliness, and all their contrivances for protection. I do not, like Rousseau, advise that we should resort to filth and barbarity in order to avoid being incommoded and invalided, but, in the name of common sense, let us understand that the going of the clock depends neither upon the paint of its cover, nor on the brightness of its face, but upon the perfection of each piece in its interior, and the nice correspondence in the movements of all!

END OF ESSAY IX.

Printed by MILLS, St. Augustine's, Bristol.

ESSAYS

ON

THE MEANS

OF

AVOIDING

HABITUAL SICKLINESS,

AND

PREMATURE MORTALITY.

ESSAY TENTH.

Printed by MILLS, St. Augustine's, Bristol,

OF IMAGINATION ALL COMPACT.

ESSAY

OF SOME OF

THE DISORDERS,

COMMONLY CALLED

NERVOUS.

PART II.

CONTAINING

OBSERVATIONS

ON

INSANITY.

Insanity more thought of in England than elsewhere, and why.

THE difficulties, belonging to the subject of the present Essay, are understood far beyond the limits of the medical profession; and it must be confessed that the ill success of those, by whom it has heretofore been treated, is enough to damp the spirit of every new enquirer. Nevertheless, considerations may be stated, which may well incite to further endeavours. There is reason to believe, that the writer, who should be happy enough to unfold some of its intricacies, would inspire a great interest among his countrymen, and that, by a proper application of his principles, he might render them a most essential service.

Foreigners, who pay attention to the turn of thought which predominates in England, are struck by the frequency of our recurrence to the idea of lunacy, and by the stress we lay upon it as the last of human calamities. Some who have published observations upon our national peculiarities, drawing, I fancy, too general an inference from particular examples, describe it as one of our customs to carry young people to Bedlam; in hopes that the sight of the objects immured within the walls of that hospital, may create a horror of the excesses that tend to reduce our nature to so humiliating a condition.

Besides the belief that insanity is a more common affliction here than in other countries, there are some remarkable circumstances in our literature which have an evident tendency to fix the imagination, at the time when it is most open to impressions, upon mental derangement. SHAKESPEARE, a name that always recals to his intelligent readers the first of poets, and the most penetrating of observers, has succeeded, by his happy use of madness as machinery, in carrying terror and compassion to a height, which they cannot perhaps be made to reach by any other means. In his lighter pieces, those who from their denomination of *fools*, must be supposed originally destitute of an average portion of understanding, are made the perpetual vehicles of remarks, whose shrewdness puts to shame the acumen of men, the most remote from folly. But it is his desponding-mad Ophelia, his raving-mad Lear, his jealous-mad Othello, his melancholy Jacques, his crafty-mad Hamlet, that awe and attach on the first exhibition, and bind the heart in a never-ceasing spell. It is in these characters, where the equilibrium of the faculties is destroyed, that he displays the force of his imagination along with that of his reason. Nor would it have been possible for people in their sober senses to have *let out* so much fancy and so much philosophy together, without seeming unnatural—even if any mortal except the delineator of these characters could have combined so much in his secret meditations.*

* "There is one circumstance altogether peculiar to Shakespeare—the singular fanciful cast of philosophy, in which he is so fond of losing himself. What distinguishes him beyond all the poets of any country previous to the

We need therefore, know nothing further, than the powerful dominion which Shakespeare exercises over our national feelings. We should have a right to conclude from this fact alone, that no subject would more deeply engage our thoughts than the one which he has addressed in this manner to our feelings. But he has been seconded with an ability beyond all reasonable expectation. Isabella, Alicia, and Clementina still display their terrible graces

---

formation of the national taste, is the depth and riches of his moral and political reflections. He does not always introduce these, when alone they are free from censure in dramatic writings—*i. e.* when they coincide with the progress of the action, suit the character of the speakers, and are explicable from their situation. He seems at times to come forward in *propriâ personâ*, and to propound his own insight into affairs and his knowledge of mankind under borrowed names. But this philosophy of his has always something richly fancied and enthusiastic, and connects best with strange and grotesque images. The more his imagination is exalted, the more profoundly he thinks and with the better success does he philosophize. No wonder therefore he assigns to his *dramatis personæ* an overstrained fancy, a character inclined to enthusiasm, a singular way of seeing things—no wonder he is so fond of placing them in the twilight between sanity and insanity—half beside themselves, half in their right wits. In displaying this peculiar state of mind of his heroes, he is at liberty to let loose the reins of his own imagination, and to utter, without reserve, all his own excentric genius suggests."—*Garve's Essays*, ii. 441.

with effect: and there are authors of later fictions, who would not have ventured forward with their lunatics or would not have succeeded with them so well, had not Shakespeare done for this class of characters what Homer did for the heroes of the Trojan war—put them in possession of the good-will of the public.

MORE OBVIOUS APPEARANCES.

In order to make an advance towards the knowledge of causes, I propose to present a concise view of the more usual phænomena of insanity, and then to endeavour to trace the chief particulars, by which it is distinguished from the sound state of intellect.

An uninformed observer, who should pass a considerable number of lunatics in review, would be struck no less by the difference between one of these unfortunate persons and another, than by the difference between them all and those, who are in full possession of themselves. At the time the disorder most strongly manifested itself, he would find a part storming with a degree of fury unknown in ordinary life, and the rest wrapped up in tenfold gloom. This observation appears to have given rise to the earliest, and what is still the usual, distribution of insanity into two species, though it was soon found impossible to adhere to it rigidly. Of these two species, one has obtained the name of MANIA, the other of MELANCHOLIA; the denomination, in the latter case, being taken from the predominance of a certain imaginary *black bile* in the constitution, but the idea of the class from the silence or dejection of the party affected.

The melancholy-mad will sometimes continue for days and weeks looking stedfastly at the same spot or with their eyes closed, without taking notice of any thing, or making any reply to questions. Nor can they often be stirred from the place or prevailed upon to take sustenance without direct force. They would be supposed by a looker-on, not acquainted with cases so much in the extreme, to be the most torpid and insensible of all mortals. And indeed I have sometimes found this opinion to be so deeply rooted in the minds of the friends of melancholic patients that the fullest force of evidence to the contrary could not shake it.

It is true the usual stimuli frequently produce none of their usual effects. The melancholy man will frequently bear the extremity of heat and of cold without complaining at least, if not without suffering. A four-fold dose of medicine is commonly enough observed to fail altogether in its effect upon him. He will look stedfastly in the face of the unclouded sun. He has a far greater chance of escaping any prevailing epidemic complaint; contagion, or the seasons having often less power to induce a change in the movements of his system.

But at other times appearances the most opposite to this seeming insensibility are to be observed. The patient trembles, shudders and starts without any obvious reason. His tears burst forth without any adequate exciting cause. His sleep is almost never composed; he often shews a degree of wildness as he is falling asleep; suddenly awakes disturbed; is haunted with the most terrific dreams; partakes with the rest of the nervous in the inconvenience of finding himself more harrassed in the morning than at bed-time.

When not totally absorbed, he complains of the whole train of uneasinesses, that are general where the sensibility is in excess, and which have been copiously described under the article, hypochondriasis. He is subject to twitchings or startings of the muscles, to obstinate contractions of the sphincters of the bladder and rectum, to spasms in the organs of deglutition. The head suffers from dreadful partial pains, particularly the hinder part; and it is sometimes a burning, sometimes cold, sometimes pressure or distension. There are pains, oppressions, anxieties also referred to the chest, heart, stomach and bowels, which have exactly the hypochondriacal character.

This analogy holds, in a most remarkable manner, of the natural functions. When the disorder has not risen to that pitch of violence, which concentrates the patient entirely within himself, the appetite shall be preternaturally voracious; and with an appetite no way immoderate, there shall exist heart-burn and a load at the stomach, with unsavoury, sour or rancid eructations. An obstinate sluggishness of the bowels attends melancholia as well as hypochondriasis; and the most drastic purgatives are at least as often unavailing.—There is an equal degree of hardness and distension of the body, with very troublesome flatulence. The skin is habitually dry; but in both complaints, a cold, clammy perspiration will burst forth by fits, particularly about the head, neck and breast; and other small irregularities of the secretory organs, as of the salivary glands and kidneys, will take place equally in one and the other.

The mutability, so remarkable in nervous complaints, and which has been described in the former part of this Essay, occurs no less in melancholy. It changes into epilepsy, catalepsy, into partial palsy and apoplexy. When it has continued for any considerable length of time, the state of the system is seen to undergo a compleat change. An atrophy, more or less compleat, supervenes. Obstinate jaundice, and often what from the depth of the bile, is termed the *black* jaundice, dropsies, cutaneous eruptions of the foulest species, infest the melancholic, particularly when intemperance in the use of intoxicating beverages has concurred in the formation of the complaint.

The dissections of the bodies of those who have died melancholic have frequently shewn a sensible disorganization in various interior parts. The alterations that respect the brain and its enclosing membranes and bones are found to agree very nearly with those already described under the article epilepsy. There is often a preternatural thickness of certain parts of the scull; and not unfrequently other parts are reduced to such a degree of thinness as to be nearly transparent. But this latter alteration seems to depend in most cases upon the unhealthy state of the parts below, which enlarging and pressing upon the bone, carry the process of *absorption* to so great a height, that more bony matter is removed than the feeding arteries supply. Interior excrescencies of some of the bones, composing the scull, have been detected as in epilepsy, as also caries, ulceration or what is frequently called *rottenness*, in speaking of the teeth.

The softer parts are observed to have undergone full as extraordinary changes. The membranes shall many times have concreted in consequence of a process, better known to anatomists since the researches of the late Mr. HUNTER. They coalesce not merely with one another, but with the brain and skull. The fine soft membranes that lie next the brain acquire a coarse texture and are inflamed. Ossifications or diseased formation of bony masses are discovered in the outer membrane, which is naturally thicker and coarser. This is particularly the case in one of its processes, which passes between the two hemispheres of the brain and is called *falciform* from its similitude to a scythe. Collections of watery liquid, effusions of blood and over-fulness of the blood-vessels are common occurrences.

This work of destruction is often distinctly seen to extend to the brain itself. Various are the relations of the examiners of the bodies of the melancholic dead concerning the hardness, dryness, and less specific gravity of this important organ. But in a much greater number of observations do we read of its changing to the opposite condition. It is there described as pultaceous, almost liquid, and yields to the slightest pressure. Hardnesses, known by the name of *schirrous* steatoms, and abscesses have been noticed. The interior of the brain has been found to contain hydalids. The cerebellum or little brain has equally lost its healthy constitution. It has contained hard tumours as also abscesses, and has been reduced to the same happy mass.

It is true that these alterations in the head have not always been found. But it should be considered that the dissecting knife, though it be the instrument to which we owe our superiority over the antients in deducing inward lesions from external signs and from particular sensations, and which therefore may be considered as the glass that shews the state of the viscera through the otherwise opake walls of skin and flesh, can be by no means adequate to lay open *all* the effects of unhealthy processes. There may be changes of parts a great deal too subtle to shew themselves upon the sides of a fresh cut. We should not therefore permit ourselves to reason negatively from dissections, which are often cursorily and ignorantly made.—Possibly chemical analysis may, one day, add to our acquaintance with the alterations of the living substance as much as the best *morbid* anatomist knows beyond one, who has never seen a dead body opened. And if means of ascertaining such effects as now escape us, should in future be discovered, those recondite effects will probably be much more universal in patients, who have suffered from a given disorder, than the grosser ones heretofore recorded, because they will lie so much nearer to the healthy condition. The stage of disorganization which our anatomy detects is probably one that many invalids never reach. It is remarkable that the scalp is affected, being looser in the mad.

As we shall find insanity consequent upon practices, that injure the stomach, and its train of auxiliary or subordinate viscera, it is satisfactory to find the appearances after death, in numerous instances confirming the opinion that these practices are its cause. Almost all the contents of the abdomen have been detected in a diseased condition—the liver for example, enlarged, or hardened; or studded with tubercles, and adhering to the contiguous parts—the gall-bladder full of concretions—the mesentery, omentum, and other parts no less deviating from their condition in health. It is related that the interior changes have sometimes corresponded to the sort of insanity; the ovaria having been indurated, when the mind had roved upon the sexual intercourse.

The mental part of melancholy is generally

considered as consisting in having the attention fixed upon a single object, and forming erroneous judgments concerning that. Thus it is remarked that the mathematician VIETA, who devoted three days and three nights uninterruptedly to an algebraic calculation, was not melancholic, because there was nothing wrong in his judgment upon the matter before him. But had he imagined the signs he used to be characters of fire, he must infallibly have been reckoned mad.

The origin of melancholy is often not easy to recognize. But when the mind adheres to one subject almost exclusively, when passion is connected with this set of ideas, and when there is at the same time great loss of sleep, the approach of the disorder may be suspected. If frequent false perceptions occur, if the behaviour undergoes a striking change, if there be an unusual degree of intractability on the one hand, or of gentleness or demonstrations of friendship on the other, if the understanding start rapidly from some matter in discussion to another quite different, and if the memory fail, and frequent absences succeed, the first stage of the complaint may be considered as having taken place.

They often distinguish MANIA from MELANCHOLIA by an universal, or at least a more general erronoeusness of judgment in the former, and by the violence of the movements, connected with these errors. Where the commencement of mania is distinct, the patient is apt to be habitually sleepless, incapable of business, averse to amusements, depressed, solitary and wrapt up in his own thoughts. Among the frequent precursors of mania, may be enumerated violent giddiness, headache, deceptions of the senses, tremblings, shiverings, twitchings of the flesh, convulsive startings, especially on falling into a dose. Great irregularity of the pulse, palpitation of the heart, oppression of the chest come on at times. Those who have too much chance of afflicting their friends with the idea of insanity infesting the family, will often undertake distant journeys at a moment's warning; engage without hesitation in the largest and most hazardous enterprizes; yield continually with more and more reluctance to argument or remonstrance; turn an equally deaf ear to the claim of a child, the entreaties of a consort, and the authority of a parent.

When confest insanity has commenced, the countenance assumes a wild, staring, frightful expression. It has been said of a celebrated public character that the most bungling drawer of a caricature never yet exhibited him so that he could not be recognized. The same may be said of maniacal lunatics in their accessions, and at times also of those that are only being worked up to madness. So clearly does tension, prominence and fire distinguish the eyes of madmen from those of persons in their sober mind—

> Ardent oculi: (says an ancient dramatic writer)
> fune opus est.

The doubt which a certain kind of behaviour often excites in common life, *whether a particular person be drunk or mad*, is justified by attentive observation of lunatics. The disease operates like spirituous liquors, insomuch that those, who are unacquainted with the symptoms of approaching mania, ascribe them to excess in drinking.

When appearances like these have gone on to shew themselves for a longer or shorter time, the frantic paroxysms commence with more or less notice. From his usual excessive and animated loquacity the patient will pass into discourse, entirely wild, incoherent and irrational. He will become agitated, assume a lofty demeanour, exhibit an aspect full of fury, express a deadly aversion or an outrageous predilection for particular persons, utter the most unprovoked threats against them, or call down the most unmerited blessings, shout, curse, pray, chaunt, laugh almost in a breath, betray feelings of the utmost lasciviousness, neglect cleanliness altogether, and exert a surprising degree of muscular power.

It may seem singular that some maniacs in the very rage of the fit shew nothing of what is called *delusion of the senses,* but very distinctly recognize objects and persons, but judge concerning them in the most erroneous manner.

The degrees of ability to recollect what has passed during an accession seem very different on different occasions. Some can recal every particular. Others are able to repeat what has been said to them, but they are apt to repeat it with false accompaniments and under circumstances different from the reality, believing themselves, for instance, to have been in quite another place and with other people. They frequently forget the coinings of their own phrenzy, but remember the sort of treatment they have undergone. Still more than partial forgetfulness occurs. They will be conscious of things having passed through their mind, but can, by no effort, call back any of the things themselves. Sometimes even this consciousness does not exist, but all their agitation passes away without leaving a trace behind.

It is, we see, therefore very much with maniacs as with those that dream. The whole picture with its finest shadings remains impressed,—or only the outlines of some of the figures are preserved,—or else the whole is obliterated; nor would it be known but for indubitable effects, that manifest themselves to the by-standers, that any striking images, or any at all, had presented themselves to the fancy. Indeed, during intervals of reason and after recovery, those who may be supposed best able to give an account of their own feelings, compare their phrenzy fits to dreams. And they agree with those who have been occupied by any interesting exhibition, in being surprised at the length of time that has elapsed since their attention was drawn to themselves. In describing the state of their intellect, they speak of an uncontroulable hurry and whirl of ideas, by the rapidity of which every endeavour to fix upon any one subject of thought in particular has been rendered abortive.

It has been often remarked that phrenetic insanity is aggravated or supervenes in the morning; as if sleep had the effect of aggravating this like other nervous complaints. A female patient, concerning whom I have the most exact information, was for a considerable time only nervous, that is, tremulous, weak, and too quickly sensible, and then she was always the better for her breakfast. But a feverish complaint having been succeeded by insanity, she was regularly for months worse after breakfast than before; and it seemed to make little difference of what the meal consisted, or whether it was cold or warm. It was enough that the stomach was any way moderately distended for the excitement to be brought on.—This *general* aggravation of the symptoms in the morning, is denied by other observers of the most extensive opportunities; and it is affirmed that the evening and great part of the night, are the seasons of most violence with incipient lunatics. Perhaps the want of sleep during the night preceding, or its variations as to tranquillity or disturbance, may make a difference; to which, if lunatics could more frequently be studied singly by those, who had already seen them much in mass and were well acquainted with all the facts that books supply, other causes of individual variations might in all likelihood be added.

In one observation the different reporters appear generally to coincide; and this is of considerable importance, when combined with the frequency of palpable disorganization in the head. The symptoms are aggravated by the horizontal position. The feelings of raving lunatics would appear to impel them to seek alleviation by keeping themselves as erect as possible. For when fastened, they are observed to sit, scarcely ever to recline. Unless indeed it should be said that the calm, recumbent state is incompatible with so much mental agitation. And we must, I believe, suppose that few persons, suddenly stimulated into anger as they were lying along, would continue to repose in the same easy manner.

It is one of the most constant consequences of derangement to impair the memory. Cases are related, in which though the diseased loquacity continued, the patient came at last to be unable to finish any one sentence; and this destruction of the associating quality often proceeds to idiotism, in which the mutterings shall be almost continual, but utterly without sense, as scarce presenting three words belonging to any one subject. Persons of good education, on being long confined, lose the power of spelling their own language, and write, like the untaught, from the sound of the words. But it is difficult to know how far this depends upon the disease, since it is not ascertained in what time want of practice in reading and writing will reduce those in a sane state to a similar degree of ignorance. Lunacy appears considerably to lessen the chance of life. Death is no uncommon consequence of violent fits, when phrenzy is succeeded by languor. Here the powers seem to be destroyed by over-exertion. The disorders above enumerated, as following upon insanity, are often fatal. But although in two sets of persons, of which one should be deranged, but otherwise similarly circumstanced, it is probable that, after a certain period, there would be more survivors among the sane; single instances of longevity are noticed in madhouses. In Bethlem Hospital, a person not long since died at 78 years of age, after being more than thirty years on the incurable establishment. This was a remarkably placid madman. Maniacs themselves sometimes live to a good old age, and shew all the signs of a good habit of body, though scarcely when very furious (*Chiarugi della Pazzia in genere e in spezie*). No lunatic, I think, occurs in the lists of those who have been most distinguished among mankind for length of years.

In our own country, as far as enquiry has hitherto determined, females are more subject to mental alienation. It is well known that this, like epilepsy and other nervous disorders, are very frequent consequences of the exercise of the functions peculiar to the female constitution. They are particularly apt to attend pregnancy and child-birth. But in return a greater number of lunatic women recover, and especially of those who have suffered from these causes.—Abroad, the case is believed to be reversed. The writer last quoted, who in four years had the opportunity of seeing eleven hundred and fifty seven lunatics in the hospitals of St. Dorothea and St. Bonifacio at Florence, states the proportion of deranged males as exceeding that of females by one-fifth.

It can hardly be doubted that a particular complexion, or rather the sort of constitution connected with it, disposes to insanity—Mr. HASLAM of Bethlem Hospital, to whom we owe interesting observations on this disease, and from whom we may one day expect still ampler information, relates that of 265 patients, only 60 had a fair skin, with light, brown, or red hair. All the rest were dark. The same fact appears to extend to other countries. Dr. PINEL remarks that those lunatics, who are most difficult to confine in a hospital, who are most remarkable for turbulence, and most subject to sudden explosions, almost all bear the following external characters. They have bolder and stronger features, sparkling eyes, an adust and often yellow complexion, hair of a jetty black, sometimes crisp; strong bones, but no fat; powerful but slender musles—a strong and hard pulse.

Some observers have found the spring most productive of alienation of mind. Others think that the seasons are all alike: but the influence of heat and cold has too distinctly appeared in various instances, to be any further questionable. In long sea-voyages, where the transition is sometimes quicker and more considerable than in the same climate, passing into colder latitudes has very regularly produced a calm, and into warmer ones all the violence of frantic accessions.

Insanity is indubitably connected with certain periods of life—a fact which, I think, will be found of considerable use in illustrating its nature and causes. Mr. HASLAM gives the following table from Bethlem Hospital.

Between the age of	Admitted.	Cured.	Not cured.
10 & 20	113	78	35
20 — 30	488	100	288
30 — 40	527	180	347
40 — 50	362	87	275
50 — 60	143	25	118
60 — 70	31	4	27

In other countries it would appear that the same law obtains. Dr. PINEL states the ages of the lunatics received during eleven successive years into the Bicêtre at Paris, as follows:

	Between 15 & 20.	20 & 30.	40 & 50.	50 & 60.	60 & 70.
In 1784	5	33	31	11	6
1785	4	39	49	14	3
1786	4	31	40	15	5
1787	12	39	41	17	7
1788	9	43	53	18	7
1789	6	38	39	14	2
1790	6	28	34	9	7
1791	9	26	32	7	3
1792	6	26	33	12	3
1793	1	13	13	4	2
1794	3	23	15	9	6

On enquiring from persons who have had the care of lunatics, what kind of persons they were accustomed to receive in the greatest proportion, I have been informed that it consisted of members of the clerical profession, but could scarce obtain any distinct opinion concerning others, except devotees. Dr. PINEL, on examining the registers of the Bicêtre, says that he found inscribed "a great many monks and priests, "as also a great many country people, who "had been driven beside themselves by "horrid pictures of futurity;—several artists, "as sculptors, painters, and musicians;—"some versifiers, in extacies with their own

"productions;—a pretty considerable num-
"ber of advocates and attorneys;—but there
"does not appear the name of a single per-
"son, accustomed to the habitual exercise
"of his intellectual faculties;—not one
"naturalist, or natural philosopher of abi-
"lity;—not a chemist or geometrician."—
It has been remarked as singular at the Charité at Berlin, that several chaplains became insane in succession. But this seems to be imputed to their situation as chaplains, though the spectacle of madness does not appear to have affected the other officers of the house, or attendants upon the insane, in the same manner.

On the Validity of the DISTINCTION *between* MELANCHOLIA *and* MANIA.

Such are the most obvious and striking facts respecting insanity. There is not one of them for which several competent vouchers may not be quoted; and as to the most important, there is a general agreement. He who would explain what sort of an affection insanity is, and how it is formed, must attend to these particulars; and by comparing them with the explanation, the reader will be in some measure enabled to judge how far it may answer his just expectations.

Modern nosologists have so entirely departed from the original principle of distinction, by which melancholy madness was characterized according to the appearances of what is vulgarly called *melancholy* in sane people, that they now give us no other criterion than the partial nature of the insanity. "Insanity" says Dr. CULLEN, "consists "in such false conceptions of the relations "of things, as lead to irrational emotions or "actions. Melancholy is partial insanity "without indigestion—Mania is universal "insanity." According to this account, the partition between the two is thin indeed. It is a mere difference in degree. It sounds much the same as splitting small-pox into *partial* and *universal*, because sometimes there happens to be only one pustule or pustules but upon one limb; while at others, the eruption is diffused over the whole body. Such a system of subdivision is contrary to that, which is followed with regard to all other disorders, and indeed contrary to what manifest propriety suggests. Nearly all disorders exhibit the same variety of gradations without being marked, for this reason, as specifically different. It is making the plant from an acorn dropped upon a rocky soil, not an oak, because it does not rise to the height, or spread to the extent, usual to the king of the forest. It answers no purpose in prognostication or practice. Accordingly, it is pitiful to see what attempts are made to maintain the distinction by writers, who have not the courage to reject it. Thus the experienced CHIARUGI says, that "mania "properly signifies raving madness. The "maniac is like a tyger or a lion, and in "this respect mania may be considered as "a state opposite to true melancholia." Nevertheless the intermixture and alteration of symptoms oblige him to create a species of raving melancholia: and he adds, "that "there are alienations of mind, attended "with a fixed hatred towards some single "object, to which, at a later period, sallies "of rage associate themselves, returning on "certain occasions. Now to determine "whether this affection deserves the name "of mania or of raving melancholy, the "behaviour of the parties must be watched. "If they be truly maniacal, they will not "for ever continue immoveable in respect "to the object of their hatred, but at times "transfer it to other objects, which shall be "exhibited to them by their excited fancy "in a very lively manner, but indistinctly "and in irregular connection with the series "of their actual sensations." So if a lunatic abhor but a single person only, he is to be set down under one head; and if he chance to take it into his head to be affronted with another, continuing otherwise just the same, he is to be removed to a different disorder. Delicate lines indeed! and which must escape not only the eyes of other people, but at times those of him who drew them! As if hatred and suspicion did not, almost in all cases of derangement, diverge from their first object, or might not at least by the smallest provocation be derived to others.—Again, the same author allows that there are maniacs, who will listen to the remonstrances of their friends, and become tranquil in consequence. That is, there is an universal insanity which is partially rational.—Again, he tells us, that "though fury may not be always present, "yet the disposition to it is always to be "observed." It is singular that a physician who had had so many hundreds of the insane under his eye, did not remember in penning this passage, how often the disposition to fury is to be observed in the melancholic, that is, in those who strictly brood over one idea, and how apt they are to burst from their meditations and offer violence to those, who injudiciously disturb them? A writer, the most remarkable of all for the multiplicity of his divisions, confesses at last, that "all "these species of insanity may be variously "combined, and frequently interchange, one "with another. It may be proper farther to "remark that the same patient sometimes "goes through several kinds of insanity— "which may be reckoned in such cases as "so many degrees or stages, during the "course of the same illness." (*Arnold on Insanity*, I. 316.) The respectable testimony of Mr. HASLAM is still more express respecting the nullity of the common distribution. "As the terms *mania* and *melancholia*," says he, "are in general use, and serve to distinguish the forms, under which insanity "is exhibited, there can be no objection to "retain them. But I would strongly oppose "their being considered as opposite diseases. "In both, the association of ideas is equally "incorrect and they appear to differ only, "from the different passions which accompany them. On dissection, the state of "the brain does not shew any appearances, "peculiar to melancholy; nor is the treatment which I have observed most successful, different from that which is employed "in mania." "We every day," says the same observer, "see the most furious maniacs "suddenly sink into a profound melancholy; "and the most depressed and miserable ob-"jects become violent and raving. We have "patients in Bethlem Hospital, whose lives "are divided between furious and melan-"cholic paroxysms, and who, under both "states, retain the same set of ideas." Let me add one more confirmation. I was very lately favoured with the perusal of a manuscript, which if it could ever appear before the public, would go far to demonstrate that this and that form of insanity are only supposed to be so limited to individuals, because cases are seldom observed long together with close attention. A medical gentleman, who had for years superintended a lunatic asylum was obliged, on account of indisposition of his own, to retire from his professional pursuits. During this secession, he lived under the same roof with an insane patient, and kept a journal for many months, in which regular entries were made morning and evening. The practitioner visited the patient daily, and sometimes repeatedly on the same day. He had almost hourly reports concerning her. The diary exhibits the utmost conceivable range of symptoms, and the author expressly assured me, that this one case shewed him every material appearance which he had observed in all the patients that had ever been under his care.

The true relation of Melancholia to Mania may, I apprehend be stated with the utmost simplicity; or if we reject these names, the true relation of the two general forms, (the *penseroso*, and sometimes the *allegro*, oftener the *furibondo*) under which insanity manifests itself; namely, abstraction and agitation. There are certain children, in whom correction or reproof is almost sure to bring on a fit of sullenness; and who, after receiving the one or the other, will stand for a considerable time as fixed as if they were only images of animated creatures. But a very trifling occasion will provoke an ebullition of the passion that is raging within. If a smaller boy happen to pass within arm's length, though without giving more offence than the wolf in the fable received from the lamb's drinking lower down the stream, the young melancholic will immediately invent some cause for quarrelling, and play off a maniacal paroxysm in miniature at the expence of his over-matched playfellow. Thus it is, I think, exactly that melancholic absorptions generate maniacal extravagancies. The storm goes on to drive for a while, and that sometimes not a short while. But in innumerable cases, the apparently dead calm, though in reality it be a season of deep retired, despondent, and sometimes pleasurable feeling returns, to be, in like manner, succeeded by the hurricane in its season.—*Torpid* melancholy! *torpid* grief! they may as well apply the epithet to the boisterous state of passion or insanity, and speak of *torpid* anger or *torpid* phrenzy. As long as it continues to be either grief or melancholy, there must be an internal agitation far beyond the average warmth of sane and complacent mortals, and this internal agitation will generally display itself to the experienced eye in some species of muscular action, as we shall see below. It is true that melancholy often ends in vacant, purring idiocy; and so does mania as well. I have already stated it to be followed by speedily fatal exhaustion.* The succession

* It may be proper to give one example. " An Austrian " prisoner remained 2 months in a state of constant agitation; crying or singing without remission, and tearing " to pieces every thing, upon which he could lay his hands. " He grew calm in the night between the 3rd and 4th of " brumaire, year 3. In the morning he appeared reasonable, but in an extreme state of debility. He took food, " and walked several turns in the courts. In the evening,

is analogous to what takes place in inflammatory disorders; the extreme of languor to the extreme of energy.

When, after a long absence, any one visits the companion of his boyish days in a state of melancholy derangement, and finds himself no more noticed than the last billet of wood which the servant laid upon the fire, it is difficult not to infer a total abolition, or at least, a great diminution of feeling. And I have, not unfrequently, been in the way of seeing such inferences instantly drawn and inexorably retained. Once or twice, I have known them corrected, either by a burst of feeling, issuing from the gloom of abstraction, like lightening from the bosom of a thunder-cloud; or by the spectator's observing quick motion of the lips and other mus-

" on withdrawing into his cell, he said he felt chilly. He " had, in consequence, an increase of blankets. In the " round which the keeper made a few hours after, he found " the man dead in his bed, preserving the same position as " on lying down."—The weather was at freezing: and Dr. Pinel mentions another death the same night. He agrees with Mr. Haslam in stating that lunatics enjoy no general privilege of insensibility to cold.—Some, however, resist its effects prodigiously (p. 31.)—The scarcity in France was dreadfully fatal in the Bicêtre, where the insane were put on a short allowance.

cular affections, associated with busy thought; or what is more decisive still, by the consequence, inevitably resulting from a whole history of ideal occurrences, which the patient, before his disease retires altogether into the interior, will sometimes relate with full belief. Now such coinings of the brain irrefragably imply action of that organ or whatever other may assist in the process. Many such processes, no doubt, are carried on with as high a *verve* or as true fervour as ever accompanied poetic fiction. Melancholia can even rekindle the embers of mental fire after all that age and residence in a court contribute to their extinction. There is now perhaps confined in the Bicêtre at Paris, there certainly was not long since, a person, once attached by his places to a prince, but whose wits perished with his fortune in the wreck of the revolution. His melancholy consists in supposing himself invested with irresistable power. On ordinary occasions, he perfectly preserves in the hospital his habits of politeness, and if contradicted, retires making a respectful bow, without heat or murmur— One day, however, when the keeper remonstrated to him on the filth that he left in his apartment, the lunatic broke out with the utmost violence, and threatened the other with annihilation.

There is, I think, one grand mistake which we perpetually commit in judging of the moral nature of one another. It extends to the insane and the sane alike. We conclude respecting the existence or non-existence of sensibility, according as it speaks the same language, and keeps the same company with our own, or the reverse. Each makes his particular inclinations the standard for the world. Scarce a person who contemplates the use of wealth or authority in a superior, but makes himself amends for the disparity by complacently imagining how much better use he could make of the same means. So far all is fair.—But fairness also requires that we grant as well as take. A perverse vine may choose to throw its tendrils round a thorn, instead of its prescriptive classical support, the elm. But can we say the vine has no tendrils, because it uses them so unworthily? We may find occasions for pronouncing that sensibility is misplaced; but it is much seldomer lost or decayed than is commonly supposed. *Where the treasure is, there will the heart be also.* But if the treasure be in a wrong place, is there therefore no heart?

Dr. JOHNSON, somewhere in his Rambler, speaks of a mathematician, who when sudden intelligence was brought him that the flames were gathering round him, instead of catching the alarm, sedately replied, that *fire naturally tends to move in a circle.* The anecdote is said to be authentic, but the relater, from this one trait, labours to make out a whole character, of which the essence may be gathered from the appellation of GELIDUS. For my part, I do not see any reason for believing that this GELIDUS had less warmth of disposition than the most frightened of those who were in haste to carry their property, or their children, or themselves, out of the reach of the conflagration. The affections of the mathematician, it is true, were bent upon none of these first. But there are such things in the world as coefficients and abscisses. And by these were his affections pre-occupied. He ought to be qualified as mad if you please, but not, for any thing that appears, as cold.

To any man, who has had great interests to meditate, the apparent or real inattention of the melancholic (consistently with profound sensibility) to the objects noticed by others, ought not to seem a strange or a puzzling doctrine. At one time, while reviewing particular ideas, we hear and see without manifesting to the bystander any tokens of our impressions. At another, we are lost in thought, and the clock strikes unnoticed. The sound cannot introduce itself among the links of the passing train. The melancholic is still more lost when the fit is on him, and he notices nothing; or else (as formerly exemplified in the case of a noctambulist), he draws every thing about him into the whirlpool of his sensations. If the minds of others may be in any measure compared to vanes, which take their direction from without, his mind is a machine, which by its rapid circumgyrations not only resists the common mover, but takes this, as it were, into tow, and forces it to become its minister.

The transition to outrageous action is nothing more difficult to comprehend.—What more natural than that feelings and ideas, strongly worked up together, should excite a corresponding energy of muscular exertion?—The following reflections may illustrate this, if it should be thought to want illustration.

Whether Madness admit of an essential Character?

There are a number of terms in language, which every man, at least till he is cross-examined, can define to his own satisfaction, but no one to the satisfaction of others. Such are *animal*, *vegetable*, and in a still more difficult degree, certain terms of moral import, as *wise*, *delicate*, *pious*, *virtuous*. They are first employed in relation to the habits of the individual, or of the community to which he belongs. Then, from a continually widening sphere of observation, they gain, by analogy, an import so extensive, that it becomes next to impossible to find a point of agreement among the objects, to which they are respectively attached. So their signification becomes almost a matter of mere feeling. To know the original meaning of each, generally affords some clue, and we can then perceive by what steps the understanding has proceeded. Thus *virtue* meant *valour*, or rather *manliness* *, before it had its present seemingly so remote and placid signification. To excel in warlike

* *Virtus*, virtue—*vir*, a man.

atchievements was at once the most difficult and useful accomplishment in a tribe of barbarians, every one ferocious and hardy, and all combating at times *ad internecionem* against other tribes. This therefore most became a man. In process of time, other qualities were felt as useful and difficult to acquire: So these were stamped as *virtuous* or *manly*.

MAD is one of those words, which mean almost every thing and nothing. At first it was, I imagine, applied to transports of rage; and when men were civilized enough to be capable of insanity, their insanity, I presume, must have been of the frantic sort, because, in the untutored, intense feelings seem regularly to carry a boisterous expression. But the frantic would, at times, fall into the opposite immoveable extreme. The conjunction of both states in the same individual might, I conceive, cause the term to be by degrees transferred to the second, where that alone appeared.

The difficulty of a definition of madness, which shall be generally accepted, is evident from another consideration. The insane have the same muscles with the sane. In both, they perform the same general office. Few need be told that in all men, discourse, look, and gesticulation, depend on fixed or alternating muscular contractions *. These alone are the outward tokens of the state within.

But the number of muscles, and the range of contraction in each, are such as to present combinations without end.—Let us attend to the countenance only. If we had a series of drawings, ascending by the closest gradations from the face of a sleeping infant (a being incapable of insanity), to that of the most furious inhabitant of Bedlam, who would undertake to point out the last among the sane, and the first among the insane? But if there be no hope of agreeing as to a simple object of sight, what chance can there be of compromise where the circumstances cannot be, in the same manner, subjected to

* The appearance of the eye, which is so striking in the maniacal state, is regulated by its muscles. Though hollow when the patient is calm, it will protrude on the commencement of the paroxysm. This arises from the rigidity of certain muscles to be seen on the back of the ball in any set of anatomical plates. The glistening is a similar operation. The dullness of the eye often arises from a sort of corrugation of the coats, though the furrows are not singly visible. But when all the moving fibres become tense, the coats are fully unfolded, and shine.—Sometimes the presence of the keeper of a madhouse shall overawe the raving patient, till his tongue and limbs become in a moment composed. But the eye will retain its characteristic expression.

steady contemplation? To what purpose does the nosologist talk of "*false conceptions of* "*things*" and of "*irrational emotions or* "*actions*", (*pathemata vel actiones præter rationem*)? I am not sure that he can explain to us what *false* is: I hold it as certain that he would be baffled by *rational.*

Too much learning has made thee mad! This is the perpetual motto of the vulgar. The more limited a man's range of information, the more readily does he attach the imputation of extravagance to any mode of conduct, varying from his own. Talking to one's self is sufficient to make one pass for insane, with those who have few or no *vivified* ideas. But one distinction will apply to these opinions, as generally as can be expected, where the circumstances, on which they are founded, are so fugitive and various, and where those, who form them, so capricious and undiscerning. They are applied to strong actions or to expressions of countenance that imply somewhat of consideration; never to mere imbecillity.

> The fool of nature stands with stupid eyes
> And gaping mouth, that testify surprize.—

Such an one will scarce any where be pronounced mad. It is something beyond, not short of, himself, that draws this sentence from the ignorant.—It is the perversion of faculties, not their privation.

The knowing do not extend the boundaries of insanity so unmercifully, but they have no exact criterion for distinguishing it. They too can only judge of others by themselves. When the mind is occupied, and the active powers employed for an end, which they cannot conceive as desirable or attainable, there the party seems no longer in his right wits. He may not yet have attained a degree of wildnesss, at which it shall be necessary to seclude him from society. But he is in a fair way to reach this point. He may not at present be dangerous to others, nor likely to walk over a precipice, or into a river. He is not yet so possessed with his delusion, but that he can strike a bargain, or attend to the items of a tradesman's bill, and ascertain its amount.

When there is no longer merely question about throwing out an imputation or cutting a joke, but depriving a person of his liberty, just the same uncertainty prevails. Those with whom it rests to decide, have nothing for it but to look into their own bosom, and on comparing the series of their own thoughts and actions with that in question, to strike the balance. If they find it considerable, they put it, of course, to the account of derangement. There are thousands who mix uncontrouled among the sane, and yet are subject to the most irrational feelings and actions from false conceptions. Such are the parents, whom we continually see wasting the inheritance of their offspring in an incessant round of wrong-headed projects for increasing it. Of the falseness of their conceptions, the injudicious adaptation of their means to the end proposed is a sufficient guarantee; and we have *primâ facie* evidence of their irrationality. May it not be doubted whether any criterion can be established upon the phænomena, exhibited by a number of the most declared lunatics, which shall exempt the great NAPOLEONE BONAPARTE from the imputation of a tainted understanding? On the eminence where he sits, and on which it is well if we may not soon have to inscribe—

——— Tyrant power
Here sits enthroned in blood——

it must be evident to every spectator that he is in a state of inordinate agitation. Whether his conceptions are false or otherwise, cannot be determined before we know exactly what they are. But if it may be presumed that he has in contemplation the happiness of the millions whom he governs and his own, their justness and the rationality of his conduct would appear very problematical. And if he *has* any other purpose in view, it would, I think, be the verdict of an enlightened jury *that he ought to be put out of harm's way.*—Similar reasonings may involve a character, lately too conspicuous in our own country, who in aspiring to rival the glory of a predecessor in the conduct of military enterprises,—in conceiving falsely of our own state and of that of the enemy, and in his perpetual vauntings, displayed all that permanence of delusion of sense, all that agitation and presumptuousness* that characterize the maniac. If with Dr. DARWIN, we make madness to arise from "excess of action of the sensorial power of "volition," we shall infallibly comprehend such cases as the preceding. But we shall be still equally at a loss for a line to divide the

* One of this man's boasts was *that he understood human nature*. Taking the pretence in a partial sense, it perhaps was well founded. He was acquainted, I believe, with all its disgraceful parts, with its corruptions, and its readiness to be gulled by mountebanks. He understood human nature, in short, just as a scavenger understands a magnificent city.

sound in mind in general from the unsound. For that term "excess" is as vague as the thing to be defined, and can only find an uncertain application, when the question arises on any particular case, according to the varying understanding and conscience of individuals. In conformity with what universally occurs on adverting to the subject of madness, Dr. DARWIN observes, "if every one, who pos-"sesses mistaken ideas or who puts false "estimates on things, was liable to confine-"ment, I know not who of my readers might "not tremble at the sight of a madhouse." But in like manner, if actions, resulting from a will, intent upon an object beyond its value, or inordinately exerted, constitute insanity, there would be cause for just the same apprehension, as it might be difficult to find one mortal, who could pass for sane with others.

Chief Particularities of the confessedly insane.

Dismissing therefore all pretensions to establish an essential character, or to give a tenable definition of madness, and throwing it out for the consideration of the intelligent, *whether it be not necessary either to confine* *insanity to one species, or to divide it into almost as many as there are cases* *, I shall extract from the history of lunatics such circumstances as principally characterise their intellectual aberrations. Afterwards, I shall endeavour to fix upon the point at which they usually begin to go astray, and follow their wanderings till they are lost. Hence it will appear to a certain extent, how they might have been kept in the path of right reason.

We have seen that imbecillity is distinguished, according to the common apprehension of men, from insanity, and that, contrary to first appearances, torpor is inconsistent with insanity in any form. It may now be added that *the mind acts with unusual energy.* The only doubt is when the patient preserves an obstinate silence. But in a large proportion of cases, the ruling ideas break out or may be elicited by entreaty. Thus in one of the entertaining articles of Zoonomia, (Class iii. Order 1, Genus 2, or vol. 2, 8vo. ed. p. 54, &c.) " a most elegant lady is said

* I have looked over the species of insanity in many writers. They have appeared to me to be conceived in this spirit. Suppose the human appetite to be divided into the appetite for animal and vegetable food: and these again into the appetite for venison, veal, mutton—pease, beans, and so forth.

"suddenly to have become melancholy, and "yet not to so great a degree but that she "could command herself to do the honours "of her table with grace and apparent ease. "After many days' entreaty she at length "told the author that she thought her mar- "rying her husband had made him unhappy "(though it was a love-match on both sides) "and that this idea she could not efface from "her mind day or night." Hundreds of similar examples of deeply retired and intense thought might be quoted. Now the induction from the majority of instances may be legitimately extended to the minority, where neither art nor authority can draw forth a confession. It seems even fair to suppose that the more unfathomable the mind, the busier are the movements going forward in its recesses.

It very generally holds *that in insanity the ideas are vivified*, or *that they are exalted to the force of impressions*. This, if any one, is the cardinal point on which the whole affair turns. When a person continues to maintain that he has been an actor in scenes, which never took place, all the world agrees in putting him down as mad. It is a fact which perpetually occurs. Thus I knew an insane officer, who believed himself to have been brought to a court-martial, though in the real history of his life there was not the least vestige of such a transaction. The insane are perpetually without cause imagining themselves infected with symptoms of the itch or of the venereal disease.

From a small germ of fact, there shoots out in a moment an extensive ramification and luxuriant foliage of imaginations, all equally distinct to the mind with the first perception. I have before me a very detailed account, by a medical insane patient, of the transactions of the three or four days, immediately preceding his confinement. He drew it up for the satisfaction of the superintendant of the lunatic asylum, into which he was received. It exhibits, in the most curiously minute manner, the workings of his fancy. The persons about him, aware of his state, keep a constant eye upon him. Of this, from the usual sagacity of invalids in the same situation, and, I suppose, from consciousness also, he becomes immediately sensible. Then, in the activity of his thoughts, he imputes motives of his own imagining to their conduct. With the help of this scaffolding, his castles run up into the air with meteor rapidity. On every countenance, he reads avowals (legible to no one else) of the schemes he conceives formed against him, and combines friends and strangers into one grand conspiracy against his life. For example—the writer supposes poison had been administered to him, but that he had escaped death by taking sparingly of the liquid with which it was mixed. On one occasion, he holds the person whom he believes to have given it him fast by the wrist. He goes on to say—" He sometimes affected to smile at " me with pity for my unhappy state of " mind. Then he would lean back on the " couch, close his eyes, then open them a " little, so that the eye could barely be seen " through the lashes, and so as to prevent " his being observed, as he thought. At " those times, he would cast the most in- " fernal looks at *me*, and afterwards round " the room, for some weapon or other to " finish what he had begun: The latter I " could see not only from his looks and the " hardness his muscles used to assume, but " also from the posture he would put himself " into—ready to jump, if he discovered " what would answer his purpose." In this style the account proceeds throughout, detailing the incidents with perfect accuracy, but forcing on each the strangest construction, and putting upon every countenance looks, the most favourable to this construction, but the most remote from the reality. The reader may find a similar example in ROUSSEAU's ingeniously mad commentary upon Mr. HUME's real or supposed expression: *Je tiens J. J. Rousseau.* (*Oeuvres compl. de Rousseau, par ordre des matières, tome xxvii.*) *

The imagination of the poets is content with personifying inanimate objects. That of lunatics frequently goes a great deal further, and strips them of their own personality. It would not be believed, were it not so perpetually experienced, that a human being could come to conceive himself made of butter; or that his legs were of straw; or that he was a barley-corn. The fact proves the force of imagination; and perhaps may be in some measure explained. An invalid of great brilliancy of parts once said to me, that but for a particular expectation, "*he would as soon be a nettle in a country* "*church-yard.*" If his spirits had declined

---

* "Les longs et funestes regards tant de fois lancès sur "moi—un ton, dont il m'est impossible de donner l'idée, "et qui correspond très-bien aux regards dont j'ai parlé." —*Rousseau.*

further, and the wish for this metamorphosis had recurred often, it might, at last, have been considered as realized. Such are the contemplations, which when seconded by external circumstances, end in melancholy. A devout person reads in scripture that *all flesh is grass.* The words strike him. He ruminates. He is doubtless flesh. But he has irrefragable authority for believing that flesh is grass. Therefore he is grass himself. What logic can be clearer! If he has gained this step, how easily may he go on to apply to his person all the qualities of grass. He may act upon it, that is, according to Dr. DARWIN, he may raise delirium into insanity, by standing out all night to receive the refreshing dews, and lurking all day in the cellar to avoid the parching sun. As the seasons revolve, he may employ literally a language much like that which WOLSEY uses metaphorically—

> "I am fallen into the sere, the yellow leaf."

Upon the preceding principle, we easily perceive why those human beings, that are merely the creatures of impression, as savages and children, are incapable of insanity. There must be combinations of ideas, and the habit of combining. Otherwise the delusive conceptions, lately exemplified, could not branch out to any length, and therefore could not maintain a footing for any time in the mind. Miss H. D. about 7 years of age, arrived from the East Indies, speaking scarce any thing but *Moors*, and, I believe, unacquainted with the letters of the alphabet, but quick and intelligent. I was desired to see her one day, when she complained in broken English that *a man had got into her stomach.* Her friends told me that this had been her complaint from the moment she appeared indisposed. I asked her how she knew that there was a man in her stomach. She said she was sure there was, because she felt him thumping very hard. Had the child harboured a previous train of superstitious ideas, ready to connect with this supposition, she might have run mad. The learned JURIEU, who wrote a commentary on the revelations, contracted, to a degree of severity, the ailments incident to sedentary scholars. The pains in his stomach gave him the idea of a hard substance pressing against it. And what should he take into his head, but that the beast of blasphemy, with seven heads and ten horns, and ten crowns upon the horns, had made good a lodgment within him? The same happens to a multitude of poor ignorant creatures, who hear of possession by devils; and when they feel an involuntary twitch, think it is the finger of the devil.

It seems to be generally understood (for there must be always a great deal conventional or arbitrary in the designation of a person as insane) that the duration of the delusion is an essential circumstance. Pending the bills, framed by two noted statesmen for placing a gag in the mouth of the English nation, I was desired to see a gentleman, whose discourse alarmed his family. He had been out in his carriage. A momentary qualm had come over him. On returning home, he spoke of having just seen persons, known to be at a distance, and mentioned occurrences as the news of the day, none of which could have happened. He appeared to read a paragraph out of the *Star*, announcing the appointment of the Duke of BEDFORD to the treasury, and of Mr. Fox to a secretaryship of State. He glanced over the debates, and when he came to the division at the end, pointed with his finger to the ministerial numbers as a small minority, though, as is well known, exactly the reverse was the case. But this was not a common blunder. He did not mistake one side for the other; but it was, on both sides, a substitution of numbers quite different from those stated in the newspaper, upon which nevertheless his eyes were intensely fixed.—The patient, it should be observed, was a quiet character and a calm politician. He had no fever, pain or other ailment. He eat and drank precisely as before; and his insanity disappeared in a couple of days without ever returning.

Exactly the same association, coalescence, or if you please, confounding of vivid ideas with impressions, arises from the same cause in fixed derangement. A lunatic, who was guarded at home, having by means of the servant got possession of a novel, set to read it with the utmost volubility, substituting the names of acquaintances for those of the author's characters, and dexterously interweaving foreign circumstances with the narrative.—Foreign feelings intruded as well as foreign ideas. For passages, not in the least pathetic, made this patient shed tears plentifully—a distinct indication of sensibility in excess. Just so do the other circumstances confirm the opinion that ideas crowd upon the mind more in this than in the sane state. These particulars add force to the positive testimony before referred to concerning the resemblance between an insane paroxysm and a dream. Indeed I can imagine no difference in the state of mind except it be the short continuance of the delusion in one case.—In attending a patient in a fit of the gout, I found one morning that he had awakened with new inflammation in his feet. He told me that his sleep had been much disturbed, and that he had dreamed, among other things, of the apothecary's putting fetters on his legs. He resisted, while it was doing, but in vain. He then insisted, with a cocked pistol in his hand, upon the removal of the fetters; and shot the apothecary dead because he refused.—A female had felt deep and frequent chagrin, because her marriage had been unproductive of children. She became lunatic. Once, while under confinement, she made unusually violent resistance to an attempt to dress her, at the same time screaming and exhibiting a degree of terror unknown before. The catamenia had just begun to flow. They probably flowed with pain. At least, it clearly appeared that she believed herself at the moment in a parturient state. Her medical superintendant, who was accustomed to act as accoucheur, thinks she suffered full as much as if she had in reality had labour-pains!—In these two cases, we find exactly the same sort of superstructure, raised by the imagination upon a painful feeling—the same spirit of resistance—*i. e.* the same efforts of volition precisely.

Considering this analogy, and the power of sleep to foster nervous complaints and to bring on their paroxysms, I have frequently endeavoured to learn, whether dreams are not sometimes continued without a break into insanity, and whether they do not increase the susceptibility to its exciting causes. My enquiries have seldom been satisfactorily answered.* Dr. PINEL, however, in his

* The authority of observant persons, having the superintendance of lunatics must be great, but there are considerations, which limit it. To the excentric movements of these animated machines, they are constant witnesses, and must therefore be good judges of their range. But does it follow that they must be equally well acquainted with their interior structure? In a madhouse at least, they can hardly gain this knowledge. The machine was laid together and mounted before its introduction there. It is one thing to behold the Nile foam down its cataracts, and spread over the face of Egypt; another to possess the secret of its springs. And those who have been most familiar with adult madness have, I believe, often failed to study it in nascent state. In the heat and bustle of their own occupations, they sometimes consider the close study of the human mind, concerning which every man of education gets a few random notions, as an occupation too retired, and too much in the shade for them; nay, one finds them disclaiming metaphysics, or professing an intention of dis-

lately published collection of facts (which, though of a gross character, will perhaps retain their value longer than many attempts at a subtler investigation of the subject) has a passage pretty much in point. "It is," he says, "extatic visions during the night that often "form the prelude to the fits of maniacal "devotion. It is also sometimes by enchant-"ing dreams and a supposed apparition of "the beloved object that madness from love "breaks out with fury, after longer or shorter "intervals of reason and tranquillity."—(*sur la manie*, p. 18. *Paris* 1801.) For the reception of so sudden a passion as that of Romeo and Juliet for each other, the heart is, I suppose, most effectually opened by nocturnal visions or day dreams. The author

---

patching this part of the subject as soon as possible. That is, they desire to forego the consideration of the essence of the thing, and to play idly about its accessories and accidents. The sane and insane mind are made up of the same stuff. A change in the hues and arrangement of their materials is the sole difference. Upon the knowledge of this change, it is probable that the power of preventing and correcting it greatly depend. The management of the insane, as far as one can judge from books, goes on too much in the gross; and without the insight to be obtained from the study I recommend, it is not easy to say how it can be more nicely adapted to the exigencies of individuals.

of Oberon and other poets, as also novelists, if their authority be worth any thing, may be quoted in favour of this doctrine.

I have met with a few instances, from which it appeared that dreams have a real influence in disposing to insanity. The following may serve as an example. A pregnant lady in the East Indies had a frightful dream, of which the day of judgment was the subject. Very soon after waking, while her nerves were still agitated by the nocturnal images, there came on one of the violent thunder-storms, usual in those climates. This recalled her scarcely dissipated delusion; and it continued for several hours. A miscarriage was the consequence; and her health received a rude shock, which it had not recovered many years afterwards when she became my patient, though she was never afterwards subject to any thing like insanity.

These facts would appear to admit of some useful application. Where we observe the constitutional tokens, above enumerated, with or without a family disposition to insanity, the hints thrown out in the former part of this essay, respecting the means of securing placid sleep, should be attended to and improved upon. A knowledge should be obtained of the nights, passed by persons for whom there are such grounds of apprehension. Their state during the days, succeeding unpleasant dreams, should be noted. On the accession of any suspicious appearances, they should be put on an abstemious diet, carried to fresh scenes, have blood taken from the head, and diligently pursue such a course of medicine as their symptoms may seem to require. A plan that succeeds but now and then in confirmed insanity would be more certainly followed by the desired consequence, if carried into effect as it is coming on. It is of much importance therefore to be put into a method of recognizing it on its first approach.

I consider delirium as having just the same relation to insanity that dreams have, that is, as being undistinguishable while it lasts. Delirium seems too to have much the same varieties. It is sometimes fierce; sometimes gentle and communicative; at others, close and solely intent upon its passing pageants. An attempt has been lately made to distinguish delirium by the absence of voluntary exertion *. But in

---

* "The ideas in delirium consist of those excited by the "sensation of pleasure or pain that precedes, and the trains "of other ideas, associated with these; and not of those "excited by external irritations or by voluntary exertion.

attending to persons, affected with delusions of sense in febrile and common nervous affections, all gradations from rambling of the thoughts without exertion to intentional violence, may sometimes be observed in the same patient, during the same attack, and in the course of the same day. And quiescent delirium passes through incompleat delirium (or that attended with weak voluntary efforts), by imperceptible shades up to full insanity, according to the definition in question.

The best account of a delirium, with which I am acquainted, is one given by a physician in his own case (*Moritz Erfahrungs-seelen-kunde*, I. 3. 44.) No medical attendant or philosophical spectator could so compleatly have delineated the various busy scenes represented before the mind. From the following short

---

" Hence the patients do not know the room which they " inhabit, or the people which surround them; nor have " they any voluntary exertion, *where the delirium is compleat:* so that their efforts in walking about a room, or " rising from their bed are unsteady, and produced by their " catenations with the immediate affections of pleasure or " pain. By the above circumstances, delirium is distin- " guished from madness, in which the patients well know " the persons of their acquaintance, and the place where " they are; and perform all the voluntary actions with " steadiness and determination." (*Zoonomia*, Vol. III. 8vo. ed. p. 495.

extract, it will appear that the varieties of steady determination of the will and passive delirious reverie succeeded each other so rapidly, and were so connected with the same disease, as not to leave any room for discriminating these affections as heterogeneous. And I fear no doctrine can stand on the slippery distinction between diseases of sensation, and diseases of volition.

The writer of the account, while ill of a low fever, had been removed into an apartment different from that in which he usually slept. The phantasy that persecuted him most was his inability to persuade himself that he was in his own house at all. The pain of this persuasion excited continual efforts and trains of thought. He seemed transported from street to street; and his imagination was active enough, to exhibit to him every moment, some different public place, in which his guards detained him on my bed. "I almost continually supplicated," says the author—and let the reader mark the steadiness of the wish, and the repetition of voluntary exertions of the voice—"I almost "continually supplicated to be only carried to "my house in *Holy-Ghost-Street*."—The persons about him, in endeavouring to pacify him without complying with his wishes, did but confirm the delusion under which he laboured.

Their cry was—*it shall be done in a few hours —or early to-morrow, as it is now night.* He lay pining in vain for the end of these few hours: and all the time his fancy created places, not the most agreeable, for his residence. Sometimes it pitched him between two walls, so close that he could not heave an arm; sometimes on a burial-ground; sometimes on the court before the hospital he attended. All the arguments of his friends availed nothing to prove to him that he was really in his own room. When they pointed out to him his own books, close beside which he lay, or the prints that hung opposite, he took it for a trick. Sometimes he did not recognize them for his own; and sometimes he conceived they had been removed to his present place of abode.—In these observations, we perceive the very nimbleness of wit, so conspicuous in confessed lunatics.— Of the effect of accidental irritations of the senses, he observes that a postilion's horn, or even the watchman, would transport him to a public place, filled with music and dancing— the neighing of a horse in the street, to a stable —the bad smell of his perspiration, or the blood coagulated in his nostrils, to a burying-ground. —When the physician in attendance consented to his removal into the wished for apartment, on the score of there being nothing to lose, his whole internal feelings underwent an instantaneous revolution. Though he had lain for days and nights without sleep, raving, supplicating, and complaining, a placid sleep, the forerunner of a rapid recovery, overtook him in ten minutes!

It was another conceit that he was hated and deserted by the whole world—that all his friends had forsaken and his patients renounced him. The foundation of fact on which this superstructure of despondency was raised, he takes to be his missing three of his most intimate friends, who were absent or incapable of attending upon him. He adds to this a mistrust towards mankind, which his friends told him they had observed in his health. The number of unpleasant things he experienced from those about him, such as their refusal to let him quit his bed, forcing him to take medicines, applying blisters, must have added force to the sentiment.

What is strongly in point is this: "my other "fancies were probably those most common in "every delirium. The flowers on my curtains and "tester I took for men in continual movement. "They all went towards the wall; and as "there were none but my acquaintance, I "often joined them. We found ourselves in "large illuminated subterraneous chambers, "where I learned such family secrets, as every "man in the world above, keeps close locked "up in the recesses of his bosom. Once I really "called my wife to my bed-side, and told her "a shocking transaction, involving two of "our friends, which I had learned in these "subterraneous assemblies. I related the story "with so much consistency and gave it such "an air of probability, as to make her take it "for a real fact, which I must have known "before my illness."

It is well understood that delirium can frequently be arrested by words uttered in a loud voice, by the introduction of a new face, or any impression strong enough to supersede the imaginative ideas. Now this happens also in madness; and the number of authentic cases, in which a cure has been thus effected, may justify a suspicion that it is an expedient not often enough resorted to. In the minute journal above mentioned as being in my possession, I find an incident, which shews the power of new and striking objects in suspending confirmed and obstinate lunacy. The patient was to be removed, through a succession of very rugged and beautiful landscapes, to another habitation. The measure occasioned great agitation; at first, a refusal and afterwards alarming suspicions. Awe of the medical superintendant made the countenance appear strangely at variance with the eyes, "the former "shewing a forced calmness, induced by the "seriousness of my observations, and the latter "a wild anxiety, demonstrative of the emo-"tions, which in fact agitated the patient's "mind." Minute examination of the open sedan chair first—then of the chairmen with reference to the medical superintendant, as if they were in league, and denunciation of the latter as a conspirator against the state, were some of the extravagancies that preceded the departure. The first part of the road led through an inhabited spot; and here frequent efforts towards calling upon the passengers for assistance, with orders to the chairmen to stop, are noted down. Next, the scenery at the foot of an ascent drew attention; but this alternated with wildness. The ascent led to immense precipices, when the patient called on the medical superintendant, notwithstanding the aversion he was held in, to walk beside the chair. The supposed danger being passed, the wildness returned. The party soon arrived at an artificial lawn encompassed by woods. Here the invalid familiarly asked the superintendant for his pencil and a card, to sketch the surrounding beauties of the landscape. No sooner was the request complied with, than an instantaneous change was remarked from serenity to great liveliness of countenance, with somewhat of a satirical cast. The eyes were, in fact, rivetted upon a human figure, bearing an umbrella before the sedan. The patient laughed heartily, and when the merry fit was a little over, committed to the card an admirable likeness of this figure; with its long bearded chin, hooked nose, and diminutive fiery eyes.*

I know not whether it will be thought to shew an affinity between the two affections, that fever, particularly where accompanied with

---

* It is probable enough that the patient could not have managed the pencil so well in the state of sanity. Thus, there are satisfactory testimonies of intellectual difficulties surmounted in dreams, which had baffled every waking effort.

The late professor Waehner of Goettingen, used to relate that in his youth, he had to put a thought into a greek couplet; he tried for two whole days, but in vain. One evening, he goes to bed on the eve of fruitless efforts. In the night, he rings for his attendant, asks for paper, pen, ink, and a light, writes down the two verses, which he had sought and found in his sleep. On awaking, without recollecting the least of what had passed, he sets again to his task, but still without fuccess, gets up chagrined and finds a very happy couplet under his own hand on his desk. Upon learning from the attendant what had happened, he could not call back the least trace of it.—Chiarugi speaks of a maniacal patient of his who, without instruction, carved admirably natural figures in wood.

delirium is one of the best ascertained causes of insanity, (Mr. *Haslam*, p. 99.)

The doctrine, constituting an exalted imagination the fundamental circumstance in lunacy, is disputed by a person, whose attention to the disorder seems equal to the advantages of his situation. A short examination of the objection may throw light on the whole subject.* The intelligent reader will probably

* "If the disease consisted in the strength of imagination, "the imagination ought to be equally strong upon all "subjects, which upon accurate observation is not found "to be the case. Had Dr. MEAD stated that together with "this increased strength of imagination, there existed an "enfeebled state of the judgment, his definition would have "been more correct. The strength or increase of any "power of the mind cannot constitute a disease of it: "strength of memory has never been suspected to produce "derangement of intellect: neither is it conceived that great "vigour of judgment can operate in that manner: on the "contrary, it will be readily granted that imbecility of "memory must create confusion by obstructing the action "of the other powers of the mind; and that if the judgment "be impaired, a man must necessarily speak, and generally "act," [what is speaking but acting?] "in a very incorrect "and ridiculous manner." *(Mr. Haslam*, p. 2.) Mr. H's definition of insanity runs thus—"an incorrect association of "familiar ideas, which is independent of the prejudices of "education, and is always accompanied with implicit belief, "and generally with either violent or depressing passions." 1. The word *incorrect* is as vague as *insane*. 2. The ideas

wonder to find it asserted that the increased force of a faculty cannot constitute a disease of the mind. To me it seems self-evident that the very belief in things unseen as seen, expressed in the objector's own definition, must be followed by mental disease. What! when the force of association lays an imaginary eruption upon the clear skin of a person, afraid of the itch, is it not all over with judgment and discrimination? Can the mad lover be less bewildered, who beholds his mistress in every female form he sees? How is it possible any judgment can consist with so busy a fancy? And what can be the fidelity of memory in that man, who is persuaded that incidents, of which he has only heard or read, have befallen himself? Neither do I perceive why the imagination, if strong as to some ideas, must be universally strong. If the memory may be

---

are often not familiar, till familiarized by the workings of the mind, that generate insanity. I shall mention instances below. 3. What is *education?* Ideas early acquired often furnish insanity with its delusions. Mr. H. mentions several patients, who said that they had seen the devil: and they described him as he stands figured in foolish books. Had they but believed that he was so formed, all might have been sound within. But carrying the idea up to the liveliness of an impression constituted lunacy. Whatever false notions be imbibed, whether as prejudices of education or otherwise, they do not impeach sanity, till exaggerated.

strong as to some ideas and weak as to others, why need it be otherwise with regard to the imagination, which is a full and vivid memory? Old people, we know, let slip the occurrences of yesterday, while they have a perfect recollection of those fifty years back. When the delirious vassal repeated the greek verses he had learned 30 years before with his lord, no other ideas of the same period recurred!—It is not necessary, for the present purpose, to annihilate these mental faculties, which they raise up, in mock majesty, to sway the thoughts for their little moment, and then to deliver the sceptre to a potentate as evanescent as themselves. But it is high time that the fiction were banished from philosophy. It was convenient enough that the modes of association, which take place most frequently, should have particular names. But to take these nominal for real essences, was the sure way to confound the whole doctrine of the intellect.

Further Remarks on the *TRANSITION to INSANITY.*

Injuries on the head, abscesses in the ear, habitual drunkenness, fevers; the cessation of natural secretions or artificial evacuations; large use of mercury, and palsy have been recorded as the chief exciting causes of madness. Of 113 patients, concerning whom Dr. PINEL obtained accurate information, 34 were reduced to insanity by domestic chagrin, 24 by disappointment in love, 30 by the events of the revolution, 25 by terrific ideas of the next world. Indulgence of grief or of any other passion, affections sudden and violent, as fright or indignation, religious terrors are observed, in this country also, to be followed by the same effect. It is remarkable that the very moral or physical causes, as blows or panic fear, that sometimes occasion insanity in grown persons, at an inferior age bring on epilepsy, hysterics, nervous headaches, atrophy. Why blows on the head should give bolder relief and brighter colours to certain ideas, must remain a mystery, till we have facts enough to unite the physical and metaphysical phænomena of human nature into one consistent body of doctrine. It is easy to give an hypothetical explanation. The brain appears to be concerned in the reproduction of ideas. Any given groupe or train may be present or passing at the moment of the injury; and the structure of the organ may be so altered, as to raise them to the utmost distinctness. Other associated ideas, by the intervention of these, may acquire a connection with the morbid actions of the brain. That frequent intoxication should, in the long run, produce a similar alteration of structure, will not surprize us, if we advert to the ordinary effects of drinking upon the head. Why a particular complexion should dispose to exaltation of ideas is likewise, at present, inexplicable. Do not states of distant parts, by sympathy, produce such changes in the brain, as to call up ideas vivid enough for madness? This may happen from a particular stroke of the heart, condition of the stomach, or tension of a remote fibre. And such exciting causes may be accompanied with obscure feelings, or none at all. Dr. DARWIN thinks a diseased liver occasions lunacy. It may be so: though both may arise from intemperance as a common cause. Nor need pain be at all concerned, as he thinks.

The action of the brain upon thoughts or impressions, and of these upon the brain, seems perfectly reciprocal. Just as in hypochondriasis, whatever lowers the spirits injures the stomach, and whatever injures the stomach, lowers the spirits. When disastrous intelligence oversets the reason, we have first the impressions on the hearing and sight, as raised by the written or spoken words and countenance; then the ideas connected with these words; lastly, the affection of the brain. This therefore one would be inclined to set down as secondary, when moral causes do the mischief. The disorganization of the brain, as an important circumstance in madness, is controverted by some, and among others, by Dr. PINEL. But accurate observers, as Mr. HASLAM, have found the head of every patient they opened, manifestly altered in its structure. And Dr. PINEL's own remark that he has never found any thing within the skull, *except* what had equally occurred after death from epilepsy, apoplexy, fever and convulsions, is in confirmation. It would be too much to expect that so clumsy a tool as the scalpel should detect specific alterations, corresponding to these several affections.—Persons, little conversant in medical disquisition, may wonder how any lucid interval should occur under a permanent lesion of the brain. But this is perpetual in pathology. Neither a stone in the bladder, nor a carious tooth, always give pain; nor does the disorganized heart always palpitate. Some external influence seems necessary to induce that condition of the parts, in which they shall be morbidly irritable. It is scarce to be believed what a trifling cause will bring on a maniacal paroxysm. A patient seeing his own face in a looking-glass, shall first smirk, then make grimaces, next gesticulate, and, in a few minutes, arrive at full frenzy.

The consideration of moral causes of slow operation is the most curious as well as most useful in this whole enquiry. A medical person, not now living, who shewed manifest signs of derangement, told me that his ill success in his profession filled him, as may well be supposed, with anxiety for his own subsistence and that of his family. He would sit at home for hours ruminating; and when he found, day after day, no summons arrive, he would saunter abroad and occupy himself with a reverie of wishes. These wishes he would sometimes arrange into a climax of events, worthy of the glassman in the Spectator. At length, he would direct his footsteps homewards under a sort of persuasion that some person of consequence had actually sent, during his absence, to call him in. This is a state full as near insanity as sanity.

Madness perpetually realises the wonders of Ariosto's magic, and prepares palaces or dungeons for those, whom it possesses. A person under disappointment or chagrin, sets about to imagine by what possibilities he may be delivered from his perplexities. They return to his thoughts by day and by night, gathering more or less body according to the intensity of his feelings. At last, after a long struggle, Imagination obtains the mastery compleatly, and ever afterwards she uses calamity as her hobby horse.

During grand political crises, when society is subverted from its foundations, insanity often takes this turn. The suddenness and magnitude of the events kindle the imagination. A total change of fortune irritates the feelings of those, who are plunged into the abyss of adversity. Grinding injustice draws forth all their indignation. No prospect of redress by ordinary means opening, they ponder upon extraordinary deliverances, till they are lost in the labyrinth of their own thoughts. For if hope deferred maketh the heart sick, hope extinguished maketh the head fanciful. In this, among other ways, does the revolution in France appear to have operated upon many intellects. It dethroned one monarch indeed, but it raised a multitude to the supreme power. Nay, the madhouses of France were peopled with gods as well as with kings. Three Louis XVIs. were seen together, disputing one another's pretensions. There were besides several kings of France, of Corsica, and other countries: there were sovereigns of the world, a Jesus Christ, a Mahomet, so many deities as to render it necessary to distinguish them by the place they came from, as the god of Lyons, the god of the Gironde.

Among a number of persons, struck with insanity, though some are seen to harbour pleasurable delusions; others, according to their character or previous habits of thought, fall into the contrary extreme. This unhappy change has been of late as largely exemplified in France as the preceding. The constant terror of the guillotine hung over some: others perpetually mourned, with aggravated ideal sorrow, in consequence of being deprived, by the military requisition, of their only child. The unfortunate father, from whom GARRICK caught the gestures and countenance of his LEAR, was probably of a temperament the reverse of sanguine, and little accustomed to the dreams of hope, even before his infant dropped out of his arms from the balcony into the street. So his imagination could but perpetually reproduce the scene in its original horrors. We know from the records of insanity that another parent of a disposition, previously more cheerful, might easily have mistaken a pillow for the child sleeping, after he had been dressed by the surgeons, and have kept eternal watch to prevent him from being disturbed by any noise.

It is no wonder that hypochondriacs, when the blue phantoms that flit at times before their fancy, become by degrees fixed and embodied, should be apt to suffer from terrific illusions. Clerical insanity, I suppose, will generally be found to have begun in hypochondriasis. There is another class of persons, exceedingly subject to insanity, and to insanity of the painful sort. Go, for instance, to the scenes of trade at London or at Bristol. Among the faces that appear at high 'Change, mark those that bespeak the cares attendant upon wealth already accumulated; and those others, where an added air of wildness characterises the speculator, too much in haste to wait for the reward of regular industry, and burning to get rich by a lucky hit. Some of these men will grow mad enough to be watched at home or sent to a lunatic asylum, where they will be haunted by the fear of coming upon the parish. Many others just sufficiently mad to be only run miserable all the rest of their lives from similar apprehension. So that the thirst of gold seems to maintain its character throughout: and there may be difficulty in saying whether it most debases the soul in sanity or insanity.

The votaries of devotion of a gloomy character often lose their wits, and have the place supplied by depressing imaginations. A poor collier heard a field-preacher rave much about damnation. He immediately felt himself encompassed by the terrors of hell.

On a dreary winter's day, he was found more than half naked, squatting in the twilight of his chimney-corner. As the wind howled over the heath, " hark !" says he to a medical person, entering to inspect his situation, " there is the devil come to fetch me in his " chariot. Did not you hear his horses " neigh ?" For keeping the intellects sound, or if it come to the worst, for merging them in cheerful madness, how much preferable is

A false religion, full of pomp and gold—

to a religion full of damning dogmas, which must necessarily be false !

The reader will easily be able to carry on this method of considering the subject to love, anger, and other affections, productive of a permanent exaltation of the imagination. I shall only add here that by some unobserved physical cause, the feelings may first be rendered intense, and then any accidental ideas will be metamorphosed into imaginations. This will in part account for changes in the delusions of mad, as well as sometimes for madness itself.

PREVENTION.

To preserve the human mind from the irregularities exemplified in this essay, in as far as they arise from moral or slowly operating causes, would seem a task of the utmost simplicity; to restore its equipoise, when that is beginning to be lost from violent gusts of passion, by no means difficult.

The first great preservative of mental, as of bodily health, is active occupation. No species of nervous disorder easily fastens upon persons, who devote a part of their time to moderate labour. It should, at the same time, be diversified and enlivened by scientific explanations. For mere labour is not enough.—The uniform operations of ingenious artisans sometimes wear a track in the brain, along which the torrent of imagination rushes with destructive violence. Hence insane projects for producing the perpetual motion. The mischief might always be avoided by contriving channels in different directions: and not only would mischief be thus avoided, but the intellectual streams, which, when collected into a single body, do so much damage, would be beneficial from their division.

It is evident that many dangerous series of reveries must be prevented by combining the thoughts and actions in efforts to produce some useful or curious purpose. It would even appear that total aberrations of the understanding are most effectually corrected upon this principle. At Saragossa in Spain, there is an hospital for the sick in general, and the insane of all countries. The patients of the latter class are divided early in the morning into parties, some of whom perform the menial offices of the house; others repair to shops, belonging to their respective trades. The majority are distributed, under the superintendance of their guards, through a large inclosure, where they are occupied in the works belonging to gardening and agriculture. Uniform experience is said to have proved the efficacy of these labours in reinstating reason in its seat. It is added that the nobles, who live in the same asylum, but in a state of idleness suitable to their rank, retain their lunacy and their privilege together, while their inferiors are restored to themselves and to society. (*Townshend's Travels*; *and Pinel*, *p.* 226.) If this be true, how shocking to think that, in other countries, lunatics are kept inactively moping, or left to torment themselves with hurtful agitations. Many of them are capable of feeling the full force of motives *.

* The overthrow of popery in France overset the mind of a young man. After practising upon himself severities equal to those of the ancient anchorites, he renounced nourishment altogether. At the end of the fourth day, the keeper repaired to his cell with a vast apparatus of terror,

As manual occupation dissipates sadness and prevents the reveries, preliminary to derangement, so does temperance preserve the mind from dangerous intensity of feeling. Repeated drunkenness brings predisposed persons nearer and nearer to insanity; and the time comes at last, when the fire once kindled by intoxicating beverages, continues to blaze on without intermission. Thus, it is satisfactory to find that the same course of life, which is attended with most cheerfulness and self-enjoyment, affords the best security against the various disorders, that result from the power acquired by man in civilized society over external things, before he knew what his own nature *(faciat vel ferat)* can either do or bear.

But though what has been said of the maintenance of health in other cases applies, with full force, to the present subject, predisposition to insanity requires a peculiar discipline of the mind. I shall not repeat what a thousand moralists have said (and Abraham Tucker

---

and laid food before him, threatening the utmost cruelty if he did not eat it before morning. He complied with the order; and during his convalescence confessed to Dr. Pinel that the most dreadful conflict had kept up for two hours between the fear of sufferings in this world and in the next.

under the name of SEARCH, more sensibly and practically than all) on restraining the passions. Let those who are interested draw all they can from this source. Only let them not follow bigots in *mortifying the inward man.* This error is calculated to multiply, not to diminish, the number of lunatics. Or what is worse, those who escape lunacy, will be wrought into hypocrites or fanatics.

In conformity, however, with an opinion I have expressed on many other occasions, I think there is much more safety in rational information, than in the most authoritative precepts or the most powerful appeals to the heart. I advise therefore that parents avail themselves of the facts dispersed through a multitude of publications, respecting the power of imagination. The facts I speak of are such as the testimony, upon which they rest, places beyond dispute. They are admitted as a common basis of speculation by writers on the mind, and are not therefore liable to the disreputable uncertainty of metaphysical opinions. It is too much to expect another Cervantes, capable of generalizing Don Quixote. Nor is it necessary. Plain narrative would suffice without distinguished dramatic powers. And

I know not how a greater service could be rendered to society than by merely making a judicious selection of such facts—from the first whims of caprice, up to the wildest ravings of amorous, avaricious, and religious insanity. Demonstrate to young persons, from their own feelings, how certain emotions affect the chest, and what sympathy there exists between the stomach and spirits. They will then know how to estimate those enthusiasts, who fancy the touch of God himself in every palpitation of the heart, and believe the glow after a good dinner to be a particular inspiration of the Holy Ghost. Whether the keepers of asylums for lunatics have availed themselves of the wise counsel of the author of Zoonomia—to force, by frequent repetition, upon the disordered mind rational ideas, exactly the reverse of those, by which they are disturbed, I am not informed. Every body knows that a dextrous method of holding up ridiculous opinions is frequently sufficient to occasion them to be renounced. And it is beyond a doubt that the same effect has been brought about by contrivances, adapted to the complexion of this and that variety of madness. So that we have proofs of the efficacy of the method in the two extremes of mental aberration, from which we may draw a favourable inference for the intermediate stages.

I believe that it will frequently be enough if a dextrous performer out-herod the very extravagancies of the patient, or take up others as similar as possible. SIMON MORIN was shamed out of the idea of his incorporation with Jesus Christ by the folly of another madman, who fancied himself God the Father. A person, who believed that he had been guillotined, and fitted with a wrong head, was cured by the following contrivance. A jocular convalescent in the Bicêtre manages to turn the conversation on the miracle of St. Denys, who carried his head under his arm, and kissed it as he walked along. The lunatic vehemently maintained the possibility of the fact, and appealed to his own case. His companion burst out into a loud laugh, and asked him in a tone of mockery: "*Why how could St. Denys contrive to "kiss his head. Was it with his heel, you "fool?*"—It is true that as you drive insanity out of one of its forts, it often retires to another. But there let it be attacked by the same arms. I perceive indeed that their use requires discretion, and that when one line of attack does not succeed, another must be tried. But none ought to meddle with the mad, who have not discretion and genius into the bargain.

As reading occupies so much of modern life, it must be a great art to lead the inclinations towards proper books. It is easy to see in general that the volatile should be fixed by one sort, and the gloomy enlivened by another. Perhaps SHAKESPEARE himself, by the beautiful soliloquy, and indeed by the whole character, of his Hamlet, has established in many minds the *tædium vitæ,* when it otherwise would have been but a slight and floating listlessness. Feelings of every kind obtain a *settlement* in the breast by being associated with harmonious language and strong images.—The difficulty, as to a choice of authors, lies in stealing away the affections from writings, congenial to a dangerous habit of thought, and attaching them to those, least likely to please at first.

The doctrine that the knowledge of the excesses (les ecarts) of the imagination, is a preservative against them, might receive illustration from an enquiry into the reason *why so few great poets have run mad?* None perhaps have derived their insanity from their poetical vein. Is it not because those, that deserve the name of poets, must have a practical knowledge at least of the way, in which imagination is affected? They are therefore in the secret. They stand in the situation, in which we would wish to place those, who seem in most danger. They are not merely worked up by passion; but however much they may work themselves up, they have a goal in view, which hinders their thoughts from going astray past recal.

END OF ESSAY X.

Printed by MILLS, St. Augustine's, Bristol.

ESSAYS

ON

THE MEANS

OF

AVOIDING

HABITUAL SICKLINESS,

AND

PREMATURE MORTALITY.

ESSAY ELEVENTH.

Printed by MILLS, St. Augustine's, Bristol.

Experiments (begun in France) and repeated at intervals during the last twenty years, and now following up with due diligence, seem to promise the extinction of the power of febrile contagion. - - - At Manchester, regulations improved from former practices and a variety of projects, have been attended with the most complete success. The observation of the propagation of fevers for years together makes one wish for an INSTITUTION like that which has been found necessary to stop the progress of conflagrations.

Int. Lecture to an Anatom. course in 1797.

ESSAY

CONTAINING

REMARKS

ON

MISCELLANEOUS TOPICS

OF

PROPHYLACTIC MEDICINE.

NOTWITHSTANDING the discipline mankind receive from adversity, there is nothing they are so slow in learning, as how to direct their foresight, and portion out their cares properly. By a fatal and almost incorrigible propensity, the fruit of a perverse method of instruction, they come to overlook what touches them nearly, but embrace what is remote with all the ardour of the soul. We are, most of us, by ignorance what the Scribes and Pharisees were by hypocrisy. If THEY *paid tithe in mint and anise and cummin, but omitted the weightier matters of the law*, WE are as assiduous about trifles and as negligent in essentials, though in the trifles that equally attract us and in the essentials we equally overlook, the vicissitudes of fashion happen to be exemplified in a remarkable manner.

It may be shewn that this game of cross purposes is repeated in every part of life; but our infatuation of attachment is most conspicuous perhaps in the superlative anxiety we discover about some things touching health or existence, and in our carelessness respecting others. In a population of three, four, or five hundred thousand souls, it may be that a single individual is drowned under such circumstances as to be recoverable, if proper pains be taken to rekindle the vital spark. Now we do not harbour the same melancholy ideas with the ancient Greeks respecting the lot of those who perish without sepulture; and from the testimony of persons, who have been restored after various kinds of suffocation, there is nothing in this manner of dying which so very particularly calls for commiseration. Yet the infrequency of the accident has not prevented societies from being formed, in behalf of those whom it befals. These societies emphatically style themselves *humane*, as if the recovery of the drowned was the paramount concern of humanity. Prizes are established, and solemnities celebrated every year with great pomp. The fine arts assist in adding a charm to the institution; and the attention of whole nations is summoned to the proceedings; and whole nations obey the summons.

Again, let but a suspicious dog make his appearance, and a hue and cry shall instantly pervade whole cities and provinces. The magistracy shall issue ordinances suitable to the emergency, and the members of the medical faculty, instructions how to proceed for the preservation of persons who have the misfortune to be bitten. Without doubt, hydrophobia is infinitely more to be deprecated than drowning. But there is nothing in the symptoms of hydrophobia that ought to impress us with much more horror than dropsy of the chest, and many other disorders that daily invade our neighbours and relations, though they hardly excite a sensation beyond the precincts of the bedchamber, which is the scene of the fearful spectacle. Nor is there above *one* of our countrymen *in a million*, I presume, that annually falls a victim to hydrophobia. Assuredly both for the improvement of the art of physic, and for the sake of the suffering parties, I think every thing ought to be done to restore suspended animation, and to prevent and cure hydrophobia. But when one looks to the alarm and the sedulous precautions, to which these accidents (the one infrequent, the other certainly not so much to be deprecated on account of any concomitant suffering), give rise, and to the supineness of society as to others equalling or exceeding them in severity and infinitely more common, is there not cause for the reproach—*these ought ye to have done, and not to leave the others undone?*

There is a time in the progress of all nations, when the regular succession of events in the material world passes without notice, and nothing has power over attention in the heavens above but an eclipse, or on the ground below, but an earth-quake. The united labours of wise men among us, have gone far towards correcting this propensity to childish wonder, and depriving the prodigies of inanimate nature of their exclusive privilege to interest. A few unquestionable facts will enable us to judge how far this may be the case in what concerns us still more nearly.

There is reason to believe that in the opulent city, near which I write, not much less than two thousand persons suffered, in a single year of the late scarcity, from pestilential contagion, with little or no medical assistance, and certainly without occasioning any public measures, for relieving the distress they felt, or repelling the danger that threatened others.*

I had estimated the annual deaths from consumption in this island, at one hundred thousand. An intelligent writer has very lately calculated in the following manner. "If the importance of a disease be in pro-"portion to the mortality, there is none so "important in Great Britain and Ireland as "the pulmonary consumption. It appears "that not fewer than one fourth, or at least "one fifth of the deaths of the inhabitants "of London are occasioned by this malady; "that of 20,000 or 25.000, in the London "bills of mortality, annually from 4000 to "5000 are in the class of consumptions. It "is allowed that these bills do not com-"prehend more than half of the present "inhabitants and that the whole number of "them is about one million. Therefore "8000 to 10,000, perish annually here by "this disease. If England and Wales only "contain ten millions of persons, it follows "then that 80,000 to 100,000 fall yearly "victims to consumption." *(Medical and*

---

* See a letter from the author to the Mayor of Bristol, on this shocking subject, reprinted in the Monthly Magazine.

Surgical Review, November 1802, p. 278.) Such statements cannot too often be placed before the public eye, because they may be expected, at last, to bring the public to bestow most pains in guarding against those evils, from which they suffer most. An excellent institution has lately been planned in London for investigating the nature and cure of cancers. But of what consequence is such an institution, relatively to one that should have for its object to ascertain the frequency of consumption, to solve the local and personal queries suggested by this disease, and, by uniting the authority of a body of medical men, to set the whole country on its guard against the sources of danger.*

---

* In the prospectus of the cancer institution, it is very properly observed that "it having pleased God that remedies "or methods of cure should be discovered for many diseases "that were once thought incurable, we ought to hope that "one may, at length be found out for CANCERS, and with "such hope, it is the duty of medical men to exert their facul-"ties for the investigation of the nature and cause of them, "and the discovery of some method or medicine, by which "they were not only to be relieved, but absolutely cured."

By parity of reasoning, it is equally the duty of medical men, individually and in combination, to exert their faculties and opportunities for investigating the nature and causes of consumption. Nor ought they to relax, as long as one general fact on the subject is in obscurity, nor till all who

To eradicate the barbarous tendency to care only for what is extraordinary, and, as far as may lie in the power of an individual, to bring feeling to a conformity with interest, I have devoted nearly the whole of the preceding essays to the elucidation of the bad consequences, resulting from the continuance of mischievous practices. And it would, I believe, make a very trifling difference in the bills of mortality or in the sum total of human misery, if we thought nothing of the events, however melancholy each singly may be, which come upon us through *the visitation of God*, provided we were emancipated from such personal misfortunes, as can be immediately traced up to our own ignorance or folly. If then, we must make the option, let us take heed to our habits, and leave casualties to shift for themselves.

There are some disorders, heretofore generally referred to the class of *incidental*, as if they were the products of the seasons or the

---

are exposed to the disease, are in possession of the knowledge, by which they may be preserved. If any new spur is wanting to our exertions on the subject of cancers, by what argument, I wonder, can it be shewn to be less necessary in respect to consumption? Is it because for every one, destroyed by cancer, a hundred are destroyed by consumption?

effects of causes out of the reach of human prevention, which for their extensive prevalence would deserve not to be passed over in a work on preventive medicine. But the more they have been studied, the less have they been found entitled to be considered in this view. They have therefore a double claim to attention in the present series of tracts. One of the most frequent of these is the disease called

CHOLERA.

It has been before stated that the liver is apt to be dangerously affected by imprudent exposure to the sun in hot climates. The summer warmth of our own climate is observed to dispose the liver to throw out its bile in excessive quantity. In consequence a complaint is formed, which is observed by physicians to mark the decline of the hot season, as faithfully as the appearance of the swallow announces the spring: and it is a complaint, sudden and distressing enough in its progress, and dangerous enough in its issue to be worthy of general attention. Its most striking symptom is a violent and repeated evacuation of a bilious fluid both upwards and downwards. But as in all considerable affections of the digestive organs, so in this, it is very little probable that the liver alone suffers. The stomach joins in an inordinate secretion of its juices, and has an impatience of its contents, which it would not shew in other circumstances, could the same materials be introduced without exciting disgust. The tract of intestines, situated below the point where the bile enters, are in the same irritable plight. No sooner are the one or the other touched by any of the bland warm liquids, which form a part of our ordinary diet, than the stomach is thrown into the action that occasions vehement vomiting, and the bowels into that, which attends purging. At the same time, the rapid manner in which the whole surface of these cavities secretes the liquids, that are proper to each, co-operates with the excitement of the liver, in exhausting the system.

When this complaint attacks in the most sudden and dangerous manner, formidable cramp in the stomach and cholicky pains are instantaneously felt, and there is a speedy puffing or enlargement of the belly, the skin of which is stretched like a drum. The patient is sensible of an extraordinary uneasiness or oppression in the pit of the stomach. The pulse becomes small, and often extremely quick and irregular. The hands and feet grow as cold as ice. It is with the greatest difficulty they can be brought to their feeling by the application of heated bricks or similar contrivances. But the inside is proportionally hot, the stools and urine scalding. The evacuations in both directions go on without ceasing. There shall be a vomiting of bile with much straining every minute. The oppression, the nausea, the cramp and cholicky pains continue.

Under these violent symptoms, the strength declines with great rapidity. The pulse in a day's time comes to creep with a vermicular motion, instead of beating. The patient does not find a moment's repose. He has soon spasms in various external and internal parts. The sense of suffocation, that distressing sensation in the throat which just precedes vomiting, is here unusually severe. Swallowing is difficult or is totally impeded by a spasmodic contraction of the top of the œsophagus. The urine is suppressed. An unintermitting hiccup comes on, with a starting of the tendons or general convulsions. The patient falls into a state of unconsciousness and perishes as if from a stroke of apoplexy.

In weak people, without timely or proper

care, the pulse sinks early; spasms and faintings take place; the interior burns; the extremities feel stone-cold, which symptoms indicate an unfavourable issue.

Where the attack is so mild as to let the patient off with some hours' harrassing discharge of bile, and other painful accompaniments of this operation, and to leave him after a few days' exhaustion, it is severe enough to make any reasonable person take precautions against it. The precautions might be thought almost too simple to be named, if one did not regularly, every year, see a whole country down with sickness, for want of knowing or observing them; which is the more to regretted as they bring with them the most absolute security.

Tropical fevers are observed to seize particularly upon persons, newly arrived in the regions where it rages. In these diseases, an excessive stimulation of the liver and the allied organs seems to bear a considerable part. And it is probable that the susceptibility which living in a colder climate confers, may be removed by gradually lowering the diet, in approaching the tropical islands. It is certain that abstaining from intemperate indulgences at the close of our summer would soon occasion cholera to cease to be a general disorder of that season. The stomach, it would seem, must in most cases be overloaded or offended before the irritation reaches the liver. In persons, particularly prone to an over-secretion of bile, sourness of the stomach will bring it on in more or less force, according to the susceptibility of the liver from the season or the constitution.

The stomach-complaints that plague children so much, and are accompanied by green stools, are among the examples of this position. Either by sympathy with the stomach, loaded with pap, biscuit, bread, or other fermentable materials, unfit for many young infants, or by actual contact between the soured mass and the mouth of the duct, a too great secretion of the gall follows; and acid eructations and purging afflict the little sufferer at the same time or in quick succession. So it is, in great measure, with more advanced patients, on a seizure of cholera, and indeed at other times with many, when the diseased affection of the digestive organs goes no farther than bilious vomiting. It is common enough for the first matter thrown up to be sufficiently acid to set the teeth on edge.

In the season of cholera, therefore, it is particularly expedient at breakfast to refrain from laying much ascescent food upon the stomach. Those who have power enough over their habits to refrain from sugared tea, would do well to use a mixture of milk and soda-water or lime-water, with a sparing portion of bread.

The temptation of fruit after dinner is now particularly dangerous. Attacks of cholera arise from no cause more commonly than from a monstrous mixture of meat, bread, fruit and wine at dinner.

Things offensive in quantity as well as quality produce the same sort of evil. Vegetables to which the stomach has not been habituated, those generally less digestible, as cucumbers, or those unsuitable to particular constitutions, as onions frequently are, rich sauces eaten in profusion, and that almost indigestible material, old, decayed cheese, are very frequently in fault.

The contrast of temperature between the noons and nights of autumn, often brings into action that susceptibility of over-stimulation, which appears to be generated by the summer-heats. Hence it is so common to find a certain degree of imprudence in regimen, after wetting the feet, or exposure to the evening damps, followed by cholera, though a greater has been committed under different circumstances with impunity. In winter both causes shall conspire without being followed by *this* bad consequence.

When oppression and distension of the stomach have come on, and there is reason from past experience or present feelings to suppose acidity, the mischief may often be stifled in its birth by draughts of alkaline water, by magnesia or other absorbent substances. But, in this case, great abstemiousness should be observed for many hours after—indeed absolute fasting, if the party be robust; otherwise small supplies of food may be taken at intervals.

When the bilious vomiting or purging has come on with the internal heat, warm liquors as tea or water gruel, do but increase the evil. Till medical assistance be obtained, small and frequent draughts of cold water, or lemonade, or nitrous acid sufficiently diluted, will alleviate the symptoms. These liquids too have the advantage of allaying the harrassing thirst and clamminess of mouth.

Another thing too, which may be innocently practised, and which therefore should never be omitted to break the force of the attack, is to warm the extremities, if cold. It is not enough to expose the chilled parts, merely to warmth, but they should be vigorously rubbed, and without too great tenderness for the invalid's feelings. It appears that the action, excited in the skin in this way, is in this case, as in others, much more beneficial than that which mere heat produces as usually applied.

FAINTNESS, FEVERISHNESS.

As during the peculiar susceptibility of one set of organs, dangerous or fatal attacks are thus liable to be incurred and thus easy to be avoided, so in all states of the system, and particularly in weak people, do very obvious oversights occasion the inconveniencies, named at the head of this section. Fatigue from travelling, or from long efforts of attention, especially with fasting, is one of the conditions under which they are most apt to occur. The fatal effects of too liberally feeding a person, who has remained without victuals for whole days and nights together, are very generally understood. But this is only an extreme case of the same kind. Persons, whose sensibility is too great and their muscular force too small, on coming off a journey into a warm room, or on taking too much food or liquids too stimulating, will immediately fall into a swoon. It is a law of the animal œconomy that the substances which, in a certain portion, but just create a more vigorous feeling of existence, will in an increased dose induce feverishness or instantaneous diminution or suspension of the vital powers or death itself. Drinking—in different quantities from a glass of rum or brandy diluted to a whole bottle pure—will produce these several degrees of affection. So also will applying solutions of opium of various strength to the inside of the heart of a frog, make it contract with more apparent vigour, or destroy its contractility in the twinkling of an eye. But the operation of these powers is in proportion to the capacity of the subject, to whom they are applied. A patient, weakened by much confinement, from fever for example, shall find the free air reviving. To another patient, weakened by long continuance of sea scurvy, the same shall be a deadly poison; if that term may be applied to a substance that kills on being respired. Therefore of a mixed party of people, who have travelled an equal distance and fasted for the same space of time, a warm room and a good dinner shall produce downright fainting in some of the females, and only oppression or physical anxiety in the others, and violent flushing of the face with general heat in stout men. Fainting depends on the contractile powers of the heart and arteries being either greatly impaired or altogether destroyed for a time. Flushing and general heat and the quick and bounding pulse, which may always, on such occasions, be detected on examination, evidently depend on the very opposite states of the whole or a part of the same organs. Both the one, therefore, and the other, may be avoided by attending to two things. It follows, in the first place, that we should be cautious, in proportion to the readiness with which our organs may be overpowered by ordinary or extraordinary agents, of suffering them to fall into a state of exhaustion. Hence the propriety of the rule, of late so frequently repeated by medical writers, not to fast too long in a state of convalescence or of debility from any cause whatever. Hence we see how it happens that people, impaired by too free use of wine, or by any species of debauchery, are often so extremely injured by the sports of the field. The consequence of many a fox-chase must be greater liability to an attack of gout in the stomach, or at least, great increase of indigestion where that troublesome affection has taken place. Sports of less severity are injurious to country gentlemen, as country gentlemen now frequently are even at the age of thirty. I shall say nothing of the injury from the heat from riding full speed, with chills between the acts of the chace.* It is sufficient to observe that between inanition and exertion, the system is too much lowered at one time of the day, as if by way of preparation for being forced the more readily into as inordinate action by the carousal of the succeeding part. Hence, with some, life is a sort of quotidian ague. They have their long cold fit in the morning; and the long hot fit in the evening and through much of the night: and as an ague, strictly so called, if it hang on, will impair digestion and disorganize the liver, or otherwise undermine the constitution, just so will the above-described sporting ague. The mode of prevention results from the mere description of the complaint. I am told indeed that some of our Nimrods upon

* There are doubtless or there have been races of men to whom these chills and heats were as nothing, or perhaps wholesome variations. But I am to speak of my contemporaries, or, at least, of a part of my contemporaries. They are not made of such well-tempered metal. Nor has their nature that range of endurance.

the wane have felt out for themselves the propriety of carrying a sandwich into the field.—The same ought to be done even by many of those whom we hear most loudly proclaiming the power of their stomach to stand the longest blockade from hunger. For their countenance and their own confessions if they are properly interrogated, and their whole habit, so flatly contradict the boast, that if the terms were not injurious, one might apply to them what Mr. POPE says of a well-known structure—

> Where London's column, pointing to the skies,
> Like a tall bully, lifts the head and lies.

There is yet another simple observance which I would strongly recommend; and that is, when the hounds are at fault and the hunters find themselves heated by exertion, to dismount and walk till the scent is recovered; as also, at the close of the chace, to provide for some additional cloathing, while their jaded steed bears them homewards as slowly as foot can fall. Indeed, earlier in the day, more cloathing would be frequently expedient. It is in vain for a man to vote himself hardy. Partaking in a diversion which requires hardihood will not confer that quality; as seems to be sometimes supposed.

How bitterly do physicians see men who vainly presume upon their power to resist the cold and damp of our winter atmosphere, rue their error.

If, on some occasions in life, it is nearly impossible to help being put into the condition of too great susceptibility, we may still be able to escape the operation of too powerful agents. This therefore is the second thing to be held in view. Large fires, high dishes, even a full meal, wine and tea, will then induce in many people, a temporary fever. The delicate, and even the robust, will often be more refreshed, after fatigue, by a sparing portion of cold meat and a glass of plain water than by any of the seemingly more efficacious restoratives. One of the most eminent and able members of the surgical profession, lately mentioned to me an instance strongly in point. After attendance on a remarkably long trial in a crouded court of justice, he went to rest on a little bread and water. A great number of those who had been present on the same occasion, gave way to the eager call of their appetite; and to his certain knowledge he almost alone of his comrades arose refreshed and alert. A large proportion of the rest were all tormented with restlessnesss and debilitating heats. Some became seriously indisposed.

FEBRILE CONTAGIOUS DISEASES.

Whatever may be the extent to which treatises on fevers, comprehending the description and treatment, the varieties produced by climate and other topics generally run, the surest rules of prevention in our own country, with the reasons of those rules, will go into a narrow compass. The researches of the moderns have done much towards illustrating this part of medicine: and the better a doctrine is understood, the more concisely can it almost always be delivered.

The first position has long been established beyond controversy, though physicians are far from being agreed concerning the exact bounds to which it extends. *Certain disorders and fevers among others, are communicated by actual contact of the person labouring under them, and even by a near approach without actual contact.* This has always been a matter of fatal experience in the febrile disorders of our climate. Every body has seen how catching the measles are. Some parents have voluntarily exposed their children to the society of persons, infected with the small-pox; and the result has justified the principle, whatever may be thought of this method of following it up into practice. A single visit, where any of the proper precautions are neglected, has sometimes been sufficient to infect physicians with the scarlet and low nervous fever. Medical practitioners, accustomed to visit the poor in their habitations, where *all* the proper precautions are apt to be neglected, have frequently fallen victims to febrile contagion. In this way, as I have been credibly informed, two young physicians of great promise have been cut off, within a short period, in the single town of Limerick in Ireland. "In Dr. HAYGARTH's letter," says Dr. STANGER, "nine physicians to hospitals "in London are enumerated, who have died "of contagious fevers. Out of the faculty "at large a greater number perish there "*annually* by this dreadful distemper." Some curious incidents shew that, other circumstances being alike, the infection will be in proportion to the nearness of approach. Thus it is stated by Dr. FENWICK, in a letter to Dr. CLARK of Newcastle upon Tyne, (who has lately exerted himself to promote an establishment for the recovery of poor fever-patients in that town, in a manner that will probably be beneficial to the kingdom at large,*) "that a beautiful girl being brought "into the clinical ward of the hospital at "Edinburgh, ill of typhus fever, some of "the pupils were particularly attentive to "her, and used to sit on her bed or loll on "her pillow. The consequence of which "was three of them were infected with the "fever."

That the power of contagion to infect extends but a little way from the patient, in whom it is generated, when he is confined where the air has free entrance and egress, is now evinced by a vast multiplicity of observations on all the diseases that are propagated by contagion. The terrors of mankind have, in all ages disposed them to imagine that the whole atmosphere was pregnant with the seeds of the reigning malady, whatever might be its nature. Thus, it appears from the earlier treatises on that complaint, which America seems unjustly accused of having transmitted to Europe, that its exciting cause was believed to be disseminated through the surrounding medium; and, as by a scarcely

* In his *collection of papers, intended to promote an institution,* &c. and the *appendix.* Newcastle, 1802. This is the most ample collection of testimonies ever brought together. I quote it freely, lest it should not be given to the public.

credible backwardness in the art of observing, the real method of its propagation continued long unascertained, it was held as no disgrace for nuns and popes to be sufferers from it. No wonder then, when the offending material is cognizable by none of the senses, and where, as is the case in some situations, it really passes from one person to another at a distance, that the same terrifying persuasion should have been harboured. The security derived to the Europeans in the Levant, from the practice of *shutting up* at the time of the prevalence of the plague failed for a long time to open our eyes to the grossness of this delusion. By degrees, the close attention, which practitioners of medicine have learned to pay to the phænomena, led them to perceive enough of the process of nature to dissipate the distressing apprehensions, transmitted to them from their forefathers. The more diligently the subject was investigated, the more confirmations were obtained of the narrow range of contagion. And now, almost every physician of an hospital and surgeon of a ship, emboldened to exert his faculties by the conviction that it is easy to keep clear of the contagious atmosphere, is able to adduce demonstrative facts.

Respecting the plague itself, Dr. Gregory,

the present professor of the practice of physic at Edinburgh, has lately furnished the following particulars.* He was informed, he says, by the Rev. Mr. CARLYLE, Arabic professor at Cambridge, who has resided at Constantinople, and had many opportunities of observing the plague, that it has scarcely ever been known to pass the narrowest streets and alleys—probably such as are not ten feet wide; and that people are safe in their own houses, while it rages in the opposite, owing to the diffusion of the effluvia in much air. Dr. RUSSEL prescribed from a window for patients, ill of the plague in the street below him at four feet distance, without receiving the smallest detriment, though on account of the tendency of the effluvia to ascend in the air, rarified by the heat of patient's bodies, that was the most dangerous position. But the second observation, communicated by Dr. GREGORY, is still more in point. Mr. TAINSH, late surgeon of the Theseus, when off the coast of Syria, in 1799, received on board Captain PHILIPPEAUX and four seamen, ill of the plague. The captain was placed in a cabbin by himself. He rejected all medicines and died on the fourth day.

* Report of the Newcastle-upon-Tyne Committee. p. 4.

For the other four, Mr. Tainsh fitted up a birth, apart from the crew (consisting of upwards of 500 men), but with no better partition than painted canvas. He kept his four patients as cool and as clean as possible. One only died. But the crew totally escaped the infection, which but for so simple a precaution, would, as Dr. Gregory observes, in all probability have destroyed one half of them.

Observations on the small-pox, which certainly ranks among our own most contagious disorders, will equally serve to demonstrate that the same security attends the same precautions, and that the contagion of this complaint does not fill the atmosphere even of a 64 gun-ship. On Christmas day, 1759, the Panther sailed for India with a crew of 420 men. Scarce had they lost sight of English land, when a feverish complaint appeared on board. It proved to be the small-pox. Within three days, 16 or 17 of the hands were attacked. Immediately on the appearance of the eruption, an enquiry was made, and it was discovered that less than half the crew, namely 193, had had the small-pox. The ventilators were kept at work from this time, night and day. Fresh air was allowed to pass through the ship wherever admittance for it could be found. Fumigations were performed and the boy well washed with vinegar twice a week. Canvas was nailed up to divide the healthy from the sick, and wetted with vinegar. The seamen liable to the infection, slept in part upon the poop under an awning, and partly in other places as remote as possible from the infected.

By these simple means was contagion rendered perfectly inert; nor was a single additional person infected during the voyage. (*Dr. Clarke's appendix*, p. 25.)

A forcible proof of the circumscribed sphere of contagion, is deducible from the history of hospitals, in which there are distinct wards or rooms for the reception of patients, who apply for admission, when ill of contagious fever. Dr. JAMES HAMILTON, physician to the Edinburgh hospital, attests that since the appointment of separate wards for persons, labouring under fever, the communication of infection from those wards is unknown to him. The arrangement is of many years standing, and Dr. HAMILTON practised in the hospital before and since. (*Collection* p. 90.) Dr. RUTHERFORD adds that when low fevers infect the other parts of the hospital, they have not originated there, but been introduced. (*ib.* p. 91.) An hospital was opened December 16th 1801, for the 4th regiment of dragoons, at Newcastle-upon-Tyne, at a time when a typhus fever of a very bad kind prevailed among the soldiers. One good airy room was set apart for fevers. The stair-case only separated this room from the entrance into the other wards, which were completely full. There have been, at one time, six patients extremely ill of typhus, and yet to the date of the report—July 1st, 1802—not one patient in the other wards was seized with any symptom of fever. (p. 111.) The work-house at Liverpool, in its general history, adds the greatest weight to the above position, while some particular fatal accidents demonstrate the danger of neglecting the due precautions. This building is so large as to have contained 1400 persons. Its fever-wards are situated in the third story: and in the next above is the nursery. The same stairs lead to both. A number of children are constantly playing upon these stairs and about the doors of the wards. Dr. CURRIE had the curiosity to follow a patient up into the ward. He counted exactly 83 children on the stairs. There was not one, by whom the patient did not pass within 3 or 4 feet; and to some he passed much nearer. When the circumstance was mentioned to the nurses, they semed to think it nothing uncommon; and the want of any actual proof or appearance of harm from the use of a common stair-case, with the prospect of a separate house for fever, rendered fruitless several consultations, respecting a distinct stair-case to the fever-wards.

The nurses live in these wards night and day, their apartment being in the centre between them, and open at top to the air of both. Several have caught the contagion at different times; but these have been nurses newly introduced; and they have seldom been affected above once; their constitutions acquiring, by habit, insensibility to the contagious impressions. Two of the present nurses have, each of them, a child living with them in this infected region, and going out to school by day. These children appear healthy, though they constantly sleep between the patients in either ward. Their mothers were satisfied that they were liable to no injury, as they never came into contact with the sick. They seemed to think that they should be themselves perfectly safe, were they not obliged to handle the sick, and immediately exposed to the contagious exhalations from their skin and lungs.

"Whatever," says Dr. Currie, to his

correspondent, "may be your sentiments, "respecting the narrowness of the sphere of "contagion, you would not, I think, have "ventured to predict so singular an exemp-"tion from fever, under such circumstances." For in the lock hospital, which is situated immediately below the fever-wards, there has been no fever since this disposition was adopted; that is, during ten years. Two years ago a very alarming fever broke out, and infected no less than 67 children. Luckily the source of the infection was clearly discovered; and the explanation not only clears away all doubt as to the inference deducible from the rest of the account, but furnishes an analogy which will extend to similar cases, where the origin of contagion, appearing on one or two occasions in situations generally exempt, remains in obscurity. "A "family" says Dr. Bostock, "residing in "a cellar in one of the most confined parts "of the town, was sent in a state of fever "to the workhouse. The parents were placed "in the (fever) wards. But, by some ne-"glect, the children were sent into the "nursery, with some degree of the disease "upon them, and without removing the "infected clothes, which they had worn "before they came into the house."

Three other persons, (the master and mistress, and a young woman,) died of low fever in 1801. The young woman had, as it appeared on enquiry, been incautiously turning over the linen from the fever-wards before it had been steeped in water. The two other fatal accidents are thus accounted for.—The apartments of the master and mistress are so near the passage, at which all persons enter the workhouse, that the door of a coach, as it brings up a patient ill of fever, must open within a foot or two of the window. In 1801, typhus being extremely prevalent, "there was an unusual pressure for admit-"tance into the fever-wards. They were "filled beyond all former precedent, and "many were obliged to be sent away. The "examination of these unhappy persons in "their unclean and contagious state, un-"washed and unventilated, and the rejection "of them, where necessary, devolved on the "master and mistress—attentive and humane "persons—who doubtless fell victims to this "dangerous and painful duty. This is the "opinion of their successors, who make a "point of never approaching within a yard "or two of suspected fevers, and have "hitherto escaped. The porter, who has "opened and shut the passage-gate for "several years has escaped also. He says "he has used the same precaution, but he "ascribes much of his safety to tobacco, "which he chews in very large quantities." (*ibid.* ps. 173—186.)

It is affirmed by Dr. CURRIE of Chester, that from the time of the opening of the fever-wards of the infirmary in that city, there never was, at any one time, reason to suspect the communication of contagion to patients in other parts of the house (*ibid.* p. 238); in this he is supported by Dr. HAYGARTH, who extends the term to nineteen years.

The preceding facts are corroborated by similar testimonies from a large proportion of the physicians in the kingdom, most conversant with febrile infection. Insomuch that the medical opposers of the fever-house, proposed to be placed near the Newcastle infirmary, "had it not in their power to "produce the authority of one modern phy-"sician of reputation in support of their "opinions." (Dr. CLARK, *Appendix*, p. 2.)

No considerate reader will require to have the cogency of such evidence pointed out to him at large. When he remembers that thousands of persons have lived under the same roof in different hospitals with patients, ill of fever, it will strike him as impossible that the influence of contagion should not have reached them, if it extend so far as from one apartment to another adjacent. And yet among a number of medical practitioners, intent upon this very subject, no one has observed an instance of the kind, where proper measures were pursued. The few straggling instances, and these for the most part loosely reported, which seem to indicate the distant action of contagion, cannot be allowed to outweigh the vast mass of opposite evidence. For if it be difficult to prevent the introduction of gross prohibited wares into a territory, must it not be nearly impossible to guard all the abodes of the healthy and sick against a material too subtle to be an object of any of our senses, and at the same time so transportable, that it may be conveyed from place to place on the skirt of a coat, in a rag, a bit of lint, or a lock of wool! "Men" says Dr. Fenwick, "are "often exposed to the near action of contagion without its being discovered; and if, "under these circumstances, they are seized "with fever, the disease may be readily attributed to some more distant exposure, of "which they have been aware. I cannot illus-

"trate my meaning better than by a case, "which Mr. Ingham read to the Special "Court on the 24th of June last. That gen-"tleman informed the governors (of the New-"castle Infirmary), that a person had caught "a fever by speaking to his servant (who had "lately recovered from one) in the open "air. A further enquiry has proved that "the person alluded to had been engaged in "a small room in paying several workmen, "many of whom resided in *Pipewellgate*— "where, at that time, a low fever was pre-"valent. In the same manner, where con-"tagion has been supposed to be conveyed "from one ship to another at a distance, a "full knowledge of all the circumstances "would probably shew that a more intimate "communication had taken place than that "which has been noticed. No attention or "sagacity can, at all times, detect such "mistakes, and it is therefore no impeach-"ment of the veracity or judgment of the "gentlemen, who give us these supposed "proofs, to consider them as inconclusive." Where therefore the examples are so rare as those of apparent infection from remote sources, no cautious reasoner would acquiesce in an affirmative conclusion, if there were no evidence on the other side. But when in trials upon a large scale, and continued for years, contagion shews no power to strike far off, it follows that it can have no such power.

Of a truth, so essential to the consolation and security of society, there yet remain proofs, if possible, stronger still. Thus at St. Thomas's hospital in London, fever-patients are indiscriminately mixed with others, but the wards are kept as sweet as the apartments of a private house. Here, says Dr. BLANE, "there was no instance, "during the last ten years of my attendance, "of any medical gentleman, patient or nurse "having caught a fever—This was not the "case before proper means of cleanliness "and ventilation were adopted." *(appendix* p. 45.*)* In another place, the same physician relates that when infectious fevers occur in private families he constantly assures the family without having been once mistaken, that if perfect cleanliness and sweet air be maintained, there is no occasion to fear infection. *(Collection* p. 102.*)* Dr. WILLAN remarks that it is well ascertained by experience in many places that if there be no close intercourse, the most malignant fever may subsist in one chamber of a house without affecting any inhabitant of the other apartments—an observation he has made a thousand times in the metropolis. *(ib.* p. 104.) Dr. John Hunter asserts that in several of the worst cases he has seen (some of which proved fatal) where the patients, being in better circumstances, were lodged in clean airy apartments, he has never known an instance of their infecting those about them; even when the patient was a married man, and his wife had slept in the same bed with him one or two nights after the commencement of the fever. In an hospital, in a ward of which there are patients, ill of this fever, provided it be well ventilated, the patients with other complaints, in the same ward, are seldom infected. It would appear that there is no great power of infection in the body *alone*, provided the air be not confined. The worst mode in which the poison can be applied seems to be by the apparel or bed-clothes of the sick. *(ib.* p. 234.)

Desgenettes *(Hist. med. de l'armee d'orient.)* mentions a woman who, without detriment, travelled sixty leagues behind the carriage of the general in chief, between two patients ill of the plague.

" Among the middle and higher ranks of
" society in Chester and its neighbourhood,
" during a period of 31 years, I scarcely recol-
" lect a single instance of the typhus fever

" being communicated to a second person—
" not even during the epidemics of 1783 and
" 1786. Fresh air and cleanliness were the
" only means employed to prevent infection.
(Haygarth's Letter to Percival.)

In a question of so great moment and of which the solution depends upon testimony, it appeared necessary to the satisfaction of the reader to bring forward a variety of witnesses. But these may suffice, when it is added that there is either a general concurrence or no contradiction, except on the part of a few opposers of the establishment of fever-wards, who have scarce succeeded in collecting any thing specious, and adduce still less from their own experience.

The reader may desire to know more precisely still the limits of contagion in ventilated places. But here we must be content with approximations. In the worst species, we may confidently believe that it does not extend a yard. Dr. C. à MERTENS physician to the foundling-hospital at Moscow, whose account of the plague in that city, I offered in vain when a student of medicine, to two London booksellers, in an English dress,* estimates the infecting distance

* It has lately, I think been translated again, and published,

of patients in the plague at a foot or under, and is most fully aware that the contagion does not reside in the air.* Dr. O'RYAN's interesting experiments shew that neither the matter of small-pox, nor a person in the eruptive fever of small-pox infect at the distance of half a yard.‡ Dr. HAYGARTH's

---

unless the translation happens to be that left by me in the hands of the late Mr. MURRAY, of Fleet-street.

* *Sola ægrotorum et rerum infectarum contactu communicabatur; atque atmosphera contagium non spargebat, sed sanissima semper fuit. Visitando tam prope adstabamus illis ut sola pedis distantia inter nos et eos sæpe vix remaneret et absque alia quâcunque cautela quam quod nec corpus nec vestes aut lectum tangeremus, a peste immunes permanserimus. Linguam propius observando, solebam linteum aceto communi imbutum naribus et ore admovere.* (*Observ. Vindobonæ.* 1778.) In a faithful history of modern attempts to ascertain the circumstances under which contagion is communicated, and to prevent infection, this physician would deserve an honourable place. Our late publications sometimes shew want of information, or a very narrow spirit of nationality.

‡ I impregnated with small-pox matter, a large ball of cotton, and placed it in the middle of an oval table, whose smaller diameter was a yard. I placed six children round the table, so that there was only half a yard distance from each of them to the infected cotton. Not one was infected on several trials. (*Dissertations sur les fiévres infecticuses.*) The experiment was repeated often within doors and without on the communication of the small-pox. The blood, mucus and tears of persons ill of measles, were similarly tried with the same result.

observations pretty nearly coincide with this result.* Dr. CURRIE supposes that contagion, issuing from a person in low fever, may not infect at more than the distance of a few inches in free air.

The history of the communication of contagion in opposite circumstances is exactly the reverse. It is in the confined and filthy habitations of the poor, that the matter which excites low, putrid, nervous fever, is generated; and there, for want of free admission of the air, it is so certainly communicated, that wherever one inhabitant sickens, others catch the disease. With regard to this fact, there is a great consent among observers; and it has been estimated that every poor person ill of fever, infects two or three.

One or two examples will illustrate the pro duction of contagion from filth and confinement. I choose first extreme cases, but the conclusion as I will afterwards prove, may be

---

* I found that the pernicious effects of the variolous matter were limited to a very narrow sphere. In the open air and in moderate cases, I discovered that the infectious distance does not exceed half a yard. Hence it is probable that, even when the distemper is malignant, the infectious influence extends but to a few yards from the poison. I soon also discovered that the contagion of fevers is confined to a narrower sphere. (*Letter to Dr. Percival.* p. 8.)

extended, in proportion, to others similar in kind, but varying in degree. It is evident that where sloth and want of cleanliness prevail on board a ship, there the effects resulting from these causes, whatever they may be, will be most strongly experienced. The two following relations strikingly confirm this reflection.

The Hankey sailed from England in 1792, with upwards of two hundred people on board, partly settlers for the island of Bulama off the coast of Africa. Both the settlers and the crew were healthy at the commencement of the voyage, nor is there any suspicion of injury from the exhalations of unhealthy ground. For reasons, which it is not necessary to repeat here, the whole number of persons carried out were obliged to live on board throughout the rainy season, when the heat is excessive. They endeavoured to shelter themselves in the best manner they could by changing the ship into a wooden house. Her sides were raised and she was covered with a roof. In so confined a situation, cleanliness was scarcely practicable; and sometime after their arrival, a dangerous fever broke out. The ship was navigated by twelve healthy seamen, then taken on board for the first time, from Bulama to Bissao; and before her return, nine of these died. Three fourths of those on board perished at Bulama.

With much difficulty, they made a passage to St. Jago, where they were furnished with four seamen, and with this help, proceeded to the West-Indies. The four new hands were all infected; two died and the others were put on shore at Grenada and St. Vincent's, in the most wretched state imaginable. On arriving at Port St. George in Grenada, no care was taken to sweeten or ventilate the vessel; nor were the clothes and bedding of those who had died on board, destroyed. In consequence, the fever, which proved one of the most fatal disorders upon record, gradually spread from the Hankey to other ships up the harbour, and thence gained the island. Of 500 sailors employed in the regular trade, the almost incredible proportion of 200 perished between the beginning of March and the end of May. Twenty-six out of twenty seven recruits, who joined the artillery in July, were infected; and twenty-one died. Nor was it till the close of 1796, that this malignant fever disappeared altogether. During the interval, it seems highly probable that it spread to other places.

In this case, there seem to be circumstances of peculiar aggravation, if the virulence of contagion be in proportion to the intensity of the generating causes. Let us, therefore, take a view of other ships under different circum stances. "Several ships, which had carried "flax-seed to Ireland, returned in June and "July to New-York, crowded with needy and "wretched emigrants from that island. They "were so thick between decks, that the air "was deprived of its usual portion of oxygen, "insomuch that on bringing the sick passengers "to shore, the common pure atmosphere was "too stimulant for their lungs; and a number "of them gasped in it, and died in a short "time. There was so much animal excrement "accumulated in one of the ships, that the "American health-officer detained her at the "quarantine ground, as poisonous and pes-"tilential, and refused to let her come up to "the city. By the pukings and purgings—"by the urinary and perspiratory discharges "of these miserable creatures, literally wallow-"ing in their own filth, the bodies of many "of them were besmeared and incrusted, "forming a layer of excrementitious grime "from head to foot. Their cloathing and "their bedding were impregnated with as "much of these excrementitious matters as "they could wipe from the bodies of the "passengers and absorb. With such coverings, "vile, offensive and pestilential in the highest "degree, were they surrounded. And these "excrements, infesting every thing in the

" neighbourhood of the sick, underwent the " usual chemical changes, in a heat nearly or " quite equal to the human body, and turned " to - - - - some poisonous matter, which " forms the exciting cause of fever. Of the " fever so excited, between thirty and forty from " one ship, died in crossing the Atlantic - - - " The survivors arrived in a state of uncleanness, " sickness and want, seldom seen in America, " but among the emigrants from that unhappy " country, who make so large a number of the " poor in the American hospitals and alms- " houses. So thoroughly contaminated with " their own corrupting excretions, were the " clothes and beds of these sufferers, that " the - - - exhalations from them poisoned the " air of the Marine hospital on Staten island, " and the medical attendants and nurses " sickened in the discharge of their humane " attentions." *(New-York Med. Repository.* Vol. 1. p. 70.)

The following fact, which we owe to Dr. ROLLO, is more analogous to what happens in common life. " One man of the horse " artillery was admitted into the hospital " with a suspicious fever, next day another. " This excited enquiry. It was found they " came from two different barrack-rooms. " These were followed by other men, in all

" amounting to eight: three of whom came " from a separate room; the rest from the " same rooms. The rooms were visited by " the commanding officer. All the rooms, " whence the infected men came, were " found to have entirely different bedding " from the rest of the barracks.—The horse " artillery being a corps in constant readi- " ness for service, and whose appointments " were always complete, had for conve- " nience of carriage hammock bedding. The " hammocks were rolled up tightly every " morning the moment the men rose; and " they were unloosed when they went into " them at night.—At this time, we had had " so much and so constant rain, that this " bedding had not been aired or opened, for " a single day, for at least two months. The " hammocks, with their bedding, were ex- " amined, and the moment they were opened " a very peculiar nauseating smell was per- " ceptible. Immediate steps were taken, and " no further mischief took place. Here, an " infectious fever evidently arose from the " confinement of the effluvia of a man's own " person, in a term of about two months." The measles and small-pox can almost always, directly or indirectly, be traced to persons who have had these distempers. But it is in low fever in thousands of instances as above. The sickness of one patient cannot be traced to any sick predecessor. It is a strong confirmation of the independent production of this contagion, that similar fevers originate among animals closely crowded together, as among sheep and hogs on board ship. If pestilence follow famine with great regularity, it may be that the effluvia from an impoverished habit are more liable to the contagious fermentation, or that the debilitated are more readily infected by floating effluvia, or that famished human creatures have not spirit to keep themselves clean, and so filth accumulates unusually under distress. Though, "in ordinary times, we are assured that the lowest class do not put clean sheets on their beds thrice in the year; and that where no sheets are used, they never wash, or scour their blankets." (*Willan's reports*).

That intermittent and remittent fevers, in most cases, spring from marsh exhalations, the concurrence of innumerable observations compels us to believe. Though it be certain that other agents, as cold applied in a certain manner, will also produce perfect agues. But what substances precisely they are, which working together engender the noxious effluvia of marshy grounds, no investigation has yet gone near towards ascertaining. So it is with the poison that excites our continued fevers, commonly but perhaps not happily termed *low*, since their first stage is probably either violent stimulation, or that suppression of the powers, which is to stimulation what the effect of swallowing a prodigious quantity of diluted spirits at once, is to ordinary intoxication. This subtle and dangerous material appears to be oftenest formed when human beings are crowded together and covered or surrounded by filth. Observers have remarked with surprize that naked negroes, jammed close together during the middle passage, (as was the custom among the dealers in human flesh, when the slave-trade was carried on with all the inhumanity, of which the uncontrouled spirit of commerce is capable,) failed to produce it.

Agues prevail in spring and autumn. Malignant fevers have been often checked by great heat and cold. Can the fermentations, from which the *febri-facient* effluvia spring, proceed through a certain range of temperature only?

If we knew the nature of the process, we should at a glance see the shortest way how to prevent and stop it. If we knew the qualities of the product, we should have the same advantage in setting about to change it. As the air has so much effect in rendering it innoxious, may one suppose it an oxydable substance? This would explain some of the phænomena better than any other conjecture. It would enable us to understand the security of a free atmosphere, and the danger from approaching patients in close rooms. Without this supposition, we may however distinguish the impropriety of the term *concentrated*, as applied to contagion in such situations. But by its help, the existence of it in its original state for want of the access of oxygen in the atmosphere surrounding the patient, and the danger of such an atmosphere become apparent. We can also better understand certain anomalies or extraordinary phænomena; and may also hope to reconcile the advocates for the narrow sphere of contagion in open places, and their opponents. When contagion settles on the clothes of a sick person, it probably assumes a more consolidated form than as it issues from the body:—in the latter it may be more expanded and more open to the action of the atmosphere. Many authors have observed that infection from clothes is more common than from contact or approach of the sick. But if the contagion be more condensed in its state of adhesion to different substances, it becomes evident that it may be blown across a court of justice, unaltered and capable of infecting, though the rarer contagious halitus from a sick person cannot be transported in this manner. What dose of the poison is sufficient to infect has lately been brought into question, and even attempted to be determined by an ingenious writer, who yet confesses that "numerous examples - - - manifestly "prove that a short and probably a momen-"tary exposure to contagion, in some in-"stances excites a fever."—(*Dr. Haygarth's Letter to Dr. Percival*, p. 42.) But perhaps the variety of phænomena would be more easily understood from the decomposition or non-decomposition of the poison, as it is inhaled, and its incapacity to break the passing trains of action, before it is altered or the contrary.*

It will be allowed in the present state of our knowledge as nothing improbable, that febri-facient contagions consist of the same

---

* If one could make experiments on such a subject, one need not despair of exciting trains of action, capable of preventing the inoculation from small-pox or cow-pox taking effect.

elements in different proportion, or perhaps with a slight addition in some cases. The variation in the morbid effects bespeak a variation in the constitution of the poison, though it must not be forgotten that difference of susceptibility will create the greatest diversity of effects. How different is the small-pox in different individuals inoculated with the same lancet! From a variety of observations, it would appear that the effluvia, which excite the small-pox, are much more widely diffused, other circumstances being equal, than those, from which our low fever springs. That is, according to the preceding supposition, they do not quite so readily combine with the oxygen of the atmosphere. It must not, in this place, be concealed that variolous matter in a liquid state does not soon lose its property of communicating the small-pox; and in a dry state it is found to resist the full action of the atmosphere for a long time, certainly for a month. "I have "tried" says Dr. CURRIE, "whether vario-"lous matter, exposed to the air, and indeed "to the wind, can communicate the disease "by inoculation, and the result has been as "I expected. The disease was communi-"cated with the usual certainty and success. "I inoculated three patients at the same time.

" in one the eruptive fever appeared on the " fifth day; in another, on the seventh; and " in the third, on the ninth. The matter " was spread on the surface of a piece of " window glass, and had been long perfectly " dry. It was discoloured by some smoke " and dust that the wind had blown over it. " I diluted it, as our practice is, with a few " drops of tepid water —— Next day, the " glass was as dry as before, and the matter " unchanged as to appearance. It now lies " before me, and the eye can distinguish no " diminution of quantity—So far as to facts. " My opinion is, that I shall be able to ino" culate with the same matter diluted in the " same way, many months or perhaps years " hence." These facts are not more unfavourable to the supposition that the air destroys contagion by acting as an oxygenant, than to any other hypothesis concerning its mode of operation. Where atmospheric air and acids perform a common office, we have few analogies, by which to conceive it, besides that of oxygenation. Marine acid may do this by the help of watry vapour. But whatever is the real alteration produced upon contagion, these reasonings will only require a change of terms. In all cases we must alike suppose that the variolous effluvia, imbedded in purulent matter, are infinitely more protected from the action of the air, than as they issue from living surfaces. By an agent, much more powerful, under the present point of view, than atmospheric air, we know that liquid variolous matter may be very speedily deprived of its power to infect. Thus, Mr. CRUIKSHANK of Woolwich, in March 1795, took two portions of recent small-pox matter from the same person, and exposed one portion to oxygenated muriatic gas for a few minutes, and with it inoculated the left arms of three drummers, while the right arms were inoculated with some of the other portion. The punctures of the left arms had no marks of inflammation, except what simple puncture produces, and they entirely disappeared in a few days. But the right arms took on the variolous action, and in two of the persons there was a general eruption. The experiment, on repetition, was attended with the same result. (*Rollo on Diabetes, first ed.* p. 61.)

These experiments would probably have led, by analogy, to a method of destroying all sorts of febrile contagion, had not such a method been already discovered with regard to some. As it is, they admirably serve to remove any of those scruples, which might have been left, where the infecting substance is not so palpable as variolous matter, and where, to common observers, the certainty of infection may not be deemed nearly so great as from the process of inoculation. Though in reality the contagion of typhus and of scarlet fever are as unfailing, when applied under favourable circumstances, as variolous matter, or variolous effluvia.

The series of observations, from which we seem entitled to draw the most important of all conclusions respecting the prevention of infection, viz. *that the contagion of low fever (and probably all contagions producing febrile disorders) may be destroyed in all their states by the fumes of some of the mineral acids; and that these fumes can be safely disseminated through the air of apartments, without removing the sick*, began with the celebrated chemist, M. de Morveau.

An attempt having been made to empty the vaults *(caves sepulchrales)* of the principal church of Dijon, there was discharged so great a quantity of putrid effluvia as to render it necessary to shut up the church; and it was the more urgent to destroy these effluvia in their source as they penetrated the neighbouring houses, where the symptoms of a contagious fever* are said to have appeared. After the vain trial of various perfumes, fumigations with aromatic vinegar, and the detonation of salt-petre, M. de MORVEAU put six pounds of kitchen salt, a little moist, into a common garden bell, imbedded in a pot full of cold ashes, and placed it over a large chafing-dish. Upon the salt he poured two pounds of oil of vitriol and immediately quitted the church, the doors of which were shut. Next day the church was found perfectly sweet; and every one who entered it declared with astonishment that he should not have entertained a suspicion of its having been so lately filled with foul effluvia. Mr. MORVEAU was induced to use marine acid on account of its expansibility and of the striking manner, in which it unites to the fumes of volatile alkali. He remarks that the nitrous acid would have been more powerful than the detonation of nitre.‡ But in these first observations, as also

* L'acide du nitre seroit plus puissant — — — — Pour purifier une masse d'air, il n'y a point une voie plus courte et plus sûre que de lâcher une acide, qui s'elevant et occupant tout l'espace, &c. &c.

‡ L'air surchargé d'exhalaisons putrides d'une odeur très-infecte, avoit causé a plusieurs personnes une mort prompte, à d'autres des maladies putrides dont elles moururent en peu de jours.

in his late work, he appears to have confounded two things essentially distinct; the offensive vapours that may accompany contagious effluvia, and these effluvia themselves, which have undoubtedly no odour, nor any quality by which they manifest themselves to the senses. This first experiment therefore could only be considered as a hopeful beginning.

In the same year, however, one of those fatal pestilences, known under the name of the *jail-fever*, broke out at Dijon; and the same method of purification was here executed with success. Notices of these transactions were published in the Philosophical Journal, which enjoyed the most extensive circulation in Europe *(Rozier, Observations sur la Physique)*; and immediately afterwards popular publications by persons of the first intelligence and weight in France, recommended this method as alone to be depended upon for destroying the contagion both of the murrain among cattle and of low fever. In 1780, the Academy of Sciences being consulted by the late French government concerning the means of destroying infection in the gaols, appointed a committee for the purpose, on which stands the great name of LAVOISIER. By this committee, the acid vapours were also pointed out as more to be depended upon than any thing else (*See Morveau des moyens de prevenir la contagion* 1801, *or the English translation by Dr. Hall.*) But it does not appear that the extrication of acid fumes was precisely mentioned as an expedient that could safely be practised in inhabited rooms in any publication before 1794, when it was done in the most express terms by the French counsel of health, and enjoined, though ineffectually, under the most binding sanctions, by the legislature.—Two years before, the oxygenated marine acid was stated to be used in dissecting-rooms for destroying putrid smells.

At a later period the marine or oxygenated marine acid fumigations, have been practised with success in France, in Spain, and in our own admirably conducted military hospital at Woolwich.

In 1795, Dr. J. C. Smith, proposed the fumes of nitrous acid for the destruction of contagion, and in the year following, a set of experiments conducted on ship-board, upon a very large scale, demonstrated the efficacy of the measure. On board the Union hospital ship, where a very fatal infectious fever was raging, not one of the crew was seized after the commencement of the fumigations. It is hardly necessary to except one of the attendants, who had a slight relapse, in consequence of his own imprudence. Similar success so distinctly attended the experiment on board some of the Russian ships, that the admiral himself applied for a fresh stock of the materials, to be used on board the rest of that fleet. All these experiments, taken together, supply the only proof of the destruction of contagion, of which the nature of the subject admits. It is not the disappearance of any odour, upon which we can place the smallest reliance. The decisive fact is the immediate cessation of infection after the fumigations, without any other change in the circumstances, under which it was taking place.

The efficacy both of nitrous and marine acid fumigations being allowed, it becomes a question which of the two is preferable, in general or on given occasions. Mr. MORVEAU has raised two objections to those by nitrous acid. He endeavours to shew by experiment, that the vapours of nitrous acid deprive that portion of the atmosphere, through which they are diffused, of part of its oxygen, and that they have less expansive force. It has been shewn at large by Dr. ODIER of Geneva, that the first objection is groundles, the appearances by which it was suggested having solely arisen from Mr. Morveau's defective manner of operating. The second objection is well founded in the abstract, but does not apply to the purpose in debate. The nitrous acid vapours are proved by the common process for distilling nitrous acid to be condensable without water, while those of the marine acid are not. The latter therefore will continue longer afloat and spread further. This, other things being alike, would, therefore, be undoubtedly preferable. But experiment has decided in favour of the sufficiency of nitrous acid fumes: and their advantage consists in their being generally little apt to provoke coughing, unless discharged in too dense a volume, and also in their being attended with but momentary inconvenience, and no danger, if this should happen.

PRACTICAL INFERENCES.

The method of preventing febrile infection, follows of itself from the preceding discussion. To speak strictly, indeed, it reduces itself to a single rule. *Destroy contagion by acid fumigations.* The fumes of nitrous acid are raised by throwing nitre into warm oil of vitriol. Let a little sand be warmed to somewhat above the heat of boiling water, that is, to a degree considerably beyond what the hand can bear.

In this sand, let a gallipot, containing an ounce of oil of vitriol, be imbedded. Then let the bowl of a tea spoon of nitre be dropped into the oil of vitriol. Let the mixture be stirred with a glass rod, or a dry stick or the barrel of a quill, well besmeared with bees wax. No metallic substance must be used in any of these processes. White fumes will be seen to ascend; and they will be sensible through a small room. A second spoonful may be required for a larger room. The measure is to be regulated by the effect. Few private houses have rooms so large that four tea-spoonfuls (half an ounce) of nitre, used at once, will not fill with fumes, too dense to be agreeable, particularly where the lungs are delicate. To fumigate thoroughly (which should be done once or twice in 24 hours,) the doors and windows should be shut. But a perpetual slow extrication of fumes is also desirable, since the contagion is perpetually generated. During this, however, the air should be freely admitted. The apparatus should be placed so near the patient, that the fumes may be just perceptible where he lies. The least practice will teach the regulation of the fumes. By throwing open a window and door, they can be lessened at once, when too strong.

When about twice the quantity of nitre has

been added to the oil of vitriol, it will expel no more fumes; and must therefore be changed.

Common salt may be used in the same way as nitre with oil of vitriol. But if the salt be well dried first, no heated sand will be necessary in the beginning; and afterwards it need not be so strongly heated.

It ought not to be concealed that the oxygenated muriatic acid, when too freely inspired, produces swooning, sneezing and a variety of disagreeable sensations. With the slightest discretion, however, it may be used for the prevention of infection; and there is only one precaution necessary, which I have found it easy to impress upon persons, perfectly unaccustomed to chemistry. It has the convenience of not requiring artificial heat; and I should undoubtedly prefer it to all known agents. Dr. ROLLO, in his excellent *account of the royal artillery hospital, at Woolwich,* directs, on the authority of Mr. CRUIKSHANK, as follows; "Take of pulverized manganese two "parts—of common salt four parts—of oil of "vitriol three parts—of water one part— "When a patient is admitted with an infectious "disease, or where there are patients with "sores having offensive discharges, one or two "gallypots are placed in the wards, with about "three ounces of the manganese and salt, to "which is added half an ounce of water, and "then is gradually poured on the whole, a part "of the ounce of oil of vitriol; the remainder "occasionally. These quantities answer to "the consumption of a day. A pot or two is "also placed on the outside of the doors of the "same wards, in the gallery. The vapour is "diffused over the whole ward and penetrates "every where, and destroys every other smell "than what itself conveys." M. MORVEAU gives the following proportions as preferable, though only on the score of œconomy: *common salt ten parts,—manganese two parts,—water four, and oil of vitriol, six parts.* If therefore we weigh out five drams of common salt, one of powdered manganese, and mix them in a mortar, and add two of water; and from a bottle containing three drams of oil of vitriol, add about a sixth part, and so on through the day, this will be sufficient to destroy contagion, for so long, in a moderate sized apartment. The vessel containing the manganese, may be at first placed in a remote part of the room, till the attendants on the sick be familiarized with the fumes and the rate of their extrication. Between the times of adding oil of vitriol, the mixture may be stirred with glass or waxed wood, whenever the extrication of fumes appears too slack.

For those who are to handle the sick, particularly in cases of the scarlet fever, it may not be a superfluous care to wash the mouth and hands, and to wet the inside of the nostrils with vitriolic, nitrous or marine acids, sufficiently diluted to be agreeable. Where glazed dresses are used to protect the skin, the mucous membrane will be liable to the effluvia; these surfaces being probably much more susceptible of contagious impressions, than those which are covered by the cuticle.

Every thing that has been in contact with the patient, as foul linen, should be thrown into water slightly acidulated with one of these acids. And where it can be conveniently done, all the evacuations of the sick should be received into acidulated water. In short, the acids should be as freely used as possible in every circumstance, belonging to persons, affected with contagious febrile disorders.

With respect to other substances, which have been used at various times as anti-contagious, such as incense, camphor, the smoke of tobacco, whatever effect they may have in rendering smells less offensive, experience has amply shewn that they are destitute of of power to destroy contagion. Of course, they could afford no security.

It may here too be noted as a dangerous error, that free use of port wine is a protection against infection during the reign of an epidemic. It is true that *habitual* drunkards have been sometimes said to escape the action of contagion, when it was fully exerted upon others. But occasional free use of wine so as to produce subsequent weakness, appears to increase susceptibility. In the fevers of our climate, the rule seems to be, that whatever enfeebles brings with it danger of infection. It has been remarked by the most experienced and intelligent observers that persons, slightly exposed to contagion, do not suffer, when afterwards exempt from debilitating causes, whereas otherwise they do. During convalescence, a relapse is occasioned by these causes. It is remarked by Dr. LIND, and the observation is strictly applicable to those that seek an excuse for indulgence in wine, that sailors after having committed such excess in drinking on shore as produced feverishness, did not return to their former health as usual, but that their feverishness put on the precise form of the contagious fever raging on board the ship. Dr. CHISHOLM found that persons, most given to the abuse of intoxicating liquors, were most liable to the malignant pestilential fever at

Grenada. There need not therefore be any particular course of regimen during the rage of fevers, than what prudence enjoins at all seasons. Intemperance and abstemiousness are alike to be shunned.

Contact of the patient himself, inhaling an atmosphere, that retains full force to infect, and the use of apparel or furniture, on which the poisonous effluvia have settled, are the grand means, by which febrile diseases are propagated. Hence, besides the necessity of airing or fumigating articles, which have touched or been near the patient, arises the propriety of excluding all persons from the sick chamber, except the necessary attendants. Even under the security of the measures above described, this rule should be sacredly observed. For not only is exemption from dangerous diseases a case, in which one would "make assurance double "sure;" but it is here in some measure as on board a ship in distress. Nothing can be so unfavourable to the regular performance of the offices required by the sick, as a promiscuous concourse, and the admixture of useless hands among those, who are occupied in their proper sphere. This restriction, however, is not to be construed to the exclusion either of a friend, whose presence is thought likely to check the agitations of delirium or of occasional assistants, where additional strength is needed. It has already appeared how small is the danger of infection in the spacious apartments of the wealthy, without a neglect of ventilation, too gross to be conceived possible in the present day under the superintendance of any medical man, or indeed in a family, possessed of the most moderate degree of intelligence*. Separation of the sick is what the condition of the poorer part of the community has long loudly called for. The call has been answered in a variety of places, for example, at Chester, Edinburgh, Manchester, Liverpool, Dublin, London, Cork, Newcastle, Waterford, Etruria. In these places wards in the infirmary have been set apart for the reception of patients, ill of fever; or distinct buildings have been

* In genteel houses I now frequently observe the contrary error, as people are apt to fly from one extreme into another. But it is a great deal more than unnecessary to invite the damps of November, through the wide-open windows, into apartments, too chilly at that season, though close shut up. It would be infinitely better to have recourse to the agent, by which pestilence has been heretofore vainly opposed: I need not say that fires should not be so fierce as those used to dissipate contagion; sufficient directions having been given, respecting the temperature of apartments.

erected or are now erecting*. In Staffordshire, the working potters themselves have contributed from their wages towards the building and the current expences of the institution. It would be highly desirable that in other places labourers, that can afford a halfpenny a week, (which is the sum contributed by the potters,) should be, in the same manner, induced to take an interest in the design. One may also hope that the incontrovertible proofs of its utility and the example of these poor manufacturers will, at length, rouse the magistrates of all our towns to assist in stopping the pestilence,

* The caprice of fortune seems to me very striking in respect to those, who have laboured to prevent infection. Dr. J. C. SMYTH, deserves much commendation, doubtless, for prosecuting the acid fumigations. Yet the most obvious analogy pointed out the use of nitre in place of common salt; and the nitrous acid was indicated by M. MORVEAU, in his first paper. Dr. SMYTH, however, has been lucky enough to be thought worthy of a national recompence, amounting to half the value of that bestowed on the introducer of the cow-pock inoculation. On the other hand, no public notice has been taken of the persons, who carried into effect the scheme of preventing infection by separation. Yet who will deny that *they* had infinitely more call for exertion of mind, and an infinitely harder conflict with prejudice to sustain? Hitherto, I fancy, they have likewise saved five hundred times as many lives.

which has been and is raging around them on all sides.

The benefits attending *houses of recovery*, into which persons affected by fever are to be carried as soon as the attack is discovered, have been ably pointed out in so many publications, calculated for extra-professional circulation, that I shall but just touch upon the subject here.* These structures accomplish two great objects. 1. *They save a number of the diseased that must otherwise perish.* During the prevalence of some epidemics, one in four of all the poor, affected with fever, have been stated to perish.‡ In London, to which the account refers, the accumulation of human beings sometimes seems almost to equal any thing on board slave-ships before the new regulations. From three to eight miserable creatures of different ages, occupy a single bed in a damp, dismal, close and filthy apartment. And it is remarked that "where two "or more persons are confined to one bed by "a fever at the same time, it almost invari-"ably happens that at least one of them "sinks under the disease." (*Murray's re-*

* See Dr. STANGER on *contagious fever in the metropolis*, and the various writers named by him.

‡ Dr. WILLAN's reports on the diseases of London for March, 1800.

marks on the poor of the metropolis.) In habitations, a degree less wretched, every sort of want—the want of air—the want of all the conveniences that have been devised for the comfort of the sick—the want of regularity in administering any medicines that may be supplied—the want of address in relieving the incidental distresses of the sick, and above all, the want of hope, so common among the indigent in their best health and so constant under disease, must occasion a most disproportionate mortality. "I have known," says Dr. FERRIAR, "nine patients in fever, crammed into three small dirty rooms, without "the regular attendance of any friend or of "a nurse. Four of these poor creatures died "—absolutely from the want of the common "offices of humanity and neglect in the administration of their medicines." The author of this essay stated to the Mayor of Bristol on good authority that eight persons were last year buried out of one house in that city from fever. And yet to listen to vulgar opinion in the place, Bristol is a place peculiarly healthy.*

---

* Those who disbelieve the accounts of last year from the Dispensary (thought dreadful even in London) and what I stated, may now convince themselves that *low fever can infest Bristol* by exploring St. Phillips'. Low fever exists in various other parts.

Such is the fate of the poor left to their own situation, and to their own care. In the house of recovery at Manchester, it results from the experience of years that a larger proportion recover than of private patients with all the accommodations that wealth can command. So decisive is the superiority of a well-combined system, regularly carried into effect! Such the chance in favour of the sick, when protected against the interference of gossips, who go about dispensing prejudice and death.

2. *The second and greatest benefit from houses of recovery is the prevention of infection.* The extent of this benefit may be, in part, appreciated from the preceding comparison between the dissemination of contagion in large and small, in airy and close apartments. But the progress of fevers among the poor is greater than can be imagined, without reference to the testimony of eye-witnesses on this head. When once introduced into the abodes of this class, fever travels with fatal certainty from individual to individual, from chamber to chamber, from floor to floor, till all is infected from the cellar below to the garret above. It not only reaches to every corps of inmates at one time, but for months together seizes the families that succeed those

that have been destroyed or disabled. Dr. FERRIAR was told by an elderly woman, that she had fifteen children, all settled in Manchester, and that all had undergone the fever within two months. The most thinly inhabited parts of the island have very little the advantage, in this respect, over the metropolis and over great manufacturing towns. In the country, a concourse of idly curious visitors nearly compensates for the moisture, darkness, and confinement of city cellars. Those, to whom the observation has not occurred, will find in a vast variety of the reports from the Scotch ministers, as published by Sir JOHN SINCLAIR, that the neighbours flock to a house, where the fever is, with as much eagerness as their superiors to a place of public amusement. So it is in country-places in England. In such places indeed they can, at present, only be admonished of the risk they themselves run, and of the mischief they do the sick. At most, the preservative of acid fumigation may be introduced by the better-informed residents. Though there is no reason why houses of recovery should not be introduced into every parish at all populous, or into an union of parishes, where the population is less. Indeed upon the evidence already accumulated*, it

* It should be stated, that the opponents of Dr. CLARKE

may be asserted that the legislature could not do a more beneficent and œconomical act than to compel a general establishment of these asylums. In all epidemics, great terror, if nothing worse, is carried into the bosom of opulent families by servants who have caught the disorder among poor acquaintances; and when the contagion is rife among the washerwomen of a town, as it is very apt to be, the apparel that has passed through their hands becomes dangerous, and the poisoned shirt of NESSUS ceases to be a fable.

That, in the present state of society, houses of recovery afford the only sure means of arresting the progress of infection by affording an opportunity to purify the habitations of the poor, is too obvious to require an explanation. The fever is not only, in many cases, communicated to other members, before the uninformed patient or his friends have any suspicion of its nature; but the apparel, the bed-clothes, the walls, and various articles of

---

at Newcastle coincided with him as to the utility of a house of recovery. They are now, I am informed, zealously labouring to raise funds for the erection of one. They objected to the situation near the infirmary, where almost all medical men think it would do much the most good. And the sum necessary for a new structure would have been saved; and what services might this sum not have rendered to Newcastle and the environs, in the hands of a board of health?

the scanty household, have received the contamination. Here is a stock of contagion, ready to continue the distemper for an indefinite time without interruption, or to give rise to it at some future period, as accident shall dispose.

The importance of removing the first person in a family, affected by fever, and of purifying the habitation and its contents, (a care that certainly cannot be confided to the poor themselves,) has suggested the idea of medical associations, whose office it is to superintend these objects, and whatever else relates to the prevention of infection. The success which has attended the endeavours of that established in Manchester, has left no doubt with intelligent judges, of the possibility of the total extinction of contagion, and of the very easy practicability of reducing its ravages within a narrow compass.

SCARLET FEVER, with Sore Throat.

Whatever may be the comparative exemption of the opulent from low or typhus fever, there is another species of contagion which they have much cause to dread. It is that which in its virulent form produces the most suddenly fatal of all our infectious disorders. It sometimes destroys in two or three days after the breaking out of the symptoms. I was once desired on a Saturday, to see a little boy affected with this complaint, and found him dying. Passing the same way on the next Monday evening, I called to enquire after the mother, and was told that a girl, whom I had seen on the first visit apparently in good health, had been seized and died in the interval. I have known other cases, almost as quickly fatal; and every physician, who has had much experience of scarlet fever, must have met with analogous examples. Another important fact, relative to this dangerous disorder, is that it is excessively infectious, and I believe that most medical practitioners would be ready to attest of the scarlatina exactly the contrary of what has been stated above, concerning typhus. If in spacious and airy apartments, the one is seldom known to be communicated, the other scarce ever fails to spread. The following are the ideas of Dr. Clarke, on the best means of proceeding for putting an end to the depredations of scarlet fever. The reflections of so judicious and experienced an observer of diseases, cannot be too widely disseminated.

He supposes a board of health established. This board, he proceeds, "should not limit its "efforts to the extermination of typhus con-"tagion, but also extend them to the preven-

"tion of the casual small-pox, and of the "scarlet fever, attended with sore throat.— "------By extending the rules of pre- "vention to the scarlet fever, and consequently "checking its progress, the lives of the chil- "dren of the middle and higher orders of "society, in large towns and in their vicinity, "will be preserved; who, at present, suffer "ten times more from this, than from any "typhus contagion.*

"The small additional expence of carrying "rules of prevention for the small-pox and "scarlet fever into execution, will scarce merit "the consideration of any board of health. "But a house of reception for poor patients, "labouring under scarlet fever, though ex- "pensive, would be attended with the great "advantage of securing cities and populous

---

* Dr. CLARKE was undoubtedly careful of exaggerating. Otherwise he might have stated a much higher proportion. In this part of England, at least, I suspect that from thirty to fifty times as many members of opulent families are affected by scarlatina, as by typhus: and adults as well as children.— One morning, some years ago, on setting out on a distant journey, I met Mr. K. a member of the Irish parliament, who shewed me in his hand, with great glee, the report of the Irish house of lords on the united Irish, and appeared perfectly well. The first news I heard on my return, was that he was dead of scarlet fever, and that it was spreading. Persons, then residing at Clifton, will remember the incident.

" towns from the destructive ravages of this " disease. Such a separate house for this com- " plaint, therefore, as soon the funds of esta- " blished boards of health will permit, should " be annexed to fever wards; taking care, at " the same time, to guard against the intermix- " ture of the respective patients, of clothes " and of servants.

" But it will be proper, in this place, to " observe, that no single house for the recep- " tion of scarlet fever, will be found adequate " for the compleat prevention of this disease; " because the subjects of it are generally " young children, nay, often infants at the " breast. To prevent this disease, therefore, " from spreading, it will be in general, neces- " sary to remove the whole family, which " would soon fill one house of reception. A " range of neat small houses, containing rooms " with opposite windows, for the reception of " poor families in succession, sending the " families back to their own purified habita- " tions, as soon as the distemper is over.

" Rooms of this kind for the reception of " whole families, employed at the breaking " out of the scarlet fever, together with the " house already mentioned to be annexed to " fever wards, for the admission of the *first* " patient seized in a poor habitation, would

" effectually extinguish the disease; and it " would be the interest of parishes to build,* " and keep in readiness, such a range of " houses." (*Supplement to collection of papers*, ps. 142—144.)

The greater the danger and fatality from the contagion of scarlet fever, with the more assiduity should fumigations, the other great preservative, be employed. That it will, in this case, be effectual, has not been demon-

---

* That parishes would find it their *pecuniary* interest to raise buildings for the recovery of their sick poor, may sound to some like a projector's paradox. And, in fact, if the scheme required a considerable uncompensated expence in addition to the burdens imposed by the poor-laws and the exhausting demands of government, the objection would be very formidable. But the parochial savings would soon much more than pay the interest of the capital, expended in building. Grown people suffer from low fever in greater proportion than children, and more men die than women. In the great epidemic at and near Lancaster, Dr. CAMPBELL observed that of men 1 died in 8; of women 1 in 19; of children 1 in 80. The death of the father therefore, which is always the more probable event, throws the family upon the parish. What a noble example, if a company of rich men, upon having secured to them the average annual charges from low fevers and their consequences, would undertake to establish houses of recovery! To demonstrate that these charges would more than pay the interest of the money expended in the building and the current cost of the establishment, would be to secure their erection every where!

strated by experiments on so large a scale, and so frequently repeated. But during the prevalence of febrile diseases of late years, among which the scarlatina has often made its appearance, I have known the oxygenated muriatic acid used; with this proof of its favour, that the infection has not spread under the circumstances in which it usually does without this precaution. Some experienced physicians have believed that scarlatina anginosa and malignant sore throat are varieties of the same disease. It is certain that the nature of the general affection is similar to that in low fever, if this be any indication of a similarity in the nature of the exciting cause.—And as the alterative power of the acid vapours has been found by precise experiments to extend to small-pox as well as typhus, we may, I think, rely upon it, that it will not fail us in either of the two former complaints.

It will, doubtless, be a part of the business of medical boards, as well as private practitioners, to determine the power of this cheap and simple method of destroying contagion not only in febrile contagious disorders, but in others not essentially so, as chin-cough. When society shall have acquired thorough confidence in their agency and perfect familiarity with their use, they must, of necessity, supersede houses of recovery. But the capital, that may have been laid out upon these edifices will not have been wasted; as they will be easily convertible to other public or to private purposes. Their general introduction will be attended with this other inestimable advantage, that it will create a spirit of vigilance to the earliest tokens of the operation of contagion, and dispose families, who have no occasion to send their sick members abroad for a cure or for the sake of stopping the dissemination of disease, to follow all the approved directions with stricter punctuality.

One very happy effect upon the general conduct of the poor at Manchester has been observed from the interposition of the board of health. By the purification of houses, out of which fever-patients have been removed, a disposition to cleanliness has been introduced. The inhabitants doubtless feeling more chearful and comfortable in their white-washed and sweetened apartments, exerted themselves to prevent a relapse into their former foul and dismal condition. So far we have a security against the future production of contagion; and those among the poor, who acquire a taste for cleanliness, will not be long before they acquire regular and sober habits.

Extension of the Plan for preventing Infection.

It is held by many that *efforts in the cause of truth are never lost*; and it must be owned that what has been done with a view to oppose the propagation of contagion tends powerfully to corroborate this opinion. Twenty years have not yet elapsed since fever-wards were first set apart in England for this purpose; and the plan of fever-houses has spread from the provinces to the metropolis, from which advantageous station they promise to extend in all directions. Wherever the question of their utility has been agitated, there the discussion has terminated compleatly in their favour, and been followed up by practical measures.

Without question, if the spirit that has been thus excited, and that seems to me to involve more of wisdom and humanity, than have ever before been combined by men acting in concert, shall receive a proper direction, its good effects will not stop at febrile infectious diseases. Other acute diseases may become objects of the attention of boards of health, and in certain cases, those who suffer from them should be equally removed from infected houses. Otherwise not only will one main intention of the establishment be de-fected; but the most valuable lives in a family be lost, while the least so are preserved. The greatest prevalence of low fever frequently occurs, when catarrhs, pleurisy, and rheumatism are most common. Though contagion be in his house, (whether of generation previous to disease or issuing from the infected), the healthy labourer shall resist: but no sooner is he debilitated by one of these other complaints than the same cause, which he before withstood, shall make a fatal impression.* Of course, the patient must be removed to a healthy atmosphere, or the contagion by which he is encompased must be destroyed.

But experience will soon suggest this and other smaller improvements. It is of more consequence to extend the views of the sup-

* Persons residing in infected apartments, sometimes by means of fresh air and the exercise of the day, continue long unaffected by contagion. But if through taking cold or any other cause, they should be confined to the house for some days, they assuredly take the fever. So it happened in the late unfavourable season. Whoever was obliged to keep his bed for a catarrh, pleurisy or inflammation of the lungs, within three or four days caught the fever; and almost every one so affected, died. The children are infected from the new source of contagion; and the mother, after closing the eyes of her husband, and perhaps of more than one of her offspring, sinks exhausted with grief, watching and fatigue, and is herself the last victim to the disease. (*Willan's reports.*)

porters of these institutions to the prevention of chronic diseases also, by which such a portion of mortality and misery are occasioned among the poor, as there must be an universal wish to see diminished. The expence and trouble, necessary to this, will make a very inconsiderable addition to the expence and trouble, inseparable from the management of a house of recovery, though the two designs may be carried on apart. The convenience of their combination will consist in this. To a house of recovery, a medical officer will be necessary, whose business it shall be to examine at their own habitations such patients as are proposed for admission, and likewise after the removal of the patients, to superintend the execution of the measures necessary for checking the progress of contagion, and preventing the renewal of its effects; such as fumigations, cleansing and white-washing the apartments; fumigating the apparel and bed clothes, and securing their destruction, if this, now we have the knowledge of the power of the acids, shall be ever necessary. Wherever this officer is called, he may at the same visit ascertain the state of the family as to chronic ailments. They are often apparent at a glance.

I shall not attempt a delineation of the wretchedness, arising among the poor from the frequency of disorders of this class. By their designation they are permanent, often lingering for years; and it is probable that they produce upon the whole a sum of pain, even more considerable, than acute diseases.—The bills of mortality attest to what a degree they are destructive. Their prevalence is beyond the conception of persons, not accustomed to the actual inspection of the poor with medical views. I do not think it too high to estimate that of *every five labourers' families where there are both parents and children, in three out of the five more than one individual will be found to suffer from some lingering malady*, DRYDEN says of his "somewhat old and very poor" widow—

> Her poverty was glad; her heart content,
> Nor knew she what the spleen or vapours meant.

But of such poverty, certain great statesmen, with whom we have been blest since DRYDEN's time, have nearly rid the land. I remember to have seen some remains of it in my youth. I meet with it no longer. From a good deal of observation and enquiry in different districts, I incline to think that a poor family of any size, where father, mother, son or daughter are not sorely harassed or dangerously affected with one or other of the following chronic disorders, is among the rarities of society:

Hysteria or the like.
Indigestion.
Chlorosis or some female complaint
Rheumatism.
Consumption incipient or confirmed.
Scrophula in its various shapes.

Now, except in far advanced consumption or schrophula, simple treatment is found almost invariably efficacious, notwithstanding the disadvantages under which the poor labour, particularly in the article of diet. How desirable then that these complaints should be detected in their nascent state, in which they are, for the most part, exceedingly obvious to an experienced eye! The principal thing necessary, therefore, is to provide for a regular domestic survey of the poor, or to fix upon some place, where parents might attend with their children at stated times. To this I am persuaded that many of them may easily be induced, for the dire distress, under which they have, of late, been labouring, seems to have had the effect of rendering them more tractable—just as a deep snow tames the birds of the air. A little conversation would, I believe, induce many to make a better disposal of their earnings in the purchase of articles of food. Tea, for example, they unanimously declare to be their only remaining comfort. They drink it twice or thrice a day. Now "the popular use of this "herb not only precludes the possibility of "purchasing more solid food, but actually "renders the stomach incapable of digesting "a more nutritious diet." *(Dr. Barry's report relative to the Cork house of recovery).* And, on these grounds, unless they practice the grossest deception, I have often succeeded in persuading them to lay it aside.

The expence of a regular system of inspection and of remedies for chronic disorders, taken when they are so distinct as to be perfectly cognizable, but before they have taken root, would be a most trifling object. From having paid the bills for drugs for a considerable number of poor chronic patients in and about Bristol, I should estimate that four hundred guineas annually would be fully enough to do the same for all such patients in the city, and for all too, that could attend at any place in it from the environs.

How advantageous to the rich would be the establishment of such a plan, in rendering the medical practitioner more familiar with the early appearance and treatment of the very complaints, to which *they* and their children are particularly subject!

ENLARGEMENT and ORGANIZATION of the preceding PLAN.

Whoever is capable of taking a comprehensive view of the great and precious, but neglected, art of PREVENTIVE MEDICINE, would desire still to reduce into order and to convert to use, many things, of which society, at present, profits or otherwise, according as chance disposes the will or abilities of individuals. It is not enough to form medical associations for the prevention of febrile infection, and for the suppression of chronic disorders on their first formation, in particular. There is wanting a general co-oporation both for these and other objects. Nor do I see in what respect the title of medicine to a national establishment, is inferior to that of agriculture. The public health appears to have as fair claims to regard, as the public sustenance. The outlines of an effectual system for the ascertainment of the state of the public health, and for its preservation, will readily occur to every person, acquainted with the established forms for conducting the great affairs of mankind. Particular regulations will present themselves as soon as the execution of the scheme shall be set about in earnest.

In the first place, it is obvious that a board must be established in the metropolis, and that there must be other boards in cities, towns and country districts, with which the first is regularly to communicate. The conduct of the business must be entrusted to select individuals, but admission to the place of meeting should be free, at stated times, to members of the medical profession without exception; and all should be invited to give information in writing, when they could not attend. A digest of the whole ought to be annually published. Should a particular emergency demand an extraordinary report, the materials would be at hand. By such an arrangement, the execution, the success, and the comparative merit of the different means of suppressing contagion and chronic disorders would come before the public; and undeniable improvements would pass with expedition from place to place. At present, the rate at which they travel, is woefully slow, and thousands perish in the mean time. For if the plan for a *fever-house* hardly reached Liverpool from Manchester in six years, how long will it be in getting to Bristol, whither the roads for useful public designs have, in all ages, been said to be heavy?

It should be among the principal cares of a national system of medical boards, to extract, from charitable institutions for the relief of the indigent sick, a quantity of medical knowledge, and by consequence, of public benefit, which hitherto they have never yielded.

By the same means, too, I conceive that the various resources, existing out of the profession and at present yielding nothing, would be called into effect. In order to illustrate this observation, I extract a passage from a paper, formerly inserted in a work of more limited circulation than the present essays. (*West-country contributions*, 1799.)

"The medical functionaries of hospitals should be required at fixed, perhaps monthly periods, to furnish an account of their respective departments, particularly noticing such phænomena, as should appear to them instructive or singular. To these meetings all the practitioners of the place and neighbourhood, together with the subscribers to the charity, should have free access. When the statement contained any thing uncommonly interesting, a commissioner or committee of verification, should be appointed to examine the circumstances. In cases, not admitting of delay, the attending physician or surgeon should call in one or more commissioners in the intervals of the sitting." - -

" After considering the stake which society has in medicine, how often in a man's life it may, according to its power and administration, wound or soothe his personal feelings and his sympathies, let him imagine himself present at one of these sittings, where the business is carried on with a spirit adequate to its importance. What motives for self congratulation and for congratulating his whole species, will the scene before him offer! The art of most immediate and most universal concern drawn out of its present darkness, in which none can distinguish whom it preserves and whom it destroys! its doubts solved! its contradictions reconciled! the cause of phænomena, where ease and existence are at issue, detected! light and order suddenly spread through trains of ideas that had long been vainly struggling in the mind of the ingenious! the stores of the most abounding in knowledge augmented! the faculties of the most acute sharpened! the interest of all classes promoted! the fortunate son of Esculapius, retiring with better informed judgment to the mansions of the opulent! the humbler practitioner carrying away comfort to the peasant and the pauper !"

" Europe abounds in private medical societies. It abounds in philosophical societies, pursuing with success every species of knowledge, but that most worthy the curiosity of the philosopher, the knowledge of man. These are the disjoined elements of the association here suggested." - - - -

" As medical philosophy gains more of the public attention, medical practitioners will become more and more the devoted servants of their art. Such as is the community, such will ever be medicine. The more intelligence in the first, the less intrigue, the less quackery, the less helplessness in the last. - - - - It is often the consciousness of the part mankind take in their exertions that stimulates the soldier in the field and the orator in the council. Under the same powerful motive the physician would also perform wonders." - - - - -

" I shall perhaps be told that our hospitals are already on the best footing and in the ablest hands, and that every thing useful to the present age and to posterity, will be recorded; consequently that no good can arise from the cumbersome regulations I propose. But I, who can name a variety of hospitals, which in a long course of years have furnished nothing or next to nothing to medical philosophy, must be slow to believe in this universal vigilance. To many of our predecessors and to many of our contemporaries, the healing art is deeply indebted for what they have recorded of their hospital practice. Nor would I desire a more conclusive argument.

For if a part is so valuable, what would the whole be? And what, according to a just estimation, are the facts that have been preserved, but fragments of a mighty wreck, demonstrating the value of the mass that has perished."

To lose one useful fact, may be to lose many lives. And does not ten thousand times more valuable information perish than is preserved? and all for want of a combination among the theatres of disease, and of a great commanding station, a national bank of medical wealth, where the private practitioner may deposit his gains of knowledge, and draw out, in return, the whole stock, accumulated by his contemporaries? In the late controversy, it was publickly declared on high authority that the infirmary at Newcastle was in a state of disorganization; and I wonder how many of our infirmaries are in the best possible state of organization. It is too obvious to require proof that the plan which brings forth, for the use of the public, all the instruction these institutions can supply, must secure to their inhabitants, the best treatment in every sense of the word—particularly if the present paralysing system of permanency in the medical functionaries be abolished, and a rotation, at due intervals, introduced in its stead.*

* The plan of the *cancer-institution*, proposes "*that a physician and surgeon be elected, not for life, but for three, five, or seven years.*" It is to be hoped that in an enlightened age, no new medical establishment will tolerate the plan of perpetuity. Objections have been drawn from the abuse of the rotation-scheme, against the thing itself. The objectors have not appeared to comprehend that there is a medium betweeen a stagnant pond and a torrent.

But *the expence! the danger of a job!*—the expence! what expence? the expence of a few thousand pounds a year for the greatest benefit that can be conferred on a people, from whom so many millions are drawn! It has been frequently said that this and that destructive disorder might be extirpated if a co-operation between governments could be attained. But of co-operation to so beneficent a purpose, there was no hope. Let the reader determine if this were a libel on governments. But whatever indifference statesmen may have towards doing good for its own sake, one would suppose that they would not be disinclined to procure for the people, at the people's own expence, a blessing, which being continually felt, and requiring no hour-long harangues to prove it such, would ensure the public gratitude to the authors—a thing particularly desirable at present, when society has been rocking like a vessel in a storm to the immiment hazard of the pilot.

But the other rub—*the job!* Against that, provision must be made by the nature of the regulations. Let there be none but moderate salaries, for hard-working clerks. The rest ought to be done from affection for the art. Those men are not worthy to be entrusted with the health of others, who will not *give* so much pains for the chance of learning how better to execute their trust. Nor can it be doubted that men, who will, can be found. During the late war, how many army physicians, in addition to their exhausting official duties, have preferred the labour of minuting down the occurrences among their sick, to that sleep, for which nature so loudly cried!—The spirit of our profession is little of a sinecure spirit. And, for the sake of the health and happiness of my brethren, I hope we shall long be content to relinquish this spirit to our betters.

With these suggestions I take my leave of an engagement, which nothing but the perpetual spectacle of the fatal effects, arising from domestic errors in every part of the country, and the hope of paving the way for more salutary practices could have induced me to undertake. From my previous professional avocations, I had no right to calculate upon the frequent and continued interruptions to which I have actually been subject. From circumstances, many of which I could not controul, the author and the printer set to work nearly at the same instant. Hence the month, previous to the publication of each number, was the utmost space allotted for preparing it; and from this, in almost every instance, distant journeys have subtracted a fourth. Such is the cause of much want of arrangement, of frequent obscurities, and many inelegancies. I know not if it will serve as their excuse. To my own palate, I confess, even, in the passages I least disapprove—

Inest amari aliquid, quod ipsis in floribus angit.

I will not, however, conceal that I often purposely deviate from a style, which I consider as suited only to the squeamish taste of an enervated age. Should improving science, as I hope, and in particular the science of human nature, restore to us the blessing of intellectual health, the niceties of those compositions, over which the artist and the amateur hang till they both sicken, may come to be referred to one class with the quaintnesses of the literary age that preceded SWIFT and ADDISON. Even in the periods of ROBERTSON and GIBBON, there may, hereafter, be discovered exactly the kind of fault that disgusted the poet in TIMON's villa—

> Grove nods at grove, each alley has a brother,
> And half the platform just reflects the other.

Wherever I could find or recollect ideas suitable to my purpose, I have freely seized them by a right, I believe, universally acknowledged to reside in the authors of elementary treatises. For I too had to deliver the elements of the most important of arts, namely, *that of living*. In such an arena, it would have been the idlest of all vanities to attempt a display of the talent of originality. Indeed, where they equally suited my purpose, I preferred facts from the experience of others, as supposing they would carry with them superior authority.—Many of our literary connoisseurs have imbibed a strong dislike of the productions of the German press; and our vile translations of vile originals, must inevitably create this feeling. Indeed, in the *belles lettres* very few of their authors will please readers familiar with those of ancient Greece and Rome, and with the best of modern Italy, France, and England. But their compilations are convenient, and their collections of original facts beginning to be really valuable to those, who will be at the pains to sift them. No other nation is equally attentive to accumulate medical materials. Nor, with the exception of a REIL and one or two other individuals, does any nation turn them to so little account. I know not whether I shall be found to have refined any of their ore, or to have transmuted German lead into a nobler metal. —In medical manuals for families their language abounds as it does in all other sorts of manuals. I should be glad, however, if any body would point me out one, in which the materials are wrought up in a manner, really fit for the service of families in this or indeed in any country. My search has been pretty extensive but vain. Yet this is the chief, or rather the only merit of such performances.

I ought, indeed, to except one point, and that among the most important, and certainly, beyond comparison, the most delicate of all. The reader will be at no loss to guess what point I mean. Here I found precious instruction both in respect to facts and their application, together with an encouraging coincidence in sentiment. I did not find these helps in the writings of physicians, but of persons long devoted to education on the amplest scale, and with an assiduity, which forsakes not the pupil at the school-door, or as soon as the cloth is drawn, but pursues him in his recreations and to his repose, ascertains his solitary habits, and refuses to leave even his inmost sentiments undetected.

What I have written upon this point has brought me so much fresh information, as